Lecture Notes in Computational Science and Engineering

31

Editors
Timothy J. Barth, Moffett Field, CA
Michael Griebel, Bonn
David E. Keyes, Norfolk
Risto M. Nieminen, Espoo
Dirk Roose, Leuven
Tamar Schlick, New York

Springer
Berlin
Heidelberg
New York
Hong Kong
London
Milan
Paris
Tokyo

Mark Ainsworth
Penny Davies
Dugald Duncan
Paul Martin
Bryan Rynne
Editors

Topics in Computational Wave Propagation

Direct and Inverse Problems

Springer

Editors

Mark Ainsworth
Penny Davies
University of Strathclyde
Department of Mathematics
Richmond Street 26
Glasgow G1 1XH
United Kingdom
e-mail: M.Ainsworth@strath.ac.uk
e-mail: penny@maths.strath.ac.uk

Paul Martin
Colorado School of Mines
Department of Mathematics
and Computer Sciences
80401-1887 Golden, USA
e-mail: pamartin@mines.edu

Dugald Duncan
Bryan Rynne
Heriot-Watt University
Department of Mathematics
Riccarton Campus
EH14 4AS Edinburgh
United Kingdom
e-mail: dugald@ma.hw.ac.uk
e-mail: B.P.Rynne@hw.ac.uk

Cataloging-in-Publication Data applied for
A catalog record for this book is available from the Library of Congress.
Bibliographic information published by Die Deutsche Bibliothek
Die Deutsche Bibliothek lists this publication in the Deutsche Nationalbibliografie; detailed bibliographic data is available in the Internet at http://dnb.ddb.de

Mathematics Subject Classification (2000):
65M, 65M06, 65M12, 65M15, 65M32, 65M60, 35L05, 65R, 65R20, 65R32

ISSN 1439-7358
ISBN 3-540-00744-X Springer-Verlag Berlin Heidelberg New York

Springer-Verlag Berlin Heidelberg New York
a member of BertelsmannSpringer Science + Business Media GmbH
http://www.springer.de

Cover Design: Friedhelm Steinen-Broo, Estudio Calamar, Spain
Cover production: *design & production*
Typeset by the authors using a Springer TEX macro package

Printed on acid-free paper 46/3142/db - 5 4 3 2 1 0

Preface

This volume consists of survey articles on current topics in computational wave propagation and inverse problems, written by leading experts in their respective fields.

The idea to compile such a volume arose in conjunction with the LMS Durham Symposium on *Computational Methods for Wave Propagation in Direct Scattering* held at the University of Durham from 15th–25th July 2002, which we jointly organised. The meeting, attended by 70 participants from the UK and overseas, was structured around a number of short, three lecture, survey courses on a range of topics on computational wave propagation and inverse problems beginning at the level of a graduate student. We were delighted to secure the participation of distinguished international researchers to present these lectures. We felt that it would be valuable to record this material for the benefit of a wider audience, and the idea was hatched that the individual lecturers should be invited to contribute a survey article. Fortunately, many of the speakers not only agreed to undertake this arduous task, but produced what we hope you will agree are the high quality contributions found in this volume.

Finally, it is a pleasure to thank the Engineering and Physical Sciences Research Council of Great Britain and the London Mathematical Society for providing the generous support that allowed the meeting to take place.

Mark Ainsworth
Penny Davies
Dugald Duncan
Paul Martin
Bryan Rynne

Glasgow, 2003

Contents

New Results on Absorbing Layers and Radiation Boundary Conditions

Thomas Hagstrom*

Department of Mathematics and Statistics, The University of New Mexico, Albuquerque, NM 87131, USA
and
Institute for Computational Mechanics in Propulsion, Ohio Aerospace Institute and NASA Glenn Research Centre, Cleveland, OH 44142, USA
hagstrom@math.unm.edu

1 Introduction

Perhaps the defining feature of waves is the fact that they propagate long distances relative to their characteristic dimension, the wavelength. This allows us to use them to probe the world around us - optically, acoustically, and now at a wide range of wavelengths in a variety of media. For numerical simulations, it is precisely this essential characteristic - the radiation of waves to the far field - that leads to the greatest difficulties. One may view this fundamental difficulty as rooted in the existence of (at least) two widely separated spatial scales. The first are the small scales associated with the wavelengths and the scatterer, and the second is the long distance between the scatterer and the observers.

The techniques discussed in this review are designed to make the accurate and efficient solution of typical wave propagation problems possible by restricting the computation to the small scale only. We do this, of course, by introducing an artificial boundary, Γ, and either imposing absorbing boundary conditions on it or surrounding it with an absorbing layer. Typical configurations are shown in Fig. 1 below.

Our focus here will be on time-domain problems for the standard equations of wave theory - the scalar wave equation, first order hyperbolic systems, and the Schrödinger equation. Although frequency-domain calculations still dominate much applied work, we believe that time-domain simulations will become increasingly important to efficiently study broadband problems and nonlinear scatterers and sources. Moreover, in the frequency domain accurate boundary conditions and integral equation methods are well-established, and the costs associated with them are fundamentally easier to control. (See, e.g. [33, 38, 85].)

* Supported in part by NSF Grant DMS-9971772 and NASA Contract NAG3-2692. Any opinions, findings, and conclusions or recommendations expressed in this paper are those of the author and do not necessarily reflect the views of NSF or NASA.

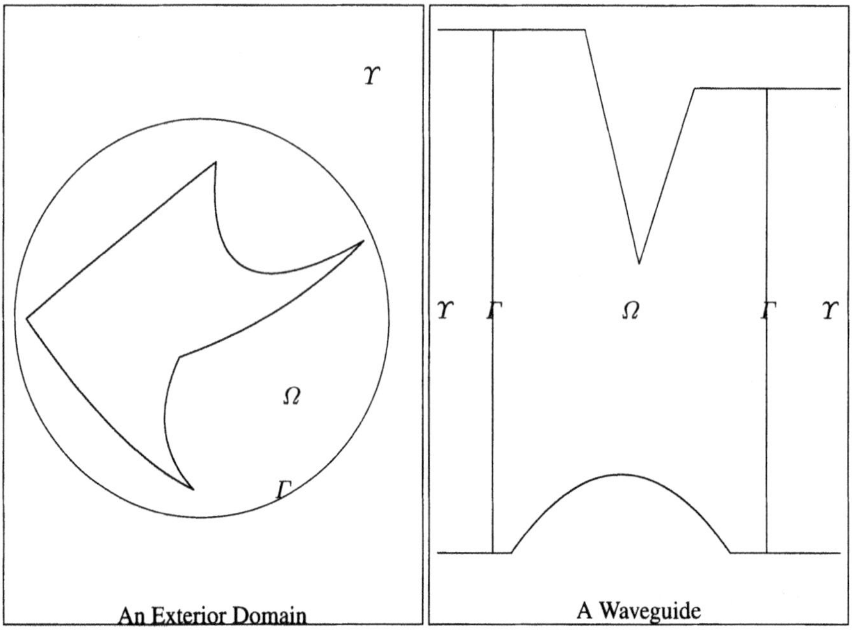

Fig. 1. Typical domain configurations. Ω is the computational domain, Γ is the artificial boundary, Υ is the tail

In 1999, the author wrote a lengthy review of the state-of-the-art in radiation boundary conditions for time-dependent problems [47]. Even then it was clear that for many important problems a number of satisfactory methods were available. Since then, new developments have provided the practitioner with a wider range of tools and raise the hope of solving the still-open problems listed in [47]. Within the narrower scope of the current work, we will review that state-of-the-art along with later contributions. Our primary goal is to explain in detail the methods that work and to illustrate their performance. We will also revisit some unsolved cases and speculate on their potential resolution.

Prior to the early nineties, the usual approach to truncating the domain was to fix some low order absorbing boundary condition. If the accuracy provided was insufficient, improvements could be made by increasing the domain size. The latter approach is both inconvenient to carry out automatically and, for three-dimensional problems in particular, rather expensive. Thus, in practice, there was no way to achieve convergence to a prescribed tolerance. The fundamental breakthroughs which changed the situation were:

i. The realisation that the boundary condition hierarchies proposed a decade earlier [16, 26, 27, 60, 61, 78] could, in fact, be stably and conveniently implemented using auxiliary functions defined only on the boundary [18, 35, 49][2];
ii. Proofs and/or numerical demonstrations of the rapid convergence with increasing order of the solutions produced by these hierarchies [45, 46, 49, 104];
iii. Low storage and fast methods to directly impose nonlocal conditions [5, 6, 28, 40, 41, 80, 86, 89];
iv. Parallel development of the perfectly matched layer (PML), an absorbing layer with a reflectionless interface [14, 17, 20, 82, 94, 96].

As mentioned earlier, the new methods provide, in many circumstances, a satisfactory solution to our problem. That is, first, they enable us to meet any prescribed error tolerance at a cost no greater than the cost of solving the interior problem. And, second, they allow us to compute on a domain which scales with the size of the scatterer, independent of solution time or error tolerance. Our hope in this paper is to explain how and why they work, focusing on the case we understand best: the scalar wave equation and its relatives.

We will illustrate many of the techniques with numerical experiments. Often these were executed with the author's research codes. The latter can be downloaded from `www.math.unm.edu/~hagstrom/downloads`. The reader is cautioned, however, that these are research codes which are not consistently documented, and that they come with no warranty whatsoever. I would also like to acknowledge the collaboration of a number of others in the work described below: Dr. Brad Alpert of NIST-Boulder, Dr. John Goodrich of NASA Glenn, Prof. Leslie Greengard of the Courant Institute, Prof. S.I. Hariharan of the University of Akron, Prof. Tim Warburton of UNM, Dr. Liyang Xu of MZA, and Igor Nazarov, currently a doctoral student at UNM.

2 Boundary Conditions

To develop useful approximate boundary conditions, it is fruitful to consider first the construction of exact boundary conditions. Plainly speaking, the condition we wish to impose is that the solution and any necessary derivatives at the boundary be the trace of an element of the set of outgoing solutions - precisely solutions of the homogeneous problem in Υ which are initially zero. One can generally derive concrete expressions for the boundary condition by performing a Laplace transformation in time with dual variable s. Taking $\Re s$ sufficiently large, the transformed problem typically has an exponential dichotomy. The exact boundary condition is then given by an appropriate projection of the solution onto the exponentially decaying subspace. Although this cookbook construction may seem difficult, we see that sometimes we can carry it through.

[2] Lindman, in the earliest of these works [78], actually suggests the auxiliary variable method.

2.1 The Wave Equation With a Cylindrical Tail

We begin with the simplest case, namely when the equation in the tail, Υ, is the scalar wave equation. Suppose we are in the first case shown in Fig. 1, so that Υ is the cylinder $(x, y) \in (0, \infty) \times \Xi$. We suppose the equation in the tail is given by:

$$\frac{1}{c^2}\frac{\partial^2 u}{\partial t^2} = \frac{\partial^2 u}{\partial x^2} + L(y, \partial/\partial y)u, \tag{1}$$

with some homogeneous boundary conditions on $\partial\Xi$. We explicitly assume that u is initially zero in Υ. We suppose L in concert with the boundary conditions is some negative, self-adjoint, linear elliptic operator with associated eigenvalues-eigenvectors:

$$L\psi_j = -\kappa_j^2\psi_j, \;\; j = 1, \ldots, \infty, \tag{2}$$

$$\int_\Xi \psi_j^2 dy = 1. \tag{3}$$

Expanding u in a Fourier series in the ψ_j and performing a Laplace transformation in t we obtain the equation:

$$\frac{\partial^2 \hat{u}_j}{\partial x^2} = \left(\frac{s^2}{c^2} + \kappa_j^2\right)\hat{u}_j, \;\; x > 0. \tag{4}$$

Clearly, the causal solution must vanish for x large and t small. Choosing $\Re s > 0$ we see that (4) has only one bounded solution, which must solve the original problem:

$$\hat{u}_j(x, s) = e^{-(c^{-2}s^2+\kappa_j^2)^{1/2}x}\hat{u}_j(0, s), \tag{5}$$

where the branch is chosen so that:

$$\Re(c^{-2}s^2 + \kappa_j^2)^{1/2} > 0, \;\; \Re s > 0. \tag{6}$$

This implies the following exact boundary condition at $x = 0$, which we write in two forms:

$$\frac{\partial \hat{u}_j}{\partial x} + c^{-1}s\hat{u}_j + \frac{\kappa_j^2}{c^{-1}s + (c^{-2}s^2 + \kappa_j^2)^{1/2}}\hat{u}_j = 0, \tag{7}$$

$$(c^{-2}s^2 + \kappa_j^2)^{-1/2}\frac{\partial \hat{u}_j}{\partial x} + \hat{u}_j = 0. \tag{8}$$

Inverting these we obtain, respectively:

$$\frac{\partial u}{\partial x} + \frac{1}{c}\frac{\partial u}{\partial t} + \mathcal{F}^{-1}\left(K_j(t) * (\mathcal{F}u(0, \cdot, \cdot))\right) = 0, \tag{9}$$

$$\mathcal{F}^{-1}\left(W_j(t) * \left(\mathcal{F}\frac{\partial u}{\partial x}(0, \cdot, \cdot)\right)\right) + u = 0. \tag{10}$$

(The first form is the one we have typically used, while the second has been used in [80].) Here, $\mathcal{F}$ is the Fourier series with respect to the eigenfunctions of L. The convolution kernels, K_j and W_j, can, in fact, be explicitly represented in terms of Bessel functions, but in general we work with their transforms directly.

Both (9) and (10) are nonlocal in space and time. However, this nonlocal operator factors into the composition of purely spatial and purely temporal operators. In many cases, for example when $L = \nabla^2$, $\mathcal{F}$ and $\mathcal{F}^{-1}$ can be applied using the FFT. Also, the temporal convolution can be treated by the algorithm in [54]. Then the cost of applying the exact nonlocal condition is acceptable. However, these algorithms require full history storage at the boundary, which is prohibitive except when the solution times are relatively short. We remark (see also [80]) that the kernels do define standard Volterra integral operators. In contrast, the retarded potential operator arising in direct integral equation formulations of the wave equation is nonstandard. The improved operators are a direct consequence of using the Dirichlet-to-Neumann or Neumann-to-Dirichlet maps.

The primary difficulty, then, is to remove the temporal and, possibly, the spatial nonlocality. The main observation is that convolution with exponential functions can performed without storing the history, but instead by solving a differential equation. If

$$\phi(t) = \int_0^t \alpha e^{-\beta(t-z)} v(z) dz, \tag{11}$$

then

$$\frac{d\phi}{dt} + \beta\phi = \alpha v, \quad \phi(0) = 0. \tag{12}$$

For the transformed variables we have:

$$\hat{\phi}(s) = \frac{\alpha}{s+\beta}\hat{v}. \tag{13}$$

The idea, then, is to construct convergent sequences of exponential approximations, $A_j^{q_j}$, to the kernels, K_j in (9) or W_j in (10), at least on finite time intervals. Working in transform space, this is equivalent to constructing rational approximations to their transforms in right half planes. To make this precise we recall the simple estimate based on Parseval's relation:

$$\|(K_j - A_j^{q_j}) * v\|_{L_2(0,T)} \leq Ce^{\eta t} \sup_{\Re s \geq \eta} |\hat{K}_j(s) - \hat{A}_j^{q_j}(s)| \cdot \|v\|_{L_2(0,T)}. \tag{14}$$

We construct $\hat{A}_j^{q_j}$ to be a rational function of degree $(q_j - 1, q_j)$. It then admits a partial fraction decomposition:

$$\hat{A}_j^{q_j} = \sum_{k=1}^{q_j} \frac{\alpha_{jk}}{s + \beta_{jk}}, \tag{15}$$

and

$$A_j^{q_j} * v = \sum_{k=1}^{q_j} \phi_{jk}(t), \tag{16}$$

$$\frac{d\phi_{jk}}{dt} + \beta_{jk}\phi_{jk} = \alpha_{jk}v, \;\; \phi_{jk}(0) = 0. \tag{17}$$

Thus we have approximated the temporal convolution by a system of differential equations for auxiliary functions defined on the boundary only. The work per time step and storage required is proportional to the number of eigenfunctions, ψ_j, the number of auxiliary functions, q_j, and the cost of computing and inverting the Fourier series.

We note that the boundary condition is also localizable in space if the j-dependence of the rational approximations can be described by polynomials in the eigenvalues, κ_j^2. We then have:

$$\hat{A}_j^{q_j} = \frac{N(s,\kappa_j^2)}{D(s,\kappa_j^2)}. \tag{18}$$

Formally (9) becomes:

$$\frac{\partial u}{\partial x} + \frac{\partial u}{\partial t} + w = 0, \tag{19}$$

$$D\left(\frac{\partial}{\partial t}, L\right) w = N\left(\frac{\partial}{\partial t}, L\right) u. \tag{20}$$

Of course we rewrite the equation for w as a first order system using additional auxiliary functions to avoid high order derivatives.

Later on we will encounter some variations of these approximations. First of all, we will often find it convenient to write $A_j^{q_j}$ as a continued fraction rather than by a partial fraction expansion. The continued fraction representation seems to allow more directly an adaptive determination of q_j, though we have not yet implemented such a scheme. Second, it is possible to compute different exponential approximations to the kernels on different time intervals - a technique which is more general than the one we've outlined. Such a method is proposed in [80] and will be discussed in more detail below.

Local Approximations

Already in [26] it was noted that Padé approximation produces boundary conditions which lead to well-posed problems. There the approximation was centred at normal incidence, which can also be thought of as an expansion of K_j valid for short time. We will follow the derivation due to Xu [104] which employs continued fractions.

To construct the approximations we note the relation:

$$\hat{K}_j(s) = \frac{\kappa_j^2}{2\frac{s}{c} + \hat{K}_j(s)}. \tag{21}$$

Then we define $\hat{A}_j^q$ recursively:

$$\hat{A}_j^q = \frac{\kappa_j^2}{2\frac{s}{c} + \hat{A}_j^{q-1}}. \tag{22}$$

In [104] a number of choices for $\hat{A}_j^0$ are considered. For example, the choice $\hat{A}_j^0 = |\kappa_j|$ leads to spatially nonlocal conditions which are exact at steady-state, generalising to high order the conditions of the type proposed in [25, 31]. Local conditions follow from the choice:

$$\hat{A}_j^0 = 0, \tag{23}$$

and may be written as the continued fraction (terminated after q terms with q independent of j):

$$\hat{A}_j^q = \frac{\kappa_j^2}{2\frac{s}{c} + \frac{\kappa_j^2}{2\frac{s}{c} + \frac{\kappa_j^2}{2\frac{s}{c} + \dots}}}. \tag{24}$$

To apply the approximate operator first define

$$w_0 = u, \quad w_k = A^{q+1-k} * w_{k-1}. \tag{25}$$

Here we have dropped the j dependence, as the multiplications by κ_j^2 may be replaced by applications of $-L$. Now the boundary condition approximating (9) is written in terms of:

$$w_1 = A^q * u. \tag{26}$$

We also have that

$$w_{q+1} = A^0 * w_q = 0. \tag{27}$$

We thus obtain the form:

$$\frac{\partial u}{\partial x} + \frac{1}{c}\frac{\partial u}{\partial t} + w_1 = 0, \tag{28}$$

$$\frac{2}{c}\frac{\partial w_k}{\partial t} = -L w_{k-1} - w_{k+1}, \quad k = 1, \dots q. \tag{29}$$

As a final step we reformulate the recursion so that only second order equations are solved. The details, along with first order reformulations suitable for incorporation into standard time-marching schemes, are found in [53]. Note that we are using the same symbols, w_k, to denote in general different auxiliary functions, except for w_1. Also, we have assumed that $q = 2P$.

$$\frac{1}{c^2}\frac{\partial^2 w_1}{\partial t^2} = \frac{1}{2} L \frac{\partial u}{\partial x} + \frac{3}{4} L w_1 - \frac{1}{4} L w_2, \tag{30}$$

$$\frac{1}{c^2}\frac{\partial^2 w_k}{\partial t^2} = -\frac{1}{4} L w_{k-1} + \frac{1}{2} L w_k - \frac{1}{4} L w_{k+1}, \quad k = 2, \dots P. \tag{31}$$

Again we emphasise the ease of applying this boundary condition to arbitrary order; it simply requires increasing the parameter P. We also note that the size of the final term, w_P, provides some measure of the error in terminating the fraction. If this measure could be made more precise, an adaptive implementation could be developed. The upcoming analysis and experiments will clarify the potential utility of an adaptive implementation. In particular we will see that the worst case analysis indicates that the necessary order could be quite large, while numerical experiments show that it is often possible to use a relatively low order.

To study the convergence of the method using (14) we must estimate:

$$\sup_{\Re s=\eta}\left|\frac{\kappa_j^2}{\frac{s}{c}+\left(\frac{s^2}{c^2}+\kappa_j^2\right)^{1/2}}-A_j^q(s)\right|. \tag{32}$$

Note that the branch points at $s=\pm ic\kappa_j$ make it difficult to derive good estimates when $\eta=0$. Thus we accept $\eta>0$. Estimates, derived in detail in [45, 104], follow recursively from the equation:

$$R^{q+1}=\frac{R^q}{(z+(z^2+1)^{1/2}+\kappa_j^{-2}R^q)(z+(z^2+1)^{1/2})}, \tag{33}$$

where

$$R^q=A_j^q(s)-\frac{\kappa_j^2}{\frac{s}{c}+\left(\frac{s^2}{c^2}+\kappa_j^2\right)^{1/2}},\quad z=\frac{s}{c|\kappa_j|}. \tag{34}$$

Using (33) we find that the error is given by:

$$|R^q|\le C\left(1+\frac{\eta}{c|\kappa_j|}\right)^{-(2q+1)}. \tag{35}$$

As we must choose $\eta=O(T^{-1})$ to obtain a meaningful estimate, we find that to achieve a relative error tolerance of ϵ for the jth harmonic we require:

$$q=O(c|\kappa_j|T\ln\frac{1}{\epsilon}). \tag{36}$$

It is relatively straightforward to turn this into an estimate depending on various Sobolev norms of u on Γ. It tells us that for a fixed problem and fixed time that the method is spectrally convergent with increasing q. However, for $c|\kappa_j|T$ large many terms may be needed.

We note that to complete the convergence argument we must derive stability estimates. This is carried through in detail in [46, 104]. A complicating feature is the fact that the exact boundary conditions themselves do not satisfy the uniform Kreiss condition [73]. However, we can show that any growth of the stability constant with q is at worst linear, so that spectral convergence is retained. In our numerical experiments here and in [48] we have used very large values of $q>100$ without observing any loss of stability.

To illustrate the stability and convergence of the local conditions we present the results of a simple numerical experiment. We solve the wave equation with $c=1$ in the two-dimensional domain $(-2,2)\times(0,1)$ for $0\le t\le 50$ with homogeneous Neumann boundary conditions on $y=0,1$. The initial data is generated by a Gaussian forcing at negative times which is shut off before $t=0$. We use the boundary conditions (28), (30), (31), an eighth order two-step method in time and eighth order differencing in space. The spatial mesh was 264×66 and the number of time steps was 20000. Evaluating to high accuracy an integral formula for the exact solution, we are able to generate precise error data, which is shown in Fig. 2. (The solution is $O(1)$ so that the absolute errors shown are comparable to relative errors.)

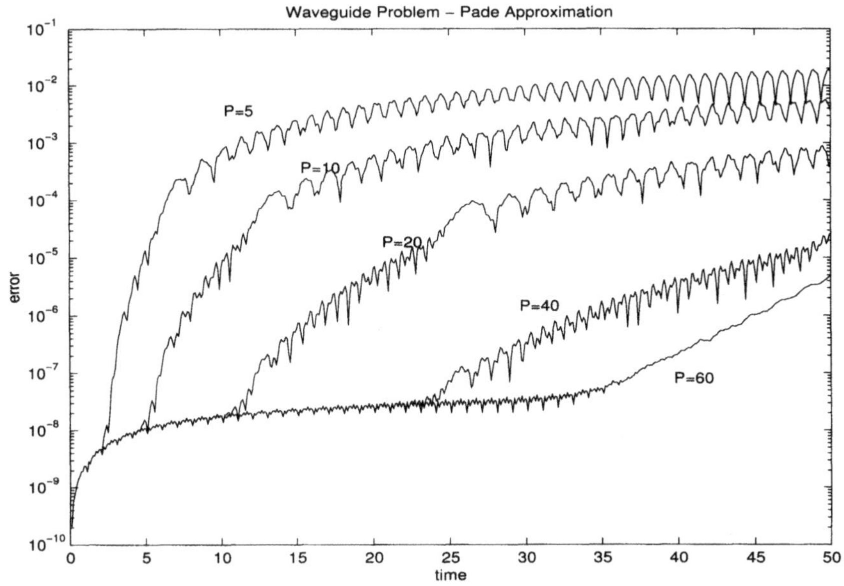

Fig. 2. Errors as a function of time and boundary condition order

The results clearly show that the Padé approximation can be used to achieve excellent accuracy at modest cost. For example, $P = 20$ suffices for a tolerance of 10^{-3} up through $t = 50$. The results are also consistent with the error analysis. Note that we are plotting the total error due both to the boundary approximations and the discretisation, which limits the gains in accuracy which can be attained by increasing P alone. The stability of the conditions for P large are also confirmed.

We finally note that a number of other local approximations are possible. In [56] various alternatives are discussed, and a reasonable case is made that they should perform better. To implement them to high order, we note from [56, 60, 61] that they are formally equivalent to:

$$\prod_{k=1}^{q} \left(\frac{1}{c_k} \frac{\partial}{\partial t} + \frac{\partial}{\partial x} \right) u = 0, \quad 0 < c_k \leq c. \tag{37}$$

Directly, (37) can be difficult to implement. However, Givoli and Neta [35] and Guddati and Tassoulas [44] have recently shown how these can be reformulated using auxiliary functions. The precise result in [35] is that (37) is equivalent to:

$$\frac{\partial u}{\partial x} + \frac{1}{c_1}\frac{\partial u}{\partial t} + w_1 = 0, \tag{38}$$

$$\left(\frac{1}{c_k} + \frac{1}{c_{k+1}}\right)\frac{\partial w_k}{\partial t} = \left(\left(\frac{1}{c^2} - \frac{1}{c_k^2}\right)\frac{\partial^2}{\partial t^2} - L\right) w_{k-1} - w_{k+1}, \quad k = 1, \ldots q. \tag{39}$$

In [44], on the other hand, a continued fraction interpolant of the wave speed function is directly computed, leading to a distinct but equivalent form, again without high order derivatives.

Clearly, (28)-(29) correspond to the special case $c_k = c$. The automatic optimisation of these parameters as well as their application to more difficult problems are subjects with great potential. (Note that Higdon has applied his boundary conditions to a variety of problems, but not to high order [63–65].)

Spatially Nonlocal Approximations

An unpleasant result of the preceding analysis is the poor behaviour of the approximations for $|\kappa_j|T$ large. As shown in [46], this sort of long time behaviour is to be expected from any homogeneous spatially local approximation. It is reasonable to ask how much better we can do if we allow spatially nonlocal conditions and construct norm-minimising rational approximations.

Results along these lines have been obtained in [6]. Theoretically it is shown that a tolerance of ϵ can be met with:

$$q = O\left(\ln \frac{1}{\epsilon} + \ln c|\kappa_j|T\right), \tag{40}$$

which is clearly superior to (36). Near-optimal approximations were also constructed numerically using a nonlinear least squares procedure. (The resulting approximations are also accurate in the maximum norm as required by the error estimate.) For example, rational approximations are constructed in [6] satisfying the tolerances in Table 1.

ϵ	$c\|\kappa_j\|T$	q
10^{-4}	10^4	21
10^{-6}	10^4	31

Table 1. Number of poles required for various tolerances and times

The rational function $\hat{A}_j^q$ is expressed in partial fraction form:

$$\hat{A}_j^q = \kappa_j^2 \sum_{k=1}^{q} \frac{\alpha_k}{\frac{s}{c} - \beta_k|\kappa_j|}, \tag{41}$$

leading to the boundary condition

$$\frac{\partial u}{\partial x} + \frac{1}{c}\frac{\partial u}{\partial t} + \sum_{k=1}^{q} \phi_k = 0, \tag{42}$$

$$\frac{\partial \hat{\phi}_{jk}}{\partial t} - \beta_k |\kappa_j| \hat{\phi}_{jk} = \alpha_k \kappa_j^2 \hat{u}_j. \tag{43}$$

Note that the coefficients α_j and β_j are independent of c and κ_j and have been computed once and for all. In particular the poles lie strictly in the left half complex plane. (Their tabulated values can be downloaded from the address mentioned above.)

We have repeated the numerical experiment described above for the wave equation in a waveguide using (42),(43) instead of (28),(30),(31). An FFT, modified for the nonuniform mesh, is used to compute the direct and inverse Fourier series. Note that we use an eighth order quadrature rule to compute the transforms, so that if we use too many Fourier modes the accuracy is degraded. The results, shown in Fig. 3, confirm the theoretical predictions. The errors when we use 16 modes are an order of magnitude smaller than the tolerances used to define the approximations themselves.

An alternative approach to constructing efficient nonlocal approximations has been proposed by Lubich and Schädle [80]. First, they reformulate the exact boundary condition using the Neumann-to-Dirichlet map (10). Second, they simplify the approximation problem somewhat by constructing local exponential approximations to the convolution kernels, $W_j(t)$, on time intervals:

$$I_l = \left(B^{l-1}\Delta t, (2B^l - 1)\Delta t\right), \tag{44}$$

where Δt is the time step employed in the discretisation and B is an integer. (They recommend $B = 10$.) The positive result of this simplification is that they can construct effective approximations with predetermined pole locations, simply approximating the inverse Laplace transform by a quadrature rule on the so-called Talbot contour. Thus they can construct the approximations themselves far more rapidly than possible with the method used in [5, 6]. This allows them to efficiently evaluate exact boundary conditions for the discrete problem, which can be more accurate when the solution is marginally resolved. On the negative side, the use of different approximations to the convolution kernel on different time intervals leads to a more complex implementation, and they require a little more work and storage than the method of [5, 6]. Precisely they require $O(\ln 1/\epsilon \cdot \ln c|\kappa_j|T)$ auxiliary functions.

2.2 The Wave Equation in an Exterior Domain

We now consider the wave equation in the second standard configuration for our class of problems, namely an exterior domain. Precisely, we suppose the tail, Υ, is such that $R^n - \Upsilon$ is bounded and that within Υ the governing equation is:

$$\frac{1}{c^2}\frac{\partial^2 u}{\partial t^2} = \nabla^2 u, \quad x \in \Upsilon. \tag{45}$$

We note that within the computational domain we may have inhomogeneities, nonlinearities, or any other perturbations.

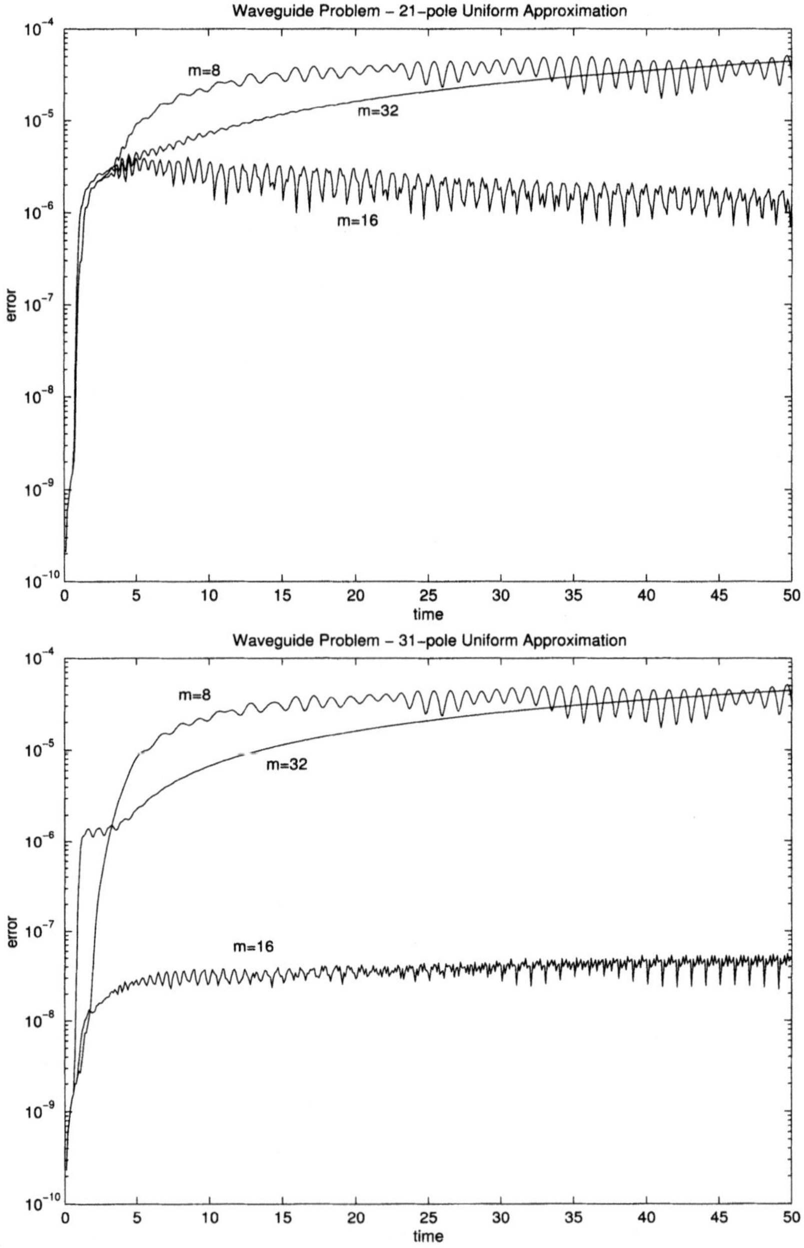

Fig. 3. Errors for the 21-pole and 31-pole least squares approximations. The parameter m is the number of Fourier modes used to evaluate the boundary condition

Performing our usual Laplace transformation in time and supposing $\Re s$ sufficiently large (typically $\Re s > 0$) we are led to the problem of describing the trace on Γ of all bounded solutions of the Helmholtz equation:

$$\nabla^2 \hat{u} - \frac{s^2}{c^2}\hat{u} = 0. \tag{46}$$

A useful format for describing this trace is the Dirichlet-to-Neumann (or Neumann-to-Dirichlet) map, which we express as:

$$\frac{\partial \hat{u}}{\partial n} = D(s)\hat{u}, \quad x \in \Gamma. \tag{47}$$

(Precisely, we may define D if Γ is sufficiently smooth by solving the exterior Dirichlet problem with data $\hat{u}$ and computing the trace of the normal derivative. Here $\partial/\partial n$ denotes the derivative into Υ, that is out of the computational domain.)

To derive concrete expressions for D it is necessary to restrict ourselves to simple boundaries, for example boundaries associated with coordinate systems in which the Helmholtz equation is separable. Even then there can be difficulties in transforming back to the time domain if the eigenfunctions of D are s-dependent. Thus the ideal case, from the point of view of analysis, is when Γ is chosen to be a sphere, and we will treat this case in great detail. However, the sphere is a wasteful choice for highly elongated or nonconvex scatterers. We will discuss some techniques which can be applied on high-aspect ratio or even nonconvex artificial boundaries. However, the problem of generalising some of the more efficient and flexible methods remains open.

Assuming now that Γ is a sphere of radius R and expanding the solution of (46) in spherical harmonics:

$$\hat{u} = \sum_{l=0}^{\infty} \sum_{m=-l}^{l} \bar{u}_{lm}(r) Y_l^m(\theta, \phi), \tag{48}$$

we find that $\bar{u}_{lm}$ satisfies:

$$\frac{\partial^2 \bar{u}_{lm}}{\partial r^2} + \frac{2}{r}\frac{\partial \bar{u}_{lm}}{\partial r} - \left(\frac{s^2}{c^2} + \frac{l(l+1)}{r^2}\right)\bar{u}_{lm} = 0. \tag{49}$$

The bounded solution of (49) is given by the modified spherical Bessel function:

$$\bar{u}_{lm}(r,s) = \frac{k_l(rs/c)}{k_l(Rs/c)}\bar{u}_{lm}(R,s), \tag{50}$$

$$k_l(z) = \frac{\pi}{2z}e^{-z}\sum_{k=0}^{l}\frac{(l+k)!}{k!(l-k)!}(2z)^{-k}. \tag{51}$$

To construct the Dirichlet-to-Neumann map we must compute the logarithmic derivative of k_l which from (51) is given by:

$$\frac{sk_l'(Rs/c)}{ck_l(Rs/c)} = -\frac{s}{c} - \frac{1}{R} - \frac{1}{R}\frac{\sum_{k=0}^{l-1}\frac{(2l-k)!}{k!(l-k-1)!}(2Rs/c)^k}{\sum_{k=0}^{l}\frac{(2l-k)!}{k!(l-k)!}(2Rs/c)^k}. \tag{52}$$

Summing over the harmonics and inverting the Laplace transform we finally find:

$$\frac{\partial u}{\partial r} + \frac{1}{c}\frac{\partial u}{\partial t} + \frac{1}{R}u + \frac{1}{R^2}\mathcal{H}^{-1}\left(S_l * (\mathcal{H}u)\right) = 0, \tag{53}$$

where $\mathcal{H}$ denotes the spherical harmonic transform and the Laplace transform of S_l is given by the last term in (52).

The primary observation to be made is that $\hat{S}_l$ is a rational function, albeit of degree $(l-1,l)$. Thus S_l can be written as a sum of exponential functions and, for each fixed harmonic, (53) can be localised in time. This fact was independently noticed by Sofronov [89, 90] and Grote and Keller [40, 41], who used it to construct temporally local boundary conditions which are exact on functions with finite harmonic content.

A second consequence of (50)-(51) is found by summing the series over l:

$$\hat{u} = \sum_{k=0}^{\infty}\frac{\pi}{(2rs/c)^{k+1}}e^{-rs/c}\sum_{l=k}^{\infty}\frac{(l+k)!}{k!(l-k)!}\sum_{m=-l}^{l}Y_l^m(\theta,\phi)\frac{\bar{u}_{lm}(R,s)}{k_l(Rs/c)}. \tag{54}$$

Defining:

$$\hat{f}_k(s/c,\theta,\phi) = \frac{\pi}{(2s/c)^{k+1}}\sum_{l=k}^{\infty}\frac{(l+k)!}{k!(l-k)!}\sum_{m=-l}^{l}Y_l^m(\theta,\phi)\frac{\bar{u}_{lm}(R,s)}{k_l(Rs/c)}, \tag{55}$$

and inverting the transform we produce the progressive wave or multipole expansion of u:

$$u = \sum_{k=0}^{\infty}\frac{f_k(ct-r)}{r^{k+1}}, \quad r \geq R. \tag{56}$$

Note again that if we assume finite harmonic content, $\bar{u}_{lm} = 0$ for $l > M$, then the series terminates with $k = M$.

From (56) we immediately derive the well-known Bayliss-Turkel conditions [16]:

$$\left(\frac{\partial}{\partial r} + \frac{1}{c}\frac{\partial}{\partial t} + \frac{2q+1}{R}\right)\cdots\left(\frac{\partial}{\partial r} + \frac{1}{c}\frac{\partial}{\partial t} + \frac{1}{R}\right)u = 0, \tag{57}$$

which we note are local and exact in the same sense as the conditions of [40, 41, 89, 90]. In the next section we will reformulate them using auxiliary functions to enable their high-order implementation.

Local Boundary Conditions

Inspired by the reformulation of the Bayliss-Turkel conditions in [15], Hagstrom and Hariharan [49] gave the first reformulation of (57) in terms of auxiliary functions on

Γ satisfying second order hyperbolic equations. Soon thereafter Huan and Thompson [70, 100] demonstrated the equivalence between the auxiliary functions of [49] and the residuals of the Bayliss-Turkel condition and also developed finite element implementations. Following them we set:

$$w_{k+1} = \left(\frac{\partial}{\partial r} + \frac{1}{c}\frac{\partial}{\partial t} + \frac{2k+1}{R}\right) \cdots \left(\frac{\partial}{\partial r} + \frac{1}{c}\frac{\partial}{\partial t} + \frac{1}{R}\right) u. \tag{58}$$

Obviously,

$$\left(\frac{\partial}{\partial r} + \frac{1}{c}\frac{\partial}{\partial t} + \frac{2k+1}{R}\right) w_k = w_{k+1}, \tag{59}$$

and (57) is equivalent to:

$$\left(\frac{\partial}{\partial r} + \frac{1}{c}\frac{\partial}{\partial t} + \frac{1}{R}\right) u = w_1, \quad w_{q+1} = 0, \tag{60}$$

combined with (59). The only problem with this formulation is the presence of the radial derivative in (59), which would force us to define the auxiliary functions in the interior. To eliminate it we note the identity derived in [70], which can easily be proven by induction:

$$\left(\frac{\partial}{\partial r} - \frac{1}{c}\frac{\partial}{\partial t} + \frac{1}{r}\right) w_k = -\frac{1}{r^2}\left(\nabla_\Gamma^2 + k(k-1)\right) w_{k-1}, \tag{61}$$

where ∇_Γ^2 denotes the Laplace-Beltrami operator on the sphere:

$$\nabla_\Gamma^2 w = \frac{1}{\sin\theta}\frac{\partial}{\partial \theta}\left(\sin\theta \frac{\partial w}{\partial \theta}\right) + \frac{1}{\sin^2\theta}\frac{\partial^2 w}{\partial \phi^2}. \tag{62}$$

Using (61) to eliminate $\frac{\partial w_k}{\partial r}$ from (59) we obtain our desired system:

$$\left(\frac{1}{c}\frac{\partial}{\partial t} + \frac{k}{R}\right) w_k = \frac{1}{2R^2}\left(\nabla_\Gamma^2 + k(k-1)\right) w_{k-1} + \frac{1}{2} w_{k+1}. \tag{63}$$

Equations (63) together with (60) are our final reformulation of (57). We again remark that they are easily implemented to any order by introducing additional auxiliary functions, that is by increasing q, and retain the property of being exact on functions described by harmonics of index up through q. We also note their obvious similarity to the planar boundary conditions based on the Padé approximation, (28)-(29). (The difference in scalings corresponds to a different scaling of the auxiliary functions.)

We have not yet carried through a complete convergence proof for these boundary conditions for fixed R and increasing q. We note that if we write:

$$u = u^{(q)} + \delta^{(q)}, \tag{64}$$

where $u^{(q)}$ is the projection of u onto the span of the harmonics of index up through q, then the consistency error is simply the result of applying the boundary condition to $\delta^{(q)}$. Moreover, for smooth solutions $\|\delta^{(q)}\|_{L_2}$ decays to zero faster than any power of q. Thus it is reasonable to conjecture that the proposed conditions are spectrally convergent uniformly in time.

An analogous sequence of boundary conditions can be constructed on circular boundaries in two space dimensions (and cylindrical boundaries in three). However, unlike the three-dimensional case, we have no expectation of time-uniform convergence. Indeed, the results of [50] essentially preclude it. In [53] we show how to rewrite these conditions so that only first order derivatives of the auxiliary functions appear, implement the conditions in a discontinuous Galerkin spectral element code for Maxwell's equations [59], and carry out numerical experiments. In Fig. 4 we plot the error over time for a two-dimensional simulation of a TE-pulse. Clearly the results are analogous to those obtained in the plane case, noting that for the waveguide experiments the solution was better resolved. (Compare with Fig. 2.)

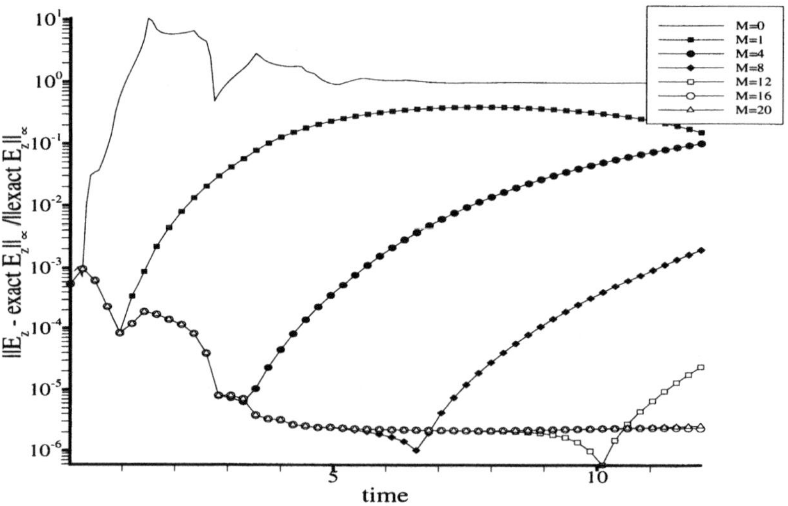

Fig. 4. Errors as a function of order, M, and time. Reformulated cylindrical Bayliss-Turkel condition. DG Maxwell solver

In addition to the applications to Maxwell's equations in [53], some other generalisations of this method have recently been completed. In [51] boundary conditions analogous to (60),(63) are constructed for the convective wave equation,

$$\left(\frac{\partial}{\partial t} + M\frac{\partial}{\partial x}\right)^2 u = \nabla^2 u, \tag{65}$$

where $0 < M < 1$ is the Mach number. These will be applicable to problems governed by the linearised subsonic Euler equations in Υ. Generalisations to the wave equation in a layered half space, motivated by problems in soil-structure interaction, have also been proposed [105].

A defect of the preceding analysis from the point of view of certain applications is the fact that the underlying expansion (56) only holds exterior to a sphere containing the problem's inhomogeneities. This could be quite wasteful of computational volume for problems with very high aspect ratio scatterers. (This defect is shared by many of the nonlocal formulations we will discuss below.) This suggests that it could be useful to find an alternative expansion with a convergence domain exterior to a high-aspect ratio boundary, which would serve as Γ. For the Helmholtz equation, Holford [67] has constructed such an expansion with Γ an oblate or prolate spheroid. Translated to the time domain, the results in [67] suggest that (56) holds exterior to a spheroid with the r coordinate replaced by its analogues in the relevant spheroidal coordinate system. On this basis we believe that a sequence of boundary conditions generalising (60),(63) could be constructed which might prove efficient for a large class of problems. We also note that related methods based on spheroidal infinite elements have been considered [8–10].

A second approach to efficiently bounding a high-aspect ratio scatterer is to use a rectangular box, applying conditions such as (28), (30),(31) or (38),(39) on each face. There are two issues which must be considered in this context. First, it is necessary to provide some boundary conditions at the edges which relate the auxiliary functions on adjacent faces. Second, it is necessary to develop some argument that the approximations will converge with increasing q.

In [102], Vacus makes substantial progress on the first issue. Considering the boundary conditions (37), he proves the existence of a unique smooth solution in the rectangular domain assuming sufficiently regular data. He also describes an algebraic procedure for deriving corner compatibility conditions. Vacus' construction involves taking high order space-time derivatives of both the equations and the boundary conditions and finding linear combinations which can be integrated to yield nontrivial constraints. It will take some additional effort to translate his results into usable compatibility relations for our preferred auxiliary variable formulations. We note that earlier Collino [18] derived corner compatibility relations for an auxiliary variable formulation of the boundary conditions based on Padé approximations. He used a sequence of exact solutions of the wave equation to determine the relations. As it is not obvious how to generalise either construction to more complex situations, a new and somewhat more straightforward approach would be very useful.

Vacus' uniqueness theorem lends some credence to the belief that the method will converge. In particular, convergence on the planes containing each face follows from the results above, and the corner compatibility conditions should lead to auxiliary functions which agree with the restrictions of the planar auxiliary functions. However, this argument is far short of a proof. Thus a definitive analysis of convergence for rectangular Γ would be an important advance. Note that it is plausible that the error estimates will be better than for the plane. The nonuniformity in time is due to the possibility of late-time glancing incidence. However, waves which impinge on one part of the boundary at glancing angles impinge on other parts nearly normally, and thus are very well-absorbed.

Nonlocal Conditions

Nonlocal boundary conditions for exterior problems have been constructed from at least three distinct formulations of exact boundary conditions. The most straightforward are based on the evaluation of (53). Directly, Grote and Keller [40, 41] compute some (typically small) number of spherical harmonic expansion coefficients and solve a system of ordinary differential equations equivalent to the order l system defined by $\hat{S}_l$. For l large, however, this method (and the companion local methods described earlier) require a large number of auxiliary variables. This number can be significantly reduced if least squares approximations are used instead. Note in this case we are approximating the degree $(l-1, l)$ rational function, $\hat{S}_l$, by a rational function of lower degree, $(q_l - 1, q_l)$. A fundamental theoretical result of [5] is that for an absolute error tolerance ϵ we can take:

$$q_l = O(\ln l + \ln 1/\epsilon). \tag{66}$$

Note that, unlike the case of a plane boundary, this estimate is uniform in time. That is, we can approximate $\hat{S}_l$ on the imaginary axis.

In either case, the nonlocal term in (53) is replaced by an expression of the form:

$$\frac{1}{R^2} \sum_{l=0}^{P} \sum_{m=-l}^{l} Y_l^m(\theta, \phi) \sum_{k=1}^{q_l} \phi_{lmk}, \tag{67}$$

with the auxiliary functions satisfying ordinary differential equations. For example, using the representation from [5, 6]:

$$\frac{1}{c} \frac{\partial \phi_{lmk}}{\partial t} + \frac{1}{R} \beta_{lk} \phi_{lmk} = \alpha_{lk} \tilde{u}_{lm}, \tag{68}$$

where $\tilde{u}_{lm}(t)$ is a coefficient in the spherical harmonic expansion of u on Γ.

The nonlinear least squares procedure has also been employed to compute the pole locations and strengths for near-optimal approximations. That is, we have computed q_l and the complex numbers α_{lk}, β_{lk} for the spherical version of (68). (They can be obtained from the website mentioned earlier.) In Table 2 we list the number of poles, and thus the number of auxiliary functions, needed for each harmonic with $\epsilon = 10^{-8}$ for both the sphere and cylinder kernels. We note that the numbers are very small - no more than 21-poles per harmonic are needed to guarantee excellent time-uniform accuracy for harmonics up through index 1024. By way of comparison, the methods of [40, 41, 49, 89, 90] would all require 1024 to be exact for such modes. Thus for a difficult problem with high harmonic content at the boundary it clearly pays to use the least squares approximations. The primary cost associated with the application of the boundary condition is the computation of the direct and inverse spherical harmonic transformations. Mohlenkamp [81], Suda and Takami [92], and Healy and coworkers [57] have devised fast algorithms for this purpose. Using them, the formal operation count for applying the boundary condition is of much lower order than the count associated with the interior solve, though for moderate size

$\hat{C}_l$		$\hat{S}_l$	
l	q_l	l	q_l
0	44		
1	15		
2	9		
3– 8	7	0– 7	l
9– 10	8	8– 10	8
11– 14	9	11– 14	9
15– 20	10	15– 19	10
21– 28	11	20– 28	11
29– 41	12	29– 40	12
42– 58	13	41– 57	13
59– 84	14	58– 83	14
85– 123	15	84– 123	15
124– 183	16	124– 183	16
184– 275	17	184– 275	17
276– 418	18	276– 418	18
419– 638	19	419– 637	19
639– 971	20	638– 971	20
972–1024	21	972–1024	21

Table 2. Number of poles needed to approximate the exact boundary condition kernels to an accuracy of 10^{-8} (taken from [5])

problems asymptotically more complex methods can be faster. Tables of poles and amplitudes can be downloaded from the web site mentioned above.

The cost of the spherical harmonic transform aside, the efficiency of the least squares approximations in the spherical case is unmatched. However, just as the sequences of local boundary conditions, their current formulation requires a spherical (cylindrical) artificial boundary, which is an expensive choice for a high-aspect ratio scatterer. Possible solutions, again just as in the local case, are to extend the construction to spheroids or boxes. A problem here is that exact conditions on these boundaries are fundamentally more complex. In particular, they can not be diagonalised by a fixed spatial basis. We feel that the development of efficient representations of boundary conditions on high-aspect ratio surfaces is an important problem which should be studied further.

The other two nonlocal formulations both allow for the flexible choice of Γ, but are otherwise more costly. The first, suggested in [101], is based on a formulation of exact boundary conditions using retarded potentials. It involves enclosing the scatterer by two surfaces, Γ_I and Γ_O, as indicated in Fig. 5.

Then we can evaluate u on Γ_O using past values of u and its derivatives on Γ_I. Precisely:

$$u(x,t) = -\frac{1}{4\pi}\int_{\Gamma_I}\left(u\frac{\partial}{\partial n}\left(\frac{1}{r}\right) - \frac{1}{r}\frac{\partial u}{\partial n} - \frac{1}{rc}\frac{\partial r}{\partial n}\frac{\partial u}{\partial t}\right)dy, \tag{69}$$

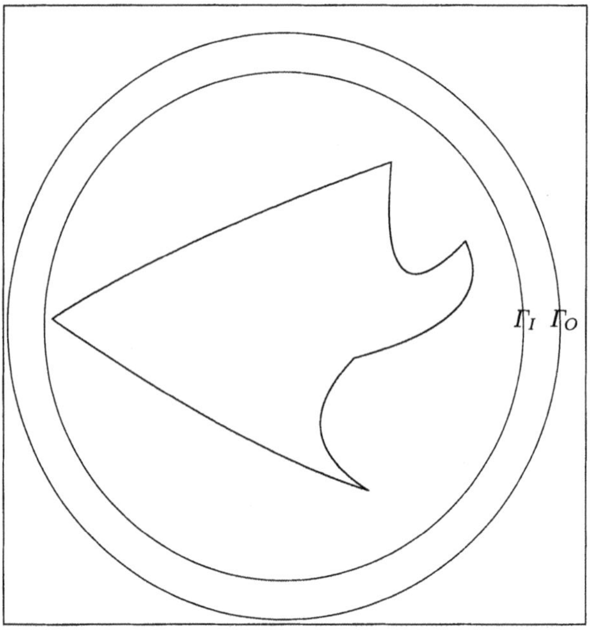

Fig. 5. Domain configurations with two boundaries

where $x \in \Gamma_O$, $\partial/\partial n$ is the normal derivative, $r = |x - y|$ and the time argument of u in the integral is $t - r/c$. We note that although the formula is nonlocal in time, the required history extends over only a finite interval determined by the maximum travel time between points on the boundary.

The first implementation of (69) appears in [34]. However, it employs a direct numerical evaluation of the integral, which implies a cost per time step proportional to the square of the number of boundary points. This is more costly by one order than the interior solve, and is thus not competitive with other methods. Recently, a fast algorithm for evaluating retarded potentials has been devised [28]. Using this algorithm, the formal operation count associated with (69) becomes comparable to the counts for the methods described above, though in practice it is still quite expensive. However, the method is competitive in certain special cases, for example for highly nonconvex scatterers.

The final method, due to Ryaben'kii and coworkers [86, 87], is based directly on the strong Huygens' principle (or the existence of lacunae) which itself follows from (69). As a first step, consider again Fig. 5 and introduce a smooth cutoff function, $\zeta(x)$, satisfying $\zeta = 1$ at and beyond Γ_O and $\zeta = 0$ inside Γ_I. Introducing the auxiliary function:

$$v = \zeta u, \tag{70}$$

we see that v satisfies a forced wave equation in all space:

$$\frac{1}{c^2}\frac{\partial^2 v}{\partial t^2} = \nabla^2 v + f, \quad x \in R^3, \tag{71}$$

where f is supported between Γ_O and Γ_I. Since $u = v$ on Γ_O, an exact Dirichlet boundary condition can be imposed if a solution of (71) can be computed. To solve (71) we note that if f is supported only in the time interval (t_j, t_{j+1}), the strong Huygens' principle would imply that $v \neq 0$ on Γ_O only on some larger time interval, (t_j, T_{j+1}). Moreover, on this time interval, v restricted to Γ_O is identical to the solution of a periodic problem in space with sufficiently large period.

The algorithm proposed in [87] is based on these observations. By a smooth partition of unity in time, f is written as the sum of functions, f_j, which are supported on time intervals of fixed duration. Then v on Γ_O is written as the sum of functions v_j which solve spatially periodic wave equations with forcing f_j. The periodic spatial domains must be chosen fairly large - larger than the domains required by a spherical artificial boundary. However, the equation for the functions v_j is simple and can be very efficiently solved using a Fourier spectral method. Thus the resulting algorithm is competitive.

We finally mention an intriguing result of Warchall [103]. He proves that if u solves the wave equation, with all inhomogeneities and initial data supported inside a convex domain, Ω, bounded by Γ, the solution at times $t \geq t_p \geq 0$ is completely determined by the data within $\Omega \cap B(x, c(t - t_p))$ at time t_p. This seems to imply that an exact boundary condition with no history dependence or auxiliary variables exists. However, to date this exact operator has not been found.

2.3 First Order Hyperbolic Systems

We now return to the case of a cylindrical tail, $\Upsilon = (0, \infty) \times \Xi$. However we assume that Ξ is rectangular and that the equation in Υ is the first order, constant coefficient hyperbolic system:

$$\frac{\partial u}{\partial t} = \Lambda \frac{\partial u}{\partial x} + \sum_j B_j \frac{\partial u}{\partial y_j} + Cu, \tag{72}$$

where the diagonal matrix Λ is partitioned into incoming and outgoing pieces. (We are assuming that $x = 0$ is noncharacteristic, but that assumption can be relaxed.)

$$\Lambda = \begin{pmatrix} \Lambda^+ & O \\ O & \Lambda^- \end{pmatrix}, \quad \Lambda^+ > O, \ \ \Lambda^- < O. \tag{73}$$

We also suppose that appropriate boundary conditions are imposed on $\partial\Xi$. After a Fourier-Laplace transformation, we derive the system of equations in x:

$$\frac{\partial \bar{u}}{\partial x} = \Lambda^{-1} \left(sI - \sum_j ik_j B_j - C \right). \tag{74}$$

For any fixed transverse mode, this system clearly possesses an exponential dichotomy for $\Re s$ sufficiently large. Moreover, the dimension of the subspace of growing solutions is that of Λ^+. An exact boundary condition is then given by:

$$P^+(s, k)\bar{u} = 0, \tag{75}$$

where P^+ is a projector into the subspace of growing solutions. Noting further that in the large s limit this equation reduces to $u^+ = 0$ we reach our final form:

$$u^+ = \mathcal{F}^{-1}\left(R(t,k) * (\mathcal{F}u)\right), \tag{76}$$

where R is now a matrix.

We note first that the projection operator is not unique and not every choice leads to a well-posed problem. For example, the obvious choice of left eigenvectors to define P^+ leads to an ill-posed problem for the linearised, subsonic Euler equations [32, 48]. Second, we have restricted the cross-section to a rectangle so that we could reduce the problem to a linear algebra problem using Fourier series. More generally the projection onto outgoing waves (decaying solutions after Laplace transformation) is more difficult. Finally, even if we can reduce the problem to linear algebra, we may be unable to find analytic representations of the projectors.

Despite these difficulties, there are a number of important examples where we can make progress. Not surprisingly, in each of these cases the basic propagating modes either satisfy the wave equation or the simple transport equation. Below we will discuss two such cases in some detail, Maxwell's equations and the linearised Euler equations. Other systems which can be discussed include the linearised shallow water equations, which are closely related to the Euler equations, and the equations of linear elasticity, which are typically formulated as a second order system. For a treatment of the latter see [39, 43, 62, 63, 83].

Maxwell's Equations

Maxwell's equations in free space using appropriate units are given by:

$$\frac{1}{c}\frac{\partial E}{\partial t} - \nabla \times B = 0, \tag{77}$$

$$\frac{1}{c}\frac{\partial B}{\partial t} + \nabla \times E = 0. \tag{78}$$

Systematic derivations of exact boundary conditions can be found in [42, 47]. Here we'll take a shortcut, noting that each of the Cartesian components of the field vectors satisfies the scalar wave equation. Thus, we can use the boundary conditions discussed above to eliminate the normal derivatives of any variable. We simply use the normal characteristic analysis to predict which variables we need to specify. Note that for Maxwell's equations there are, at any boundary, two incoming, two outgoing and two characteristic variables. Thus we will impose two boundary conditions. See [53] for more details and numerical experiments with local conditions.

In our waveguide geometry, where x is the normal coordinate and (y, z) the tangential coordinates, the incoming normal characteristic equations are:

$$\left(\frac{1}{c}\frac{\partial}{\partial t} - \frac{\partial}{\partial x}\right)(E_y - B_z) + \frac{\partial E_x}{\partial y} - \frac{\partial B_x}{\partial z} = 0, \tag{79}$$

$$\left(\frac{1}{c}\frac{\partial}{\partial t} - \frac{\partial}{\partial x}\right)(E_z + B_y) + \frac{\partial E_x}{\partial z} + \frac{\partial B_x}{\partial y} = 0. \tag{80}$$

Using (9) to eliminate the x-derivatives yields the exact boundary condition:

$$\left(\frac{2}{c}\frac{\partial}{\partial t} + \mathcal{F}^{-1} \circ (K_j *) \circ \mathcal{F}\right)(E_y - B_z) + \frac{\partial E_x}{\partial y} - \frac{\partial B_x}{\partial z} = 0, \tag{81}$$

$$\left(\frac{2}{c}\frac{\partial}{\partial t} + \mathcal{F}^{-1} \circ (K_j *) \circ \mathcal{F}\right)(E_z + B_y) + \frac{\partial E_x}{\partial z} + \frac{\partial B_x}{\partial y} = 0. \tag{82}$$

Finally, we can replace the exact nonlocal term by either the local sequences or the nonlocal least squares approximations.

A similar procedure can be carried out on a sphere. Each Cartesian component of the fields satisfies (53). One simply writes the equations for the incoming variables at each point and replaces r derivatives using (53) and whatever local or nonlocal approximation to the nonlocal term, $\mathcal{H}^{-1}(S_l * (\mathcal{H}u))$, is chosen. As described, this method would require the definition of auxiliary functions for all six components. However, the more complex derivations in [42, 47] show that only two sets are required.

Linearised Euler Equations

As a second example, consider the compressible Euler equations linearised about a uniform, subsonic flow:

$$\frac{D\rho}{Dt} + \nabla \cdot u = 0, \tag{83}$$

$$\frac{Du}{Dt} + \frac{1}{\gamma}\nabla p = 0, \tag{84}$$

$$\frac{Dp}{Dt} + \gamma \nabla \cdot u = 0. \tag{85}$$

Here the material derivative is defined by:

$$\frac{D}{Dt} \equiv \frac{\partial}{\partial t} + U \cdot \nabla, \tag{86}$$

and U is the uniform flow field about which we've linearised. Note that units have been chosen so that the sound speed is one. Thus:

$$\sum_{j=1}^{3} U_j^2 = M^2 < 1. \tag{87}$$

We also assume $U_1 > 0$.

The linearised Euler equations support waves of differing types. Combining (84) and (85) we find that the pressure, p, satisfies the convective wave equation:

$$\frac{D^2 p}{Dt^2} = \nabla^2 p. \tag{88}$$

On the other hand, the entropy, $S = p - \gamma\rho$, and the vorticity, $\omega = \nabla \times u$ satisfy the transport equation:

$$\frac{DS}{Dt} = 0, \ \frac{D\omega}{Dt} = 0. \tag{89}$$

Thus the boundary condition problem for the pressure is similar to the boundary condition problem for the wave equation. Indeed, we have shown in [48, 51] that the local boundary condition sequences discussed above for both waveguide and exterior geometries can be generalised to the convective wave equation. In addition, exact boundary conditions for (89) are extremely simple; namely:

$$S = 0, \ \ \omega = 0, \ \text{ if } \ U \cdot n < 0, \tag{90}$$

where U is the outward normal - that is if the flow is incoming. The primary difficulty is to combine these relationships to produce a well-posed problem at inflow.

In [48] the full range of well-posed formulations of exact boundary conditions is displayed. (See also [84, 91].) At outflow, where there is a single incoming acoustic mode, the boundary condition is essentially uniquely defined:

$$\frac{D_{\tan}}{Dt}(p - u_1) + \mathcal{K}p + U_1 \left(\frac{\partial u_2}{\partial x_2} + \frac{\partial u_3}{\partial x_3} \right) = 0, \tag{91}$$

At inflow there are many possibilities. The simplest realisation, which is used in the numerical experiments, is:

$$\gamma\rho - p = 0, \tag{92}$$

$$\frac{D_{\tan}}{Dt}(p + u_1) + \frac{1}{2}\mathcal{K}(p + u_1) + \frac{1 - U_1}{2} \left(\frac{\partial u_2}{\partial x_2} + \frac{\partial u_3}{\partial x_3} \right) = 0, \tag{93}$$

$$\frac{D_{\tan} u_2}{Dt} + U_1 \frac{\partial u_1}{\partial x_2} + \frac{\partial p}{\partial x_2} = 0, \tag{94}$$

$$\frac{D_{\tan} u_3}{Dt} + U_1 \frac{\partial u_1}{\partial x_3} + \frac{\partial p}{\partial x_3} = 0. \tag{95}$$

Here, $\mathcal{K}$ denotes a nonlocal operator very closely related to the operator appearing in (9). Precisely:

$$\mathcal{K}w = \mathcal{F}^{-1} \left((1 - U_1)^2 k^2 K(\sqrt{1 - U_1^2} kt) * (\mathcal{F}w) \right), \tag{96}$$

where $K(\kappa_j t)$ is the kernel we had before. In particular, both the local and nonlocal approximations we have discussed can be used here. In [48], for example, we use the Padé sequence for a problem involving a periodic array of pressure pulses. The results are similar to those reported in Fig. 2.

As mentioned above, it is also possible to generalise the local boundary condition sequence we have discussed from the wave equation to the convective wave equation [51]. However, we have yet to complete the construction of a well-posed inflow condition. We note that for the exterior problem a smooth artificial boundary must contain a point where $U \cdot n = 0$. Thus the switch between inflow and outflow conditions may require some additional compatibility conditions for the auxiliary variables. Given the importance of compressible flow problems, the resolution of this issue is certainly important.

Variable Coefficients and General Systems

We've now completed the list of hyperbolic problems which we can satisfactorily solve. Although it contains many of the more important equations arising in applications, it certainly doesn't contain them all. Particularly glaring is our inability to treat variable coefficient problems and general hyperbolic systems.

We note that in their original work [27], Engquist and Majda do treat the variable coefficient case from the perspective of the reflection of singularities. More recently, practical algorithms for constructing the Engquist-Majda conditions to high order have been proposed [7]. However, the underlying expansions are only convergent up to smooth errors, which may not be small. Thus, unless a stronger form of convergence can be established for some cases, it is not obvious that high order constructions will be useful. (We note that a sort of convergence has been established in the highly oscillatory limit in [55] for the lowest order Engquist-Majda conditions.)

A second approach to these issues is the reformulation of Higdon's boundary conditions in [35, 36]. Already in their original version applications to problems in stratified media have been undertaken [64]. For problems with travelling waves only, the input to the method is simply a range of wave speeds which can, in principle, be estimated from the coefficients. However, the reformulations currently require a second order system and need to be generalised to the first order case. Moreover, modal solutions to variable coefficient problems may be growing or decaying and it is unclear if the conditions need to be reformulated in that case. In any event, we believe the method has great promise, but it is clear that much more experimentation and analysis is needed.

2.4 The Schrödinger Equation

Waves, particularly dispersive waves, can also be described by higher order equations. Work on these systems is less well-developed, though low order conditions of Higdon type have certainly been used. In recent years, however, there have been advances in the treatment of the most important equation from this class, the Schrödinger equation. In particular, two recent dissertations, by Jiang at NYU [72] and Schädle in Tübingen [88] deal with the construction, analysis and testing of highly accurate nonlocal conditions. (See also [80].)

We begin again with the case of a cylindrical tail:

$$-i\frac{\partial u}{\partial t} = \frac{\partial^2 u}{\partial x^2} + L(y, \partial/\partial y)u. \tag{97}$$

Expanding in a Fourier series in the cross-section based on the eigenfunctions of L and performing a Laplace transform in time we obtain a formula for an exact boundary condition:

$$\frac{\partial \hat{u}_j}{\partial x} + (-is + \kappa_j^2)^{1/2} \hat{u}_j = 0. \tag{98}$$

We immediately note an important difference between (98) and its wave equation analogue. To derive boundary conditions for the wave equation, we simply removed the large s behaviour from the nonlocal term until the remaining operator was equivalent to convolution with a bounded function. As for large s the symbol was linear, the total operator was the sum of a differential operator and the convolution. Here, in contrast, the symbol behaves like $(is)^{1/2}$ for s large. To desingularise this somewhat, we can use the Neumann-to-Dirichlet map as in [80, 88],

$$\mathcal{F}^{-1}\left(V_j * \left(\mathcal{F}\frac{\partial u}{\partial x}\right)\right) + e^{-i\frac{\pi}{4}} u = 0, \tag{99}$$

or, as in [72], write:

$$(-is + \kappa_j^2)^{1/2} = e^{-i\frac{\pi}{4}} \frac{(s + i\kappa_j^2)}{(s + i\kappa_j^2)^{1/2}}, \tag{100}$$

leading to:

$$\frac{\partial u}{\partial x} + e^{-i\frac{\pi}{4}} \mathcal{F}^{-1}\left(V_j * \left(\mathcal{F}\left(\frac{\partial u}{\partial t} - iLu\right)\right)\right) = 0. \tag{101}$$

In either case, the boundary condition is expressed in terms of convolution with:

$$V_j(t) = \mathcal{L}^{-1}(s + i\kappa_j^2)^{-1/2} = \frac{e^{-i\kappa_j^2 t}}{\sqrt{\pi t}}. \tag{102}$$

Unlike the kernels we dealt with earlier, the singularity at $t = 0$ precludes the uniform approximation of V_j on $(0, \infty)$. Thus both authors split the convolution into two pieces, a local piece on $(0, \delta)$ to be handled by direct discretisation, and a global piece where V_j is replaced by exponential or piecewise exponential functions. Jiang's global approximation is based on approximating the integral formula:

$$\frac{1}{\sqrt{\pi t}} = \frac{2}{\pi} \int_0^\infty e^{-z^2 t} dz, \tag{103}$$

and leads to the estimate:

$$q_l = O\left(\ln 1/\epsilon \cdot (\ln T/\delta + \ln\ln 1/\epsilon)\right). \tag{104}$$

Lubich and Schädle, on the other hand, use the same algorithm they suggest for kernels arising from the wave equation. They require about the same number of poles, though the implementation is more complex. They do treat the discrete problem, however.

Similarly, exact boundary conditions can be formulated on spherical and cylindrical boundaries. Transforms of the resulting kernels are still expressed by logarithmic derivatives of modified Bessel functions, but now the arguments are proportional to $s^{1/2}$. Theoretically, Jiang shows that the approximation theorems of [5] can be applied in this case, precisely to:

$$\mathcal{L}^{-1}\left(\frac{\sqrt{is}K_\nu'(\sqrt{is})}{(s-s_\nu)K_\nu(\sqrt{is})}\right). \tag{105}$$

(A multiplication and division by $s-s_\nu$ has been performed to regularise the kernel.) Choosing δ as above he shows that for harmonics of index l the number of poles required is:

$$q_l = O\left((\ln 1/\epsilon + \ln 1/\delta)\cdot(\ln l + \ln 1/\epsilon + \ln 1/\delta)\right). \tag{106}$$

Actual approximations are also computed in [72] by a least squares procedure. Generally, more poles are required than for the simpler case of the wave equation. Nonetheless, these nonlocal approximations do provide arbitrary accuracy at relatively small cost. Moreover, high-order local alternatives do not as yet exist.

As an example of the difficulties in constructing high-order local boundary conditions for the Schrödinger equation, we point to the recent paper of Allonso-Mallo and Reguera [4], where they study the stability of Higdon-type boundary conditions proposed for the Schrödinger equation in [24, 29]. They find mild instabilities which worsen with increasing order for semi-discretisation of the problem. Their analysis ends at rather low order by our standards - a rational approximation of degree $(3, 2)$. It leads to doubt about the utility of very high order conditions, though this still needs to be tested.

3 Absorbing Layers

The alternative to domain truncation by accurate boundary conditions is the use of an absorbing or sponge layer. The idea is simple - extend the computational domain from the boundary of the physical domain, Γ_I, to a second boundary, Γ_O, and change the equations in the new buffer zone so that waves decay. It is obviously not difficult to write down dissipative wave equations whose solutions decay rapidly as they propagate. However, impinging waves generally reflect off the interface or transition zone and thus return to Ω. As a result, early methods of this type (e.g. [71]) required a gradual increase in the absorption parameters, leading to fairly thick layers. All of this changed with Bérenger's introduction of the perfectly matched layer (PML) for Maxwell's equations [14].

The new property of the PML is a reflectionless interface with the physical domain. In the original paper, the origins of this property are somewhat obscured. However, soon thereafter it was noted that the PML was in fact equivalent to a continuation into the complex plane of the real, normal coordinate, for example x for our cylinder domains [17]. This observation has led to a number of important generalisations to curvilinear coordinates [20, 82, 94, 96], which we'll discuss below, and to more complex media [95, 97, 98]. We will present a nonstandard derivation of the PML which generalises the coordinate transformation approach. A result of this generalisation has been the extension of the PML to the linearised Euler equations. Interestingly, in this framework the reflectionless interface property is easily satisfied, but the dissipativity condition is not.

3.1 Perfectly Matched Layers for Hyperbolic Systems

Consider our usual cylindrical tail and suppose we are solving a first order hyperbolic system which we allow to have varying coefficients in the cross section;

$$\frac{\partial u}{\partial t} + A(y)\frac{\partial u}{\partial x} + \sum_j B_j(y)\frac{\partial u}{\partial y_j} + C(y)u = 0. \tag{107}$$

Just as before, the physical domain is located in $x < 0$, the interface, Γ, between the physical domain and the absorbing region is located at $x = 0$, and the absorbing region itself lies between $x = 0$ and $x = L$.

Performing a Laplace transformation in time we find modal solutions of (107):

$$\hat{u} = e^{\lambda x}\phi, \tag{108}$$

where

$$\left(sI + \lambda A + \sum_j B_j(y)\frac{\partial}{\partial y_j} + C(y)\right)\phi = 0. \tag{109}$$

We label incoming and outgoing solutions by looking at $\Re\lambda$ for $\Re s$ sufficiently large. However, in the limit $\Re s \to 0$, we typically have $\Re\lambda \to 0$. The idea behind the layer construction is to modify the equation so that modes are damped as they propagate, meaning that $\Re\lambda$ is bounded away from zero for $\Re s \geq 0$. We also want to assure that there is no reflection at the interface between the physical and absorbing regions. The most direct way to do this is to design the problem in the layer so that its modal solutions for any s have the same eigenfunctions, ϕ, as (109).

Thus our starting point for constructing the layer is the formal modal solution:

$$\hat{u} = e^{\lambda x + \left(\lambda \hat{R}^{-1} - \mu\right)\int_0^x \sigma(z)dz}\phi. \tag{110}$$

In the examples we have carried through so far, R is a first order differential operator in time and the transverse variables with Laplace transform $\hat{R}$,

$$R = \frac{\partial}{\partial t} + \sum_j \beta_j \frac{\partial}{\partial y_j} + \alpha, \tag{111}$$

μ is some number, and $\sigma \geq 0$ is the absorption parameter. We note that more complex choices for R and μ are certainly possible, for example R and μ could be defined by higher degree rational functions of s with operator or matrix coefficients. This may prove to be necessary to extend the method to other problems. We also note that if $\mu = 0$ and $\hat{R}$ is represented by a function of s (and tangential wave numbers for constant coefficient problems), we can view the transformation as the result of extending x into the complex plane, as in [17] and its descendants.

It is now a simple matter to write down a system of equations satisfied by u in the layer. In particular, when we substitute (110) into the new system we want (109) to be the result. This yields:

$$\left(sI + (I - (R+\sigma)^{-1}\sigma)A\left(\frac{\partial}{\partial x} + \sigma\mu\right) + \sum_j B_j \frac{\partial}{\partial y_j} + C \right) \hat{u} = 0. \tag{112}$$

Introducing auxiliary functions w we finally obtain:

$$\frac{\partial u}{\partial t} + A(y)\left(\frac{\partial u}{\partial x} + \sigma\mu u\right) + \sum_j B_j(y)\frac{\partial u}{\partial y_j} + C(y)u + w = 0. \tag{113}$$

$$Rw + \sigma w + \sigma A(y)\left(\frac{\partial u}{\partial x} + \sigma\mu u\right) = 0. \tag{114}$$

We emphasise that the interface, $x = 0$, is nonreflecting for *any* choice of R, μ and σ. The difficulty is to choose them so that all waves are damped. Below we will examine some special constant coefficient cases where this can be achieved, albeit by inspection rather than by some systematic process.

A practical difficulty in implementing the PML is the choice of the absorption profile, σ. This is purely a numerical issue, as theoretically we can choose it to be constant and arbitrarily large. In our experiments we make it linear, and as our methods are high order we treat the interface as an internal boundary and use characteristic matching across it. However, we typically need to experiment to find near optimal slopes for the linear profile. The only serious analysis of this question which we are aware of is [19], where numerically optimal profiles for a specific discretisation in the frequency domain are found.

Maxwell's Equations

We return to (77)-(78), which we recall was the first system for which a PML was constructed [14]. Reordering the unknowns into the 6-vector given by $u = (E_x, B_x, E_y, B_y, E_z, B_z)^T$ we note that the matrix A is given in block form by:

$$A = c\begin{pmatrix} O & O & O \\ O & O & T \\ O & -T & 0 \end{pmatrix}, \quad T = \begin{pmatrix} 0 & 1 \\ -1 & 0 \end{pmatrix}. \tag{115}$$

From (114) and the structure of A we conclude that the first two components of w are zero. Thus we will take w to be a 4-vector appearing in the last four equations of

(113) and (114). As the system has constant coefficients we may additionally perform a Fourier transformation in y and z. This leads to explicit expressions for λ:

$$\lambda = 0, \pm \left(\frac{s^2}{c^2} + k_x^2 + k_y^2 \right)^{1/2}, \tag{116}$$

which, of course, are by now quite familiar. Letting $\bar{R}$ denote the symbol of R our problem is to guarantee:

$$\Re \left(\frac{\left(\frac{s^2}{c^2} + k_x^2 + k_y^2 \right)^{1/2}}{\bar{R}} - \mu \right) > 0. \tag{117}$$

A simple choice which by an easy calculation can be shown to satisfy (117) is:

$$R = \frac{s}{c} + \alpha, \quad \alpha \geq 0, \quad \mu = 0, \tag{118}$$

leading to the equations:

$$\frac{1}{c} \frac{\partial E}{\partial t} - \nabla \times B + \begin{pmatrix} 0 \\ w_y^{(E)} \\ w_z^{(E)} \end{pmatrix} = 0, \tag{119}$$

$$\frac{1}{c} \frac{\partial B}{\partial t} + \nabla \times E + \begin{pmatrix} 0 \\ w_y^{(B)} \\ w_z^{(B)} \end{pmatrix} = 0, \tag{120}$$

$$\left(\frac{1}{c} \frac{\partial}{\partial t} + \alpha + \sigma \right) w^{(E)} + \sigma \begin{pmatrix} \frac{\partial B_z}{\partial x} \\ -\frac{\partial B_y}{\partial x} \end{pmatrix} = 0, \tag{121}$$

$$\left(\frac{1}{c} \frac{\partial}{\partial t} + \alpha + \sigma \right) w^{(B)} + \sigma \begin{pmatrix} -\frac{\partial E_z}{\partial x} \\ \frac{\partial E_y}{\partial x} \end{pmatrix} = 0. \tag{122}$$

Although the equations themselves look quite different, a study of the eigenvalues reveals the formal equivalence between solutions of (119)-(122) and solutions in the layer constructed in [13]. In particular, the choice of $\alpha > 0$ corresponds to the complex frequency shift discussed in [13].

Linearised Euler Equations

We now consider the linearised Euler equations, (83)-(85), again assuming a subsonic flow with $U_1 \neq 0$. Then A is nonsingular so that we require the full complement of auxiliary variables. Again we perform a Fourier-Laplace transformation and find that λ is given by:

$$\lambda = -\frac{\tilde{s}}{U_1}, \quad \frac{U_1\tilde{s} \pm \left(\tilde{s}^2 + (1-U_1^2)(k_y^2+k_z^2)\right)^{1/2}}{1-U_1^2}, \tag{123}$$

$$\tilde{s} = s + ik_yU_2 + ik_zU_3. \tag{124}$$

We note the two distinct forms for λ, corresponding to the two types of waves.

We first choose $\bar{R}$ in analogy with the case of Maxwell's equations,

$$\bar{R} = \tilde{s}, \tag{125}$$

is suggested. Then $\lambda/\bar{R}$ is given by:

$$-\frac{1}{U_1}, \quad \frac{U_1}{1-U_1^2} \pm \frac{\left(\tilde{s}^2 + (1-U_1^2)(k_y^2+k_z^2)\right)^{1/2}}{(1-U_1^2)\tilde{s}}. \tag{126}$$

Clearly when the square root is zero we can't have the correct sign on both terms. Thus we choose:

$$\mu = \frac{U_1}{1-U_1^2}. \tag{127}$$

Now all terms now have the correct sign. We note that, as in (118), we could add a positive lower order term, α, to R. However, it would then be necessary to make μ dependent on $\tilde{s}$, leading to additional auxiliary variables.

We have carried out preliminary numerical experiments with the new layer for a problem defined by a periodic array of pressure pulses in the two-dimensional uniform flow $U_1 = 0.3$, $U_2 = 0.4$. The numerical method is eighth order in space and time. Relative errors as a function of time and average absorption parameter are listed in Table 3. Here the interior mesh is 258×64 while we have 34×64 points in each layer. The absorption profile is linear, and the layer is terminated by characteristic end conditions. Clearly, the accuracy is acceptable. Moreover, the insensitivity of the results to changes in $\bar{\sigma}$ indicates that the primary error source is the discretisation itself. We note that the same problem has been solved in [48] using the Padé sequence of local boundary conditions, but there the solutions were better resolved so that the dominant error was due to the boundary condition. We plan to redo these computations so that direct comparisons can be made.

t	$\bar{\sigma} = 50$	$\bar{\sigma} = 75$	$\bar{\sigma} = 100$
1	1.6(−3)	1.6(−3)	1.6(−3)
2	1.4(−3)	1.4(−3)	1.4(−3)
4	1.1(−3)	1.1(−3)	1.1(−3)
8	1.3(−3)	1.4(−3)	1.4(−3)
16	2.4(−3)	2.5(−3)	2.9(−3)
32	3.4(−3)	2.4(−3)	2.4(−3)

Table 3. Relative errors - periodic array of pressure pulses, Euler PML $L = 1/2$

There have been other earlier attempts to generalise the PML to the Euler equations, beginning with Hu [68]. However, these all had stability problems [37, 58, 93].

Recent fixes in [2, 69] are restricted to flows aligned with the layer coordinate, while others [79] sacrifice the perfect matching property. Thus the general formulation developed here seems to have provided a convenient framework for constructing a true PML. We note that similar PMLs for advective problems have been constructed in [11, 23, 69].

Returning to the variable coefficient formulation, we have also begun experiments with the PML for linearisations about jet flow profiles [52]. A confounding factor in this case is the instability of the flow profile itself. Now the parameter μ is chosen numerically to suppress instabilities in the layer. So far, we have been unable to construct an absolutely stable layer, but we have been able to make the growth rate small enough that accurate results can be obtained over long time intervals.

Exterior Problems

There are a variety of ways that the PML can be applied to problems in exterior domains. In particular, rectangles or convex regions of arbitrary aspect ratio can be used. This is, in our view, the primary advantage of the PML over high-order absorbing boundary conditions.

The standard approach is to use a rectangular box. Then, besides unidirectional layers attached to each face, there must be corner layers attached to the layers themselves. In these we use multiple sets of auxiliary variables. In a three dimensional corner, for example, we use three sets of auxiliary functions. Thus our general formulation would be:

$$\frac{\partial u}{\partial t} + \sum_j A_j \left(\frac{\partial}{\partial x_j} + \mu_j \right) u + Cu + \sum_j w^{(j)} = 0, \tag{128}$$

$$R_j w^{(j)} + \sigma_j A_j \left(\frac{\partial}{\partial x_j} + \mu_j \right) u + \sigma_j w^{(j)} = 0. \tag{129}$$

(Here we assume constant coefficients and rename the coefficient matrices.)

An alternative to a rectangular box is a curved boundary. Using the coordinate mapping technique, the PML has been extended to spheres and cylinders [20, 82, 94]. The idea is simply to apply the complex change of variables to the radial coordinate. For example, consider Maxwell's equations for a TE mode in two space dimensions using polar coordinates. Following [20] we make the change of variables:

$$r \to r\left(1 + \frac{c}{s}\bar{\sigma}\right), \quad \bar{\sigma} = r^{-1} \int_R^r \sigma(\rho) d\rho, \tag{130}$$

the equations for the Laplace transforms become:

$$\frac{s}{c}\hat{E}_r - \left(1 + \frac{c}{s}\bar{\sigma}\right)^{-1} \frac{1}{r}\frac{\partial \hat{B}_z}{\partial \theta} = 0, \tag{131}$$

$$\frac{s}{c}\hat{E}_\theta + \left(1 + \frac{c}{s}\sigma\right)^{-1} \frac{\partial \hat{B}_z}{\partial r} = 0, \tag{132}$$

$$\frac{s}{c}\hat{B}_z + \left(1+\frac{c}{s}\sigma\right)^{-1}\frac{\partial \hat{E}_\theta}{\partial r} + \left(1+\frac{c}{s}\bar{\sigma}\right)^{-1}\frac{1}{r}\hat{E}_\theta - \left(1+\frac{c}{s}\bar{\sigma}\right)^{-1}\frac{1}{r}\frac{\partial \hat{E}_r}{\partial \theta} = 0. \quad (133)$$

Returning to the time domain we see that only a single auxiliary variable is needed:

$$\left(\frac{1}{c}\frac{\partial}{\partial t} + \bar{\sigma}\right) E_r - \frac{1}{r}\frac{\partial B_z}{\partial \theta} = 0, \quad (134)$$

$$\left(\frac{1}{c}\frac{\partial}{\partial t} + \sigma\right) E_\theta + \frac{\partial B_z}{\partial r} = 0, \quad (135)$$

$$\left(\frac{1}{c}\frac{\partial}{\partial t} + \bar{\sigma}\right) B_z + \frac{\partial E_\theta}{\partial r} + \frac{1}{r}E_\theta - \frac{1}{r}\frac{\partial E_r}{\partial \theta} + w = 0, \quad (136)$$

$$\left(\frac{1}{c}\frac{\partial}{\partial t} + \sigma\right) w + (\sigma - \bar{\sigma})\frac{\partial E_\theta}{\partial r} = 0. \quad (137)$$

For scatterers which are well-fit by a sphere, this method has the advantage that it avoids the expensive corner regions where multiple sets of auxiliary functions are needed. Even more general curvilinear coordinates are considered in [96, 99]. Here the only restriction is that the local radii of curvature of the coordinate system remain positive.

Stability and Convergence

To prove the convergence of the solutions obtained using the PML to the restriction to Ω of the unbounded domain solution one must establish the consistency of the method, as we have done in some cases for the boundary condition sequences, and its stability. In the frequency domain this has been carried out [66, 74, 75], but we know of no analogous results in the time domain. Stability, however, has been the focus of a great deal of analysis, which we will discuss below.

As first analysed in [1], Bérenger's original formulation is not strongly well-posed, suggesting the possibility of instability and ill-posedness under perturbation. Later formulations, such as the ones presented in [82], are strongly well-posed. Strong well-posedness for (113)-(114) is easily analysed. Freezing coefficients and performing a Fourier expansion in all spatial variables leads to the symbol of the PML equations:

$$\hat{P} = -\begin{pmatrix} ik_1 A + \sum_j ik_j B_j & O \\ \sigma i k_1 A & \sum_j ik_j \beta_j \end{pmatrix}. \quad (138)$$

Clearly, $\hat{P}$ is diagonalisable assuming the strong hyperbolicity of the original problem whenever $\sum_j ik_j\beta_j$ is *not* an eigenvalue of $ik_1 A + \sum_j ik_j B_j$. If it is, the associated eigenvectors must be null vectors of A or we must have $k_1 = 0$. These conditions can be checked for our two examples.

Beyond strong well-posedness, which we see is independent of the sign of the damping term, σ, we require that the layer be asymptotically stable; that is we require that no growing modes exist. This analysis is more difficult. However, for the

Maxwell PML Bécache and Joly have recently developed an energy method which establishes this stronger form of stability [12, 13]. Interestingly the method does not apply when σ is variable, which is standard in practice. Therefore, the analysis of the asymptotic stability of the layer equations is an important open problem. One might hope that a new technique for treating this issue would lead to new insights into the construction of stable layers for variable coefficient problems.

3.2 Other Absorbing Layer Techniques

Although the PML has dominated the attention of the computational community in recent years, there are important problems to which it has not been applied. In many of these problems, older, ad hoc methods are still in use. A case of particular interest arises in the attempt to simulate the aeroacoustics of shear flows such as jets. Due to flow instabilities, both fine scale turbulence and large vortical structures are present at the outflow boundary. These certainly invalidate the analysis used to derive the PML and radiation boundary condition for the linearised Euler equations. Moreover, in practice the accuracy is also degraded, as shown in [21]. These authors suggest a different method, which seems to have become the method of choice at present for direct simulations of aeroacoustic phenomena [30, 76].

The basic description of the method of [21] is as follows; apply a smooth grid stretching in the absorbing region, which has the effect of making the propagating waves of shorter wavelength relative to the grid, and use some form of artificial viscosity or filtering to damp them. The computational experiments of [21] indicate that the method can be effective so long as the buffer zone is sufficiently wide.

In comparison with the PML, even more parameters need to be chosen without much theoretical guidance: grid stretching profiles, buffer zone lengths, and low-pass filters. As the method is clearly useful, we believe it would be of interest to develop some theory. An interesting move in this direction is the work in [22], which casts such procedures as a composition of grid mapping and filtering using the language of supergrid models. Although we emphasise that no hard error estimates directly follow, the formulation may be a good starting point.

3.3 PML for the Schrödinger Equation

Finally, we consider the construction of a PML for the Schrödinger equation. In [3, 77] they have been constructed using the complex coordinate stretching technique in more complicated settings than we will consider here, including systems describing excitons and one-way wave equations.

After Fourier-Laplace transformation, solutions of (97) are of exponential form with:

$$\lambda = \pm \left(-is + \kappa_j^2\right)^{1/2} . \tag{139}$$

As before, we seek to replace this in the layer by:

$$\lambda + \frac{\lambda}{\bar{R}} \int_0^x \sigma(z) dz. \tag{140}$$

We note that we must now also require continuity of $\frac{\partial u}{\partial x}$, so we will impose the additional condition:

$$\sigma(0) = 0. \tag{141}$$

Also, as the system is isotropic we don't require $\mu \neq 0$.

Unlike the exponents which we dealt with in the previous cases, the argument of λ now varies only between 0 and $-\frac{\pi}{2}$ when $\Re s > 0$. Thus it can made to have a nonvanishing real part by a simple rotation. That is we can choose $\bar{R}$ to be any complex number whose argument lies between 0 and $-\frac{\pi}{2}$. We thus obtain:

$$-i\frac{\partial u}{\partial t} = \frac{1}{1+\sigma e^{i\gamma}}\frac{\partial}{\partial x}\left(\frac{1}{1+\sigma e^{i\gamma}}\frac{\partial u}{\partial x}\right) + Lu, \quad 0 < \gamma < \frac{\pi}{2}. \tag{142}$$

Equation (142) has a clear interpretation. We simply make the system parabolic in the layer. However, note that increasing the absorption parameter σ corresponds to decreasing the diffusion coefficient, thus spawning a boundary layer at the interface. We also emphasise that, unlike the PML for hyperbolic equations, no additional variables are needed in the layer.

We have performed some simple one-dimensional numerical experiments with the Schrödinger PML. Choosing $\gamma = \frac{\pi}{4}$ we use second order differences and a second order BDF method in time. The layers contained twenty-five points, the absorption profile was quadratic, and the termination was with Dirichlet conditions. The solution, which we compute up to $T = 50$, was given initially by $(32/\pi)^{1/4}\exp(i\eta x - 16x^2)$, $\eta = 1, 10$. The errors as a function of $\bar{\sigma}$ and t are plotted in Fig. 6. The results are quite good, though the reader should be cautioned that this is a one-dimensional experiment. They do show a difficult-to-predict dependence on $\bar{\sigma}$.

4 Conclusions and Open Problems

In summary, for a small but important list of problems domain truncation methods capable of delivering arbitrary accuracy at acceptable cost are now available to the computational scientist. These include the equations of acoustics, electromagnetics and elastodynamics in homogeneous media and waveguide or exterior geometries. Certainly there is room for improvement in the efficiency and the mathematical analysis of these successful methods, but we believe they are already well-grounded, reliable tools.

As we've presented a number of different techniques, it is natural to try to compare them. At present, it would be premature to definitively claim that one or the other is best. For a fixed domain configuration, it is the author's experience that high-order boundary condition methods are somewhat more efficient and easy to use than the PML. This is due to the fact that the PML depends on the choice of the absorption profile and on adding points in a volume distribution to achieve convergence, whereas the boundary condition methods only require the addition of auxiliary functions. On the other hand, the PML can be used in arbitrary convex domains, and

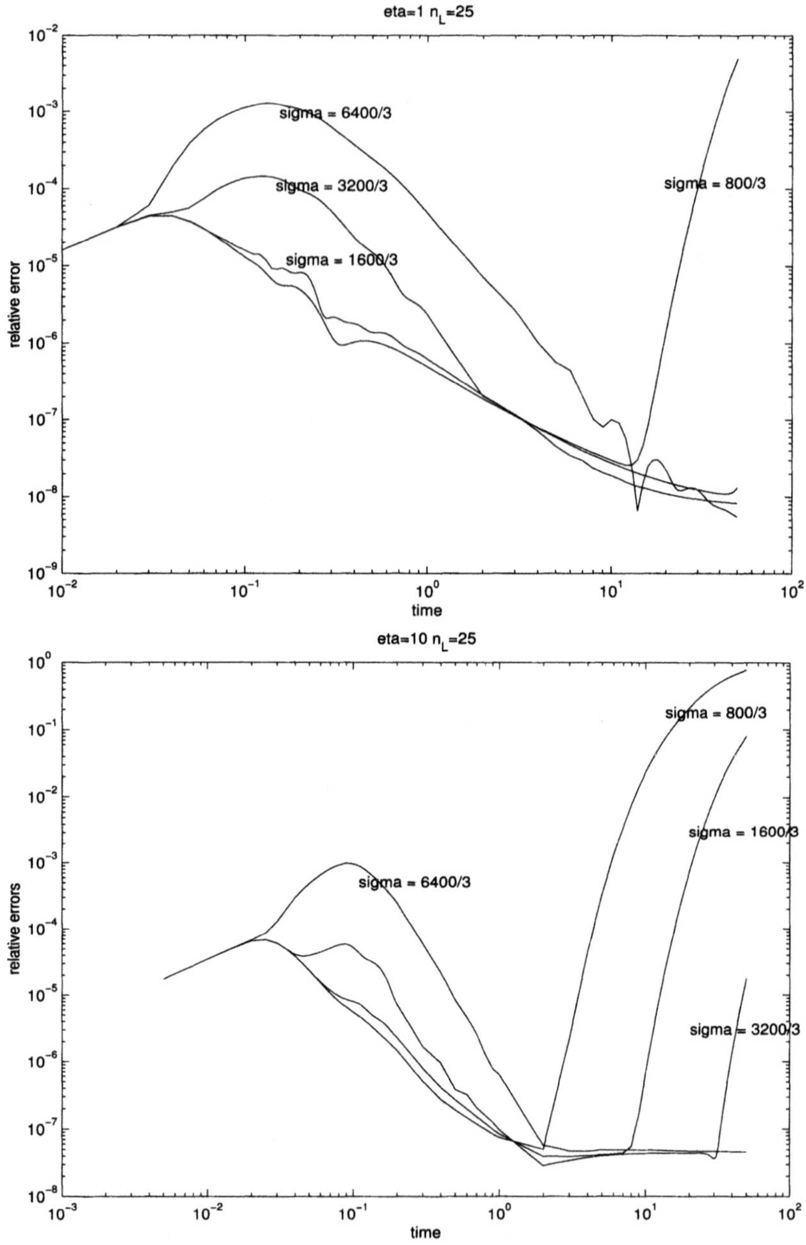

Fig. 6. Errors for the Schrodinger PML, 25-point layer, $\eta = 1, 10$

thus is currently more efficient for high-aspect ratio scatterers. Comparing boundary conditions, the nonlocal approximations are the clear winners from the point of view of complexity analysis. However, the local sequences are easier to implement, and often require many fewer auxiliary functions than the error analyses suggest. Thus they can be competitive if a good adaptive strategy can be found.

We have identified a number of important open problems whose resolution is likely to have the final say on the future importance of the various accurate techniques. These include:

i. Extension of the high-order local and/or nonlocal boundary conditions to high-aspect ratio boundaries such as boxes or spheroids.
ii. Sharp error analysis for the time-domain PML, particularly in high-aspect ratio domains.
iii. Convergence analysis of Higdon-type boundary conditions for problems with variable coefficients.
iv. Improved understanding of the stability of PMLs leading to their extension to a wider range of problems including those with variable coefficients.
v. Adaptive order determination for local boundary condition sequences.
vi. Optimisation of absorption parameters in the PML.
vii. Mathematical analysis of alternative absorbing layer methods.

We believe that progress on these issues is possible, and that it will lead to substantial improvements in our ability to simulate waves.

References

1. S. Abarbanel and D. Gottlieb. A mathematical analysis of the PML method. *J. Comput. Phys.*, 134:357–363, 1997.
2. S. Abarbanel, D. Gottlieb, and J. Hesthaven. Well-posed perfectly matched layers for advective acoustics. *J. Comput. Phys.*, 154:266–283, 1999.
3. A. Ahland, D. Schulz, and E. Voges. Accurate mesh truncation for Schrödinger equations by a perfectly matched layer absorber: Application to the calculation of optical spectra. *Phys. Rev. B*, 60:5109–5112, 1999.
4. I. Alonso-Mallo and N. Reguera. Weak ill-posedness of spatial discretizations of absorbing boundary conditions for Schrödinger-type equations. *SIAM J. Numer. Anal.*, 40:134–158, 2002.
5. B. Alpert, L. Greengard, and T. Hagstrom. Rapid evaluation of nonreflecting boundary kernels for time-domain wave propagation. *SIAM J. Numer. Anal.*, 37:1138–1164, 2000.
6. B. Alpert, L. Greengard, and T. Hagstrom. Nonreflecting boundary conditions for the time-dependent wave equation. *J. Comput. Phys.*, 180:270–296, 2002.
7. X. Antoine and H. Barucq. Microlocal diagonalization of strictly hyperbolic pseudodifferential systems and application to the design of radiation boundary conditions in electromagnetism. *SIAM J. Appl. Math.*, 61:1877–1905, 2001.
8. R.J. Astley. Transient spheroidal elements for unbounded wave problems. *Computer Meth. Appl. Mech. Engrg.*, 164:3–15, 1998.
9. R.J. Astley. Infinite elements for wave problems: A review of current formulations and an assessment of accuracy. *Int. J. for Numer. Meth. Engrg.*, 49:951–976, 2000.

10. R.J. Astley and J. Hamilton. Infinite elements for transient flow acoustics. Technical Report 2001-2273, AIAA, 2002.
11. E. Bécache, A.-S. Bonnet-Ben Dhia, and G. Legendre. Perfectly matched layers for the convected Helmholtz equation. In preparation, 2002.
12. E. Bécache and P. Joly. On the analysis of Bérenger's perfectly matched layers for Maxwell's equations. *Math. Model. and Numer. Anal.*, 36:87–119, 2002.
13. E. Bécache, P. Petropoulos, and S. Gedney. On the long-time behavior of unsplit Perfectly Matched Layers. Preprint, 2002.
14. J.-P. Bérenger. A perfectly matched layer for the absorption of electromagnetic waves. *J. Comput. Phys.*, 114:185–200, 1994.
15. A. Barry, J. Bielak, and R.C. MacCamy. On absorbing boundary conditions for wave propagation. *J. Comput. Phys.*, 79:449–468, 1988.
16. A. Bayliss and E. Turkel. Radiation boundary conditions for wave-like equations. *Comm. Pure and Appl. Math.*, 33:707–725, 1980.
17. W. Chew and W. Weedon. A 3-D perfectly matched medium from modified Maxwell's equations with stretched coordinates. *Microwave Optical Technol. Lett.*, 7:599–604, 1994.
18. F. Collino. High order absorbing boundary conditions for wave propagation models. Straight line boundary and corener cases. In R. Kleinman et al., editor, *Proceedings of 2nd Int. Conf. on Math. and Numer. Aspects of Wave Prop. Phen.*, pp. 161–171. SIAM, 1993.
19. F. Collino and P. Monk. Optimizing the perfectly matched layer. *Computer Meth. Appl. Mech. Engrg.*, 164:157–171, 1998.
20. F. Collino and P. Monk. The perfectly matched layer in curvilinear coordinates. *SIAM J. Sci. Comput.*, 19:2061–2090, 1998.
21. T. Colonius, S. Lele, and P. Moin. Boundary conditions for direct computation of aerodynamic sound generation. *AIAA J.*, 31:1574–1582, 1993.
22. T. Colonius and H. Ran. A super-grid scale model for simulating compressible flow on unbounded domains. *J. Comput. Phys.*, 2002. To appear.
23. J. Diaz and P. Joly. Stabilized perfectly matched layers for advective wave equations. In preparation, 2002.
24. L. DiMenza. Transparent and absorbing boundary conditions for the Schrödinger equation in a bounded domain. *Numer. Funct. Anal. Optim.*, 18:759–775, 1997.
25. B. Engquist and L. Halpern. Far field boundary conditions for computation over long time. *Appl. Numer. Math.*, 4:21–45, 1988.
26. B. Engquist and A. Majda. Absorbing boundary conditions for the numerical simulation of waves. *Math. Comp.*, 31:629–651, 1977.
27. B. Engquist and A. Majda. Radiation boundary conditions for acoustic and elastic wave calculations. *Comm. Pure and Appl. Math.*, 32:313–357, 1979.
28. A. Ergin, B. Shanker, and E. Michielssen. Fast evaluation of three-dimensional transient wave fields using diagonal translation operators. *J. Comput. Phys.*, 146:157–180, 1998.
29. T. Fevens and H. Jiang. Absorbing boundary conditions for the Schrödinger equation. *SIAM J. Sci. Comput.*, 21:255–282, 1999.
30. J. Freund and S. Lele. Computer simulation and prediction of jet noise. In *High Speed Jet Flows: Fundamentals and Applications*. Taylor Francis, 2001.
31. T. Geers. Singly and doubly asymptotic computational boundaries. In *Computational Methods for Unbounded Domains*, pp. 135–142, Dordrecht, the Netherlands, 1998. Kluwer Academic Publishers.
32. M. Giles. Nonreflecting boundary conditions for Euler equation calculations. *AIAA Journal*, 28:2050–2058, 1990.

33. D. Givoli. Exact representations on artificial interfaces and applications in mechanics. *Appl. Mech. Rev.*, 52:333–349, 1999.
34. D. Givoli and D. Kohen. Non-reflecting boundary conditions based on Kirchoff-type formulae. *J. Comput. Phys.*, 117:102–113, 1995.
35. D. Givoli and B. Neta. High-order nonreflecting boundary scheme for time-dependent waves. *J. Comput. Phys.*, 2002. To appear.
36. D. Givoli, B. Neta, and I. Patlashenko. Finite element solution of exterior time-dependent wave problems with high-order boundary treatment. Submitted, 2002.
37. J. Goodrich and T. Hagstrom. A comparison of two accurate boundary treatments for computational aeroacoustics. In *3rd AIAA/CEAS Aeroacoustics Conference*, 1997.
38. L. Greengard and V. Rokhlin. A new version of the fast multipole method for the Laplace equation in three dimensions. *Acta Numerica*, 6:229–269, 1997.
39. M. Grote. Nonreflecting boundary conditions for elastodynamic scattering. *J. Comput. Phys.*, 161:331–353, 2000.
40. M. Grote and J. Keller. Exact nonreflecting boundary conditions for the time dependent wave equation. *SIAM J. Appl. Math.*, 55:280–297, 1995.
41. M. Grote and J. Keller. Nonreflecting boundary conditions for time dependent scattering. *J. Comput. Phys.*, 127:52–81, 1996.
42. M. Grote and J. Keller. Nonreflecting boundary conditions for Maxwell's equations. *J. Comput. Phys.*, 139:327–342, 1998.
43. M. Grote and J. Keller. Exact nonreflecting boundary conditions for elastic waves. *SIAM J. Appl. Math.*, 60:803–818, 2000.
44. M. Guddati and J. Tassoulas. Continued-fraction absorbing boundary conditions for the wave equation. *J. Comput. Acoust.*, 8:139–156, 1998.
45. T. Hagstrom. On the convergence of local approximations to pseudodifferential operators with applications. In E. Bécache, G. Cohen, P. Joly, and J. Roberts, editors, *Proc. of the 3rd Int. Conf. on Math. and Numer. Aspects of Wave Prop. Phen.*, pp. 474–482. SIAM, 1995.
46. T. Hagstrom. On high-order radiation boundary conditions. In B. Engquist and G. Kriegsmann, editors, *IMA Vol. on Computational Wave Propagation*, pp. 1–22, New York, 1996. Springer-Verlag.
47. T. Hagstrom. Radiation boundary conditions for the numerical simulation of waves. *Acta Numerica*, 8:47–106, 1999.
48. T. Hagstrom and J. Goodrich. Accurate radiation boundary conditions for the linearized Euler equations in Cartesian domains. *SIAM J. Sci. Comput.*, 2002. To appear.
49. T. Hagstrom and S.I. Hariharan. A formulation of asymptotic and exact boundary conditions using local operators. *Appl. Numer. Math.*, 27:403–416, 1998.
50. T. Hagstrom, S.I. Hariharan, and R. MacCamy. On the accurate long-time solution of the wave equation on exterior domains: Asymptotic expansions and corrected boundary conditions. *Math. Comp.*, 63:507–539, 1994.
51. T. Hagstrom, S.I. Hariharan, and D. Thompson. High-order radiation boundary conditions the convective wave equation in exterior domains. Submitted.
52. T. Hagstrom and I. Nazarov. Absorbing layers and radiation boundary conditions for jet flow simulations. Technical Report AIAA 2002-2606, AIAA, 2002.
53. T. Hagstrom and T. Warburton. High-order radiation boundary conditions for time-domain electromagnetics using unstructured discontinuous Galerkin methods. In preparation, 2002.
54. E. Hairer, C. Lubich, and M. Schlichte. Fast numerical solution of nonlinear Volterra convolutional equations. *SIAM J. Sci. Stat. Comput.*, 6:532–541, 1985.

55. L. Halpern and J. Rauch. Error analysis for absorbing boundary conditions. *Numer. Math.*, 51:459–467, 1987.
56. L. Halpern and L. Trefethen. Wide-angle one-way wave equations. *J. Acoust. Soc. Am.*, 84:1397–1404, 1988.
57. D. Healy, D. Rockmore, P. Kostelec, and S. Moore. FFTs for the 2-sphere - Improvements and variations. *Adv. Appl. Math.*, 2002. To appear.
58. J. Hesthaven. On the analysis and construction of perfectly matched layers for the linearized Euler equations. *J. Comput. Phys.*, 142:129–147, 1998.
59. J. Hesthaven and T.Warburton. High-order/spectral methods on unstructured grids. I. Time-domain solution of Maxwell's equations. *J. Comput. Phys.*, 181:186–221, 2002.
60. R. Higdon. Absorbing boundary conditions for difference approximations to the multi-dimensional wave equation. *Math. Comp.*, 47:437–459, 1986.
61. R. Higdon. Numerical absorbing boundary conditions for the wave equation. *Math. Comp.*, 49:65–90, 1987.
62. R. Higdon. Radiation boundary conditions for elastic wave propagation. *SIAM J. Numer. Anal.*, 27:831–870, 1990.
63. R. Higdon. Absorbing boundary conditions for elastic waves. *Geophysics*, 56:231–254, 1991.
64. R. Higdon. Absorbing boundary conditions for acoustic and elastic waves in stratified media. *J. Comput. Phys.*, 101:386–418, 1992.
65. R. Higdon. Radiation boundary conditions for dispersive waves. *SIAM J. Numer. Anal.*, 31:64–100, 1994.
66. T. Hohage, F. Schmidt, and L. Zschiedrich. Solving time-harmonic scattering problems based on the pole condition: Convergence of the PML method. Technical Report ZIB-Report 01-23, Zuse Institut Berlin, 2001.
67. R. Holford. A multipole expansion for the acoustic field exterior to a prolate or oblate spheroid. Preprint, 1998.
68. F. Hu. On absorbing boundary conditions for linearized Euler equations by a perfectly matched layer. *J. Comput. Phys.*, 129:201–219, 1996.
69. F. Hu. A stable, perfectly matched layer for linearized Euler equations in unsplit physical variables. *J. Comput. Phys.*, 173:455–480, 2001.
70. R. Huan and L. Thompson. Accurate radiation boundary conditions for the time-dependent wave equation on unbounded domains. *Int. J. Numer. Meth. Engrg.*, 47:1569–1603, 2000.
71. M. Israeli and S. Orszag. Approximation of radiation boundary conditions. *J. Comput. Phys.*, 41:115–135, 1981.
72. S. Jiang. *Fast Evaluation of Nonreflecting Boundary Conditions for the Schrödinger Equation*. PhD thesis, New York University, 2001.
73. H.-O. Kreiss and J. Lorenz. *Initial-Boundary Value Problems and the Navier–Stokes Equations*. Academic Press, New York, 1989.
74. M. Lassas and E. Somersalo. On the existence and convergence of the solution of PML equations. *Computing*, 60:228–241, 1998.
75. M. Lassas and E. Somersalo. Analysis of the PML equations in general convex geometry. *Proc. Roy. Soc. Edinburgh A*, 131:1183–1207, 2001.
76. S. Lele. Direct numerical simulation of compressible turbulent flows: fundamentals and applications. In A. Hanifi, P. Alfredsson, A. Johansson, and D. Henningson, editors, *Transition, Turbulence and Combustion Modelling*, Chap. 7. Kluwer, Dordrecht, 1999.
77. M. Levy. Perfectly matched layer truncation for parabolic wave equation models. *Proc. Roy. Soc. Lond. A*, 457:2609–2624, 2001.

78. E. Lindman. Free space boundary conditions for the time dependent wave equation. *J. Comput. Phys.*, 18:66–78, 1975.
79. J.-L. Lions, J. Métral, and O. Vacus. Well-posed absorbing layer for hyperbolic problems. *Numer. Math.*, 2001. To appear.
80. C. Lubich and A. Schädle. Fast convolution for non-reflecting boundary conditions. *SIAM J. Sci. Comput.*, 24:161–182, 2002.
81. M. Mohlenkamp. A fast transform for spherical harmonics. *J. of Fourier Anal. and Applic.*, 5:159–184, 1999.
82. P. Petropoulos. Reflectionless sponge layers as absorbing boundary conditions for the numerical solution of Maxwell's equations in rectangular, cylindrical and spherical coordinates. *SIAM J. Appl. Math.*, 60:1037–1058, 2000.
83. C. Randall. Absorbing boundary condition for the elastic wave equation. *Geophysics*, 53:611–624, 1988.
84. C. Rowley and T. Colonius. Discretely nonreflecting boundary conditions for linear hyperbolic systems. *J. Comput. Phys.*, 15:500–538, 2000.
85. V. Ryabe'nkii. *Method of Difference Potentials and its Applications*. Springer-Verlag, New York, 2001.
86. V. Ryaben'kii, S. Tsynkov, and V. Turchaninov. Global discrete artificial boundary conditions for time-dependent wave propagation. *J. Comput. Phys.*, 174:712–758, 2001.
87. V. Ryaben'kii, S. Tsynkov, and V. Turchaninov. Long-time numerical computation of wave-type solutions driven by moving sources. *Appl. Numer. Math.*, 38:187–222, 2001.
88. A. Schädle. *Ein Schneller Faltungsalgorithmus für Nichtreflektierende Randbedingungen*. PhD thesis, Eberhard-Karls-Universität Tübingen, 2002.
89. I. Sofronov. Conditions for complete transparency on the sphere for the three-dimensional wave equation. *Russian Acad. Sci. Dokl. Math.*, 46:397–401, 1993.
90. I. Sofronov. Artificial boundary conditions of absolute transparency for two- and three-dimensional external time-dependent scattering problems. *Euro. J. Appl. Math.*, 9:561–588, 1998.
91. I. Sofronov. Non-reflecting inflow and outflow in wind tunnel for transonic time-accurate simulation. *J. Math. Anal. Appl.*, 221:92–115, 1998.
92. R. Suda and M. Takami. A fast spherical harmonic transform algorithm. *Math. Comput.*, 71:703–715, 2002.
93. C. Tam, L. Auriault, and F. Cambuli. Perfectly matched layer as an absorbing boundary condition for the linearized Euler equations in open and ducted domains. *J. Comput. Phys.*, 144:213–234, 1998.
94. F. Teixeira and W. Chew. PML-FDTD in cylindrical and spherical grids. *IEEE Microwave and Guided Wave Lett.*, 7:285–287, 1997.
95. F. Teixeira and W. Chew. Systematic derivation of anisotropic PML absorbing media in cylindrical and spherical coordinates. *IEEE Microwave and Guided Wave Lett.*, 7:371–373, 1997.
96. F. Teixeira and W. Chew. Analytical derivation of a conformal perfectly matched absorber for electromagnetic waves. *Microwave and Optical Tech. Lett.*, 17:231–236, 1998.
97. F. Teixeira and W. Chew. General closed-form PML constitutive tensors to match arbitrary bianisotropic and dispersive linear media. *IEEE Microwave and Guided Wave Lett.*, 8:223–225, 1998.
98. F. Teixeira and W. Chew. Finite-difference computation of transient electromagnetic waves for cylindrical geometries in complex media. *IEEE Trans. on Geosci. and Remote Sensing*, 38:1530–1543, 2000.

99. F. Teixeira, K.-P. Hwang, W. Chew, and J.-M. Jin. Conformal PML-FDTD schemes for electromagnetic field simulations: A dynamic stability study. *IEEE Trans. on Ant. Prop.*, 49:902–907, 2001.
100. L. Thompson and R. Huan. Implementation of exact non-reflecting boundary conditions in the finite element method for the time-dependent wave equation. *Comput. Methods Appl. Mech. Engrg.*, 187:137–159, 2000.
101. L. Ting and M. Miksis. Exact boundary conditions for scattering problems. *J. Acoust. Soc. Am.*, 80:1825–1827, 1986.
102. O. Vacus. Mathematical analysis of absorbing boundary conditions for the wave equation: The corner problem. *Math. Comput.*, 2002. To appear.
103. H. Warchall. Wave propagation at computational domain boundaries. *Commun. in Part. Diff. Eq.*, 16:31–41, 1991.
104. L. Xu. *Applications of High-Order Radiation Boundary Conditions.* PhD thesis, The University of New Mexico, 2001.
105. C. Zhao and T. Liu. Non-reflecting artificial boundaries for modelling scalar wave propagation problems in two-dimensional half space. *Comput. Meth. Appl. Mech. Engrg.*, 191:4569–4585, 2002.

Fast, High-Order, High-Frequency Integral Methods for Computational Acoustics and Electromagnetics

Oscar P. Bruno

Applied and Computational Mathematics, Caltech, Pasadena, CA 91125, USA
bruno@acm.caltech.edu

1 Introduction

We review a set of algorithms and methodologies developed recently for the numerical solution of problems of scattering by complex bodies in three-dimensional space. These methods, which are based on integral equations, high-order integration, Fast Fourier Transforms and highly accurate high-frequency integrators, can be used in the solution of problems of electromagnetic and acoustic scattering by surfaces and penetrable scatterers — even in cases in which the scatterers contain geometric singularities such as corners and edges. All of the solvers presented here exhibit high-order convergence, they run on low memories and reduced operation counts, and they result in solutions with a high degree of accuracy. In particular, our approach to direct solution of integral equations results in algorithms that can evaluate accurately in a personal computer scattering from hundred-wavelength-long objects — a goal, otherwise achievable today only by super-computing. The high-order high-frequency methods we present, in turn, are efficient where our direct methods become costly, thus leading to an overall computational methodology which is applicable and accurate throughout the electromagnetic spectrum.

With regards to engineering relevance of the accuracy exhibited by high order numerical methods, it has been correctly argued that often an accuracy better than 1%–0.1% is not significant in engineering practice. (An important class of problems in which significantly higher accuracies are needed relate to low-observable applications, where the quantities of interest are small residuals of large incident fields.) In any case, it is our contention that high-order accuracy is extremely valuable in all applications, since it allows one to produce accurate estimates of the errors incurred in a given calculation. Indeed, estimation of the accuracy of a certain approximation requires, say, the possibility of evaluation of the solution to at least one additional digit of accuracy. With first order convergence (which necessarily results from a direct integration method or without use of high-order surface representations) this requires a refinement of a one dimensional mesh by a factor of ten. For the two-dimensional integrals we consider this translates into a factor of $10^2 = 100$ in the number of discretisation points — and thus, in the number of unknowns in the problem! Clearly, refinement of a first order algorithm is generally not a viable approach for evaluation of the accuracy of a numerical solution. A high-order algorithm such as that used in Table 1 right, in turn, can produce an additional digit of accuracy (and, therefore, a good measure of the accuracy of a numerical solution) by a small increase in the overall size of the numerical problem. This is an extremely valuable feature in a numerical method; as indicated throughout the following text, our approaches produce such high-order accuracy by exploiting the high-order convergence of both trapezoidal rule integration and approximation via Fourier series for smooth periodic functions. Fast numerics in our algorithms results from use of $\mathcal{O}(N\log(N))$ Fast Fourier Transforms as well as the novel high-frequency integrators described in Sect. 6.

The goal of this paper is to provide a concise review of the main lines of the methodologies mentioned above. The presentation is organised as follows: after a brief general discussion in Sect. 2 on integration and interpolation, in Sect. 3 we describe methods that can be used to produce high order representation of regular and singular surfaces from a given triangulation. In Sects. 4 and 5 we then present our direct integral solvers for surface and volumetric scattering, respectively. In Sect. 6, finally, we introduce our high-order high-frequency integral solvers, and we illustrate their capabilities through applications to problems of scattering in two-dimensional space.

2 High-Order Integration and Interpolation of Smooth Functions

Many aspects of our methods result from consideration of the remarkable properties exhibited by the trapezoidal rule for integration of smooth periodic functions over d-dimensional cubes ($d = 1, 2, \dots$). Of course the integrals arising in integral equation formulations involve surfaces which are much more complex than a square in 2-dimensions or a cube in 3-dimensions, and they require consideration of highly non-smooth integrands–which arise from both singularities in the Green functions and geometric singularities of the scatterer: edges, corners, etc. We have shown, however, that, in spite of these singularities, appropriate transformations permit one to obtain high-order integrators for scattering problems from trapezoidal rules and Fourier series. Further, use of Fast Fourier Transforms for evaluation of Fourier series and convolutions allows for fast numerics in addition to high order accuracy. To demonstrate these facts in simple settings we preface our discussion with some considerations on the properties of the trapezoidal rule and Fourier series for one-dimensional functions under various periodicity and smoothness assumptions.

2.1 Trapezoidal-Rule Integration

Let us thus consider the problem of integration of a function $y = f(x)$ over a one-dimensional interval $[a, b]$. Table 1 displays relative errors and convergence ratios exhibited by the trapezoidal rule as applied to three problems — which encapsulate some relevant issues under consideration: $\int_0^{1/2} \sqrt{x}\, dx$; $\int_0^{\pi/4} e^{\cos^2(x)}\, dx$ and $\int_0^{\pi} e^{\cos^2(x)}\, dx$. From Table 1 we see that integration of the non-smooth function $\sqrt{x}$ results in errors larger than $\mathcal{O}(h^2)$ for an N interval grid of mesh-size $h = (b-a)/N$: refinement of the mesh by a factor of 2 leads to error reductions by a factor smaller than 4. When applied to integration of the smooth function $f(x) = e^{\cos^2(x)}$ over the interval $[0, \pi/4]$, in turn, the trapezoidal rule results in quadratic errors: refinement of the mesh by a factor of 2 leads to error reductions by a factor of 4. In the last case in which the *smooth and periodic* function $f(x) = e^{\cos^2(x)}$ shown in Fig. 1 is integrated over its period, we see an enormously higher, exponential convergence rate.

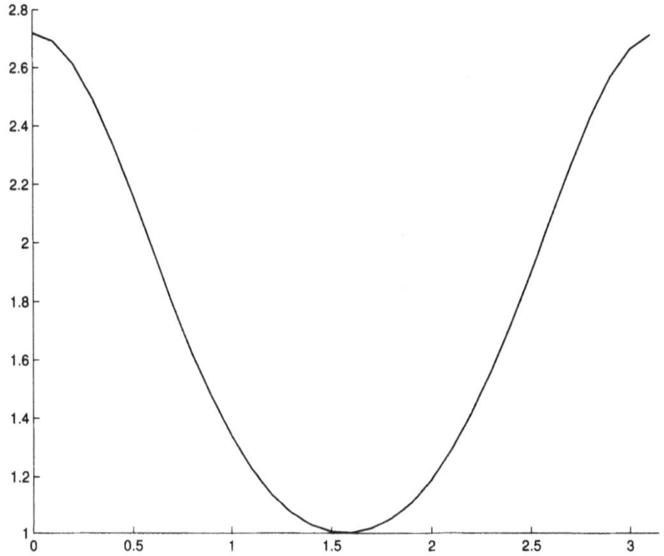

Fig. 1. $f(x) = e^{\cos^2(x)}$

The behaviour exhibited by the trapezoidal rule is easy to explain. As is well known, for a general smooth function, the trapezoidal rule gives rise to quadratic errors $\mathcal{O}(h^2)$ (see Table 1 centre) since the error in the approximation of a function by its trapezoidal approximation is of the order of h^3 within each one of the $\mathcal{O}(1/h)$ individual integration elements. For the non-smooth function considered here, the basic-element approximation error for the ℓ-th integration element is of the order of $\mathcal{O}(\ell^{-3/2}h^{3/2})$. Thus, summing over ℓ we see that the overall errors are of the order of $h^{3/2}$: larger than the $\mathcal{O}(h^2)$ errors that result in the smooth case. A remarkable phenomenon takes place as the trapezoidal rule is applied to functions such as $e^{\cos^2(x)}$ which are periodic on the integration domain, and smooth for $-\infty < x < \infty$: in this

N	Rel. Error	Ratio
1	2.5(-1)	
2	9.5(-2)	2.6
4	3.5(-2)	2.7
8	1.3(-2)	2.7
8192	4.2(-7)	

N	Rel. Error	Ratio
1	4.8(-2)	
2	1.2(-2)	4.0
4	2.9(-3)	4.0
8	7.4(-4)	4.0
8192	7.0(-10)	

N	Rel. Error	Ratio
1	5.5(-1)	
2	6.0(-2)	9.1
4	3.1(-4)	1.9(+2)
8	7.2(-10)	4.3(+5)
16	2.1(-23)	3.4(+13)

Table 1. Relative errors in trapezoidal-rule approximations for $\int_a^b f(x)\,\mathrm{d}x$ using N intervals ($N+1$ discretisation points). **Left:** $f(x) = \sqrt{x}$, $a = 0$, $b = 1/2$. **Centre:** $f(x) = \mathrm{e}^{\cos^2(x)}$, $a = 0$, $b = \pi/4$. **Right:** $f(x) = \mathrm{e}^{\cos^2(x)}$, $a = 0$, $b = \pi$

case the integration errors decrease exponentially fast: see Table 1 right. Two basic facts make this behaviour possible, namely

1. The extremely fast convergence of Fourier series of smooth periodic functions [29], and,
2. The fact that, as can be easily verified, the N–interval trapezoidal rule produces the integral of the Fourier harmonics $\mathrm{e}^{\mathrm{i}\ell x}$ between 0 and 2π *exactly* for $\ell = -(N-1)\dots(N-1)$.

From Table 1 we thus see that use of 8 integration intervals for integration of the smooth periodic function $f(x) = \mathrm{e}^{\cos^2(x)}$ in its periodicity interval results in errors of the order of 10^{-10}. For the non-periodic integration problem in Table 1 centre, an equivalent accuracy requires 8192 intervals. And, use of this highly refined mesh gives an accuracy of only 10^{-7} when the integration problem for the non-smooth function $\sqrt{x}$ is considered.

As discussed above, the integrands occurring in integral equations of the type (3) below generally are neither one dimensional, nor smooth or periodic. The methods presented in the rest of this text, however, do reduce these general integration problems to problems of evaluation of sequences of one-dimensional integrals of smooth periodic functions.

2.2 High-Order Interpolation

In addition to supporting a rationale for the convergence properties of the trapezoidal rule, Point 1 above suggests an efficient interpolation method: once problems have been reduced to consideration of smooth and periodic functions, as indicated in the following sections, any interpolations that may be needed can be produced with high order accuracy through use of Fourier series — which can be computed with reduced operation counts by means of Fast Fourier Transforms. Further, as discussed in the following section, unequally spaced FFTs can be used for efficient high-order treatment of the surface representation problem.

3 High-Order Surface Representation
(***Joint work with M. Pohlman*** [16])

Clearly, a high-order surface integrator must use high-order representations of the integration surfaces. Such high-order representations are generally not directly available, however: usually only triangulations are provided, so that first order derivatives are discontinuous, or, at most, spline representation with continuous first order derivatives and discontinuous curvatures are given. Since the curvature of the surface occurs as part of the double-layer integrands (see equation (2) below), a high order integration method requires several derivatives *of the curvature* to exist — thus our efforts to produce very smooth representations of a given surface from triangulations or other low order representations.

The problem of producing smooth representations from a triangulation has been considered often in recent years; see [30] and references therein. Previous methods have utilised piecewise polynomial approximations, and, under certain assumptions concerning regularity of a given triangulation, have provided representations with continuous derivatives of first order. Higher order differentiability has not been produced by these means as yet, and the prospects for the feasibility of such extensions do not seem favourable: it appears that piecewise polynomial approximations may not be helpful in our context.

Our approach to high order geometry representation is based on use of Fourier series for certain periodic functions associated with the given surface. We need to treat separately the cases of *regular representations* — in regions away from boundaries, edges, corners or other singularities *and* containing "large" numbers of data points —, and the case of *singular representations* — in regions *either* near boundaries or singularities *or* containing "small" numbers of data points. These are discussed, in turn, in Sects. 3.1 and 3.2 below. A number of examples of application to general geometries are then presented in Sect. 3.3.

3.1 Surface Representation Around Regular Points

We introduce our method for representation of surfaces around regular points through a one dimensional problem: interpolation of smoothly varying, irregularly spaced data on the curve $(x - 1/2)^2 + y^2 = 1/4$, as shown in the upper-left portion of Fig. 2. In the present regular interpolation problem we seek to produce representations (interpolations) of this data in a region away from the end points of the interval of definition.

To produce the required interpolation we begin by multiplying the given data by the smooth windowing function depicted in the upper right portion of Fig. 2, all of whose derivatives vanish at the interval end points. The result of this operation is shown in the middle left portion of this figure. Clearly, the windowed data is as smooth as the original data. But, unlike the original points, the windowed points constitute a discretisation of the *smooth and periodic function* that results by periodic continuation. As noted in Sect. 2.2, the convergence of such Fourier series is of high-

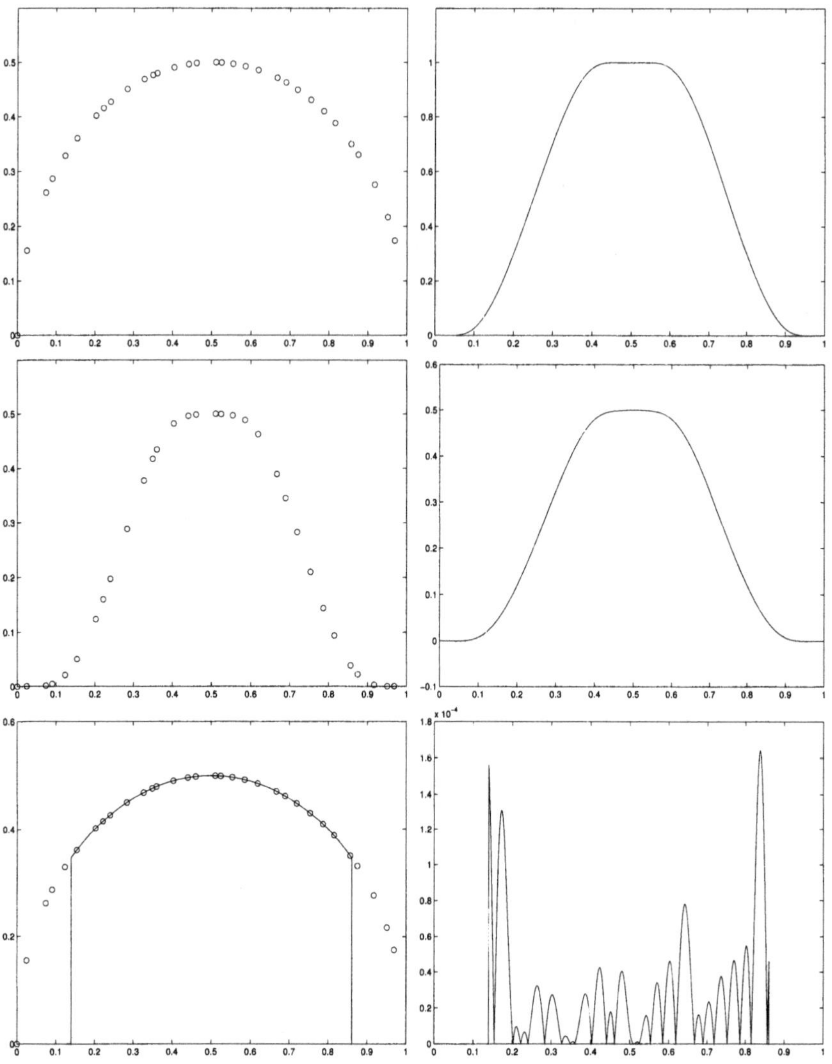

Fig. 2. Upper left: given data; 32 points on the curve $(x - 1/2)^2 + y^2 = 1/4$. **Upper right:** windowing function. **Centre left:** windowed data. **Centre right:** interpolation of the windowed data in an interior region, away from the end points of the interval of definition. **Lower left:** interpolation of the given data in the interior region after division by the widowing function. **Lower right:** interpolation error ($\mathcal{O}(10^{-4})$)

order in the number of discretisation points — and thus provides an ideal high order interpolator.

The needed Fourier series can be obtained efficiently by means of the unevenly spaced FFT algorithm (USFFT) [5, 33–35]. To eliminate potentially oscillatory behaviour in the interpolation (that could arise if more Fourier modes are used than data points are available) we use an under-determined USFFT, and we penalise high-order coefficients by means of an appropriate least squares procedure [16]. Once the Fourier series is obtained, interpolation of the windowed data within the interval of definition and away from the interval end-points can be obtained by evaluation of the Fourier series and division by the corresponding value of the windowing function; the results of these operations in the example under consideration are shown in Fig. 2. Thus, an ordinary FFT can be used to produce an equi-spaced, high-order discretisation of the given surface; differentiation of the Fourier series, in turn, produces tangents, curvatures, and other derivatives that may be needed in a given application. The accuracies rendered by these procedures in the one-dimensional case under consideration are given in Table 2; similar results are obtained in the two-dimensional case.

N	Max. $\Delta x_{\max}$	Func. Err.	Ratio	d_x Err.	Ratio	d_x^2 Err.	Ratio
8	2.1e-01	5.5e-02		9.7e-01		4.8e+01	
16	1.2e-01	2.9e-03	1.9e+1	9.0e-02	1.1e+1	1.2e+01	3.8e+0
32	5.0e-02	2.4e-04	1.2e+1	1.2e-02	7.5e+0	9.5e-01	1.3e+1
64	3.0e-02	8.7e-06	2.8e+1	1.3e-03	9.6e+0	2.2e-01	4.3e+0
128	1.4e-02	3.1e-09	2.8e+3	7.2e-07	1.8e+3	2.2e-04	9.7e+2
256	7.5e-03	4.8e-13	6.4e+3	3.3e-10	2.2e+3	2.0e-07	1.2e+3

Table 2. Convergence of interpolations of the function $y = \sqrt{1/4 - (x - 1/2)^2}$ using N unevenly spaced data points: Maximum errors in the interpolated values of the function (Func.) and its first and second derivatives (d_x and d_x^2) using N unequally spaced points $x_1 \dots x_N$ with maximum separation $\Delta x_{\max}$ ($x_1 = 0$, $x_N = 1$, $x_i = ih + \text{rand}(-h/2, h/2)$ for $1 < i < N$, where $h = 1/(N-1)$ and where $\text{rand}(-h/2, h/2)$ is a random number in the interval $(-h/2, h/2)$.) Maximum errors were evaluated through comparison of exact and interpolated values at large numbers of points in the interval $[0, 1]$

Clearly, the methods described above are applicable to a triangulated surface as soon as a projection of the triangulation vertices is available, as indicated in Fig. 3. The existence of a planar projection is not necessary, however: projection surfaces of any shape can be used when convenient. This is particularly important for representation of highly curved portions of a surface. The representation for the canopy of the F15 airplane depicted in Fig. 6, for example, was obtained through projection of the triangle vertices on a paraboloidal surface. For highly curved or otherwise complex surfaces our algorithm uses the so called "Intrinsic Parametrisation" to produce adequate projection or parametrisation surfaces, see [32]. Intrinsic parametrisation provides a projection of an arbitrary triangulation, however complex, onto a triangulation of a planar region. An application of our interpolation algorithm to the result produced by the intrinsic parametrisation yields a smooth interpolated surface which lies very close to the given surface and which, therefore, serves as an ideal projection surface. In many cases the surface resulting from interpolation of the intrinsic

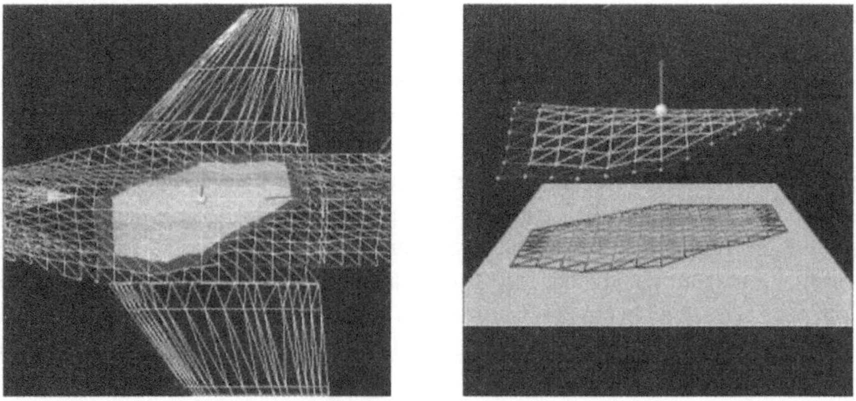

Fig. 3. The first step in the surface representation algorithm: selection of patches and a two-dimensional parameter space such that the given points on each patch can be viewed as a discretisation of a smooth *function*. A planar parameter space is shown here, but, as indicated towards the end of Sect. 3.1 parametrisation other than projections can (and should) be used as well

parametrisation itself provides an excellent smooth interpolation for the original data; see [16] for details.

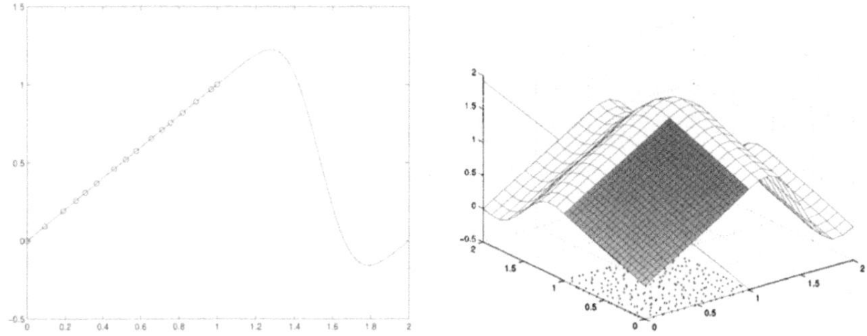

Fig. 4. Surface representation around edges and corners. **Left:** one dimensional case. **Right:** Two dimensional case

3.2 Surface Representation Around Singular Points

In most engineering applications, geometric singularities in surfaces arise as *intersections of two or more smooth surfaces*. In the interest of brevity we confine our discussion of the singular interpolation problem to such cases, although singularities of other types (e.g. of conical type) can be treated by methods similar to those described in what follows. The methods described in this section are also useful when description of interior portions of a surface must be provided, for which the sampling available is somewhat sparse, and which is therefore insufficient to resolve a windowing function such as that depicted in the upper right portion of Fig. 2.

Focusing on singularities arising from intersection of smooth surfaces, then, it suffices to provide smooth representations of surfaces whose triangulations are abruptly terminated at an edge, although the original surface could be continued smoothly — just like each face of a cube can be continued smoothly, beyond its edges and corners, as a larger planar surface. Fig. 4 provides examples in one- and two-dimensional contexts: the points with abscissae in the interval $[0, 1]$ in the left figure and the points on the (planar) square with abscissae shown in $[0, 1] \times [0, 1]$ represent data from which parametrisations for the line and the square are to be obtained. The nature of the singular surface-representation problem, then, is that use of smooth windowing functions to induce periodicity is not appropriate. In such cases we resort to a different strategy, in which the smooth non-periodic data is parametrised by a smooth periodic function with period larger than the domain in which the data is given.

To accomplish this for the straight segment in the left portion of Fig. 4, for example, let the column vectors $x = (x_0, \ldots, x_{N-1})^T$ and $p = (p_0, \ldots, p_{N-1})^T$ contain the discrete abscissae (in the interval $[0, 1]$) and ordinates of the sample points, respectively. To find a Fourier series periodic in the interval $[0, 2]$, say, that interpolates this data, coefficients $\hat{p}$ must be found which satisfy the equation

$$p_l = \sum_{k=-N/2}^{N/2-1} \hat{p}_k e^{\pi i k x_l}, \quad l = 0, \ldots, N-1 . \tag{1}$$

The interpolator resulting from direct solution of a system of the form (1) is likely to be highly oscillatory, and, thus, to provide a poor representation of the original surface and its derivatives. To correct this problem, conditions additional to those in (1) are imposed on the coefficients $\hat{p}_k$ to ensure that, in some appropriate sense, they decay fast — as behoves the smooth function they interpolate. This can be accomplished by requiring that these coefficients multiplied by certain growing factors vanish in the least-square sense. The factors equal one for low order coefficients, and various rates of growth, from polynomial to exponential, can be used for higher order modes; see [16] for a discussion of theoretical and numerical issues associated with such strategies.

In general contexts, finally, it is necessary to provide a *smooth curve* which delimits the portion of the parametrised surface that corresponds to the actual given triangulation. Such a *bounding curve* curve can be obtained by the same methods described in this and the previous section, this time applied to a one dimensional "triangulation": the sequence of edges on the boundary of the two-dimensional triangulation. An example of such a boundary curve is shown in the lower right portion of Fig. 6.

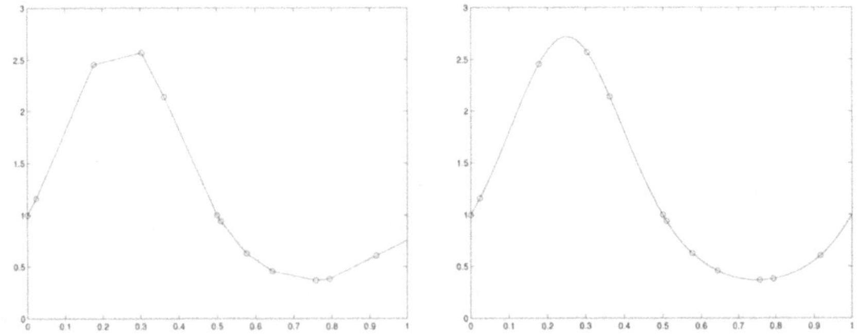

Fig. 5. A sparsely sampled surface (left) and its high-order approximation (right)

3.3 Surface Representation: Additional Examples

We present two additional examples which demonstrate some points of interest concerning the character of our surface representation algorithms.

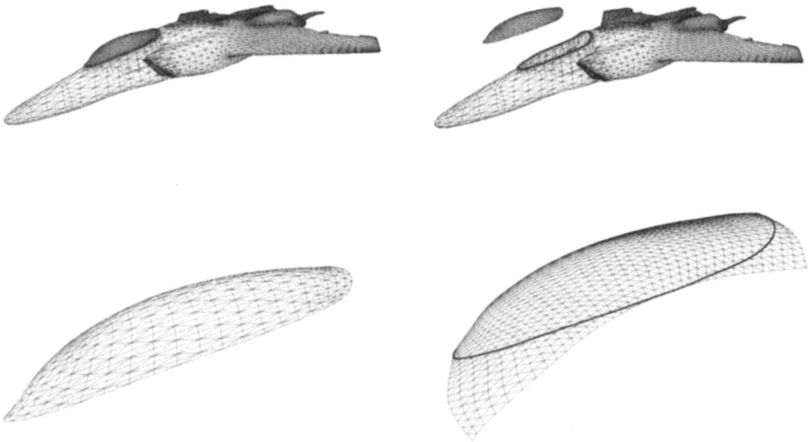

Fig. 6. F-15 aircraft. **Upper left:** original triangulation. **Upper right:** the canopy is separated from the rest of the triangulation, for treatment by the smooth parametrisation algorithm. **Lower left:** Close-up on the separated canopy. **Lower right:** Smooth parametrisation of extended surface, including bounding curve (in solid black), all produced automatically by the algorithm. The parametrised surface passes through all the vertices of the given triangulation. The bounding curve passes through all the boundary-vertices of the given triangulation

In the first of these examples the given, rather sparse discretisation is displayed in Fig. 5 left. We note that the high order representation produced by our algorithm, which is presented on the right portion of that figure, might not be a faithful representation of the original surface. Of course, this is also true of the linear interpolation shown on the left graph. The representation errors arising from the sparsity of the mesh are caused by the uncertainty implicit in the discretisation provided. In any case, with the information available the representation algorithm produces a smooth surface which can be utilised in a given application, and which allows for high-order numerics. If the high-order surface obtained is not faithful to the true physical surface then the predictions might be incorrect. But, at least, use of a high-order algorithm in such a situation allows a correct solution with accurate error estimates to be obtained for the incorrect surface, so that discrepancies in comparison with experiment can safely be attributed to either problems in the experimental setup or on the original surface representation, and thus a clear path for remedial action results: 1) to refine the surface sampling, and/or 2) to recheck the experimental setup.

Our second example, illustrated in Fig. 6, demonstrates the performance of our algorithms to produce smooth parametrisations from actual engineering triangulations. The parameter space used for this geometry was a paraboloid close to the canopy. Other portions of the airplane such as the wing require much more curved parameter surfaces. Such parametrisations (not displayed here) have been obtained by means of smoothed intrinsic parametrisations, as described towards the end of Sect. 3.1.

4 Fast, High-Order Surface Scattering Solvers

(*Joint work with L. Kunyansky* [10, 11, 15, 46] *and R. Paffenroth* [17])

In what follows we present our Fourier-based high-order algorithms for the numerical solution of problems of scattering by surfaces in three-dimensional space. This algorithm evaluates scattered fields through fast, high-order solution of the corresponding boundary integral equations. The high-order accuracy of our solver is achieved through use of *partitions of unity* (that is, a set of windowing functions which add up to one throughout the surface), together with *analytical* resolution of kernel singularities. The acceleration, in turn, results from use of a novel approach which, based on high-order *"two-face" equivalent source* approximations, reduces the evaluation of far interactions to evaluation of 3-D FFTs. This approach is faster, substantially more accurate, and it runs on dramatically lower memories than other FFT-based methods. The present overall algorithm computes one matrix-vector multiply in $\mathcal{O}\left(N^{6/5}\log N\right)$ to $\mathcal{O}(N^{4/3}\log N)$ operations, where N is the number of surface discretisation points. The latter estimate applies to smooth surfaces, for which our high order algorithm provides accurate solutions with small values of N; the former, more favourable count is valid for highly complex surfaces requiring significant amounts of sub-wavelength sampling. Further, our approach exhibits super-algebraic convergence, it can be applied to smooth and non-smooth scatterers, and, unlike other accelerated schemes it does not suffer from accuracy breakdowns of any kind (compare [31,53] and [47, p. 576]). In what follows we introduce the main algorithmic components in our approach, and we demonstrate its performance with a variety of numerical results. In particular, we show that the present algorithm can evaluate accurately in a personal computer scattering from bodies of acoustical sizes of several hundreds.

4.1 Fast High-Order Surface Integration Algorithm

For simplicity we restrict our presentation to the problem of acoustic scattering by a sound-soft obstacle. (In Sect. 4.2, however, numerical results for both electromagnetic and acoustic applications of our methods are given.) The relevant "combined field" integral equation is given by the appropriate combination of a single- and a double-layer potential (see e.g. [28])

$$(S\varphi)(\mathbf{r}) = \int_{\partial D} \Phi(\mathbf{r},\mathbf{r}')\,\varphi(\mathbf{r}')\,\mathrm{d}s(\mathbf{r}') \quad \text{and} \quad (K\varphi)(\mathbf{r}) = \int_{\partial D} \frac{\partial\Phi(\mathbf{r},\mathbf{r}')}{\partial\nu(\mathbf{r}')}\,\varphi(\mathbf{r}')\,\mathrm{d}s(\mathbf{r}'), \tag{2}$$

Here $\Phi(\mathbf{r},\mathbf{r}') = \mathrm{e}^{\mathrm{i}k|\mathbf{r}-\mathbf{r}'|}/4\pi\,|\mathbf{r}-\mathbf{r}'|$ is the Green function for the Helmholtz equation $\Delta u + k^2 u = 0$, and $\nu(\mathbf{r}')$ is the external normal to the (closed) surface ∂D at point $\mathbf{r}'$. Explicitly, given the values of the incoming wave $\psi^i(\mathbf{r})$ on ∂D, the scattered field can be obtained easily once the integral equation

$$\frac{1}{2}\varphi(\mathbf{r}) + (K\varphi)(\mathbf{r}) - \mathrm{i}\gamma\,(S\varphi)(\mathbf{r}) = \psi^i(\mathbf{r}), \qquad \mathbf{r} \in \partial D \tag{3}$$

has been solved for the unknown density $\varphi(\mathbf{r})$. Naturally, the possibility of producing fast and accurate solutions for our problems hinges on our ability to evaluate the integrals (2) accurately and efficiently. In attempting to develop such accurate and efficient integrators one faces two main problems, namely, accurate evaluation of the singular *adjacent interactions* — without undue compromise of speed — and fast evaluation of the voluminous number of *non-adjacent interactions* — without compromise in accuracy. In what follows we present a solution to these problems.

Partitions of Unity

In order to deal with topological characteristics of closed surfaces, which are given in terms of the local parametrisation discussed in Sect. 3, we utilise partitions of unity. In detail, we use a covering of the surface ∂D by a number K of overlapping two-dimensional patches $\mathcal{P}^j, j = 1, \cdots, K$ (called local charts in differential geometry), together with smooth mappings to coordinate sets $\mathcal{H}^j$ in two-dimensional space, where actual integrations are performed. Further, we utilise a partition of unity subordinated to this covering of ∂D, i.e. we introduce a set of non-negative smooth functions $\{w^j, j = 1, \ldots, K\}$, such that (i) w^j is defined, smooth and non-negative in ∂D, and it vanishes outside $\mathcal{P}^j$, and (ii) $\sum_{j=1}^{K} w^j = 1$ throughout ∂D. This allows us to reduce the problem of integration of the density $\varphi(\mathbf{r})$ over the surface to a calculation of integrals of smooth functions φ^j compactly supported in the planar sets $\mathcal{H}^j$.

Adjacent Integration

Substantial difficulties in the high-order evaluation of *adjacent interactions* are caused by the singular nature of the integral kernels. While, certainly, the well-known strategy of "singularity subtraction" gives rise to bounded integrands, integration of such bounded functions by means of classical high-order methods does not exhibit high-order accuracy — since the subsequent derivatives of the integrand are themselves unbounded. The new basic high-order integrator we present is based on analytical resolution of singularities. The resolution is achieved by integration in polar coordinates centred around each singular point. The Jacobian of the corresponding change of variables has the effect of cancelling the singularity, so that high order integration in the both radial and angular directions can be performed using the trapezoidal rule. Since the corresponding radial quadrature points do not lie on the Cartesian grid, a high-order, fast interpolation technique has been developed for evaluation of the necessary function values at the radial integration points. Efficiency is of utmost importance here, since we use one such polar coordinate transformation *at each target point.* Our high order integrator exhibits super-algebraic convergence for smooth and non-smooth scattering surfaces [10, 15, 46].

Non-Adjacent Integration and Acceleration

Our accelerator is closely related to two of the most advanced FFT methods developed recently [7, 51]. An important common element between these two methods and our technique is a concept of equivalent (or auxiliary) sources, located on a subset of a 3-D Cartesian grid. In all three cases, the intensities of these sources are chosen to approximate the field radiated by the scatterer, which allows for fast computation of the "non-adjacent interactions" through the use of 3-D FFTs. Surface problems like the ones we consider are treated in [7, 51] by means of equivalent sources located in a *volumetric* grid — in such a way that equivalent sources with non-zero intensities occupy *all Cartesian nodes adjacent to the scatterer.* Since the spacing of this Cartesian grid cannot be coarsened beyond some threshold for surface problems such a scheme requires an $\mathcal{O}(N^{3/2})$ FFT. Therefore, previous FFT surface scattering solvers require $\mathcal{O}(N^{3/2})$ units of RAM and they run in $\mathcal{O}\left(N^{3/2}\log N\right)$ operations. Our algorithm, in contrast, subdivides the volume occupied by the scatterer into a number of (relatively large) cubic cells, and it places equivalent sources *on the faces* of those cells. As we have shown, such a design reduces significantly the sizes of the required FFTs — to as little as $\mathcal{O}\left(N^{6/5}\right)$ to $\mathcal{O}\left(N^{4/3}\right)$ points — with proportional improvement in storage requirements and operation count. Further, it results in super-algebraic convergence of the equivalent source approximations *as the size of the scatterer is increased.*

Resolution of Singularities

To obtain resolution of the singular integrands around the ogive's conical singularities, for example, a combination of two changes of variables were used: a polar change of variables similar to that described in the section "Adjacent integration" above, followed by a polynomial change of variables which regularises the Hölder-type singularity of the underlying density; see [15] for details.

4.2 Surface Scattering: Numerical Results

We present results for well known and widely used test geometries: large spheres, ellipsoids, cubes and ogives; solutions with analogous accuracies and computing times for non-exact geometries have been presented elsewhere. In particular, we present comparisons with the FMM solver FISC [56]. The following caveat should be taken into account when considering these comparisons: our results for large spheres correspond to solutions of three-dimensional acoustic scattering problems — solutions of the Helmholtz equation — whereas the FISC data corresponds to solutions of the Maxwell equations. There are of course some differences between the Helmholtz and Maxwell problems; in particular the unknowns in the Maxwell integral equations are two-dimensional vectors, as opposed to the single scalar unknown arising in the Helmholtz integral equation. However, our methods apply to the full Maxwell problem — although our implementations do not yet include Maxwell solvers with equivalent source acceleration. Results provided by our non-accelerated Maxwell solver have already been obtained: for example, results for an electromagnetic cube of about one wavelength in diagonal were obtained with errors of the order of 10^{-4}, while results for the electromagnetic flying saucer of Fig. 7 of two wavelengths in diameter were obtained with errors 10^{-5}. Implementations of our fully accelerated EM solvers for general, possibly singular surfaces are currently being produced.

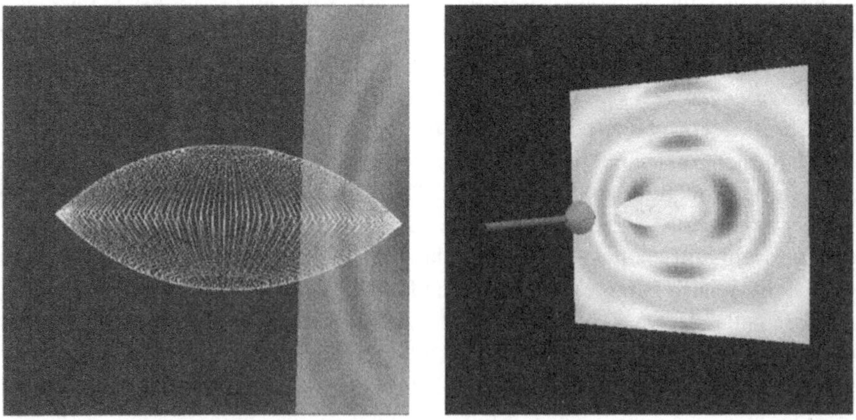

Fig. 7. Left: Flying saucer. **Right:** Electromagnetic scattering from a two-wavelength diameter flying saucer, with maximum far field errors of $3.0 \cdot 10^{-5}$

Results for problems of scattering by large spheres are presented in Table 3. We see that the performance of the present methods compares very well with that of lead-

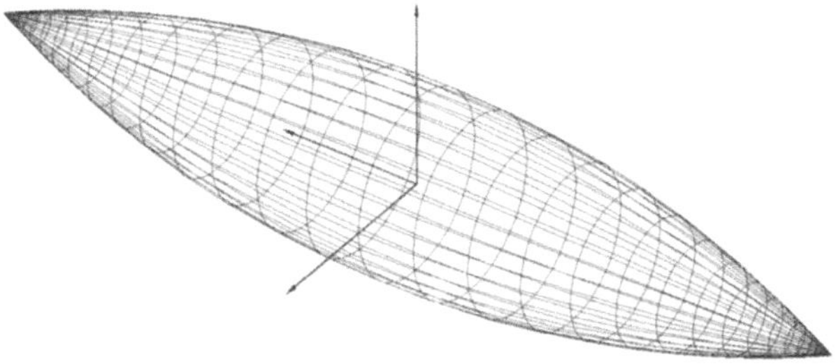

Fig. 8. Ogive geometry presented in [60]

ing solvers; note the excellent accuracies provided by the algorithm in competitive running times.

Algorithm	Diameter	Time	RAM	Unknowns	RMS Error	Computer
FISC	120λ	$32 \times 14.5h$	$26.7Gb$	9,633,792	4.6%	SGI Origin 2000 (32 proc.)
Present	80λ	$55h$	$2.5Gb$	1,500,000	0.005%	AMD 1.4GHz (1 proc.)
Present	100λ	$68h$	$2.5Gb$	1,500,000	0.03%	AMD 1.4GHz (1 proc.)

Table 3. Scattering from large spheres

Table 4 displays a set of results obtained for scattering from a singular surface: the ogive depicted in Fig. 8, for acoustical sizes (distances between tips) equal to 1λ, 10λ and 20λ. For the larger sizes we used the accelerator described above; note the substantial improvements in computing times resulting from the acceleration algorithm.

5 Fast, High-Order Volumetric Scattering Solvers

In this section we describe a class of high-order integral equation methods for the evaluation of electromagnetic scattering by bounded penetrable scatterers with variable indexes of refraction. Such problems find application in a wide range of fields, including biology and medicine (tomography, ultrasound) as well as plasma physics, neutron scattering, etc. The evaluation of useful numerical solutions for such scattering problems remains a highly challenging problem, requiring novel mathematical approaches and powerful computational tools.

Our approach to the problem of scattering by large volumetric scatterers is based, in part, on a concept of cancellation of errors: certain large errors in integrands arising by approximation of discontinuous functions by their Fourier series can result in

Type	Size	Unknowns	Iterations	Time/It.	ε_∞	ε_2
Non Accelerated	1λ	1568	20	69s	2.5×10^{-3}	$1.4 \cdot 10^{-3}$
Non Accelerated	1λ	6336	17	12m 45s	3.8×10^{-5}	$2.2 \cdot 10^{-5}$
Non Accelerated	1λ	25472	17	3h 27m	9.8×10^{-7}	$4.8 \cdot 10^{-7}$
Accelerated	10λ	34112	13	26m	3.8×10^{-4}	$2.1 \cdot 10^{-4}$
Accelerated	20λ	34112	14	14m	6.0×10^{-3}	$2.4 \cdot 10^{-3}$
Accelerated	20λ	72320	19	67m	5.4×10^{-5}	$2.1 \cdot 10^{-5}$

Table 4. Scattering by an ogive. Results produced on a single processor 400MHz PC with 1Gb of RAM. The column labelled "Iterations" displays the number of iterations required by the linear algebra solver GMRES [55] to produce results with the errors quoted. The quantities ε_∞ and ε_2 are, respectively, the relative mean-square norm and the absolute maximum norm of the error in the far field

Geometry	Diameters	Time	Unknowns	RMS Error	Computer
Cube (Present work)	$10\lambda \times 10\lambda \times 10\lambda$	$21h$	96,774	0.049%	AMD 1.4GHz (1 proc.)
Flying Saucer (Present work)	$42\lambda \times 42\lambda \times 17\lambda$	$53h$	290,874	0.0045%	AMD 1.4GHz (1 proc.)

Table 5. Acoustic scattering by geometries containing edges

small errors in values of integrals — provided integrands and integrals are set up appropriately; see Sect. 5.1 below and the references therein for details. Since the ideas of error cancellation have proven somewhat controversial we preface our discussion on volumetric scattering by a one dimensional example (given in the following paragraph), which illustrates this phenomenon in a simple setting. Then in Sect. 5.1 we present our volumetric scattering solver. As it happens, the behaviour of this solver can be improved by smoothly partitioning the integration problem into the "large" interior integration problem, and the "thin" volumetric boundary integration problem. Our approach for thin volume integration follows in Sect. 5.2.

The proposed concept of error cancellation may be readily illustrated by means of an elementary numerical example: use of Fourier expansions to produce high-order numerical evaluations of an integral of the form $\int_0^1 f(\theta)g(\theta)\,\mathrm{d}\theta$, where (i) f is a discontinuous periodic function, and where (ii) g is a function which is continuous together with its derivative, but whose second derivative is discontinuous. For this example we use the functions

$$f(\theta) = \begin{cases} 1 & 0 \le \theta < 1/4 \\ -1 & 1/4 \le \theta < 3/4 \\ 1 & 3/4 \le \theta \le 1 \end{cases}$$

and

$$g(\theta) = \begin{cases} \theta^2/2 & 0 \le \theta < 1/4 \\ -\theta^2/2 + \theta/2 - 0.0625 & 1/4 \le \theta < 3/4 \\ (\theta - 1)^2/2 & 3/4 \le \theta \le 1 \end{cases}$$

see Fig. 9; we have $\int_0^1 f(\theta)g(\theta)\,d\theta = -1/48$. The discontinuities of f and the degree of smoothness of g correspond, respectively, to those of the refractive index n and the scattered field u *for a soft acoustic scatterer with a discontinuous refractive index.*

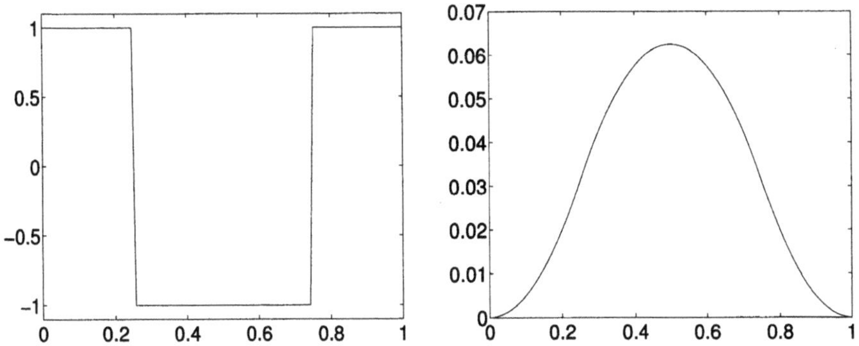

Fig. 9. The functions f and g

In analogy with some aspects of our method, here we proceed to evaluate the integral of fg by replacing f and g by their respective Fourier series truncated to order F (including modes between $-F$ and F only), evaluating their *pointwise* product, and integrating the result by means of the trapezoidal rule with N_θ points and mesh-size $h = 1/(N_\theta - 1)$. [We emphasise here that the Fourier coefficients of the discontinuous function f should be accurate, and could not therefore be produced by a simple integration rule. Fortunately, it is not hard to produce accurate Fourier coefficients for f, as it is generally the case for the Fourier coefficients of a given distribution n of refractive indexes: either closed expressions or simple one-dimensional high-order integration rules "by-pieces" can be used.] The accuracies resulting from these operations are displayed in Table 6. We see that the error in the approximate integral is of the order h^3, in spite of the Gibbs phenomenon and the low order convergence of the series for the discontinuous function f. This cancellation of errors can be explained through consideration of the error arising in the zero-th order coefficient of the function fg as a result of the truncations used, see [13]. Naturally, even higher order convergence results for smoother functions f and g (or n and u). We point out, however, that, even in the most singular case considered here, it suffices to use 64 points to produce results with an accuracy better than full single precision. Tight error estimates for the numerical method presented in this paper are given in [13].

5.1 Large Penetrable Bodies
(*Joint work with McKay Hyde* [13, 14, 43] *and A. Sei* [21–23])

We consider frequency-domain acoustic scattering by bounded, inhomogeneous media with refractive index $n(\mathbf{r})$. Given an incident field u^i and calling u^s the scattering

F	N_θ	Absolute Error	Ratio
2	4	3.0 (-4)	
4	8	4.8 (-5)	6.3
8	16	6.5 (-6)	7.4
16	32	8.3 (-7)	7.8
32	64	1.0 (-7)	8.3

Table 6. Convergence test for the evaluation of $\int_0^1 f(\theta)g(\theta)\,\mathrm{d}\theta$ as a function of the number of Fourier modes used for f and g

field to be calculated, the total field $u = u^i + u^s$ solves both the Helmholtz equation [28, p. 2],

$$\Delta u + k^2 n^2(\mathbf{r})u = 0, \quad \mathbf{r} \in \mathbb{R}^3,$$

and the Lippmann-Schwinger integral equation [28, p. 214]

$$\begin{aligned} u(\mathbf{r}) &= u^i(\mathbf{r}) - k^2 \int_\Omega \Phi(\mathbf{r},\mathbf{r}')\, m(\mathbf{r}')\, u(\mathbf{r}')\,\mathrm{d}\mathbf{r}' \\ &= u^i(\mathbf{r}) - K[u](\mathbf{r}), \end{aligned} \tag{4}$$

where $m = 1-n^2$ and, again, $\Phi(\mathbf{r},\mathbf{r}') = \mathrm{e}^{ik|\mathbf{r}-\mathbf{r}'|}/4\pi\,|\mathbf{r}-\mathbf{r}'|$ is the Green's function for the Helmholtz equation in three dimensions.

As is well known, the complexity of the integration associated with this integral equation can be reduced to $\mathcal{O}(N \log N)$ operations through use of the fast Fourier transform (FFT). Perhaps the most familiar such method is the k-space or conjugate-gradient FFT method (CG-FFT) [8, 61, 62], in which the convolution with the Green's function is computed via a Fourier transform (computed with an FFT) and multiplication in Fourier space. Although this method provides a reduced complexity, it is only first-order accurate. This low-order accuracy arises because the FFT provides a poor approximation to the Fourier transform when, as in this case, the function is not smooth and periodic.

Although our methods also use FFTs to achieve a reduced complexity, they yield, in addition, *high-order accuracy*. To our knowledge, only limited attempts have been made at devising high-order methods for this problem. Liu and Gedney [39] proposed an $\mathcal{O}(N^2)$ locally corrected Nyström scheme for scattering in two dimensions. This method provides high-order convergence rates that are not limited by the regularity of the scatterer, but it does not possess the desirable low operation counts of Fourier based methods. Vainikko [58], on the other hand, proposed a method for smooth scatterers that is related to our approach. In this method, the integral equation is modified to produce a periodic solution by cutting off the Green's function (either smoothly or discontinuously) outside a cube that is at least twice as large as the scatterer. The solution to the modified integral equation is smooth and periodic on this larger cube and, furthermore, agrees with the true solution on the support of the scatterer. Thus, for smooth scatterers, the solution is smooth and periodic and can, therefore, be approximated to high-order with a truncated Fourier series. However, the convergence rates of this approach lag significantly behind those of our approach — producing only first-order convergence in the case of discontinuous scatterers. Vainikko introduces a completely different approach for piecewise-smooth scatterers that produces $\mathcal{O}(h^2(1 + \log h))$ convergence in both the near and far fields, where h is the discretisation spacing in each direction. This approach requires that for each level of discretisation, one must approximate the volume fraction of each cell that lies on each side of a discontinuity in the refractive index. This seems rather difficult to obtain, especially for complicated scatterers in three dimensions. This contrasts with the limited geometrical requirements of our method, in which various portions of a scatterer can be treated separately to finally assemble the Fourier series of the overall refractive index distribution as a sum of the Fourier series for the refractive indexes of the components.

Our goal is to obtain an FFT-based method (thereby yielding $\mathcal{O}(N \log N)$ complexity) that is also high-order accurate. As is well known, the trapezoidal rule can be used to evaluate convolution integrals and Fourier coefficients and, in these cases, is algorithmically equivalent to the FFT. Also, FFTs can efficiently evaluate a truncated Fourier series on a set of equally-spaced grid points. However, the trapezoidal rule yields high-order convergence only for *smooth and periodic* integrands and similarly, truncated Fourier series exhibit high-order convergence only when approximating *smooth and periodic* functions.

For these reasons, a primary obstacle to the development of a high-order, FFT-based method is the lack of regularity in the scatterer. Hence, perhaps the most important aspect of our approach is the substitution of the scatterer by an appropriate Fourier-smoothed approximation. This approach counters conventional wisdom: a Fourier approximation of a discontinuous function necessarily yields low-order convergence. However, as shown in the example in the introduction to Sect. 5, through our computational examples and as we have proven rigorously [13], an *appropriate* numerical integration of such a Fourier approximated function does indeed yield high-order accuracy.

Another obstacle to the development of a high-order method concerns the Green's function. There are two ways to evaluate the convolution by means of FFT-based methods. First, one could evaluate the convolution by means of the trapezoidal rule. However, even with the Fourier-smoothed scatterer, the polar singularity in the Green's function produces first-order convergence. Alternatively, one could approximate the convolution operator $K[u](\mathbf{r})$ (see (4)) itself by means of a truncated Fourier series. However, although $K[u](\mathbf{r})$ is smooth, because of the slowly decaying tail of the Green's function, it is not periodic. Hence, the Fourier series converges to first-order only.

To overcome these difficulties, we decompose the Green's function by means of a smooth partition of unity into 1) a smooth part with infinite support and 2) a singular part with compact support. The convolution with the smooth part of the Green's function is computed to high-order by means of the trapezoidal rule. Finally, the convolution with the singular part is computed to high-order by approximating the (now smooth and periodic) operator by a truncated Fourier series. Each one of these convolutions is computed using FFTs yielding high-order accuracy and a total complexity of $\mathcal{O}(N \log N)$. We thereby obtain a method that is as simple and efficient as the CG-FFT method, but, which, unlike the CG-FFT, yields high-order accuracy.

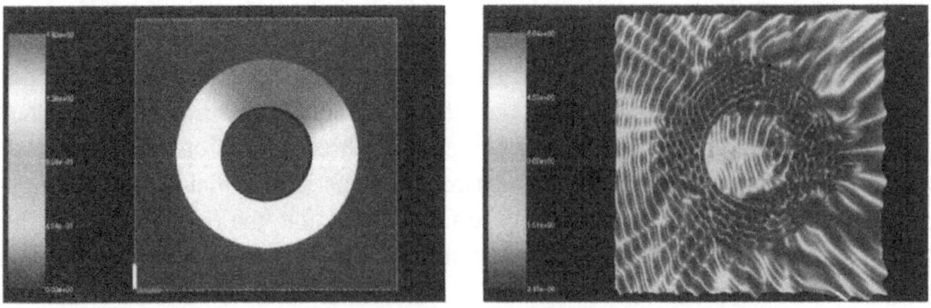

(a) Scatterer ($q = -m = n^2 - 1$) (b) Near Field Intensity ($|u|^2$)

N	Memory	Iter.	Time	NF Error	Ratio	FF Error	Ratio
12K	19Mb	54	36s	2.16e-5		7.33e-9	
25K	39Mb	54	72s	4.81e-7	44.91	1.06e-11	691.51
50K	75Mb	54	160s	1.05e-8	45.81	4.50e-12	Conv.
99K	150Mb	54	331s	4.76e-10	22.06	4.52e-12	Conv.
198K	305Mb	54	561s	1.36e-11	35.0	4.61e-12	Conv.
396K	609Mb	54	1172s	1.94e-12	Conv.	4.72e-12	Conv.

(c) Convergence Results

Fig. 10. Two-dimensional scatterer. Diameter = 10λ. Computations using N unknowns

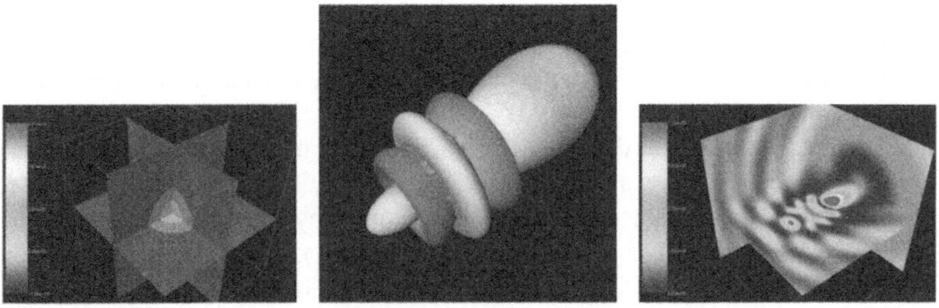

(a) Scatterer ($q = n^2 - 1$) (b) Far Field Intensity (c) Near Field Intensity

Discretisation	Time (s) × #CPUs	FF Error
10 × 10 × 10	2.15 × 1	0.146
20 × 20 × 20	15.6 × 1	4.56(-3)
40 × 40 × 40	125 × 1	9.55(-4)
80 × 80 × 80	1119 × 1	5.43(-5)
160 × 160 × 160	475 × 32	7.11(-6)

(d) Convergence Results

Fig. 11. Layered Sphere of radius a– $ka = 4$

(**a**) Scatterer ($q = n^2 - 1$) (**b**) Far Field Intensity (**c**) Near Field Intensity

N	Time (s) × #CPUs	NF Error	Ratio	FF Error	Ratio
10 × 10 × 10	2.15 × 32	3.70		43.0	
20 × 20 × 20	15.6 × 32	1.35	2.73	10.6	4.05
40 × 40 × 40	125 × 32	4.80(-2)	28.2	8.66(-2)	122
80 × 80 × 80	1119 × 32	8.28(-3)	5.79	4.47(-2)	1.94
160 × 160 × 160	475 × 32	6.48(-5)	128	7.76(-5)	576

(**d**) Convergence Results

Fig. 12. Array of Smooth Scatterers (Potentials) – $6\lambda \times 6\lambda \times 6\lambda$

Large Penetrable Bodies: Numerical Results

We illustrate the capabilities of this method by means of three computational examples. In our first example we consider a problem of two-dimensional scattering by a scatterer containing the refractive index distribution depicted in Fig. 10. We then present a problem of scattering for a piecewise-constant (discontinuous) layered sphere in three dimensions (shown in Fig. 11), for which the analytical solution is known. Finally, we consider in Fig. 12 a $5 \times 5 \times 5$ array of scattering potentials. In the Tables of this section the columns labelled "Iter." display the number of iterations required by the linear algebra solver GMRES [55] to produce results with the errors quoted, the quantities "NF Error" and "FF Error" are the absolute maximum norm of the error in the near field and the far field, respectively, and the quantity "Ratio" is the quotient between two consecutive error quotes. Except in the case in which the exact solution is known, errors were evaluated through comparison with the solution obtained for a significantly finer discretisation.

The numerical results of this section demonstrate both the $\mathcal{O}(N \log N)$ complexity and the high-order convergence rate of our method. In particular, the method seems to yield significantly more than second-order convergence in the near field and third-order convergence in the far field for discontinuous scatterers.

5.2 Thin Penetrable Scatterers (*Joint work with F. Reitich* [18])

As mentioned earlier, the convergence of the solver described in the previous section can be improved significantly by smoothly partitioning the integration problem into a "large" interior integration problem and a "thin" volumetric boundary integration problem. Further, the problem of scattering by thin volumetric bodies is of independent interest, and, because of its proximity to a surface scattering problem admits a special treatment — as discussed in what follows. For simplicity we restrict this discussion to the two-dimensional case.

Let us thus consider a thin volumetric scatterer such as that depicted in Fig. 13. It is natural to integrate in directions parallel and transverse to the discontinuity surface. The problem of integration in the direction parallel to the discontinuity surface is related to (but different from) the problem of surface integration in the surface scattering case. Indeed the present problem requires, in addition to integration on the singular surface, integration in near singular surfaces, for which the treatment of Sect. 4 is not appropriate. Indeed, the near singular integrals contain smooth but nearly singular integrands, whose nearly singular behaviour is not appropriately resolved by the polar changes of variables introduced in Sect. 4.

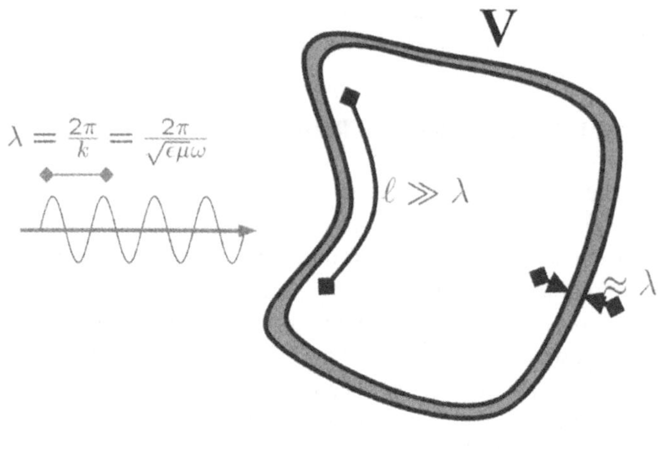

Fig. 13. A thin volumetric scatterer

We thus resort to use of a different type of change of variables, namely, one given by a function of the type depicted in Fig. 14, several (or all) of whose derivatives vanish at the origin. Such a change of variables regularises every one of the nearly singular integrands, and thus allows for high order integration in the tangential direction. We note, however, that such a change of variables implies a refinement of data around each target point. Since this refined data is not available, we must resort to interpolation — which we do by means of Fourier series, as discussed in Sect. 2. A similar change of variables and interpolation method is used for integration in the transverse direction. Acceleration of the integrator, finally, is obtained through the equivalent source method described in Sect. 4.1.

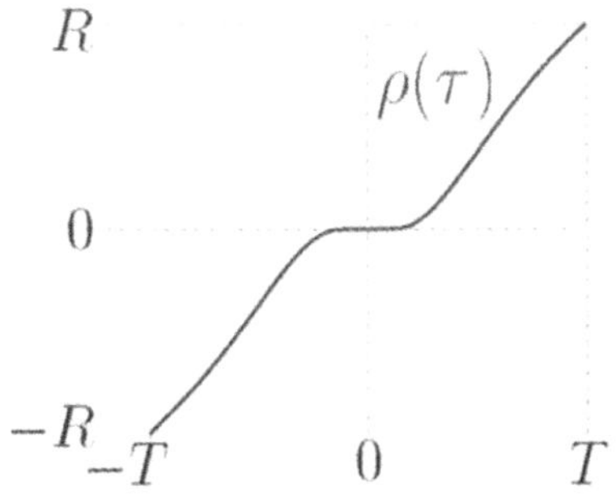

Fig. 14. Change of variables

In Fig. 15 we present the field within a thin volumetric scatterer, and in Fig. 16 we give the result of a convergence study, showing perfect agreement with the theoretical error estimates associated with the changes of variables used.

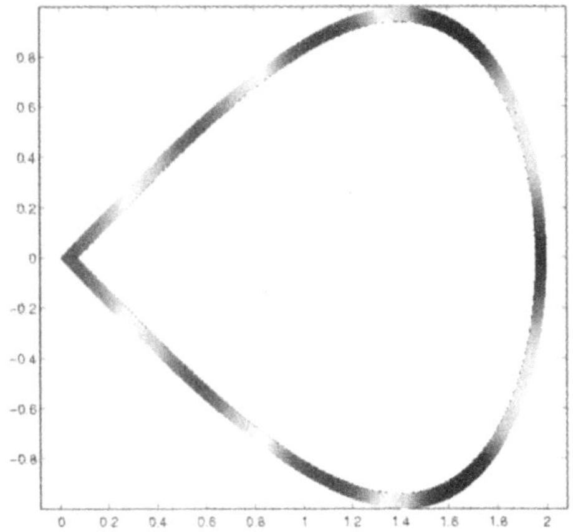

Fig. 15. Field within a thin drop-shaped volumetric scatterer

6 Rigorous High-Order, High-Frequency Integral Methods

However efficient, direct numerical methods are necessarily capped in the sizes of problems they can treat on a given computer. It is therefore desirable to produce numerical methods which become more efficient as the frequencies (and, thus, the size of the problem) grow. If accurate high-frequency solvers are made available with a bounded computational complexity as the frequency tends to infinity, (that is, methods with an asymptotic $\mathcal{O}(1)$ computational complexity) then one can envision the

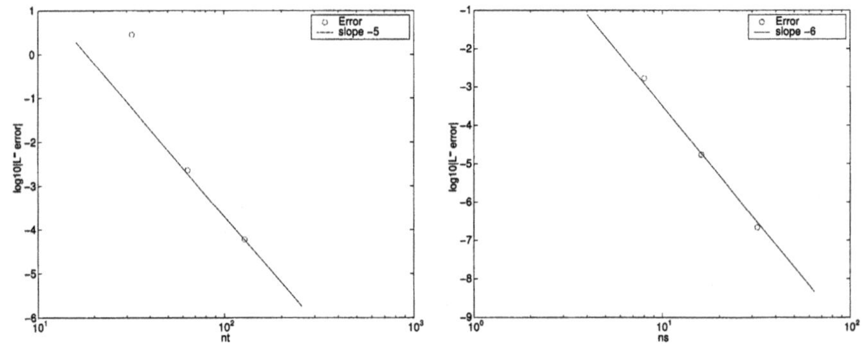

Fig. 16. Convergence of the thin volume scattering solver for the scatterer of Fig. 15. **Left:** Error in the parallel integrator. **Right:** Error in the transverse integrator. The changes of variables used have five and six vanishing derivatives at the origin, respectively, and thus, give rise to the 5-th and 6-th order convergence shown

development of a computational capability able to solve essentially arbitrarily scattering problems. Our recent efforts in these regards have resulted in a class of high-order high-frequency surface-scattering solvers which we present in the following three sections.

Versions of these solvers, based on Nystrom discretisation and applicable to scattering by bounded obstacles are described in Sects. 6.2 and 6.3 below. We begin our discussion in Sect. 6.1 with a different method for scattering by rough surfaces which is based on high-frequency perturbation series. Although the perturbation series approach does not generalise easily to scattering configurations giving rise to shadowing (that is, to regions of the scattering surface which, from a geometrical optics standpoint, are not illuminated by the incident field), a discussion of the perturbation method does provide a useful link between classical non-convergent high-frequency methods such as the Kirchhoff approximation and the general, convergent high frequency solvers presented in the subsequent two sections.

6.1 HF-Scattering by Periodic Surfaces (*Joint work with A. Sei and M. Caponi* [24])

Let us consider the problem of scattering by a periodic rough surface $y = f(x)$ in two-dimensional space. The scattered field can be computed from the surface current density ν induced on the scattering surface by an incident plane wave [59]. Assuming plane wave incidence and calling θ the angle between the incident wave-vector and the vertical, the function ν solves the integral equation

$$\nu(x,k) - \frac{\mathrm{i}}{2}\int_{-\infty}^{\infty} h\,(kR(x,x'))\, g(x,x')\,\nu(x',k)\,\mathrm{d}x' = -\mathrm{e}^{\mathrm{i}\alpha x - \mathrm{i}\beta f(x)} \tag{5}$$

where

$$\begin{aligned} R(x,x') &= \sqrt{(x'-x)^2 + (f(x') - f(x))^2}\,, \\ h(t) &= tH_1^1(t) \quad \text{(where } H_1^1 \text{ is the Hankel function)}\,, \\ g(x,x') &= (f(x') - f(x) - (x'-x)f'(x'))/\left(R(x,x')\right)^2\,, \\ \alpha &= k\sin\theta\,, \quad \beta = k\cos\theta\,. \end{aligned}$$

To solve this equation we use an asymptotic expansion of high order around $k = \infty$. Our asymptotic expansion results from the geometrical optics type ansatz

$$\nu(x,k) \sim \mathrm{e}^{\mathrm{i}\alpha x - \mathrm{i}\beta f(x)} \sum_{n=0}^{+\infty} \frac{\nu_n(x)}{k^n}\,. \tag{6}$$

From (5) and (6) we obtain

$$\sum_{n=0}^{+\infty} \frac{1}{k^n}\left(\nu_n(x) - \frac{\mathrm{i}}{2} I^n(x,k)\right) = -1 \tag{7}$$

where

$$\begin{aligned} I^n(x,k) &= \int_{-\infty}^{+\infty} h(kR(x,x'))\, g(x,x')\, \mathrm{e}^{-\mathrm{i}k\psi(x,x')}\, \nu_n(x')\,\mathrm{d}x' \\ \psi(x,x') &= (x-x')\sin\theta - (f(x) - f(x'))\cos\theta\,. \end{aligned}$$

To obtain the coefficients $\nu_n(x)$ we must produce the asymptotic expansion of $I^n(x,k)$ in powers of $1/k$. To do this we use two separate integration regions and we define

$$\begin{aligned} I_-^n(x,k) &= \int_{-\infty}^{x} h(kR(x,x'))\, g(x,x')\, \mathrm{e}^{-\mathrm{i}k\psi(x,x')}\, \nu_n(x')\,\mathrm{d}x' \\ I_+^n(x,k) &= \int_{x}^{+\infty} h(kR(x,x'))\, g(x,x')\, \mathrm{e}^{-\mathrm{i}k\psi(x,x')}\, \nu_n(x')\,\mathrm{d}x'\,. \end{aligned}$$

We focus first on an asymptotic expansion for $I_+^n(x,k)$. Using $t = x' - x$ we can write

$$I_+^n(x,k) = \int_0^{+\infty} h(k\phi_+(x,t))\, g(x,x+t)\, \mathrm{e}^{-\mathrm{i}k\psi(x,t)}\, \nu_n(x+t)\,\mathrm{d}t\,, \tag{8}$$

where $\phi_+(x,t) = \sqrt{t^2 + (f(x+t) - f(x))^2}$ and $\psi(x,t) = t\sin\theta - (f(x+t) - f(x))\cos\theta$. For the simplest treatment we present here we assume that $f(x)$ is such that ϕ_+ satisfies the condition

$$\frac{\partial\phi_+}{\partial t}(x,t) > 0 \quad \text{for } t \geq 0$$

so that the map $t \mapsto \phi_+(x,t)$ is invertible; this condition is generally satisfied by rough surfaces relevant to the applications under consideration. Then setting

$$u = \phi_+(x,t) \iff t = \phi_+^{-1}(x,u)$$

we can rewrite (8) in the form

$$I_+^n(x,k) = \int_0^{+\infty} h(ku)\, F_n^+(x,u)\, \mathrm{e}^{-\mathrm{i}k\psi^+(x,u)}\, du$$

with

$$\begin{aligned} F_n^+(x,u) &= \frac{g(x, x+\phi_+^{-1}(x,u))}{\phi_+'(x,\phi_+^{-1}(x,u))}\, \nu_n(x+\phi_+^{-1}(x,u)) \\ \psi^+(x,u) &= \big(f(x+\phi_+^{-1}(x,u)) - f(x)\big)\cos\theta \\ &\quad - \phi_+^{-1}(x,u)\sin\theta\,. \end{aligned} \tag{9}$$

We can now use the Taylor series of $u \mapsto F_n^+(x,u)$ around $u=0$

$$\begin{aligned} F_n^+(x,u) &= \sum_{m=0}^{+\infty} \frac{\partial^m F_n^+(x,0)}{\partial u^m} \frac{u^m}{m!} = \sum_{m=0}^{+\infty} p_{n,m}^+(x) u^m \\ p_{n,m}^+(x) &= \frac{1}{m!} \frac{\partial^m F_n^+(x,0)}{\partial u^m} \end{aligned}$$

to express $I_+^n(x,k)$ in the form

$$\begin{aligned} I_+^n(x,k) &= \sum_{m=0}^{+\infty} p_{n,m}^+(x) \int_0^{+\infty} u^m\, h(ku)\, \mathrm{e}^{-\mathrm{i}k\psi^+(x,u)}\, du \\ &= \sum_{m=0}^{+\infty} \frac{p_{n,m}^+(x)}{k^{m+1}} \int_0^{+\infty} v^m\, h(v)\, \mathrm{e}^{-\mathrm{i}k\psi^+(x,\frac{v}{k})}\, dv \\ &= \sum_{m=0}^{+\infty} \frac{p_{n,m}^+(x)}{k^{m+1}} A^+(k,m,x) \end{aligned} \tag{10}$$

where we have set

$$A^+(k,m,x) = \int_0^{+\infty} v^m\, h(v)\, \mathrm{e}^{-\mathrm{i}k\psi^+(x,\frac{v}{k})}\, dv\,. \tag{11}$$

(These non-convergent integrals must be re-interpreted by means of analytic continuation — in a manner similar to that use in the definition and manipulation of Mellin transforms. We do not provide details about this analytic continuation procedure here; see [6] for a complete treatment in the case of the Mellin transform.)

To complete our expansion of $I_+^n(x,k)$ we need to produce a corresponding expansion of the quantity $A^+(k,m,x)$ in powers of $1/k$. With $\varepsilon = 1/k$ we call

$$\tilde{A}^+(\varepsilon,m,x) = A^+(k,m,x) = \int_0^{+\infty} v^m\, h(v)\, \mathrm{e}^{-\mathrm{i}\psi^+(x,\varepsilon v)/\varepsilon}\, dv\,. \tag{12}$$

By evaluation of the successive derivatives of $\tilde{A}^+(\varepsilon,m,x)$ with respect to ε at $\varepsilon = 0$ it is easy to check that the coefficients of the Taylor series of $\tilde{A}^+(\varepsilon,m,x)$ with respect to ε can be obtained directly if all the integrals in the sequence

$$\tilde{A}^+(m,x) = \int_0^{+\infty} v^m\, h(v)\, \mathrm{e}^{-\mathrm{i}\frac{\partial\psi^+}{\partial u}(x,0)}\, dv$$

are known; for example, the first derivative is given by the following expression (where $\psi_n^+(x) = \dfrac{\partial^n \psi^+}{\partial u^n}(x, u = 0)$)

$$\left.\frac{\partial \tilde{A}^+}{\partial \varepsilon}(\varepsilon, m, x)\right|_{\varepsilon=0} = -\mathrm{i}\frac{\psi_2^+(x)}{2}\tilde{A}^+(m+2, x)\,.$$

Using (10) and (12) the expansion for $I_+^n(x,k)$ results; clearly, the expansion of $I_-^n(x,k)$ can be obtained through a similar derivation. The combined expansion for $I^n(x,k)$ involves combinations of quantities such as $p_{n,m}^+(x)$, $\psi_\ell^+(x)$, $p_{n,m}^-(x)$, $\psi_\ell^-(x)$, $\tilde{A}^+(m,x)$, $\tilde{A}^-(m,x)$. It can be shown that all of these quantities — and therefore the integrals $\tilde{A}^\pm(m,x)$ for all m— can be obtained from the integrals

$$\int_0^{+\infty} v^{m+1} H_1^1(v)\cos(av)\,\mathrm{d}v \qquad \text{if } m \text{ is even}$$

$$\int_0^{+\infty} v^{m+1} H_1^1(v)\sin(av)\,\mathrm{d}v \qquad \text{if } m \text{ is odd}$$

for which closed form formulas are available [2].

To test the accuracy of our numerical procedures we compare our results for the high-frequency method with those given by a well tested approach [20] (based on high-order perturbation theory on boundary perturbations) in an "overlap" wavelength region — in which, as we show, both algorithms are very accurate. Note that these two methods are substantially different in nature: one is a high order expansion in λ whereas the other is a high order expansion in the height h of the profile. In the examples that follow we list relative errors for the computed values of the corresponding scattered energy shown; the results given in the columns denoted by Order 0–11 are the relative errors for the values of the scattered energy calculated from the high frequency code to orders 0–11 in the scattering direction listed.

For brevity we only present a classical test example: a sinusoidal profile $f(x) = h/2\cos(2\pi x)$; other applications of this method can be found in [24]. In this example we have taken $h = 0.025$, $\lambda = 0.025$ and an incidence angle of $40°$. The errors given for this problem in Table 7 were computed through comparison with the results given by the boundary variations code. The convergence of the high frequency method is nicely illustrated by this example which, in fact, validates both the high frequency and the boundary variations calculation. We note that an approximation of order 11 in powers of $1/k$ is accurate to machine precision. For reference it is useful to indicate the times required by these computations: the calculation of order 11 shown in this table resulted from a 17 second run on a DEC Alpha workstation (500MHz).

6.2 HF-Scattering by Convex Bodies: Shadowing
(*Joint work with C. Geuzaine and A. Monro* [12])

As mentioned earlier, the methods presented in the previous section do not generalise directly to configurations including shadow boundaries. Our high-frequency

Scattering Direction #	Order 0	Order 3	Order 7	Order 11
0	9.7e-4	1.3e-8	4.0e-13	3.7e-16
1	2.5e-3	1.8e-8	5.6e-14	1.8e-16
2	2.4e-3	1.5e-8	5.4e-13	5.2e-16
3	1.2e-3	2.0e-8	9.1e-14	5.8e-16
4	5.3e-3	1.6e-8	9.5e-13	2.3e-15
5	5.2e-4	2.4e-8	5.8e-14	4.5e-16
6	1.9e-2	6.0e-9	3.4e-12	5.3e-15

Table 7. Errors in high frequency approximations of various orders

algorithms for bounded scatterers thus avoid use of expansions in powers of $1/k$ and, instead, use pointwise discretisation of certain slowly varying densities. This approach is described in what follows under the assumption of *convexity* of the scattering surface. A discussion of the non-convex case is given in Sect. 6.3.

Our rigorous (convergent) high-frequency algorithm for bounded scatterers in the convex case relies on two main elements. The first of these elements is a transformation of a boundary integral equation which allows it to explicitly capture, with coarse discretisation, the rapidly oscillating phase progression of the surface currents. For this purpose, an ansatz is used for the unknown currents in the boundary integral equation, which takes the form of a product of a highly-oscillatory exponential and a slowly varying amplitude. The slowly varying amplitude can then be represented by a number of degrees of freedom independent of the frequency. This idea is related to those presented in [48] for partial differential equations and in [1, 44] for integral equations; unlike the previous approaches, however, the present treatment accounts rigorously for the fact that the ansatz is only valid in certain regions of the scattering surface.

The second main element in the present algorithm is a localised integration method related to the method of stationary phase. This localised integration scheme can be viewed as a natural link between high-frequency approximate, *non-convergent* methods such as the Kirchhoff approximation and a direct method of moments. As discussed below, the size of the reduced integration support is such that the overall method can solve scattering problems within a prescribed error tolerance for arbitrarily small wavelengths within a fixed computing time.

In addition to these main elements, our solver uses high order discretisation schemes for accuracy: the Nystrom method of [28] in two dimensions, and the method described in Sect. 4 in three dimensions. In all cases, the high-order nature of the high-frequency solver is achieved through use of Fourier interpolation and the trapezoidal rule for integration of periodic functions. The overall result is a high-order *convergent* integral method that can solve accurately in a personal computer scattering problems throughout the electromagnetic spectrum. The examples presented at the end of this section illustrate the efficiency of the method. In particular, those results show the high order convergence of the solver as well as its asymptotically bounded computational complexity as the frequency increases; see Tables 8 and 9.

Boundary Integral Formulation

To introduce some of the issues arising in our high-frequency integral method we begin by considering two different integral formulations for the Helmholtz equation $\Delta u + k^2 u = 0$ under Dirichlet boundary conditions

$$u|_{\partial D} = -u^i .$$

(For the sake of simplicity we treat a scalar scattering problem — acoustic or TE-electromagnetic; the full electromagnetic problem can be handled in a similar way.) The first of these formulations takes as unknown function the boundary values of the normal derivative:

$$\frac{1}{2}\frac{\partial u}{\partial \boldsymbol{\nu}_{\mathbf{r}}}(\mathbf{r}) = \left(\frac{\partial u^i}{\partial \boldsymbol{\nu}_{\mathbf{r}}}(\mathbf{r}) + \mathrm{i}\gamma u^i(\mathbf{r})\right) + \int_{\partial D} \frac{\partial \Phi(\mathbf{r},\mathbf{r}')}{\partial \boldsymbol{\nu}_{\mathbf{r}}} \frac{\partial u}{\partial \boldsymbol{\nu}_{\mathbf{r}'}}(\mathbf{r}')\,\mathrm{d}s(\mathbf{r}') + \mathrm{i}\gamma \int_{\partial D} \Phi(\mathbf{r},\mathbf{r}') \frac{\partial u}{\partial \boldsymbol{\nu}_{\mathbf{r}'}}(\mathbf{r}')\,\mathrm{d}s(\mathbf{r}') . \qquad (13)$$

The second formulation we consider, which takes as unknown a certain density φ, is given by (3) — which we rewrite here as

$$\frac{\varphi(\mathbf{r})}{2} = u^i(\mathbf{r}) - \int_{\partial D} \frac{\partial \Phi(\mathbf{r},\mathbf{r}')}{\partial \boldsymbol{\nu}_{\mathbf{r}'}} \varphi(\mathbf{r}')\,\mathrm{d}s(\mathbf{r}') + \mathrm{i}\gamma \int_{\partial D} \Phi(\mathbf{r},\mathbf{r}')\varphi(\mathbf{r}')\,\mathrm{d}s(\mathbf{r}') . \qquad (14)$$

In these equations γ is an arbitrary positive constant; appropriate choices of this parameter can be very advantageous in practice — see [10].

As mentioned above, our high-frequency approach is based on a high-frequency ansatz for the unknowns of the problem. For a convex scatterer, for example, the ansatz for an unknown surface function v (in the illuminated portions of the scatterer) takes the form

$$v(\mathbf{r}) = v_{slow}(\mathbf{r})\mathrm{e}^{\mathrm{i}\mathbf{k}\cdot\mathbf{r}}, \qquad (15)$$

where v_{slow} is a slowly oscillatory function (e.g. it can be represented with a fixed accuracy by Fourier series with a fixed number of terms *for arbitrarily large wave numbers* k). The validity of such an ansatz indicates that, on illuminated surfaces, the unknown v oscillates along with the incident field.

As it happens, only the solution of certain integral equations can be represented through an ansatz of the form (15). As a rule, an integral equation whose unknown is a *physical quantity* can be represented by an ansatz of the form (15); for example, the unknown in (13) (the normal derivative of the solution!) can be represented in this manner. We have verified, further, that the density φ in (14) for a convex scatterer does not admit such a representation.

The difference in character between the unknown densities arising in these two types of equations can be understood easily through consideration of an example for a simple scattering surface: a pair of parallel planes. It is easy to check that the combination of integrals in (13) *integrated over the illuminated plane only* produces field values on the non-illuminated surface which equal, precisely, the value of the inhomogeneous term in (13) on the non-illuminated boundary. It follows that the unknown function vanishes on the non-illuminated boundary, and therefore the integral over that boundary does not give rise to additional fields on the illuminated boundary. Thus, a solution of the equation can be obtained, in this case, by consideration of scattering by illuminated surface alone. In the case of (14) this is not true: further corrections on the illuminated surface must be introduced, as the non-illuminated surface "scatters" a field into the illuminated surface, which then gives rise to additional fields on the non-illuminated surface, and so on, so that the simplest ansatz (15) is not valid in this case.

Considerations related to these can be used to determine whether, for general, non-planar surfaces, the solutions of a given integral equation satisfy an ansatz of the form (15). Indeed, while such a discussion would generally not be exact for finite wave numbers and curved surfaces, these arguments can be used *asymptotically* as $k \to \infty$ — which suffices to determine the validity (or lack of validity) of our integral ansatz for a given integral equation.

High-Frequency Integral Equations

Using the integral formulation (13) and the ansatz (15) for the unknown function

$$v(\mathbf{r}) = \frac{\partial u}{\partial \boldsymbol{\nu}_{\mathbf{r}}}(\mathbf{r})$$

we obtain the equation

$$\frac{1}{2}v_{slow}(\mathbf{r}) - (\tilde{K}'v_{slow})(\mathbf{r}) - \mathrm{i}\gamma(\tilde{S}v_{slow})(\mathbf{r}) = \mathrm{i}\boldsymbol{\nu}\cdot\boldsymbol{\alpha}k + \mathrm{i}\gamma, \quad \mathbf{r} \in \partial D, \tag{16}$$

where $\tilde{S}$ and $\tilde{K}'$ denote the integral operators

$$(\tilde{S}w)(\mathbf{r}) = \int_{\partial D} \Phi(\mathbf{r},\mathbf{r}')\mathrm{e}^{\mathrm{i}k\boldsymbol{\alpha}\cdot(\mathbf{r}'-\mathbf{r})}w(\mathbf{r}')\,\mathrm{d}s(\mathbf{r}'), \tag{17}$$

$$(\tilde{K}'w)(\mathbf{r}) = \int_{\partial D} \frac{\partial\Phi(\mathbf{r},\mathbf{r}')}{\partial\boldsymbol{\nu}(\mathbf{r})}\mathrm{e}^{\mathrm{i}k\boldsymbol{\alpha}\cdot(\mathbf{r}'-\mathbf{r})}w(\mathbf{r}')\,\mathrm{d}s(\mathbf{r}')\,. \tag{18}$$

The kernels in (17) and (18) are now the only highly-oscillatory functions in the boundary integral formulation (16). Being a slowly varying function, the unknown density v_{slow} can be represented, to within any prescribed tolerance, by a *fixed* set of discretisation points, independently of frequency.

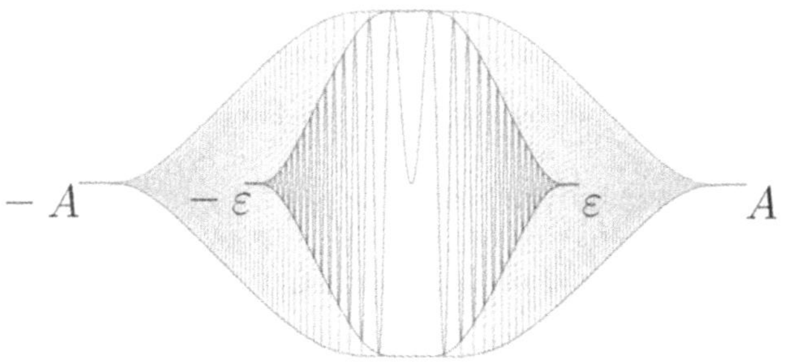

Fig. 17. Functions $f_A(x)\mathrm{e}^{\mathrm{i}kx^p}$ and $f_\varepsilon(x)\mathrm{e}^{\mathrm{i}kx^p}$ with envelopes f_A and f_ε, respectively

Localised Integration

To apply the ideas of the method of stationary phase [4] we obtain the critical points of the integrals (17) and (18). The details of such an evaluation depend on the particular kernels under consideration, but in the present case, for $\mathbf{r} \neq \mathbf{r}'$, both kernels in (17) and (18) behave asymptotically as

$$\mathrm{e}^{\mathrm{i}k\left[|\mathbf{r}-\mathbf{r}'|+\boldsymbol{\alpha}\cdot(\mathbf{r}'-\mathbf{r})\right]} = \mathrm{e}^{\mathrm{i}k\phi}, \tag{19}$$

i.e. as the kernel of a generalised Fourier integral with phase ϕ. The critical points are

- the target (observation) point $\mathbf{r}$ itself;
- the stationary points, i.e. the points where the phase ϕ in the integrals has a vanishing gradient. Note that these stationary points vary as a function of the target point; and
- the shadow boundaries, where the second derivative of the phase vanishes.

(In Appendix A we present, as an example, the details of the evaluation of the corresponding stationary points for a TE integral equation.) In view of the method of stationary phase we know that, *asymptotically*, the only significant contributions to the integrals (17) and (18) arise from values of the slow integrands and their derivatives at the critical points. In order to construct a *convergent* method for arbitrary frequencies, we introduce a integration procedure based on localisation *around* critical points.

As an example, let us consider the problem of integration of the one-dimensional smooth function $f_A(x)\mathrm{e}^{\mathrm{i}kx^p}$ depicted in Fig. 17. For the smooth cutoff $f_\varepsilon(x)\mathrm{e}^{\mathrm{i}kx^p}$ in the interval $[-\varepsilon, \varepsilon]$, one can show that [12]:

$$\int_{-A}^{A} f_A(x)\mathrm{e}^{\mathrm{i}kx^p} = \int_{-\varepsilon}^{\varepsilon} f_\varepsilon(x)\mathrm{e}^{\mathrm{i}kx^p} + \mathcal{O}((k\varepsilon^p)^{-n}),\ \forall n,\ p \geq 2.$$

(A similar error estimate for the integrals (17) and (18) can be obtained by expanding the phase ϕ in (19) around the critical points.) Estimates of this type provide our criteria for localised integration: For each target point the corresponding set of critical points is covered by a number of small regions, as indicated in what follows:

- the target point is covered by a region U_t of radius proportional to the wavelength λ;
- the ℓ-th stationary point is covered by a region U^{ℓ}_{st} of radius proportional to $\sqrt{\lambda}$;
- the shadow-boundary set is covered by U_{sb}: the set of points whose distance to the boundary is less than a radius proportional to $\sqrt[3]{\lambda}$.

A partition of unity (POU) (see Sect. 4 and [10]) is used to smoothly split the integral over ∂D into a number of integrals over subsets of ∂D. This POU is taken to be subordinated to the covering by open sets U_t, U^{ℓ}_{st}, U_{sb} and the complement V of a closed set which is contained and closely approximates the union $U_t \cup U^{\ell}_{st} \cup U_{sb}$. (In other words, the set where each of the functions making up the partition of unity is not zero is contained in one of the sets U or V.) The integral over all of ∂D is then split as a sum of integrals over V and each one of the U sets, with integrands which include the corresponding partitions of unity. The integral in the outside region V is neglected.

It can rigorously be shown that setting $v_{slow} \equiv 0$ in the deep shadow region does indeed give rise to a convergent numerical method. As it happens, this implies that the only critical points that need to be considered for convex scatterers are the target point and the shadow boundary [12].

Spectral Implementation

As it happens, the unknown density v_{slow} in (16) can be obtained *within a prescribed error tolerance* through interpolation from the corresponding values on a fixed of set of discretisation points *independent of frequency*. Interpolations of very high order can be obtained by means of smooth cutoffs and Fourier series, both in two and three dimensions.

The integral in the region U_t is evaluated by means of a discretisation with a mesh-size proportional to λ. Our choice of local integrator is that described in the text of [28] in the two-dimensional case, and that of [10] in the three-dimensional case. The integral in the region U^{ℓ}_{st}, in turn, is evaluated by the trapezoidal rule with a discretisation with a mesh size proportional to $\sqrt{\lambda}$, while the integral in the region U_{sb} is evaluated by means of a discretisation with a mesh size proportional to $\sqrt[3]{\lambda}$ in the direction orthogonal to the shadow boundary, and a constant (frequency-independent) mesh size along the shadow boundary. This constant mesh size is that needed to correctly represent the slow density v_{slow}. In all cases, the values of the slow densities at the integration points are obtained through interpolation from the fixed discretisation mesh mentioned above. Note that, because of the smooth cutoffs used, all integrands are smooth periodic functions — for which the trapezoidal rule gives rise to high-order convergence. Also note that a special procedure is necessary to guarantee that the non-empty intersections occurring between the various U sets defined above do not cause difficulties [12].

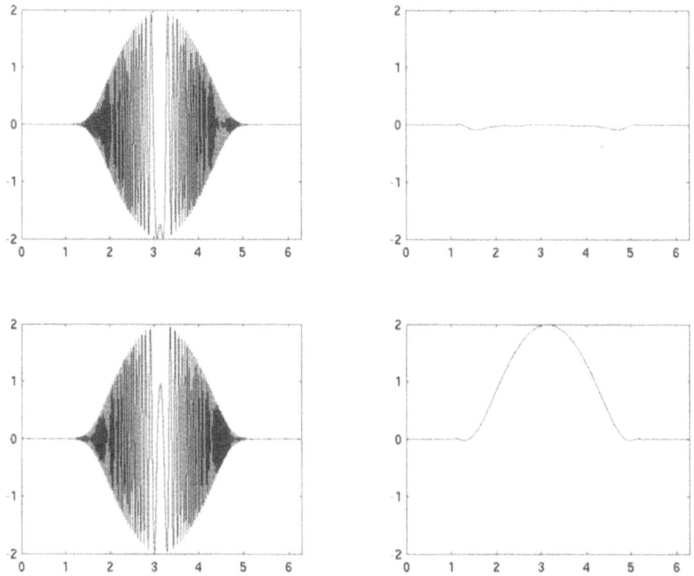

Fig. 18. Scattering by a circular cylinder of radius a, with $ka = 200$: real and imaginary part of $v/k = \frac{1}{k}\frac{\partial u}{\partial \nu_{\mathbf{r}}}(\mathbf{r})$ as functions of the angular coordinate (top left and bottom left); real and imaginary part of v_{slow}/k as functions of the angular coordinate (top right and bottom right)

The numerical method is then completed through use of an iterative method such as GMRES [55] to build a matrix-free solver. According to these prescriptions, for a given error tolerance, the evaluation of all the relevant integrals is performed within that tolerance through a fixed number of operations—independently of frequency.

Numerical Results

ka	GMRES Iter.	Max. Error	Mean Square Err.	CPU (s)
1	9	1.8e-12	8.8e-12	< 1
10	17	2.0e-12	9.2e-12	< 1
100	31	5.0e-5	2.5e-5	8
1000	30	7.8e-4	2.1e-4	84
10000	33	2.6e-3	6.6e-4	83

Table 8. Scattering by a circular cylinder of radius a, using 100 unknowns

Unknowns	GMRES Iter.	Max. Error
25	13	4.4e-3
50	23	1.2e-3
100	31	1.2e-4
200	34	4.4e-6
400	39	1.0e-9
800	44	1.0e-12

Table 9. Scattering by a circular cylinder of radius a, with $ka = 150$: error convergence

Tables 8 and 9 and Fig. 18 show results of a preliminary version of the two-dimensional high-frequency integral algorithm, as applied to a circular cylinder of radius a. Errors were computed by comparison with an exact solution for the integral equation. This example illustrates the asymptotic bounded complexity (the error for $k = 1000$ is roughly equal to the error for $k = 10000$, using the same number of unknowns and the same number of integration points, leading to identical computation times) as well as the high order convergence of the solver. As illustrated by Table 8, the high-frequency solver is well conditioned and requires a small number of GMRES iterations for arbitrarily large wave numbers.

Extensions of this high-frequency method to non-smooth geometries (containing singularities such as corners and edges) can be found in [12]. For non-convex obstacles, the ansatz introduced in Sect. 6.2 may not be valid. Generalisations of the ansatz (15) which are valid for non-convex scatterers are given in the next section.

6.3 HF-Scattering by Non-Convex Bodies: Multiple Scattering (*Joint work with F. Reitich* [19])

For a non-convex obstacle the ansatz (15) may not be valid; indeed, this ansatz is not valid if the incident field is such that its ray theory approximation gives rise to multiple reflections, as shown Fig. 19. This fact can be understood easily: in the case of Fig. 19, for example, the reflections in the lower part of the concave region act as sources of an "incident" field for the upper portion of the concavity which should be accounted for in the ansatz for the density μ.

A correct version of the ansatz for the scatterer of Fig. 19 is

$$\mu(x) = \mu^0_{slow}(x)\mathrm{e}^{\mathrm{i}k\cdot x} + \mu^1_{slow}(x)\mathrm{e}^{\mathrm{i}(k^1(x)\cdot x+\phi^1)} + \mu^2_{slow}(x)\mathrm{e}^{\mathrm{i}(k^2(x)\cdot x+\phi^2)} . \quad (20)$$

Here $\mu^1_{slow}(x)$ and $\mu^2_{slow}(x)$ are defined in the upper part of the concavity, $k^1(x)$ and $k^2(x)$ are vectors of magnitude equal to $|k|$ and direction given by the geometrical optics rays, and where ϕ^1 and ϕ^2 are initial phases. Note that, indeed, two slow densities arise as a result of the reflections from the lower portions of the concavity, since points in the upper part of the concavity are illuminated by two reflections from the lower part. The algorithm for the non-convex case can now be completed by simply applying the algorithm of Sect. 6.2 to this modified type of ansatz. Note,

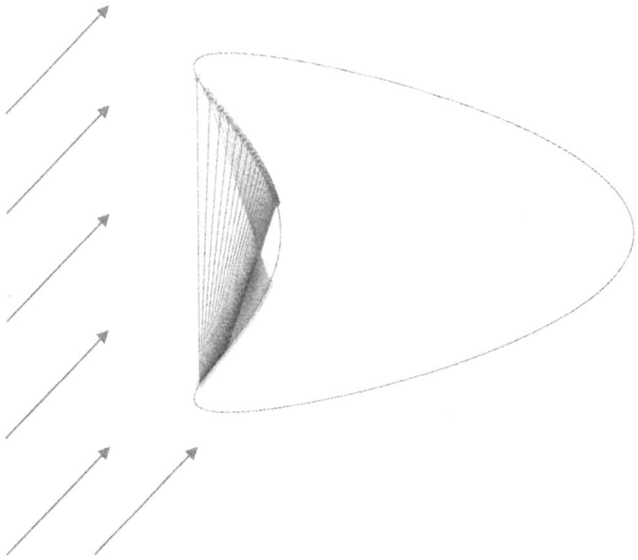

Fig. 19. A multiple scattering configuration

Unknowns	Max. Err. in μ^j_{slow}
512	5.0 e(-3)
768	2.0 e(-4)
925	3.0 e(-6)

Table 10. Convergence of the multiple slow densities with increasing number of unknowns for the multiple-scattering geometry shown in Fig. 19

for example, that reflections usually give rise to new shadow boundaries, which need to be treated as detailed in the algorithm of Sect. 6.2.

Table 10 provides a verification of the multiple scattering ansatz for the geometry of Fig. 19. Here the scattered field for $ka = 800$ was obtained by means of a well-known high-order solver [28], and then, slow densities μ^j_{slow} ($j = 1, 2, 3$) were obtained in such a way that (20) is satisfied. Note that the problem of obtaining the slow densities μ^j_{slow} is under-determined, since three functions need to be obtained from the single density μ. And, further, a mechanism needs to be introduced to guarantee that the slow densities are indeed slow — that is, that they do not contain oscillations on the level of the wavelength. These two problems are addressed in our context through use of least squares: the slow densities are obtained in such a way that (20) is satisfied *in the least squares sense* together with an equation that sets to zero an appropriate multiple of the L^2 norm of a derivative of its Fourier series. This makes the solution unique, and appropriately penalises oscillations: in Fig. 20 we display slow densities in various portions of the scatterer together with the highly oscillatory density μ. The results in Table 10 show the convergence obtained for the

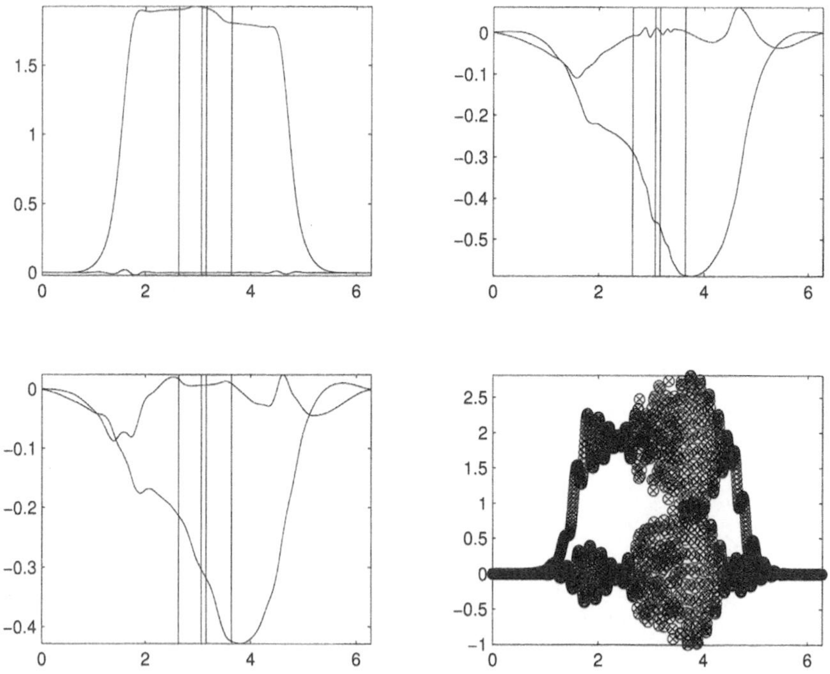

Fig. 20. Densities μ^j_{slow} in the upper multiple scattering region for $j = 0, 1$ and 2 are displayed in the upper-left, upper-right and lower-left portions of this figure, respectively. The lower-right graph displays the full density μ in the same region

functions μ^j_{slow} obtained in this manner and, thus, validate our assumption of an ansatz that accounts for multiple reflections through use of corresponding multiple slow densities.

A Appendix: Evaluation of Stationary Points

For this example we consider *convex* obstacles in two dimensions whose boundaries admit polar parametrisation

$$r = r(\theta), \quad 0 \leq \theta \leq 2\pi. \tag{21}$$

Let the phase of the incident wave be given by

$$\mathrm{i}k(\alpha, \beta) = \mathrm{i}k(\cos\varphi, \sin\varphi)$$

and consider the total phase

$$\mathrm{i}k\phi = \mathrm{i}k\left[\,|y - x| + (\alpha, \beta) \cdot y\,\right]$$

where x and y are arbitrary points on the boundary of the obstacle, the *target* and *source* points respectively. (The total phase is obtained as the phase in the $k \to \infty$ asymptotic expression of $G(x,y)\,\mathrm{e}^{\mathrm{i}k\cdot y}$). Without loss of generality we may assume $\varphi = 0$. Using polar coordinates

$$x = r(\theta_0)\mathrm{e}^{\mathrm{i}\theta_0}, \quad y = r(\theta)\mathrm{e}^{\mathrm{i}\theta} \quad \text{and} \quad y - x = \rho\mathrm{e}^{\mathrm{i}\psi},$$

we have

$$\begin{aligned}\phi &= \phi(\theta)\\ &= \rho + r(\theta)\cos\theta\\ &= \sqrt{r(\theta_0)^2 + r(\theta)^2 - 2r(\theta_0)r(\theta)\cos(\theta-\theta_0)} + r(\theta)\cos\theta\,.\end{aligned}$$

The stationary points then correspond to the solutions of

$$0 = \phi'(\theta) = \frac{r(\theta)\dfrac{\mathrm{d}r}{\mathrm{d}\theta}(\theta) - r(\theta_0)\dfrac{\mathrm{d}r}{\mathrm{d}\theta}(\theta)\cos(\theta-\theta_0) + r(\theta_0)r(\theta)\sin(\theta-\theta_0)}{\sqrt{r(\theta_0)^2 + r(\theta)^2 - 2r(\theta_0)r(\theta)\cos(\theta-\theta_0)}} + \frac{\mathrm{d}r}{\mathrm{d}\theta}(\theta)\cos\theta - r(\theta)\sin\theta \qquad (22)$$

in the interval $[0, 2\pi)$. The solution of this nonlinear equation can be obtained in $\mathcal{O}(N)$ operations by means of Newton's method.

References

1. K. R. Aberegg and A. F. Peterson, Application of the integral equation-asymptotic phase method to two-dimensional scattering. *IEEE Transactions on Antennas and Propagation* **43** 534–537 (1995).
2. M. Abramowitz and I. Stegun *Handbook of mathematical functions with formulas, graphs, and mathematical tables*, US Dept Commerce (June 1964).
3. C. R. Anderson, An implementation of the fast multipole method without multipoles, *SIAM J. Sci. Stat. Comput.* **13**, 923–947 (1992).
4. C. M. Bender and S. A. Orszag, *Advanced Mathematical Methods for Scientists and Engineers*, McGraw-Hill (1978).
5. G. Beylkin, On the fast fourier transform of functions with singularities, *Applied Computational Harmonic Analysis*, **2** 363–381 (1995).
6. N. Bleistein and R.A. Handelsman, *Asymptotic expansions of integrals*, Dover Publications, New York (1986).
7. E. Bleszynski, M. Bleszynski, and T. Jaroszewicz, AIM: Adaptive integral method for solving large-scale electromagnetic scattering and radiation problems, *Radio Science* **31**, 1225–1251 (1996).
8. N. Bojarski, The k-space formulation of the scattering problem in the time domain, *J.Acoust. Soc. Am.* **72**, 570–584 (1982).
9. A. Brandt and A. A. Lubrecht, Multilevel matrix multiplication and fast solution of integral equations, *J. Comput. Phys.* **90**, 348–370 (1990).

10. O. P. Bruno and L. A. Kunyansky, A Fast, High-Order Algorithm for the Solution of Surface Scattering Problems: Basic Implementation, Tests, and Applications *J. Comput. Phys.* **169**, 80–110 (2001).
11. O. P. Bruno and L. A. Kunyansky, Surface scattering in 3-D: an accelerated high order solver, *Proc. R. Soc. Lond. A* **457**, 2921–2934 (2001).
12. O. Bruno, C. Geuzaine and A. Monro, Rigorous high-frequency solvers for electromagnetic and acoustic scattering: convex scatterers. In preparation.
13. O. Bruno and M. Hyde, High order solution of scattering by penetrable bodies. In preparation.
14. O. Bruno and M. Hyde, A fast high-order method for scattering by three-dimensional inhomgogeneous media. In preparation.
15. O. P. Bruno and L. A. Kunyansky, High-Order Fourier approximation in Scattering by two-dimensional inhmogeneous media. In preparation.
16. O. Bruno and M. Pohlman, Fast, High-order surface representation of smooth and singular surfaces. In preparation.
17. O. Bruno and R. Paffenroth, Fasster: A fast, parallel, high-order surface scattering solver. In preparation.
18. O. Bruno and F. Reitich, A fast high-order algorithm for evaluation of electromagnetic scattering from thin penetrable bodies. In preparation.
19. O. Bruno and F. Reitich, Rigorous high-frequency solvers for electromagnetic and acoustic scattering: non-convex scatterers and multiple scattering. In preparation.
20. O. Bruno and F. Reitich, *Numerical solution of diffraction problems: a method of variation of boundaries I, II, III*, J. Opt. Soc. A **10**, 1168–1175, 2307–2316, 2551–2562 (1993).
21. O. Bruno and A. Sei, A high order solver for problems of scattering by heterogeneous bodies, *Proc. of the 13-th Annual Review of Progress in Applied Computational Electromagnetism* (Applied Computational Electromagnetics Society), 1296–1302 (1997).
22. O. Bruno and A. Sei, A fast high-order solver for EM scattering from complex penetrable bodies: TE case; *IEEE Trans. Antenn. Propag.* **48**, 1862–1864 (2000).
23. O. Bruno and A. Sei, A fast high-order solver for problems of scattering by heterogeneous bodies; To appear in IEEE Trans. Antenn. Propag..
24. O. Bruno, A. Sei and M. Caponi, High-order high-frequency solution of rough surface scattering problems, *Radio Science* **37**, 2-1—2-13 (2002).
25. L. Canino, J. Ottusch, M. Stalzer, J. Visher and S. Wandzura, Numerical solution of the Helmholtz equation using a High-Order Nystrom Discretization, *J. Comp. Phys.* **146**, 627–663 (1998).
26. M. F. Catedra, E. Cago, and L. Nuno, A Numerical Scheme to Obtain the RCS of Three-Dimensional Bodies of Resonant Size Using the Conjugate Gradient Method and the Fast Fourier Transform, *IEEE Trans. Antennas Propag.* **37**, 528–537 (1989).
27. R. Coifman, V. Rokhlin, and S. Wandzura, The Fast Multipole Method for the Wave Equation: A Pedestrian Prescription, *IEEE Antennas and Propagation Magazine* **35**, 7–12, (1993).
28. D. Colton and R. Kress, *Inverse acoustic and electromagnetic scattering theory*, Springer-Verlag, Berlin/Heidelberg (1998).
29. R. Courant and D. Hilbert, *Methods of mathematical physics*, Wiley (1953).
30. I. Daubechies, I. Guskov, P. Schröder and W. Sweldens, Wavelets on irregular point sets *Phil. Trans. R. Soc. Lond. A* **357**, 2397–2413 (1999).
31. B. Dembart and E. Yip, The accuracy of Fast Multipole Methods for Maxwell's Equations, *IEEE Computational Science and Engineering* **4**, 48–56 (1998).
32. M. Desbrun, M. Meyer and P. Alliez, Intrinsic Parametrizations of Surface Meshes, *Eurographics 2002* **21**, G. Drettakis and H. P. Seidel, Eds. (2002).

33. J. W. Duijndam and M. A. Schonewille, Nonuniform fast Fourier transform, *Geophysics* **64** (1999).
34. A. Dutt and V. Rokhlin, Fast fourier transforms for nonequispaced data, *SIAM Journal of Scientific Computing*, **14** 1368–1393 (1993).
35. A. Dutt and V. Rokhlin. Fast Fourier transforms for nonequispaced data, ii, *Applied and Computational Harmonic Analysis*, **2** 85–100 (1995).
36. M. Epton and B. Dembart, Multipole Translation Theory for the Three-Dimensional Laplace and Helmholtz Equations, *SIAM J. Sci. Comput.* **16**, 865–897, (1995).
37. L. B. Felsen and N. Marcuvitz, *Radiation and Scattering of Waves*, Prentice-Hall (1973).
38. V. A. Fock, *Electromagnetic Diffraction and Propagation Problems*. Elmsford, NY, Pergamon (1965).
39. G. Liu and S. Gedney, High-order Nyström solution of the volume EFIE for TM-wave scattering, *Microwave and Optical Technology Letters* **25**, 8–11 (2000).
40. G. H. Golub and C. F. Van Loan, *Matrix Computations* (Second Ed.), John Hopkins (1993).
41. A. Greenbaum, L. Greengard and G. McFadden, Laplace equation and the Dirichlet-Neumann map in multiply connected domains, *J. Comput. Phys.* **105**, 267–278 (1993).
42. L. Greengard, J. F. Huang, V. Rokhlin, and S. Wandzura, Accelerating fast multipole methods for the Helmholtz equation at low frequencies, *IEEE Comput. Sci. Eng.* **5**, 32–38 (1998).
43. M. Hyde and O. Bruno, Two-dimensional scattering by an inhomogeneous medium. In preparation.
44. R. M. James, A contribution to scattering calculation for small wavelengths—the high frequency panel method, *IEEE Transactions on Antennas and Propagation*, 38 1625–1630 (1990).
45. J. B. Keller and R. M. Lewis, *Asymptotic methods for partial differential equations: the reduced wave equation and Maxwell's equations*, Vol. 1 of *Surveys in Applied Mathematics*, pp. 1–82, Plenum Press, New York (1995).
46. L. A. Kunyansky and O. P. Bruno, A Fast, High-Order Algorithm for the Solution of Surface Scattering Problems II. Theoretical considerations. Submitted.
47. C. Labreuche, A convergence theorem for the fast multipole method for 2 dimensional scattering problems, *Mathematics of computation* **67**, 553–591 (1998).
48. R. T. Ling, Numerical solution for the scattering of sound waves by a circular cylinder, *AIAA Journal*, **25** 560–566 (1987).
49. E. Martensen, Über eine methode zum räumlichen Neumannschen problem mit einer anwendung für torusartige berandungen, *Acta Math.* **109**, 75–135 (1963).
50. J. R. Mautz and R. F. Harrington, A combined-source solution for radiation and scattering from a perfectly conducting body, *IEEE Transactions on Antennas and Propagation*, AP-27 445–454 (1979).
51. J. R. Phillips and J. K. White, A Precorrected-FFT Method for Electrostatic Analysis of Complicated 3-D Structures, *IEEE Trans. Computer-Aided Design of Integrated Circuits and Systems* **16**, 1059–1072 (1997).
52. V. Rokhlin, Rapid solution of integral equations of classical potential theory, *J. Comput. Phys.* **60**, 187–207 (1985).
53. V. Rokhlin, Rapid solution of integral equations of scattering theory in two dimensions, *J. Comput. Phys.* **86**, 414–439 (1990).
54. V. Rokhlin, Diagonal Form of Translation Operators for the Helmholtz equation in Three Dimensions, *Applied and Computational Harmonic Analysis* **1**, 82–93 (1993).
55. Y. Saad and M. H. Schultz, GMRES: a generalized minimal residual algorithm for solving non-symmetric linear systems, *SIAM J. Sci. Statist. Comput.* **7**, 857–869 (1986).

56. J. M. Song, C. C. Lu, W. C. Chew, and S. W. Lee, Fast Illinois Solver Code (FISC), *IEEE Antenna and Propagation Magazine* **40**, 27–34 (1998).
57. J. M. Song, C. C. Lu, and W. C. Chew, Multilevel Fast Multipole Algorithm for Electromagnetic Scattering by Large Complex Objects, *IEEE Trans. Antennas Propag.* **45**, 1488–1493 (1997).
58. G. M. Vainikko, Fast solvers of the Lippmann-Schwinger equation, in *Direct and Inverse Problems of Mathematical Physics*, R. P. Gilbert and J. Kajiwara and Y. S. Xu Eds. (2000).
59. A.G. Voronovich, *Wave scattering from rough surfaces*, Springer-Verlag, Berlin (1994).
60. A. C. Woo, H. T. G. Wang, M. J. Schuh, and M. L. Sanders, Benchmark Radar Targets for the Validation of Computational Electromagnetics Programs, *IEEE Antennas and Propagation Magazine* **35**, 84–89 (1993).
61. X. M. Xu and Q. H. Liu, Fast spectral-domain method for acoustic scattering problems, *IEEE Trans. Ultrasonics, Ferroelectrics, and Frequency Control* **48**, 522–529 (2001).
62. P. Zwamborn and P. Van den Berg, Three dimensional weak form of the conjugate gradient FFT method for solving scattering problems, *IEEE Trans. Microwave Theory Tech.* **40**, 1757–1766 (1992).

Galerkin Boundary Element Methods for Electromagnetic Scattering

Annalisa Buffa[1] and Ralf Hiptmair[2]

[1] Istituto di Matematica applicate e tecnologie informatiche del CNR, Pavia, Italy
annalisa@ian.pv.cnr.it
[2] Seminar für Angewandte Mathematik, ETH Zürich, CH-8092 Zürich, Switzerland
hiptmair@sam.math.ethz.ch

Summary. Methods based on boundary integral equations are widely used in the numerical simulation of electromagnetic scattering in the frequency domain. This article examines a particular class of these methods, namely the Galerkin boundary element approach, from a theoretical point of view. Emphasis is put on the fundamental differences between acoustic and electromagnetic scattering. The derivation of various boundary integral equations is presented, properties of their discretised counterparts are discussed, and a-priori convergence estimates for the boundary element solutions are rigorously established.

Key words: Electromagnetic scattering, boundary integral equations, boundary element methods

1 Introduction

The numerical simulation of electromagnetic scattering aims at computing the interaction of electromagnetic waves with a physical body, the so-called scatterer. The scatterer occupies a bounded domain Ω_s in three-dimensional affine space $\mathbb{R}^3$. In general, Ω_s will have Lipschitz-continuous boundary $\Gamma := \partial\Omega_s$ [41, Sect. 1.2], which can be equipped with an exterior unit normal vector field $\mathbf{n} \in \boldsymbol{L}^\infty(\Gamma)$. With boundary element methods in mind, we do not lose generality by considering only piecewise smooth Ω_s, i.e. curvilinear Lipschitz polyhedra in the parlance of [35].

We only consider linear materials and time-harmonic electromagnetic fields of angular frequency $\omega > 0$. Excitation is provided by the fields $\mathbf{e}_i, \mathbf{h}_i$ of an incident (plane) wave. Under these circumstances we can derive the following transmission problem from Maxwell's equations [31, Chap. 6]:

$$\begin{aligned}
\mathbf{curl}\,\mathbf{e} = -i\omega\underline{\mu}\mathbf{h}\,, \quad \mathbf{curl}\,\mathbf{h} = i\omega\underline{\epsilon}\mathbf{e} \quad \text{in } \Omega_s \cup \Omega'\,,\\
\gamma_{\mathbf{t}}^+\mathbf{e} - \gamma_{\mathbf{t}}^-\mathbf{e} = -\gamma_{\mathbf{t}}^+\mathbf{e}_i\,, \quad \gamma_{\mathbf{t}}^+\mathbf{h} - \gamma_{\mathbf{t}}^-\mathbf{h} = -\gamma_{\mathbf{t}}^+\mathbf{h}_i \quad \text{on } \Gamma\,,\\
\int_{\partial B_r} \left|\gamma_{\mathbf{t}}\mathbf{h} \times \mathbf{n} + \boldsymbol{\gamma}\mathbf{e}\right|^2 \,\mathrm{d}S \to 0 \quad \text{for } r \to \infty\,.
\end{aligned} \tag{1}$$

Here and below, B_r denotes a ball of radius r centred at the origin, $\gamma_{\mathbf{t}}\mathbf{u}$ stands for the tangential trace $\mathbf{u}\times\mathbf{n}$, and superscripts $-$ and $+$ tag traces onto Γ from Ω_s and $\Omega' := \mathbb{R}^3 \setminus \overline{\Omega}$, respectively. The vector fields $\mathbf{e} = \mathbf{e}(\mathbf{x})$, $\mathbf{h} = \mathbf{h}(\mathbf{x})$ represent the unknown complex amplitudes (phasors) of the electric and magnetic field, respectively. The material parameters $\underline{\boldsymbol{\mu}} = \underline{\boldsymbol{\mu}}(\mathbf{x})$ (permeability tensor), $\underline{\boldsymbol{\epsilon}} = \underline{\boldsymbol{\epsilon}}(\mathbf{x})$ (dielectric tensor), $\mathbf{x} \in \mathbb{R}^3$, are uniformly positive definite and bounded. In fact, information on the scatterer is completely contained in $\underline{\boldsymbol{\mu}}$ and $\underline{\boldsymbol{\epsilon}}$: inside Ω_s they may vary, but in the "air region" Ω' both material parameters agree with the constants $\mu_0 > 0$ and ϵ_0, respectively. At ∞ the so-called Silver-Müller radiation conditions are imposed.

This system of equations can always be reduced to a second order wave equation in terms either of the electric or the magnetic field, e.g., $\mathbf{e}$ satisfies the *electric wave equation*

$$\mathbf{curl}\,\underline{\boldsymbol{\mu}}^{-1}\,\mathbf{curl}\,\mathbf{e} - \omega^2\underline{\boldsymbol{\epsilon}}\mathbf{e} = 0 \quad \text{in } \Omega_s \cup \Omega' \,. \tag{2}$$

Note that the uniqueness of solutions of the system (1) is a direct consequence of Rellich's Lemma [24, 53].

Apart from generic dielectric and even lossy scatterers the following special situations are of practical interest.

- The scatterer is assumed to be a "perfect conductor" in which no electric field can exist. This leads to an exterior Dirichlet problem for the electric wave equations in Ω', because the transmission conditions in (1) are replaced by the boundary condition $\gamma_{\mathbf{t}}^+\mathbf{e} = -\gamma_{\mathbf{t}}^+\mathbf{e}_i$ on Γ for the electric field.
- If the scatterer is a thin perfectly conducting sheet, we arrive at a *screen problem*. In this case $\Omega_s = \emptyset$ and Γ becomes a compact piecewise smooth two-dimensional surface with boundary. As before, we demand $\gamma_{\mathbf{t}}\mathbf{e} = -\gamma_{\mathbf{t}}\mathbf{e}_i$ on both sides of Γ. For screen problems Ω' does not possess a Lipschitz boundary any more. Moreover, the screen Γ itself might not even be a Lipschitz surface itself, in case it branches. The resulting mathematical complications are treated in [16].
- If the scatterer is a good conductor with smooth surface, its impact on the fields can be modelled by *impedance boundary conditions* (Leontovich boundary conditions) [3, 10, 54]

  $$\gamma_{\mathbf{t}}^+\mathbf{e} - \underline{\boldsymbol{\eta}}(\gamma_{\mathbf{t}}^+\mathbf{h}\times\mathbf{n}) = \underline{\boldsymbol{\eta}}(\gamma_{\mathbf{t}}^+\mathbf{h}_i\times\mathbf{n}) - \gamma_{\mathbf{t}}^+\mathbf{e}_i \quad \text{on } \Gamma\,.$$

  The surface impedance $\underline{\boldsymbol{\eta}}$ is a complex tensor with uniformly positive definite real part and non-zero imaginary part.

All these problems have in common that scattered fields on the unbounded domain Ω' have to be determined. As the material coefficients are constant in Ω', boundary integral equation methods are perfectly suited for this job. In addition, they are posed on the two-dimensional surface Γ, which relieves us from meshing (a part of) Ω'. In the case of complicated geometries this is a strong point in favour of boundary integral equation methods, compared to volume based schemes with absorbing boundary conditions (*cf.* the contribution of T. Hagstrom on absorbing layers and radiation boundary conditions in this collection) at an artificial cut-off boundary.

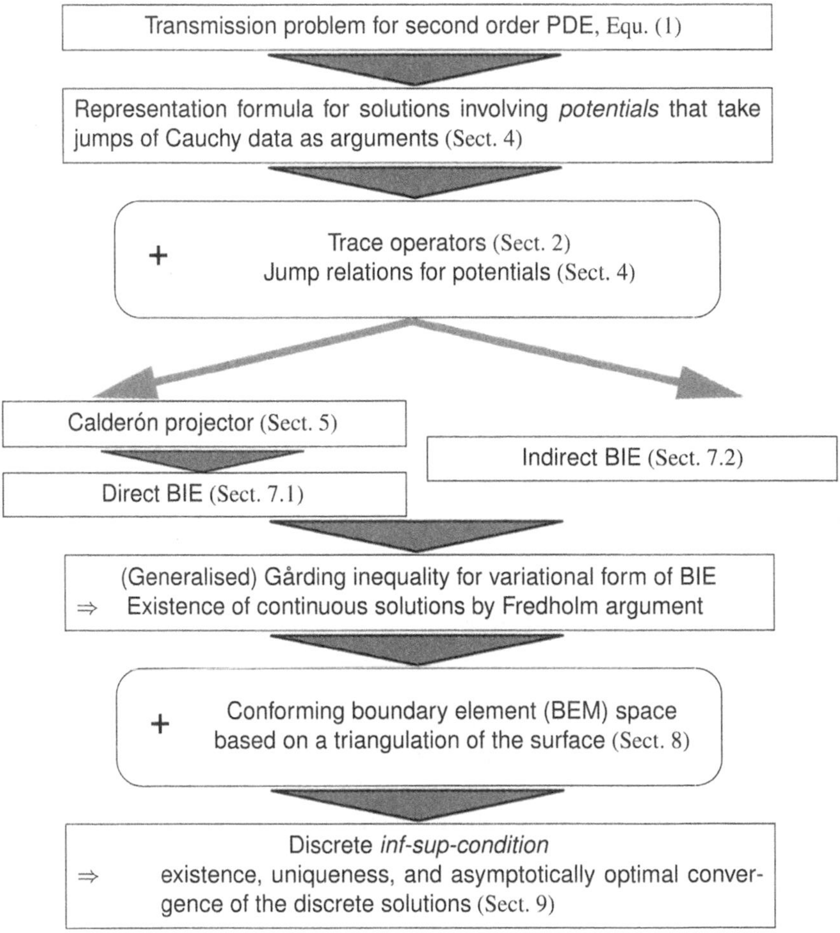

Fig. 1. "Road map" for the derivation and analysis of Galerkin boundary element methods for electromagnetic scattering (BIE = boundary integral equation)

In this article we will deal exclusively with the Galerkin method for the discretisation of the boundary integral equations. It is based on variational formulations in suitable trace spaces. This permits us to use powerful tools from functional analysis. They pave the way for a rigorous and comprehensive convergence theory. We acknowledge that several other numerical methods based on boundary integral formulations exist and are widely used alternatives to Galerkin schemes:

- the collocation method, which can be regarded as a special Petrov-Galerkin approach [42, Sect. 4.4]
- the method of source potentials, which requires a second surface away from Γ, on which a source distribution is sought. An example of this method for electromagnetic scattering is analysed in [43].

- Nyström methods, which directly tackle the boundary integral equations by means of a quadrature rule. For an exposition we refer to the contribution of O. Bruno in this volume and to [49, Chap. 12].

Unfortunately, the theoretical understanding of these methods is rudimentary in comparison with Galerkin schemes. For this reason we restrict the presentation to Galerkin methods.

As far as Galerkin boundary element methods are concerned, there is a fairly canonical approach to their construction and theoretical examination. This standard procedure is depicted in the flowchart of Fig. 1. The plan of this paper closely follows these lines.

We point out that issues of implementation and efficient solution of the resulting linear systems of equations are not covered by this article. We will also skip quite a few proofs, which the reader may look up in the research papers that underly this survey. In particular, we mention [15, 20] as main references for Sect. 2.1, [45, Chap. 5] for Sect. 3, [31, Chap. 6] and [16, 21] as regards Sects. 4-9, and [44] as source for Sect. 10.

2 Function Spaces and Traces

In order to write problem (1) or equivalent formulations of it in a mathematically rigorous way, we need a precise characterisation of the function spaces on which the equations are posed. This section is devoted to definitions and main properties of function spaces which are concerned with the rigorous formulation of the problem (1). The first section concerns spaces on the domain, either Ω_s, $\Omega' = \mathbb{R}^3 \setminus \overline{\Omega_s}$ or $\mathbb{R}^3$, while in the second we define and characterise suitable spaces on the manifold Γ, which will be of key importance for the definition of integral operators.

2.1 Function Spaces in the Domain

Let $\Omega \subseteq \mathbb{R}^3$ be any of the sets Ω_s , Ω' , $\mathbb{R}^3$ and define the Fréchet space $\boldsymbol{L}^2_{\text{loc}}(\Omega)$ of complex, vector valued, locally square integrable functions $\mathbf{u} : \Omega \to \mathbb{C}^3$. We also make use of the Sobolev spaces $\boldsymbol{H}^s_{\text{loc}}(\Omega)$, $s \geq 0$ with the convention $\boldsymbol{H}^0 \equiv \boldsymbol{L}^2$ (see, e.g., [1] for definitions). The sub-fix $_{\text{loc}}$ is systematically removed when Ω is bounded: in this case, the $\boldsymbol{H}^s(\Omega)$ are Hilbert spaces endowed with the natural graph-norm $\|\mathbf{u}\|_{\boldsymbol{H}^s(\Omega_s)}$ and semi-norm $|\mathbf{u}|_{\boldsymbol{H}^s(\Omega_s)}$, respectively [1]. Round brackets will consistently be used to express inner products.

With $\mathbf{d}$ a first order differential operator, we define for any $s \geq 0$

$$\boldsymbol{H}^s_{\text{loc}}(\mathbf{d}, \Omega) := \{\mathbf{u} \in \boldsymbol{H}^s_{\text{loc}}(\Omega) : \ \mathbf{du} \in \boldsymbol{H}^s_{\text{loc}}(\Omega)\} \ , \tag{3}$$

$$\boldsymbol{H}^s_{\text{loc}}(\mathbf{d}0, \Omega) := \{\mathbf{u} \in \boldsymbol{H}^s_{\text{loc}}(\Omega) : \ \mathbf{du} = 0\} \ . \tag{4}$$

When $s = 0$, we simplify the notation by setting $\boldsymbol{H}^0 = \boldsymbol{H}$. If Ω is bounded, $\boldsymbol{H}^s_{\text{loc}}(\mathbf{d}, \Omega)$ is endowed with the graph norm $\|\cdot\|^2_{\boldsymbol{H}^s(\mathbf{d},\Omega)} := \|\cdot\|^2_{\boldsymbol{H}^s(\Omega)} + \|\mathbf{d}\cdot\|^2_{\boldsymbol{H}^s(\Omega)}$

and seminorm $|\cdot|^2_{\boldsymbol{H}^s(\mathbf{d},\Omega)} := |\cdot|^2_{\boldsymbol{H}^s(\Omega)} + |\mathbf{d}\cdot|^2_{\boldsymbol{H}^s(\Omega)}$. This defines the spaces $\boldsymbol{H}^s(\mathbf{curl},\Omega)$, $\boldsymbol{H}^s(\mathrm{div},\Omega)$ and $\boldsymbol{H}^s(\mathbf{curl}\,0,\Omega)$, $\boldsymbol{H}^s(\mathrm{div}\,0,\Omega)$.

From Gauß' theorem we obtain integration by parts formulae for the spaces $\boldsymbol{H}(\mathbf{curl},\Omega_s)$, $\boldsymbol{H}(\mathrm{div},\Omega_s)$. If $\mathbf{u}$, $\mathbf{v} \in C^\infty(\overline{\Omega_s})^3$ and $p \in C^\infty(\overline{\Omega_s})$, then we have $\mathrm{div}(\mathbf{u}p) = \mathrm{div}\,\mathbf{u}\,p + \mathbf{u}\cdot\nabla p$ and $\mathrm{div}(\mathbf{u}\times\mathbf{v}) = \mathbf{curl}\,\mathbf{u}\cdot\mathbf{v} - \mathbf{curl}\,\mathbf{v}\cdot\mathbf{u}$, and, finally, $(\mathbf{u}\times\mathbf{v})\cdot\mathbf{n} = -(\mathbf{u}\times\mathbf{n})\cdot\mathbf{v}$ on the boundary Γ. These imply the following formulae:

$$\int_{\Omega_s} \mathrm{div}(\mathbf{u}p) = \int_{\Omega_s} (\mathrm{div}\,\mathbf{u}\,p + \mathbf{u}\cdot\nabla p)\,\mathrm{d}\mathbf{x} = \int_{\Gamma} p\,\mathbf{u}\cdot\mathbf{n}\mathrm{d}S\,, \tag{5}$$

$$\int_{\Omega_s} (\mathbf{u}\cdot\mathbf{curl}\,\mathbf{v} - \mathbf{curl}\,\mathbf{u}\cdot\mathbf{v})\,\mathrm{d}\mathbf{x} = \int_{\Gamma} (\mathbf{u}\times\mathbf{n})\cdot\mathbf{v}_{|\Gamma}\,\mathrm{d}S\,. \tag{6}$$

These formulae suggest the definitions of the mappings $\gamma_{\mathbf{t}} : \mathbf{u} \mapsto \mathbf{u}_{|\Gamma}\times\mathbf{n}$ and $\gamma_{\mathbf{n}} : \mathbf{u} \mapsto \mathbf{u}_{|\Gamma}\cdot\mathbf{n}$, $\mathbf{u} \in C^\infty(\overline{\Omega_s})^3$.

The trace theorem for $\boldsymbol{H}^1(\Omega)$ [40, Theorem 1.5.1.1] shows that the tangential trace $\gamma_{\mathbf{t}} : \boldsymbol{C}^\infty(\bar{\Omega}) \mapsto \boldsymbol{L}^\infty(\Gamma)$ and the normal trace: $\gamma_{\mathbf{n}} : \boldsymbol{C}^\infty(\bar{\Omega}) \mapsto L^\infty(\Gamma)$ are continuous as mappings $\boldsymbol{H}(\mathbf{curl};\Omega) \mapsto \boldsymbol{H}^{-\frac{1}{2}}(\Gamma)$ and $\boldsymbol{H}(\mathrm{div};\Omega) \mapsto H^{-\frac{1}{2}}(\Gamma)$, respectively. Here, $H^{-\frac{1}{2}}(\Gamma)$ and $\boldsymbol{H}^{-\frac{1}{2}}(\Gamma)$ are the dual spaces of $H^{\frac{1}{2}}(\Gamma)$ and $\boldsymbol{H}^{\frac{1}{2}}(\Gamma) := (H^{\frac{1}{2}}(\Gamma))^3$, respectively, with respect to the pivot spaces $L^2(\Gamma)$ and $\boldsymbol{L}^2(\Gamma)$. Consequently, the traces can be extended to $\boldsymbol{H}(\mathbf{curl};\Omega)$ and $\boldsymbol{H}(\mathrm{div};\Omega)$, respectively.

In the sequel we will consider the electric wave equation (2). Now, since the field $\mathbf{u}$ is a locally square-integrable function satisfying $\mathbf{curl}\,\mathbf{curl}\,\mathbf{u} - \mathbf{u} = 0$, we can conclude that $\mathbf{curl}\,\mathbf{curl}\,\mathbf{u}$ is locally square-integrable, too. Hence, the space

$$\boldsymbol{H}_{\mathrm{loc}}(\mathbf{curl}^2,\Omega) := \{\mathbf{u} \in \boldsymbol{H}_{\mathrm{loc}}(\mathbf{curl};\Omega),\ \mathbf{curl}\,\mathbf{curl}\,\mathbf{u} \in \boldsymbol{L}^2_{\mathrm{loc}}(\Omega)\}$$

comes into play as the natural space for the solutions of the electric/magnetic wave equation with constant coefficients. It will be crucial for meaningful strong formulations of electromagnetic transmission problems.

2.2 Function Spaces on the Manifold Γ

Recall that through local charts one defines standard Sobolev spaces on the manifold $\Gamma = \partial\Omega_s$. We denote them as $H^s(\Gamma), \mathbf{H}^s(\Gamma)$, $s \in [-1,1]$, for scalars and vectors, respectively. We saw that the tangential trace operator $\gamma_{\mathbf{t}}$ possesses an interpretation as a continuous mapping $\boldsymbol{H}(\mathbf{curl};\Omega) \mapsto \boldsymbol{H}^{-\frac{1}{2}}(\Gamma)$. This is actually sufficient for the understanding of homogeneous boundary conditions for fields in the Hilbert space context. However, in order to impose meaningful non-homogeneous boundary conditions or, even more important, to lay the foundations for boundary integral equations we need to identify a proper trace space "$X(\Gamma)$" of $\boldsymbol{H}(\mathbf{curl};\Omega)$, $\Omega \subset \mathbb{R}^3$ a "generic" domain. It has to meet two essential requirements:

1. The inner product on $X(\Gamma)$ has an *intrinsic* definition that does not rely on the embedding of Γ into $\mathbb{R}^3$, i.e, $X(\Gamma)$ should have an interpretation as sections of the tangent bundle to $T\Gamma$ of Γ.
2. We demand that $\gamma_{\mathbf{t}} : \boldsymbol{H}(\mathbf{curl};\Omega) \mapsto X(\Gamma)$ is *continuous* and *surjective.*

Note that the same issue for the operator $\gamma_{\mathbf{n}} : \boldsymbol{H}(\mathrm{div};\Omega) \to H^{-\frac{1}{2}}(\Gamma)$ was resolved a long time ago [39, Sect. I.2.2].

We emphasise that for the discussion of traces it hardly matters whether Ω is bounded or not. We assume in this section that Ω is bounded (this allows for integration on Ω), but with this slight change, the results of this section remain true also for unbounded domains, in particular, the open complement of Ω.

Smooth boundaries.

To illustrate ideas, we first consider a C^∞*-smooth* Γ. Then the Sobolev spaces $\boldsymbol{H}^s(\Gamma)$ of functions and $\boldsymbol{H}^s_{\mathbf{t}}(\Gamma)$ of tangential vector-fields, as well as differential surface operators (we shall use the self evident notation div_Γ, $\mathbf{curl}_\Gamma, \mathrm{curl}_\Gamma, \ldots$) can be defined for all $s \in \mathbb{R}$ using local charts and transformations [24, Sect. 3.1, Appendix] [53, Sect. 2.5.2]. It is a classical result that smooth functions on Γ are dense in all these spaces. Standard trace and tangential trace generate continuous and surjective operators $\gamma : H^{s+\frac{1}{2}}(\Omega) \mapsto H^s(\Gamma)$ and $\gamma_{\mathbf{t}} : \boldsymbol{H}^{s+\frac{1}{2}}(\Omega) \mapsto \boldsymbol{H}^s_{\mathbf{t}}(\Gamma)$ for all $s > 0$, where

$$\boldsymbol{H}^s_{\mathbf{t}}(\Gamma) \cong \{\boldsymbol{\phi} \in \boldsymbol{H}^s(\Gamma),\ \boldsymbol{\phi} \cdot \mathbf{n} = 0\} \subset \boldsymbol{L}^2_{\mathbf{t}}(\Gamma) \tag{7}$$

are Sobolev spaces of tangential vector-fields. We denote by $\boldsymbol{H}^{-s}_{\mathbf{t}}(\Gamma)$ the dual space of $\boldsymbol{H}^s_{\mathbf{t}}(\Gamma)$ with $\boldsymbol{L}^2_{\mathbf{t}}(\Gamma)$ as a pivot space. Angle brackets will designate the duality pairings. Now, since for any $\mathbf{u} \in \boldsymbol{H}(\mathbf{curl};\Omega)$ we have $\gamma_{\mathbf{t}}\mathbf{u} \cdot \mathbf{n} = 0$, thus $\gamma_{\mathbf{t}}\mathbf{u} \in \boldsymbol{H}^{-1/2}_{\mathbf{t}}(\Gamma)$. Moreover, using both (5) and (6), we can easily see that

$$\mathrm{div}_\Gamma(\gamma_{\mathbf{t}}\mathbf{u}) = \gamma_{\mathbf{n}}(\mathbf{curl}\,\mathbf{u}) \quad \forall \mathbf{u} \in \boldsymbol{H}(\mathbf{curl};\Omega), \tag{8}$$

which implies $\mathrm{div}_\Gamma(\gamma_{\mathbf{t}}\mathbf{u}) \in H^{-1/2}(\Gamma)$.

Now, it is natural to define the space

$$\boldsymbol{TH}^s(\mathrm{div}_\Gamma;\Gamma) := \{\boldsymbol{\mu} \in \boldsymbol{H}^s_{\mathbf{t}}(\Gamma),\ \mathrm{div}_\Gamma\boldsymbol{\mu} \in H^s(\Gamma)\}. \tag{9}$$

The tangential trace $\gamma_{\mathbf{t}} : \boldsymbol{H}(\mathbf{curl};\Omega) \mapsto \boldsymbol{TH}^{-\frac{1}{2}}(\mathrm{div}_\Gamma;\Gamma)$ turns out to be continuous and surjective [53, Theorem 5.4.2]. For smooth surfaces the issue of tangential traces in $\boldsymbol{H}(\mathbf{curl};\Omega)$ was investigated in the papers of L. Paquet [55] and Alonso/Valli [2]. A survey of the results is also given in the monographs by M. Cessenat [24, Sect. 2.1] and J.-C. Nédélec [53, Sect. 5.4.1].

Moreover, if we define the anti-symmetric pairing

$$\langle \boldsymbol{\mu}, \boldsymbol{\eta} \rangle_{\tau,\Gamma} := \int_\Gamma (\boldsymbol{\mu} \times \mathbf{n}) \cdot \boldsymbol{\eta}\, \mathrm{d}S\,, \quad \boldsymbol{\mu}, \boldsymbol{\eta} \in \boldsymbol{L}^2_{\mathbf{t}}(\Gamma)\,, \tag{10}$$

then we can rewrite (6) as

$$\int_{\Omega} (\mathbf{curl}\,\mathbf{u}\cdot\mathbf{v} - \mathbf{u}\cdot\mathbf{curl}\,\mathbf{v})\,\mathrm{d}\mathbf{x} = \langle \gamma_{\mathbf{t}}\mathbf{v}, \gamma_{\mathbf{t}}\mathbf{u}\rangle_{\boldsymbol{\tau},\Gamma}, \tag{11}$$

which suggests that the space $\boldsymbol{TH}^{-\frac{1}{2}}(\mathrm{div}_\Gamma;\Gamma)$ coincides with its dual when using $\langle\cdot,\cdot\rangle_{\boldsymbol{\tau},\Gamma}$ as duality pairing. This statement will be clarified in the case of non-smooth surfaces at the end of this section.

Piecewise smooth and Lipschitz boundaries.

Only recently have results been obtained for non-smooth boundaries. We owe it to the pioneering work of one of the authors together with P. Ciarlet jr., who first examined piecewise smooth boundaries in [14, 17, 18]. The issue of traces of $\boldsymbol{H}(\mathbf{curl};\Omega)$ for general Lipschitz-domains was finally settled jointly by one of the authors, M. Costabel and D. Sheen in [20]. These articles and Sect. 2 of [19] supply the main references for the current section.

The challenges faced in the case of piecewise smooth boundaries are highlighted by the simple observation: even if $\mathbf{u}\in \boldsymbol{C}^\infty(\bar{\Omega})$ we do **not** have $\gamma_{\mathbf{t}}\mathbf{u}\in \boldsymbol{H}^{\frac{1}{2}}(\Gamma)$, because the tangential trace is inevitably discontinuous across edges of Γ. The first consequence of this fact is that $\gamma_{\mathbf{t}}\boldsymbol{H}^1(\Omega)\times\mathbf{n}\not\subseteq \gamma_{\mathbf{t}}\boldsymbol{H}^1(\Omega)$, although the two objects are both good candidates to be "tangential" vector fields of "regularity" $\frac{1}{2}$. Thus we have to resort to the following definition:

Definition 1. *We introduce the Hilbert space* $\boldsymbol{H}^s_\times(\Gamma) := \gamma_{\mathbf{t}}(\boldsymbol{H}^{s+1/2}(\Omega))$, $s\in(0,1)$, *equipped with an inner product that renders* $\gamma_{\mathbf{t}} : \boldsymbol{H}^{s+1/2}(\Omega)\mapsto \boldsymbol{H}^s_\times(\Gamma)$ *continuous and surjective. Its dual space with respect to the pairing* $\langle\cdot,\cdot\rangle_{\boldsymbol{\tau},\Gamma}$ *is denoted by* $\boldsymbol{H}^{-s}_\times(\Gamma)$.

The dual space is well defined due to the density of $\boldsymbol{H}^{\frac{1}{2}}_\times(\Gamma)\subset \boldsymbol{L}^2_{\mathbf{t}}(\Gamma)$. The case of smooth and non-smooth surfaces differ considerably, which we aim to highlight by different notations: $\boldsymbol{H}^s_{\mathbf{t}}(\Gamma)$ for smooth Γ and $\boldsymbol{H}^s_\times(\Gamma)$ for non-smooth Γ.

For curvilinear polyhedra this space can be given a more concrete meaning. To that end, write $\Gamma^1,\dots,\Gamma^P$, $P\in\mathbb{N}$, for the finitely many curved polygonal faces of Γ, i.e. $\Gamma := \bigcup_{j=1}^P \bar{\Gamma}^j$, meeting at non-degenerate edges. For any tangential vector $\boldsymbol{\mu}$, we denote by $\boldsymbol{\mu}^j$ the restriction of $\boldsymbol{\mu}$ to Γ^j. Then, according to [17, Proposition 1.6] an equivalent norm on $\boldsymbol{H}^{\frac{1}{2}}_\times(\Gamma)$ can be expressed as

$$\|\boldsymbol{\mu}\|^2_{\boldsymbol{H}^{\frac{1}{2}}_\times(\Gamma)} := \sum_{j=1}^P \|\boldsymbol{\mu}^j\|^2_{\boldsymbol{H}^{\frac{1}{2}}_{\mathbf{t}}(\Gamma^j)} + \sum_{j=1}^P\sum_{i\in\mathcal{I}_j}\int_{\Gamma^j\times\Gamma^i}\frac{|\boldsymbol{\mu}^i\cdot\boldsymbol{\nu}^{ij}(\mathbf{x}) - \boldsymbol{\mu}^j\cdot\boldsymbol{\nu}^{ji}(\mathbf{y})|^2}{|\mathbf{x}-\mathbf{y}|^3}\,\mathrm{d}S^2\,.$$

where $\mathcal{I}_j$ is the set of indices of smooth components abutting Γ^j, and $\boldsymbol{\nu}^{ij}$ denotes the tangential outer normal to Γ^i restricted to the edge $\overline{\Gamma^j}\cap\overline{\Gamma^i}$. Loosely speaking,

$\boldsymbol{H}_{\times}^{\frac{1}{2}}(\Gamma)$ contains vector-fields that are in $\boldsymbol{H}_{\mathbf{t}}^{\frac{1}{2}}(\Gamma^j)$ for each face Γ^j and feature a "weak normal continuity" enforced by the second term in the definition of the norm.

Using (6) and the same reasoning as for regular surfaces, we have that $\gamma_{\mathbf{t}}$: $\boldsymbol{H}(\mathbf{curl};\Omega) \to \boldsymbol{H}_{\times}^{-\frac{1}{2}}(\Gamma)$ is linear and continuous. In view of (8), we also know that this operator does not admit a right inverse. In order to repeat the argument sketched above for regular domains, we need a theory of differential operators on non-smooth manifolds. We do not want to delve into the details of these developments, and we refer the reader to [14, 16, 20] for a discussion on the subject. We need only the following definition: for $\mathbf{u} \in C^\infty(\overline{\Omega})$ set

$$\operatorname{div}_\Gamma \gamma_{\mathbf{t}}\mathbf{u} := \begin{cases} \operatorname{div}_j(\gamma_{\mathbf{t}}\mathbf{u})^j & \text{on } \Gamma^j\,, \\ \left((\gamma_{\mathbf{t}}\mathbf{u})^j \cdot \boldsymbol{\nu}^{ij} + (\gamma_{\mathbf{t}}\mathbf{u})^i \cdot \boldsymbol{\nu}^{ji}\right)\delta_{ij} & \text{on } \overline{\Gamma^j} \cap \overline{\Gamma^i}; \end{cases} \tag{12}$$

where δ_{ij} is the delta distribution (in local coordinates) whose support is the edge $\overline{\Gamma^j} \cap \overline{\Gamma^i}$ and div_j denotes the 2D-divergence computed on the face Γ^j. By density, this differential operator can be extended to less regular distributions and, in particular, to functionals in $\boldsymbol{H}_{\times}^{-\frac{1}{2}}(\Gamma)$. Moreover, (8) holds true in the appropriate sense. Thus, we set

$$\boldsymbol{H}_{\times}^{-\frac{1}{2}}(\operatorname{div}_\Gamma, \Gamma) := \{\boldsymbol{\mu} \in \boldsymbol{H}_{\times}^{-\frac{1}{2}}(\Gamma),\ \operatorname{div}_\Gamma \boldsymbol{\mu} \in H^{-\frac{1}{2}}(\Gamma)\}\,.$$

Finally, we denote by $\mathbf{curl}_\Gamma$ the operator adjoint to $\operatorname{div}_\Gamma$ with respect to the scalar product $\langle \cdot, \cdot \rangle_{\boldsymbol{\tau},\Gamma}$, i.e.

$$\langle \mathbf{curl}_\Gamma q, \mathbf{p} \rangle_{\boldsymbol{\tau},\Gamma} = \langle \operatorname{div}_\Gamma \mathbf{p}, q \rangle_{\frac{1}{2},\Gamma}\,, \quad \mathbf{p} \in \boldsymbol{H}_{\times}^{-\frac{1}{2}}(\operatorname{div}_\Gamma, \Gamma)\,,\ q \in H^{\frac{1}{2}}(\Gamma)\,. \tag{13}$$

The following theorem proves that the space $\boldsymbol{H}_{\times}^{-\frac{1}{2}}(\operatorname{div}_\Gamma, \Gamma)$ fits the criterion announced at the beginning of this section:

Theorem 1. *The operator $\gamma_{\mathbf{t}} : \boldsymbol{H}(\mathbf{curl};\Omega) \mapsto \boldsymbol{H}_{\times}^{-\frac{1}{2}}(\operatorname{div}_\Gamma, \Gamma)$ is continuous, surjective, and possesses a continuous right inverse.*

Proof. See Theorem 4.4 in [18] for the case of Lipschitz polyhedra. The more general assertion for Lipschitz domains is shown in [20, Sect. 4].

In the case of Maxwell's equations the role of Cauchy data is played by $\gamma_{\mathbf{t}}\mathbf{e}$ and $\gamma_{\mathbf{t}}\mathbf{h}$. By the fundamental symmetry of electric and magnetic fields, $\boldsymbol{H}_{\times}^{-\frac{1}{2}}(\operatorname{div}_\Gamma, \Gamma)$ is the right trace space for both fields. Everything fits because this space is its own dual, as is confirmed by the following theorem [20, Lemma 5.6]

Theorem 2 (Self-duality of $\boldsymbol{H}_{\times}^{-\frac{1}{2}}(\operatorname{div}_\Gamma, \Gamma)$). *The pairing $\langle \cdot, \cdot \rangle_{\boldsymbol{\tau},\Gamma}$ can be extended to a continuous bilinear form on $\boldsymbol{H}_{\times}^{-\frac{1}{2}}(\operatorname{div}_\Gamma, \Gamma)$. With respect to $\langle \cdot, \cdot \rangle_{\boldsymbol{\tau},\Gamma}$ the space $\boldsymbol{H}_{\times}^{-\frac{1}{2}}(\operatorname{div}_\Gamma, \Gamma)$ becomes its own dual.*

When we want to examine the convergence of boundary element methods quantitatively, extra smoothness of the functions to be approximated is indispensable. For any $s > \frac{1}{2}$, we define $\mathbf{H}^s_-(\Gamma) := \{\mathbf{u} \in \mathbf{L}^2_{\mathbf{t}}(\Gamma) \; : \; \mathbf{u}_{|\Gamma^j} \in \boldsymbol{H}^s_{\mathbf{t}}(\Gamma^j)\}$ and $\boldsymbol{H}^s_\times(\Gamma) := \boldsymbol{H}^{\frac{1}{2}}_\times(\Gamma) \cap \mathbf{H}^s_-(\Gamma)$. The corresponding space of scalar functions will be denoted by $H^s_-(\Gamma)$. To characterise the smoothness we resort to the family of Hilbert spaces

$$\boldsymbol{H}^s_\times(\mathrm{div}_\Gamma, \Gamma) := \begin{cases} \boldsymbol{H}^{-\frac{1}{2}}_\times(\mathrm{div}_\Gamma, \Gamma) \,, & \text{if } s = -\frac{1}{2} \,, \\ \{\boldsymbol{\mu} \in \boldsymbol{H}^s_\times(\Gamma), \, \mathrm{div}_\Gamma \boldsymbol{\mu} \in H^s(\Gamma)\} \,, & \text{if } -\frac{1}{2} < s < \frac{1}{2} \,, \\ \{\boldsymbol{\mu} \in \boldsymbol{H}^s_\times(\Gamma), \, \mathrm{div}_\Gamma \boldsymbol{\mu} \in H^s_-(\Gamma)\} \,, & \text{if } s > \frac{1}{2} \,. \end{cases}$$

As demonstrated in [16, Appendix 2], these spaces can be obtained through complex interpolation for $-\frac{1}{2} \leq s < \frac{1}{2}$. From this fact we conclude the following trace theorem (see [16]).

Theorem 3. *The tangential trace mapping $\gamma_{\mathbf{t}}$ can be extended to a continuous mapping $\gamma_{\mathbf{t}} : \boldsymbol{H}^s(\mathbf{curl}, \Omega) \mapsto \boldsymbol{H}^{s-\frac{1}{2}}_\times(\mathrm{div}_\Gamma, \Gamma)$ for all $0 \leq s < 1$.*

3 Maxwell Versus Helmholtz

There is a striking similarity between the electric wave equation (2) and the scalar Helmholtz equation

$$-\,\mathrm{div}(\underline{\boldsymbol{\mu}}^{-1}\,\mathbf{grad}\, p) - \omega^2 \epsilon p = 0 \,, \quad \text{in } \Omega_s \cup \Omega' \,. \tag{14}$$

In fact, the relationship between (2) and (14) runs much deeper than mere appearance: both equations emerge from a single equation for differential forms on $\mathbb{R}^3$, where (14) involves 0-forms, whereas (2) is the version for 1-forms [45, Sect. 2]. Hardly surprisingly, the theories of boundary integral equation methods for the related boundary value problems largely rely on the same principles. Nevertheless, the technical difficulties encountered in the treatment of the electric wave equation and related boundary element methods are significantly bigger than in the case of (14).

To appreciate what accounts for the fundamental difference between electromagnetism and acoustics, let us temporarily consider the variational *source problem* in a bounded Lipschitz domain Ω, cf. [45, Sect. 5]. For (2) this reads: for $\mathbf{j} \in \boldsymbol{L}^2(\Omega)$ find $\mathbf{e} \in \boldsymbol{H}(\mathbf{curl}; \Omega)$ such that for all $\mathbf{v} \in \boldsymbol{H}(\mathbf{curl}; \Omega)$

$$a_M(\mathbf{e}, \mathbf{v}) := \left(\underline{\boldsymbol{\mu}}^{-1}\,\mathbf{curl}\,\mathbf{e}, \mathbf{curl}\,\mathbf{v}\right)_0 - \omega^2 \left(\underline{\epsilon}\mathbf{e}, \mathbf{v}\right)_0 = -i\omega \left(\mathbf{j}, \mathbf{v}\right)_0 \,, \tag{15}$$

where $(u, v)_0 := \int_\Omega uv \,\mathrm{d}\mathbf{x}$. The related problem for the Helmholtz equation and $f \in L^2(\Omega)$ seeks $p \in H^1(\Omega)$ such that

$$a_H(p, q) := (\mathbf{grad}\, p, \mathbf{grad}\, q)_0 - \omega^2 \,(p, q)_0 = (f, q)_0 \quad \forall q \in H^1(\Omega) \,. \tag{16}$$

Investigations of the convergence of Galerkin schemes for (16) usually centre on the concept of coercivity of the underlying bilinear form $a_H(\cdot, \cdot)$, that is, the fact that the zero order term is a *compact perturbation* of the second order term, the principal part, and that a Gårding inequality of the form

$$|a_H(p, \overline{p}) + c_H(p, \overline{p})| \geq C \|p\|^2_{H^1(\Omega)} \quad \forall p \in H^1(\Omega) \tag{17}$$

holds with $C > 0$ and a bilinear form $c_H(\cdot, \cdot)$, which is compact in $H^1(\Omega)$. As has been demonstrated by Schatz [58], *cf.* also [62], this is the key to a priori asymptotic error estimates for Galerkin finite element methods. Evidently, we cannot expect an analogue of (17) from a_M. The blame lies with the infinite dimensional kernel of the $\mathbf{curl}$-operator, which foils compactness of the imbedding $\boldsymbol{H}(\mathbf{curl}; \Omega) \hookrightarrow \boldsymbol{L}^2(\Omega)$.

The issue of coercivity can also be discussed from the point of view of "energies": both acoustic and electromagnetic scattering are marked by an incessant conversion of energies. In acoustics, potential and kinetic energy of the fluid are converted into each other, in electromagnetism the same roles are played by the electric and magnetic energy. In acoustics the potential energy (with respect to the bounded control volume Ω) is a compact perturbation of the kinetic energy[3]. Therefore we can clearly single out the Laplacian as the principal part of the Helmholtz operator. Conversely, in electromagnetism the electric and magnetic energies of a field are perfectly symmetric. Neither is a compact perturbation of the other. This means that no part of the electric wave equation is "principal". Formally speaking, the operator of the electric wave equation lacks the essential property of strong ellipticity. A concise summary is given in Table 1.

Table 1. Acoustics vs. electromagnetics in terms of dominant energies

Acoustic wave equation	Electric wave equation
$-\Delta p - \kappa^2 p = 0$	$\mathbf{curl}\,\mathbf{curl}\,\mathbf{e} - \kappa^2 \mathbf{e} = 0$
Energies entering the Lagrangian:	
Kinetic "energy" $\int_\Omega \lvert \mathbf{grad}\, p\rvert^2 \,\mathrm{dx}$ Potential "energy" $\int_\Omega \lvert p\rvert^2 \,\mathrm{dx}$	Magnetic "energy" $\int_\Omega \lvert \mathbf{curl}\,\mathbf{e}\rvert^2 \,\mathrm{dx}$ Electric "energy" $\int_\Omega \lvert \mathbf{e}\rvert^2 \,\mathrm{dx}$
Potential energy is a compact perturbation of kinetic energy	Symmetry between electric and magnetic quantities

The lack of a principal part can be overcome by the *splitting* of the fields into two components. One set of components, called the electric, will feature dominant electric energy. With the other set, the magnetic quantities, the situation is reversed. This will promote either $\mathbf{curl}\,\mathbf{curl}$ or Id to the role of a principal part. As a consequence, on each component the electric wave equation should be amenable to the

[3] Roles might be reversed depending on the formulation of the acoustic equations.

same treatment as the Helmholtz equation. In the context of electromagnetic problems the splitting idea has been pioneered by Nédélec and was first applied to integral operators in [37]. Since then it has emerged as a very powerful theoretical tool, see [6, 19, 26] and, in particular, the monograph [53]. Three features of a splitting prove essential:

1. one subspace in the splitting agrees with the kernel of **curl**,
2. the compact embedding of the other subspace (complement space) into $\boldsymbol{L}^2(\Omega)$,
3. the extra smoothness of vector-fields in the complement space.

This makes it possible to opt for the Helmholtz-type *regular splitting* provided by the next lemma.

Lemma 1 (Regular decomposition lemma). *There exists a continuous projector* $\mathsf{R} : \boldsymbol{H}(\mathbf{curl};\Omega) \mapsto \boldsymbol{H}^1(\Omega) \cap \boldsymbol{H}(\mathrm{div}\,0;\Omega)$ *such that* $\mathrm{Ker}(\mathsf{R}) = \boldsymbol{H}(\mathbf{curl}\,0;\Omega)$.

The proof is given in [45, Sect. 2.4] and makes use of the existence of regular vector potentials, *cf.* Lemma 3.5 in [7].

Evidently, the three requirements are satisfied by the decomposition

$$\boldsymbol{H}(\mathbf{curl};\Omega) = \boldsymbol{\mathcal{X}}(\Omega) \oplus \boldsymbol{\mathcal{N}}(\Omega)\,,\quad \boldsymbol{\mathcal{X}}(\Omega) := \mathsf{R}(\boldsymbol{H}(\mathbf{curl};\Omega)) \subset \boldsymbol{H}^1(\Omega)\,, \tag{18}$$

where we write $\boldsymbol{\mathcal{N}}(\Omega) := \boldsymbol{H}(\mathbf{curl}\,0;\Omega)$. The continuity of the projectors guarantees the stability of this decomposition. Now, we can consider the variational problem (15) with respect to (18): thanks to the compact embedding of $\boldsymbol{H}^1(\Omega)$ into $\boldsymbol{L}^2(\Omega)$ we see that the second term of the bilinear form

$$(\mathbf{e}_\perp,\mathbf{v}_\perp) \mapsto \left(\underline{\mu}^{-1}\,\mathbf{curl}\,\mathbf{e}_\perp,\mathbf{curl}\,\mathbf{v}_\perp\right)_0 - \omega^2\,(\underline{\epsilon}\mathbf{e}_\perp,\mathbf{v}_\perp)_0\,,\qquad \mathbf{e}_\perp,\mathbf{v}_\perp \in \boldsymbol{\mathcal{X}}(\Omega)\,,$$

is a compact perturbation of the first: a_M is coercive on $\boldsymbol{\mathcal{X}}(\Omega)$. Coercivity on $\boldsymbol{\mathcal{N}}(\Omega)$ is trivial. In addition, terms like

$$(\mathbf{e}_0,\mathbf{v}_\perp) \mapsto (\underline{\epsilon}\mathbf{e}_0,\mathbf{v}_\perp)_0\,,\qquad \mathbf{e}_0 \in \boldsymbol{\mathcal{N}}(\Omega),\ \mathbf{v}_\perp \in \boldsymbol{\mathcal{X}}(\Omega)\,,$$

which effect the coupling of $\boldsymbol{\mathcal{X}}(\Omega)$ and $\boldsymbol{\mathcal{N}}(\Omega)$ with respect to a_M, can also be dismissed as compact perturbations. Using the isomorphism $\mathsf{X}_\Omega : \boldsymbol{H}(\mathbf{curl};\Omega) \mapsto \boldsymbol{H}(\mathbf{curl};\Omega)$, defined by $\mathsf{X}_\Omega := \mathsf{R} - \mathsf{Z}$, where $\mathsf{Z} := \mathrm{Id} - \mathsf{R}$ is the complementary projector to R, to "flip signs", we arrive at

$$|a_M(\mathbf{u},\mathsf{X}_\Omega\overline{\mathbf{u}}) - \mathbf{c}_M(\mathbf{u},\overline{\mathbf{u}})| \geq C\,\|\mathbf{u}\|^2_{\boldsymbol{H}(\mathbf{curl};\Omega)} \quad \forall \mathbf{u} \in \boldsymbol{H}(\mathbf{curl};\Omega)\,, \tag{19}$$

with some $C > 0$ and a compact bilinear form $\mathbf{c}_M$ on $\boldsymbol{H}(\mathbf{curl};\Omega)$. The *generalised Gårding inequality* (19) is the crucial assumption in the following fundamental theorem:

Theorem 4. *If a bilinear form* $a : V \times V \mapsto \mathbb{C}$ *on a reflexive Banach space* V *satisfies*

$$|a(u,\mathsf{X}_\Omega\overline{u}) - c(u,\overline{u})| \geq C\,\|u\|_V^2 \quad \forall u \in V\,,$$

with $C > 0$, *a compact bilinear form* $c : V \times V \mapsto \mathbb{C}$, *and an isomorphism* $\mathsf{X}_\Omega : V \mapsto V$, *then the associated operator* $A : V \mapsto V'$ *is Fredholm with index 0.*

In particular, for a bilinear form meeting the requirements of the theorem, injectivity of the associated operator implies its surjectivity by the *Fredholm alternative* [11].

It is hardly surprising that the splitting idea also plays a pivotal role in the analysis of boundary integral equations arising from the electric wave equation. Here, it is applied to the trace space $\boldsymbol{H}_{\times}^{-\frac{1}{2}}(\mathrm{div}_{\Gamma},\Gamma)$:

Lemma 2. *There exists a projection* $\mathsf{R}^{\Gamma} : \boldsymbol{H}_{\times}^{-\frac{1}{2}}(\mathrm{div}_{\Gamma},\Gamma) \mapsto \boldsymbol{H}_{\times}^{\frac{1}{2}}(\Gamma)$ *such that* $\mathrm{Ker}(\mathsf{R}^{\Gamma}) = \boldsymbol{H}_{\times}^{-\frac{1}{2}}(\mathrm{div}_{\Gamma}0,\Gamma)$ *and*

$$\|\mathsf{R}^{\Gamma}\boldsymbol{\mu}\|_{\boldsymbol{H}_{\times}^{\frac{1}{2}}(\Gamma)} \leq C\|\mathrm{div}_{\Gamma}\boldsymbol{\mu}\|_{H^{-\frac{1}{2}}(\Gamma)}. \tag{20}$$

Proof. Pick $\boldsymbol{\lambda} \in \boldsymbol{H}_{\times}^{-\frac{1}{2}}(\mathrm{div}_{\Gamma},\Gamma)$ and set $\mu := \mathrm{div}_{\Gamma}\boldsymbol{\lambda} \in H^{-\frac{1}{2}}(\Gamma)$. Solve the Neumann problem

$$w \in H^{1}(\Omega_s)/\mathbb{R}: \quad \Delta w = 0 \quad \text{in } \Omega_s\,, \quad \gamma_{\mathbf{n}}^{-}\,\mathbf{grad}\,w = \mu \quad \text{on } \Gamma\,.$$

We find that $\mathbf{v} := \mathbf{grad}\,w \in \boldsymbol{H}(\mathrm{div}\,0;\Omega_s)$. Using Lemma 3.5 in [7], there exists $\mathbf{w} \in \boldsymbol{H}^{1}(\Omega_s)$ such that $\mathbf{v} = \mathbf{curl}\,\mathbf{w}$, $\mathrm{div}\,\mathbf{w} = 0$. This defines an operator $\mathsf{J} : H^{-\frac{1}{2}}(\Gamma) \mapsto \boldsymbol{H}^{1}(\Omega_s)$ by $\mathsf{J}\mu := \mathbf{w}$. Its continuity is elementary

$$\|\mathsf{J}\,\mu\|_{\boldsymbol{H}^{1}(\Omega_s)} \leq C\,\|\mathbf{v}\|_{\boldsymbol{L}^{2}(\Omega_s)} \leq C\,\|\mu\|_{H^{-\frac{1}{2}}(\Gamma)}\,,$$

and inherited by the mapping $\mathsf{R}^{\Gamma} := \gamma_{\mathbf{t}} \circ \mathsf{J} \circ \mathrm{div}_{\Gamma} : \boldsymbol{H}_{\times}^{-\frac{1}{2}}(\mathrm{div}_{\Gamma},\Gamma) \mapsto \boldsymbol{H}_{\times}^{\frac{1}{2}}(\Gamma)$. Moreover, we see that $\mathrm{div}_{\Gamma}\mathsf{R}^{\Gamma}\boldsymbol{\lambda} = \gamma_{\mathbf{n}}^{-}\,\mathbf{curl}\,\mathsf{L}\mathbf{v} = \gamma_{\mathbf{n}}^{-}\mathbf{v} = \mathrm{div}_{\Gamma}\boldsymbol{\lambda}$. □

As before, the projector complementary to R^{Γ} will be denoted by Z^{Γ}. We arrive at a stable decomposition of the trace space

$$\boldsymbol{H}_{\times}^{-\frac{1}{2}}(\mathrm{div}_{\Gamma},\Gamma) := \boldsymbol{\mathcal{X}}(\Gamma) \oplus \boldsymbol{\mathcal{N}}(\Gamma)\,, \tag{21}$$

where $\boldsymbol{\mathcal{X}}(\Gamma) := \mathsf{R}^{\Gamma}(\boldsymbol{H}_{\times}^{-\frac{1}{2}}(\mathrm{div}_{\Gamma},\Gamma))$ and $\boldsymbol{\mathcal{N}}(\Gamma) = \boldsymbol{H}_{\times}^{-\frac{1}{2}}(\mathrm{div}_{\Gamma}0,\Gamma)$. Both components inherit the norm of $\boldsymbol{H}_{\times}^{-\frac{1}{2}}(\mathrm{div}_{\Gamma},\Gamma)$.

Corollary 1. *The embedding* $\boldsymbol{\mathcal{X}}(\Gamma) \hookrightarrow \boldsymbol{L}_{\mathbf{t}}^{2}(\Gamma)$ *is compact.*

It is illuminating to give a physical interpretation of the decomposition. First, view $\boldsymbol{H}_{\times}^{-\frac{1}{2}}(\mathrm{div}_{\Gamma},\Gamma)$ as a space of tangential components of electric fields. Then, we encounter traces of "static" irrotational fields in $\boldsymbol{\mathcal{N}}(\Gamma)$, whereas traces of "dynamic" field components, whose **curl**s do not vanish, are associated with $\boldsymbol{\mathcal{X}}(\Gamma)$. By Faraday's law the latter are linked with magnetic fields. All in all, we can attribute an "electric nature" to the space $\boldsymbol{\mathcal{N}}(\Gamma)$, and a "magnetic nature" to $\boldsymbol{\mathcal{X}}(\Gamma)$. The arguments are simply reversed when considering magnetic traces $\gamma_{N}\mathbf{e}$, because given the absence of source currents the magnetic field is irrotational in the stationary case. This means that components of $\gamma_{N}\mathbf{e}$ that belong to $\boldsymbol{\mathcal{N}}(\Gamma)$ are "magnetic", whereas components in $\boldsymbol{\mathcal{X}}(\Gamma)$ are "electric", *cf.* Table 2.

Table 2. Physical nature of components occurring in the splitting of fields and traces

Field	Space	Magnetic components	Electric components
$\mathbf{e}$	$\boldsymbol{H}(\mathbf{curl};\Omega)$	$\boldsymbol{\mathcal{X}}(\Omega)$	$\boldsymbol{\mathcal{N}}(\Omega)$
$\gamma_{\mathbf{t}}\mathbf{e}$	$\boldsymbol{H}_{\times}^{-\frac{1}{2}}(\mathrm{div}_\Gamma,\Gamma)$	$\boldsymbol{\mathcal{X}}(\Gamma)$	$\boldsymbol{\mathcal{N}}(\Gamma)$
$\gamma_N\mathbf{e}$	$\boldsymbol{H}_{\times}^{-\frac{1}{2}}(\mathrm{div}_\Gamma,\Gamma)$	$\boldsymbol{\mathcal{N}}(\Gamma)$	$\boldsymbol{\mathcal{X}}(\Gamma)$

4 Representation Formulae

In this section we start from the electric wave equation (2) in the air region Ω', where $\underline{\mu}$ and $\underline{\epsilon}$ can be regarded as scalar constants μ_0 and ϵ_0. Then, the partial differential equation (2) can be recast as

$$\mathbf{curl}\,\mathbf{curl}\,\mathbf{e} - \kappa^2\mathbf{e} = 0\,. \tag{22}$$

The constant $\kappa := \omega\sqrt{\epsilon_0\mu_0} > 0$ is called the *wave number*, because $\kappa/2\pi$ tells us the number of wavelengths per unit length. Henceforth, κ will stand for a fixed positive wave number[4].

Definition 2. *A distribution* $\mathbf{e} \in \boldsymbol{H}_{\mathrm{loc}}(\mathbf{curl}^2,\Omega)$ *is called a* Maxwell solution, *on some generic domain* Ω, *if it satisfies* (22) *in* Ω, *and the Silver–Müller radiation conditions at* ∞, *if* Ω *is not bounded.*

It is our objective to derive a boundary integral representation formula for Maxwell solutions. In order to handle transmission conditions in the calculus of distributions, we introduce *currents*, that is, distributions supported on Γ. For a function $\varphi \in H^{-\frac{1}{2}}(\Gamma)$, a tangential vector-field $\boldsymbol{\xi} \in \boldsymbol{H}_{\times}^{-1}(\Gamma)$, and test functions $\Phi \in \mathcal{D}(\mathbb{R}^3)$, $\boldsymbol{\Phi} \in \boldsymbol{\mathcal{D}}(\mathbb{R}^3) := (\mathcal{D}(\mathbb{R}^3))^3$, we define

$$(\varphi\delta_\Gamma)(\Phi) := \langle\varphi,\gamma\Phi\rangle_{\frac{1}{2},\Gamma}\,,\quad (\boldsymbol{\xi}\boldsymbol{\delta}_\Gamma)(\boldsymbol{\Phi}) := \quad \langle\boldsymbol{\xi},\gamma_{\mathbf{t}}\boldsymbol{\Phi}\rangle_{\boldsymbol{\tau},\Gamma} = \langle\boldsymbol{\xi},\boldsymbol{\gamma}\boldsymbol{\Phi}\rangle_{-1,\Gamma}\,.$$

Recall the notation used in the introduction: the superscripts $-$ and $+$ tag traces onto Γ from Ω_s and Ω' respectively. Now, in the sense of distributions, integration by parts yields, *cf.* [16, Sect. 2.3],

$$\begin{aligned}
&\text{for } \mathbf{u} \in \boldsymbol{H}_{\mathrm{loc}}(\mathrm{div};\Omega_s\cup\Omega') &&: \quad \mathrm{div}\,\mathbf{u} = \mathrm{div}\,\mathbf{u}_{|\Omega_s\cup\Omega'} + [\gamma_{\mathbf{n}}]_\Gamma(\mathbf{u})\,\delta_\Gamma\,,\\
&\text{for } \mathbf{u} \in \boldsymbol{H}_{\mathrm{loc}}(\mathbf{curl};\Omega_s\cup\Omega') &&: \quad \mathbf{curl}\,\mathbf{u} = \mathbf{curl}\,\mathbf{u}_{|\Omega_s\cup\Omega'} - [\gamma_{\mathbf{t}}]_\Gamma(\mathbf{u})\,\boldsymbol{\delta}_\Gamma\,,\\
&\text{for } \boldsymbol{\xi} \in \boldsymbol{H}_{\times}^{-\frac{1}{2}}(\mathrm{div}_\Gamma,\Gamma) &&: \quad \mathrm{div}(\boldsymbol{\xi}\,\boldsymbol{\delta}_\Gamma) = (\mathrm{div}_\Gamma\boldsymbol{\xi})\,\delta_\Gamma\,.
\end{aligned}$$

For the sake of brevity, we have used the jump operator $[\cdot]_\Gamma$ defined by $[\gamma]_\Gamma := \gamma^+ - \gamma^-$ for some trace γ onto Γ. For notational simplicity it is also useful to resort to the average $\{\gamma\}_\Gamma = \frac{1}{2}(\gamma^+ + \gamma^-)$. Both operators can only be applied to functions defined in $\Omega_s\cup\Omega'$. Moreover, we set $\gamma_N^\pm := \kappa^{-1}\gamma_{\mathbf{t}}^\pm \circ \mathbf{curl}$.

[4] We point out that all considerations remain true if $\mathrm{Im}\,\kappa > 0$ (lossy media)

Now, let $\mathbf{u}$ be a Maxwell solution in $\Omega_s \cup \Omega'$, which, of course, satisfies $\operatorname{div} \mathbf{u} = 0$ in $\Omega_s \cup \Omega'$. Then the following identity holds in the sense of distributions,

$$\begin{aligned}
-\boldsymbol{\Delta}\mathbf{u} - \kappa^2\mathbf{u} &= \mathbf{curl}\,\mathbf{curl}\,\mathbf{u} - \mathbf{grad}\operatorname{div}\mathbf{u} - \kappa^2\mathbf{u} \\
&= \mathbf{curl}\left(\mathbf{curl}\,\mathbf{u}_{|\Omega_s\cup\Omega'} - [\gamma_{\mathbf{t}}]_\Gamma(\mathbf{u})\,\boldsymbol{\delta}_\Gamma\right) - \mathbf{grad}\left([\gamma_{\mathbf{n}}]_\Gamma(\mathbf{u})\,\delta_\Gamma\right) - \kappa^2\mathbf{u} \\
&= \mathbf{curl}\,\mathbf{curl}\,\mathbf{u}_{|\Omega_s\cup\Omega'} - \kappa\,[\gamma_N]_\Gamma(\mathbf{u})\,\boldsymbol{\delta}_\Gamma - \mathbf{curl}([\gamma_{\mathbf{t}}]_\Gamma(\mathbf{u})\,\boldsymbol{\delta}_\Gamma) - \\
&\quad - \mathbf{grad}([\gamma_{\mathbf{n}}]_\Gamma(\mathbf{u})\,\delta_\Gamma) - \kappa^2\mathbf{u} \\
&= -\kappa\,[\gamma_N]_\Gamma(\mathbf{u})\,\boldsymbol{\delta}_\Gamma - \mathbf{curl}([\gamma_{\mathbf{t}}]_\Gamma(\mathbf{u})\,\boldsymbol{\delta}_\Gamma) - \mathbf{grad}([\gamma_{\mathbf{n}}]_\Gamma(\mathbf{u})\,\delta_\Gamma)\,.
\end{aligned}$$

As far as the differential operator $\mathbf{curl}\,\mathbf{curl} - \kappa^2\,\mathrm{Id}$ is concerned, the integration by parts formula (11) suggests the distinction between *Dirichlet trace* $\gamma_{\mathbf{t}}$ and *Neumann trace* γ_N. The trace γ_N can be labelled "magnetic", because it actually retrieves the tangential trace of the magnetic field solution. It has much in common with the Neumann trace operator $\gamma_{\mathbf{n}} \circ \mathbf{grad}$ for the Helmholtz equation: for instance, it fails to be defined on $\boldsymbol{H}_{\mathrm{loc}}(\mathbf{curl}; \Omega_s \cup \Omega')$, but the weak definition

$$-\frac{1}{\kappa}\int_\Omega \mathbf{curl}\,\mathbf{u}\cdot\mathbf{curl}\,\mathbf{v} - \mathbf{curl}\,\mathbf{curl}\,\mathbf{u}\cdot\mathbf{v}\,\mathrm{d}\mathbf{x} = \langle\gamma_N\mathbf{u}, \gamma_{\mathbf{t}}\mathbf{v}\rangle_{\boldsymbol{\tau},\Gamma}\,, \qquad (23)$$

$\mathbf{v} \in \boldsymbol{\mathcal{D}}(\mathbb{R}^3)$, renders it meaningful on $\boldsymbol{H}_{\mathrm{loc}}(\mathbf{curl}^2, \Omega_s \cup \Omega')$ [46, Lemma 3.3]:

Lemma 3. *The trace γ_N furnishes a continuous and surjective mapping γ_N : $\boldsymbol{H}_{\mathrm{loc}}(\mathbf{curl}^2, \Omega' \cup \Omega_s) \mapsto \boldsymbol{H}_\times^{-\frac{1}{2}}(\operatorname{div}_\Gamma, \Gamma)$.*

Definition 3. *Pairs $(\boldsymbol{\zeta}, \boldsymbol{\mu}) \in \boldsymbol{H}_\times^{-\frac{1}{2}}(\operatorname{div}_\Gamma, \Gamma) \times \boldsymbol{H}_\times^{-\frac{1}{2}}(\operatorname{div}_\Gamma, \Gamma)$ are called interior/exterior* Maxwell Cauchy data, *if there is a Maxwell solution $\mathbf{u}$ in Ω_s and Ω', respectively, such that $\boldsymbol{\zeta} = \gamma_{\mathbf{t}}^{\pm}\mathbf{u}$, $\boldsymbol{\mu} = \gamma_N^{\pm}\mathbf{u}$.*

We know from [31, Theorem 6.7] that the Cartesian components of Maxwell solutions will satisfy the Sommerfeld radiation condition and the scalar Helmholtz equation in $\Omega_s \cup \Omega'$. Using the results from [52, Chap. 9], we can apply componentwise convolution with the outgoing fundamental solution of the Helmholtz equation $E_\kappa(\mathbf{x}) := \exp(i\kappa|\mathbf{x}|)/4\pi|\mathbf{x}|$, $\mathbf{x} \neq 0$, and we find that almost everywhere in $\mathbb{R}^3$ the components of $\mathbf{u} = (u_1, u_2, u_3)^T$ satisfy

$$\begin{aligned}
u_j(\mathbf{x}) = &- \kappa([\gamma_N]_\Gamma(\mathbf{u})\,\boldsymbol{\delta}_\Gamma)(E_\kappa(\mathbf{x}-\cdot)\mathbf{e}_j) - ([\gamma_{\mathbf{t}}]_\Gamma(\mathbf{u})\,\boldsymbol{\delta}_\Gamma)(\mathbf{curl}(E_\kappa(\mathbf{x}-\cdot)\mathbf{e}_j)) + \\
&+ ([\gamma_{\mathbf{n}}]_\Gamma(\mathbf{u})\,\delta_\Gamma)(\operatorname{div}(E_\kappa(\mathbf{x}-\cdot)\mathbf{e}_j))\,, \quad j = 1,2,3\,.
\end{aligned}$$

Using $\mathbf{grad}_{\mathbf{x}}\, E_\kappa(\mathbf{x}-\mathbf{y}) = -\,\mathbf{grad}_{\mathbf{y}}\, E_\kappa(\mathbf{x}-\mathbf{y})$, we arrive at the famous Stratton–Chu representation formula for the electric field in $\Omega_s \cup \Omega'$ [60], *cf.*[31, Sect. 6.2], [53, Sect. 5.5], [24, Chap. 3, Sect. 1.3.2],

$$\mathbf{u} = -\kappa\boldsymbol{\Psi}_{\mathbf{V}}^{\kappa}([\gamma_N]_\Gamma(\mathbf{u})) - \mathbf{curl}\,\boldsymbol{\Psi}_{\mathbf{V}}^{\kappa}([\gamma_{\mathbf{t}}]_\Gamma(\mathbf{u})) - \mathbf{grad}\,\Psi_V^{\kappa}([\gamma_{\mathbf{n}}]_\Gamma(\mathbf{u}))\,. \qquad (24)$$

Here, $\boldsymbol{\Psi}_{\mathbf{V}}^{\kappa}(\cdot)$ and $\Psi_V^{\kappa}(\cdot)$ are *potentials*, that is, mappings of boundary data to analytic functions defined everywhere off the boundary. In detail, Ψ_V^{κ} and $\boldsymbol{\Psi}_{\mathbf{V}}^{\kappa}$ are the scalar and vectorial single layer potential, whose integral representation is given by ($\mathbf{x} \notin \Gamma$)

$$\Psi_V^\kappa(\phi)(\mathbf{x}) := \int_\Gamma \phi(\mathbf{y}) E_\kappa(\mathbf{x}-\mathbf{y})\, dS(\mathbf{y}), \boldsymbol{\Psi}_{\mathbf{V}}^\kappa(\boldsymbol{\mu})(\mathbf{x}) := \int_\Gamma \boldsymbol{\mu}(\mathbf{y}) E_\kappa(\mathbf{x}-\mathbf{y})\, dS(\mathbf{y}).$$

A simplification of (24) is possible by observing that, by (8),

$$\mathrm{div}_\Gamma(\gamma_N^\pm \mathbf{u}) = \kappa^{-1}\gamma_{\mathbf{n}}^\pm(\mathbf{curl}\,\mathbf{curl}\,\mathbf{u}) = \kappa(\gamma_{\mathbf{n}}^\pm \mathbf{u}) \quad \text{in } H^{-\frac{1}{2}}(\Gamma)\,. \tag{25}$$

This enables us to get rid of the normal components trace in (24). We end up with the pointwise identity

$$\mathbf{u}(\mathbf{x}) = -\boldsymbol{\Psi}_{DL}^\kappa([\gamma_{\mathbf{t}}]_\Gamma(\mathbf{u}))(\mathbf{x}) - \boldsymbol{\Psi}_{SL}^\kappa([\gamma_N]_\Gamma(\mathbf{u}))(\mathbf{x})\,, \quad \mathbf{x} \in \Omega_s \cup \Omega'\,, \tag{26}$$

where we have introduced the (electric) *Maxwell single layer potential* according to

$$\boldsymbol{\Psi}_{SL}^\kappa(\boldsymbol{\mu})(\mathbf{x}) := \kappa \boldsymbol{\Psi}_{\mathbf{V}}^\kappa(\boldsymbol{\mu})(\mathbf{x}) + \frac{1}{\kappa}\,\mathbf{grad}_{\mathbf{x}}\,\Psi_V^\kappa(\mathrm{div}_\Gamma \boldsymbol{\mu})(\mathbf{x})\,, \quad \mathbf{x} \notin \Gamma\,, \tag{27}$$

and the (electric) *Maxwell double layer potential*

$$\boldsymbol{\Psi}_{DL}^\kappa(\boldsymbol{\mu})(\mathbf{x}) := \mathbf{curl}_{\mathbf{x}}\,\boldsymbol{\Psi}_{\mathbf{V}}^\kappa(\boldsymbol{\mu})(\mathbf{x})\,, \quad \mathbf{x} \notin \Gamma\,. \tag{28}$$

We have chosen these names in order to underscore the similarity of (26) with the representation formula for solutions of the Helmholtz equation [31, Sect. 3.1], [52, Chap. 9].

Next, we aim to fit the potentials into the functional framework devised in Sect. 2. To this end we have to show that the potentials $\boldsymbol{\Psi}_{DL}^\kappa$ and $\boldsymbol{\Psi}_{SL}^\kappa$ are continuous operators between the canonical function spaces for traces and the appropriate function spaces for Maxwell solutions. An important result from [19] and [33] will be useful

Lemma 4. *The single layer potentials Ψ_V^κ and $\boldsymbol{\Psi}_{\mathbf{V}}^\kappa$ give rise to continuous mappings*

$$\Psi_V^\kappa : H^{-\frac{1}{2}+s}(\Gamma) \mapsto H_{\mathrm{loc}}^{1+s}(\mathbb{R}^3)\,, \quad \boldsymbol{\Psi}_{\mathbf{V}}^\kappa : \boldsymbol{H}_\times^{-\frac{1}{2}+s}(\Gamma) \mapsto \boldsymbol{H}_{\mathrm{loc}}^{1+s}(\mathbb{R}^3)\,,$$

for any s, $-\frac{1}{2} < s \leq \frac{1}{2}$.

From this lemma we conclude that both $\boldsymbol{\Psi}_{SL}^\kappa$ and $\boldsymbol{\Psi}_{DL}^\kappa$ are well defined for arguments in the trace space $\boldsymbol{H}_\times^{-\frac{1}{2}}(\mathrm{div}_\Gamma, \Gamma)$. To gain deeper insights into the continuity property of the Maxwell single layer and double layer potentials, we have to make some preparations, *cf.* Lemma 2.3 in [51].

Lemma 5. *For $\boldsymbol{\mu} \in \boldsymbol{H}_\times^{-\frac{1}{2}}(\mathrm{div}_\Gamma, \Gamma)$ we have $\mathrm{div}\,\boldsymbol{\Psi}_{\mathbf{V}}^\kappa(\boldsymbol{\mu}) = \Psi_V^\kappa(\mathrm{div}_\Gamma \boldsymbol{\mu})$ in $\boldsymbol{L}^2(\mathbb{R}^3)$.*

By definition and $\mathbf{curl} \circ \mathbf{grad} = 0$, it is immediate that $\mathbf{curl} \circ \boldsymbol{\Psi}_{SL}^\kappa = \kappa \boldsymbol{\Psi}_{DL}^\kappa$ on $\boldsymbol{H}_\times^{-\frac{1}{2}}(\mathrm{div}_\Gamma, \Gamma)$. On the other hand, using the previous lemma, we get for $\boldsymbol{\mu} \in \boldsymbol{H}_\times^{-\frac{1}{2}}(\mathrm{div}_\Gamma, \Gamma)$,

$$\begin{aligned}\mathbf{curl}\,\boldsymbol{\Psi}_{DL}^\kappa(\boldsymbol{\mu}) &= \mathbf{curl}\,\mathbf{curl}\,\boldsymbol{\Psi}_{\mathbf{V}}^\kappa(\boldsymbol{\mu}) = (-\boldsymbol{\Delta} + \mathbf{grad}\,\mathrm{div})\boldsymbol{\Psi}_{\mathbf{V}}^\kappa(\boldsymbol{\mu}) \\ &= \kappa^2 \boldsymbol{\Psi}_{\mathbf{V}}^\kappa(\boldsymbol{\mu}) + \mathbf{grad}\,\Psi_V^\kappa(\mathrm{div}_\Gamma \boldsymbol{\mu}) = \kappa \boldsymbol{\Psi}_{SL}^\kappa(\boldsymbol{\mu})\,.\end{aligned}$$

Here, we have used $-\boldsymbol{\Delta}\boldsymbol{\Psi}^{\kappa}_{\mathbf{V}}(\boldsymbol{\mu}) = \kappa^2\boldsymbol{\Psi}^{\kappa}_{\mathbf{V}}(\boldsymbol{\mu})$. Altogether, both potentials are Maxwell solutions, that is, for $\boldsymbol{\mu} \in \boldsymbol{H}_{\times}^{-\frac{1}{2}}(\mathrm{div}_\Gamma, \Gamma)$ they fulfil

$$(\mathbf{curl}\,\mathbf{curl} - \kappa^2\,\mathrm{Id})\boldsymbol{\Psi}^{\kappa}_{SL}(\boldsymbol{\mu}) = 0 \quad , \quad (\mathbf{curl}\,\mathbf{curl} - \kappa^2\,\mathrm{Id})\boldsymbol{\Psi}^{\kappa}_{DL}(\boldsymbol{\mu}) = 0 \,, \tag{29}$$

off the boundary Γ in a pointwise sense, and, globally, in $\boldsymbol{L}^2_{\mathrm{loc}}(\mathbb{R}^3)$. In addition, they comply with the Silver–Müller radiation conditions. From these relationships and Lemma 4 we infer the desired continuity properties.

Theorem 5. *The following mappings are continuous*

$$\begin{aligned}
\boldsymbol{\Psi}^{\kappa}_{SL} &: \boldsymbol{H}_{\times}^{-\frac{1}{2}}(\mathrm{div}_\Gamma, \Gamma) \mapsto \boldsymbol{H}_{\mathrm{loc}}(\mathbf{curl}^2, \Omega_s \cup \Omega') \cap \boldsymbol{H}_{\mathrm{loc}}(\mathrm{div}\,0; \Omega_s \cup \Omega')\,,\\
\boldsymbol{\Psi}^{\kappa}_{DL} &: \boldsymbol{H}_{\times}^{-\frac{1}{2}}(\mathrm{div}_\Gamma, \Gamma) \mapsto \boldsymbol{H}_{\mathrm{loc}}(\mathbf{curl}^2, \Omega_s \cup \Omega') \cap \boldsymbol{H}_{\mathrm{loc}}(\mathrm{div}\,0; \Omega_s \cup \Omega')\,.
\end{aligned}$$

Now, we are in a position to extract the desired identities from (26).

Theorem 6 (Stratton-Chu representation formula). *Any Maxwell solution* $\mathbf{u}$ *in* Ω_s *possesses the representation*

$$\mathbf{u} = \boldsymbol{\Psi}^{\kappa}_{DL}(\gamma_{\mathbf{t}}^{-}\mathbf{u}) + \boldsymbol{\Psi}^{\kappa}_{SL}(\gamma_N^{-}\mathbf{u}) \quad \text{in } \boldsymbol{H}(\mathbf{curl}^2, \Omega_s)\,.$$

If $\mathbf{u}$ *is a Maxwell solution in* Ω' *that satisfies the Silver–Müller radiation conditions, it can be written as*

$$\mathbf{u} = -\boldsymbol{\Psi}^{\kappa}_{DL}(\gamma_{\mathbf{t}}^{+}\mathbf{u}) - \boldsymbol{\Psi}^{\kappa}_{SL}(\gamma_N^{+}\mathbf{u}) \quad \text{in } \boldsymbol{H}_{\mathrm{loc}}(\mathbf{curl}^2, \Omega')\,.$$

5 Boundary Integral Operators

By Lemma 3, Theorem 5 provides the foundation for applying both the Dirichlet trace $\gamma_{\mathbf{t}}$ and the Neumann trace γ_N to the potentials $\boldsymbol{\Psi}^{\kappa}_{SL}$ and $\boldsymbol{\Psi}^{\kappa}_{DL}$. This is the canonical way of constructing boundary integral operators [52, Chap. 7]. In the case of second order elliptic problems, four different boundary integral operators arise. Yet, due to the fact that $\mathbf{curl} \circ \boldsymbol{\Psi}^{\kappa}_{SL} = \kappa\boldsymbol{\Psi}^{\kappa}_{DL}$, $\mathbf{curl} \circ \boldsymbol{\Psi}^{\kappa}_{DL} = \kappa\boldsymbol{\Psi}^{\kappa}_{SL}$ implies

$$\gamma_N^{\pm}\boldsymbol{\Psi}^{\kappa}_{SL} = \gamma_{\mathbf{t}}^{\pm}\boldsymbol{\Psi}^{\kappa}_{DL}\,, \quad \gamma_N^{\pm}\boldsymbol{\Psi}^{\kappa}_{DL} = \gamma_{\mathbf{t}}^{\pm}\boldsymbol{\Psi}^{\kappa}_{SL}\,, \tag{30}$$

two different boundary integral operators are sufficient for electromagnetic scattering: we obtain the *boundary integral operators*

$$\mathbf{S}_\kappa := \{\gamma_{\mathbf{t}}\}_\Gamma \circ \boldsymbol{\Psi}^{\kappa}_{SL} = \{\gamma_N\}_\Gamma \circ \boldsymbol{\Psi}^{\kappa}_{DL}\,, \quad \mathbf{C}_\kappa := \{\gamma_{\mathbf{t}}\}_\Gamma \circ \boldsymbol{\Psi}^{\kappa}_{DL} = \{\gamma_N\}_\Gamma \circ \boldsymbol{\Psi}^{\kappa}_{SL}\,.$$

The continuity of $\mathbf{S}_\kappa$ and $\mathbf{C}_\kappa$ is immediate from Theorem 5, in conjunction with Lemma 3 and Theorem 1.

Corollary 2. *The operators* $\mathbf{S}_\kappa, \mathbf{C}_\kappa : \boldsymbol{H}_{\times}^{-\frac{1}{2}}(\mathrm{div}_\Gamma, \Gamma) \mapsto \boldsymbol{H}_{\times}^{-\frac{1}{2}}(\mathrm{div}_\Gamma, \Gamma)$ *are continuous.*

As auxiliary boundary integral operators, which supply building blocks for $\mathbf{S}_\kappa$ and $\mathbf{C}_\kappa$, we introduce the two single layer boundary integral operators

$$V_\kappa := \{\gamma\}_\Gamma \circ \Psi_V^\kappa \quad , \quad \mathbf{V}_\kappa := \{\gamma_\mathbf{t}\}_\Gamma \circ \boldsymbol{\Psi}_\mathbf{V}^\kappa \, .$$

By combining Lemma 4 with continuity properties of the traces, we obtain the following result

Corollary 3. *The boundary integral operators $V_\kappa : H^{-\frac{1}{2}}(\Gamma) \mapsto H^{\frac{1}{2}}(\Gamma)$ and $\mathbf{V}_\kappa$: $\boldsymbol{H}_\times^{-\frac{1}{2}}(\Gamma) \mapsto \boldsymbol{H}_\times^{\frac{1}{2}}(\Gamma)$ are continuous.*

By inspecting the potential $\boldsymbol{\Psi}_{SL}^\kappa$, and recalling $\gamma_\mathbf{t} \circ \mathbf{grad} = \mathbf{curl}_\Gamma \circ \gamma$, it is clear that we can write

$$\mathbf{S}_\kappa = \kappa \mathbf{V}_\kappa + \kappa^{-1} \mathbf{curl}_\Gamma \circ V_\kappa \circ \mathrm{div}_\Gamma \, . \tag{31}$$

For the sake of implementation, more concrete *boundary integral representations* of the boundary integral operators are indispensable. It takes subtle theory to establish them, but here we only cite the result. A comprehensive treatment for second-order elliptic operators is given in [52, Sect. 7.2]. As variational formulations are our primary concern, expressions for the bilinear forms associated with $\mathbf{S}_\kappa$ and $\mathbf{C}_\kappa$ will be given: for tangential vector fields $\boldsymbol{\mu}, \boldsymbol{\xi} \in \boldsymbol{L}^\infty(\Gamma)$ we obtain

$$\begin{aligned} \langle \mathbf{S}_\kappa \boldsymbol{\mu}, \boldsymbol{\xi} \rangle_{\boldsymbol{\tau},\Gamma} = & -\kappa \int_\Gamma \int_\Gamma E_\kappa(\mathbf{x}-\mathbf{y}) \, \boldsymbol{\mu}(\mathbf{y}) \cdot \boldsymbol{\xi}(\mathbf{x}) \, \mathrm{d}S(\mathbf{y},\mathbf{x}) + \\ & + \frac{1}{\kappa} \int_\Gamma \int_\Gamma E_\kappa(\mathbf{x}-\mathbf{y}) \, \mathrm{div}_\Gamma \boldsymbol{\mu}(\mathbf{y}) \, \mathrm{div}_\Gamma \boldsymbol{\xi}(\mathbf{x}) \, \mathrm{d}S(\mathbf{y},\mathbf{x}) \, , \end{aligned} \tag{32}$$

$$\langle \mathbf{C}_\kappa \boldsymbol{\mu}, \boldsymbol{\xi} \rangle_{\boldsymbol{\tau},\Gamma} = - \int_\Gamma \int_\Gamma \mathbf{grad}_\mathbf{x} \, E_\kappa(\mathbf{x}-\mathbf{y}) \cdot (\boldsymbol{\mu}(\mathbf{y}) \times \boldsymbol{\xi}(\mathbf{x})) \, \mathrm{d}S(\mathbf{y},\mathbf{x}). \tag{33}$$

The first integral arises from (31) through integration by parts. Its kernel $E_\kappa(\mathbf{x}-\mathbf{y})$ is weakly singular, because $E_\kappa(\mathbf{x}-\mathbf{y}) = O(|\mathbf{x}-\mathbf{x}|^{-1})$ for $\mathbf{y} \to \mathbf{x}$. Thus, the integral makes sense as an improper integral. The second integral has a strongly singular kernel behaving like $O(|\mathbf{x}-\mathbf{y}|^{-2})$ for $\mathbf{y} \to \mathbf{x}$, and has to be read as a Cauchy principal value.

A fundamental tool for deriving boundary integral equations are *jump relations* describing the behaviour of the potentials when crossing Γ. For the Maxwell single and double layer potential they closely resemble those for conventional single and double layer potentials for second order elliptic operators [52, Chap. 6]. For smooth domains these results are contained in [31, Thm. 6.11], [53, Thm. 5.5.1], and [57].

Theorem 7 (Jump relations). *The interior and exterior Dirichlet- and Neumann-traces of the potentials $\boldsymbol{\Psi}_{SL}^\kappa$ and $\boldsymbol{\Psi}_{DL}^\kappa$ are well defined and, on $\boldsymbol{H}_\times^{-\frac{1}{2}}(\mathrm{div}_\Gamma, \Gamma)$, satisfy*

$$[\gamma_\mathbf{t}]_\Gamma \circ \boldsymbol{\Psi}_{SL}^\kappa = [\gamma_N]_\Gamma \circ \boldsymbol{\Psi}_{DL}^\kappa = 0 \, , \quad [\gamma_N]_\Gamma \circ \boldsymbol{\Psi}_{SL}^\kappa = [\gamma_\mathbf{t}]_\Gamma \circ \boldsymbol{\Psi}_{DL}^\kappa = -\mathrm{Id} \, .$$

Proof. The jump condition for the Dirichlet trace of the single layer potential is immediate from its regularity asserted in Lemma 4. By (30) we get the continuity of the Neumann trace $\boldsymbol{\Psi}^{\kappa}_{DL}$. Then, the jump of the Neumann trace of $\boldsymbol{\Psi}^{\kappa}_{SL}$ can be determined from (26). Finally, by (30), this also settles the contention for the Dirichlet trace of the double layer potential. □

Now, with the jump relations in mind, let us apply the exterior and interior trace operators to the representation formulae of Theorem 6:

$$\gamma_{\mathbf{t}}^{-}\mathbf{u} = \tfrac{1}{2}\gamma_{\mathbf{t}}^{-}\mathbf{u} + \mathbf{C}_{\kappa}(\gamma_{\mathbf{t}}^{-}\mathbf{u}) + \mathbf{S}_{\kappa}(\gamma_{N}^{-}\mathbf{u}), \quad \gamma_{\mathbf{t}}^{+}\mathbf{u} = \tfrac{1}{2}\gamma_{\mathbf{t}}^{+}\mathbf{u} - \mathbf{C}_{\kappa}(\gamma_{\mathbf{t}}^{+}\mathbf{u}) - \mathbf{S}_{\kappa}(\gamma_{N}^{+}\mathbf{u}),$$
$$\gamma_{N}^{-}\mathbf{u} = \mathbf{S}_{\kappa}(\gamma_{\mathbf{t}}^{-}\mathbf{u}) + \tfrac{1}{2}\gamma_{N}^{-}\mathbf{u} + \mathbf{C}_{\kappa}(\gamma_{N}^{-}\mathbf{u}), \quad \gamma_{N}^{+}\mathbf{u} = -\mathbf{S}_{\kappa}(\gamma_{\mathbf{t}}^{+}\mathbf{u}) + \tfrac{1}{2}\gamma_{N}^{+}\mathbf{u} - \mathbf{C}_{\kappa}(\gamma_{N}^{+}\mathbf{u}).$$

A concise way to write these formulae relies on the *Calderon projectors*, *cf.* [21, Sect. 3.3], [37, Formula (29)], and [53, Sect. 5.5],

$$\mathbb{P}_{\kappa}^{-} := \begin{pmatrix} \frac{1}{2}\,\mathrm{Id} + \mathbf{C}_{\kappa} & \mathbf{S}_{\kappa} \\ \mathbf{S}_{\kappa} & \frac{1}{2}\,\mathrm{Id} + \mathbf{C}_{\kappa} \end{pmatrix}, \quad \mathbb{P}_{\kappa}^{+} := \begin{pmatrix} \frac{1}{2}\,\mathrm{Id} - \mathbf{C}_{\kappa} & -\mathbf{S}_{\kappa} \\ -\mathbf{S}_{\kappa} & \frac{1}{2}\,\mathrm{Id} - \mathbf{C}_{\kappa} \end{pmatrix}. \tag{34}$$

By Theorem 6 the operators $\mathbb{P}_{\kappa}^{-}$, $\mathbb{P}_{\kappa}^{+} : \boldsymbol{H}_{\times}^{-\frac{1}{2}}(\mathrm{div}_{\Gamma}, \Gamma)^2 \mapsto \boldsymbol{H}_{\times}^{-\frac{1}{2}}(\mathrm{div}_{\Gamma}, \Gamma)^2$ are projectors, that is,

$$\mathbb{P}_{\kappa}^{-} \circ \mathbb{P}_{\kappa}^{-} = \mathbb{P}_{\kappa}^{-} \quad , \quad \mathbb{P}_{\kappa}^{+} \circ \mathbb{P}_{\kappa}^{+} = \mathbb{P}_{\kappa}^{+}. \tag{35}$$

Also note that $\mathbb{P}_{\kappa}^{-} + \mathbb{P}_{\kappa}^{+} = \mathrm{Id}$ and that the range of $\mathbb{P}_{\kappa}^{+}$ coincides with the kernel of $\mathbb{P}_{\kappa}^{-}$ and vice versa. The next result promotes Calderon projectors to a pivotal role in the derivation of boundary integral equations, *cf.* [61, Thm. 3.7].

Theorem 8. *The pair of functions* $(\boldsymbol{\zeta}, \boldsymbol{\mu}) \in \boldsymbol{H}_{\times}^{-\frac{1}{2}}(\mathrm{div}_{\Gamma}, \Gamma) \times \boldsymbol{H}_{\times}^{-\frac{1}{2}}(\mathrm{div}_{\Gamma}, \Gamma)$ *are suitable interior or exterior Maxwell Cauchy data, if and only if they lie in the kernel of* $\mathbb{P}_{\kappa}^{+}$ *or* $\mathbb{P}_{\kappa}^{-}$*, respectively.*

For the subsequent analysis it is convenient to examine the operator

$$\mathbb{A}_{\kappa} := \begin{pmatrix} \mathbf{C}_{\kappa} & \mathbf{S}_{\kappa} \\ \mathbf{S}_{\kappa} & \mathbf{C}_{\kappa} \end{pmatrix} : \boldsymbol{H}_{\times}^{-\frac{1}{2}}(\mathrm{div}_{\Gamma}, \Gamma)^2 \mapsto \boldsymbol{H}_{\times}^{-\frac{1}{2}}(\mathrm{div}_{\Gamma}, \Gamma)^2 .$$

It is linked with the Calderon projectors by $\mathbb{P}_{\kappa}^{-} = \frac{1}{2}\,\mathrm{Id} + \mathbb{A}_{\kappa}$, $\mathbb{P}_{\kappa}^{+} = \frac{1}{2}\,\mathrm{Id} - \mathbb{A}_{\kappa}$.

The operators $\mathbf{C}_{\kappa}$ enjoy a hidden symmetry, made precise in the next lemma, see [21, Thm. 3.9].

Lemma 6. *We have* $\langle \mathbf{C}_{\kappa}\boldsymbol{\zeta}, \boldsymbol{\mu}\rangle_{\boldsymbol{\tau},\Gamma} = \langle \mathbf{C}_{\kappa}\boldsymbol{\mu}, \boldsymbol{\zeta}\rangle_{\boldsymbol{\tau},\Gamma}$ *for all* $\boldsymbol{\zeta}, \boldsymbol{\mu} \in \boldsymbol{H}_{\times}^{-\frac{1}{2}}(\mathrm{div}_{\Gamma}, \Gamma)$.

6 Compactness and Coercivity

The ultimate goal is to establish the coercivity of bilinear forms occurring in weak formulations of boundary integral equations. To achieve this we need to identify compact perturbations, *cf.* Lemma 3.2 of [47] and the proof of Theorem 3.12 in [21].

Lemma 7. *The integral operators* $\delta V_\kappa := V_\kappa - V_0 : H^{-\frac{1}{2}}(\Gamma) \mapsto H^{\frac{1}{2}}(\Gamma)$ *and* $\delta \mathbf{V}_\kappa := \mathbf{V}_\kappa - \mathbf{V}_0 : \boldsymbol{H}_\times^{-\frac{1}{2}}(\Gamma) \mapsto \boldsymbol{H}_\times^{\frac{1}{2}}(\Gamma)$ *are compact.*

Slightly abusing notation, we define

$$\mathbf{S}_0 := \kappa \mathbf{V}_0 + \kappa^{-1} \mathbf{curl}_\Gamma \circ V_0 \circ \operatorname{div}_\Gamma . \tag{36}$$

From Lemma 7 and (31) we find that switching from $\mathbf{S}_\kappa$ to $\mathbf{S}_0$ amounts to a compact perturbation.

Corollary 4. *The operator* $\mathbf{S}_\kappa - \mathbf{S}_0 : \boldsymbol{H}_\times^{-\frac{1}{2}}(\operatorname{div}_\Gamma, \Gamma) \mapsto \boldsymbol{H}_\times^{-\frac{1}{2}}(\operatorname{div}_\Gamma, \Gamma)$ *is compact.*

The significance of this can be appreciated in light of the following result, *cf.* Theorem 3 in [36, Vol. IV, Chap. XI, § 2], and Theorem 6.2 in [46].

Lemma 8 (Ellipticity of single layer potentials). *The operators* V_0 *and* $\mathbf{V}_0$ *are continuous, self-adjoint with respect to the bilinear pairings* $\langle \cdot, \cdot \rangle_{\frac{1}{2},\Gamma}$ *and* $\langle \cdot, \cdot \rangle_{\boldsymbol{\tau},\Gamma}$, *respectively, and satisfy*

$$\langle \mu, V_0 \overline{\mu} \rangle_{\frac{1}{2},\Gamma} \geq C \|\mu\|^2_{H^{-\frac{1}{2}}(\Gamma)} \qquad \forall \mu \in H^{-\frac{1}{2}}(\Gamma) ,$$

$$\langle \boldsymbol{\mu}, \mathbf{V}_0 \overline{\boldsymbol{\mu}} \rangle_{\boldsymbol{\tau},\Gamma} \geq C \|\boldsymbol{\mu}\|^2_{\boldsymbol{H}_\times^{-\frac{1}{2}}(\Gamma)} \qquad \forall \boldsymbol{\mu} \in \boldsymbol{H}_\times^{-\frac{1}{2}}(\operatorname{div}_\Gamma 0, \Gamma) .$$

with constants $C > 0$ *only depending on* Γ.

Again, it proves highly instructive to recall facts about boundary integral operators related to the Helmholtz equation: Table 3 lists similarities and differences of the situations faced in the case of the Helmholtz equation and Maxwell's equations, respectively. The lack of ellipticity of the off-diagonal operators in $\mathbb{A}_\kappa$ has the same roots as the absence of a direct compact embedding in the case of the Maxwell source problem, *cf.* Sect. 3.

The roots of the difficulties being the same as for the Maxwell source problem, the same ideas should provide remedies: we have to employ stable splittings that target the trace space $\boldsymbol{H}_\times^{-\frac{1}{2}}(\operatorname{div}_\Gamma, \Gamma)$ and decompose it into the kernel of $\operatorname{div}_\Gamma$ and a suitable more regular complement. The decomposition (21) introduced in Sect. 3 meets all the requirements and will be used below.

Please recall the discussion in Sect. 3 of the coercivity of the bilinear form associated with the Maxwell source problem. The same considerations will now be applied to the bilinear form spawned by the boundary integral operator $\mathbf{S}_\kappa$ through $(\boldsymbol{\xi}, \boldsymbol{\mu}) \mapsto \langle \mathbf{S}_\kappa \boldsymbol{\mu}, \boldsymbol{\xi} \rangle_{\boldsymbol{\tau},\Gamma}$. To begin with, the "lower order" term $\mathbf{V}_\kappa$ in the operator $\mathbf{S}_\kappa$, *cf.* Formula (31), becomes compact on the "regular component" $\boldsymbol{\mathcal{X}}(\Gamma)$ of the decomposition (21).

Lemma 9. *The bilinear forms* $\langle \mathbf{V}_\kappa \cdot, \cdot \rangle_{\boldsymbol{\tau},\Gamma} : \boldsymbol{\mathcal{X}}(\Gamma) \times \boldsymbol{H}_\times^{-\frac{1}{2}}(\Gamma) \mapsto \mathbb{C}$ *and* $\langle \mathbf{V}_\kappa \cdot, \cdot \rangle_{\boldsymbol{\tau},\Gamma} : \boldsymbol{H}_\times^{-\frac{1}{2}}(\Gamma) \times \boldsymbol{\mathcal{X}}(\Gamma) \mapsto \mathbb{C}$ *are compact.*

Table 3. Comparison of analytical aspects of the acoustic and electromagnetic boundary integral operators, supplementing Table 1. The symbols K_κ and D_κ denote the double layer integral operator, and the hypersingular integral operator for the Helmholtz operator $-\Delta - \kappa^2 \,\mathrm{Id}$, respectively. Details can be found in [52, Chap. 9]

Helmholtz equation	Maxwell equations
$-\Delta p - \kappa^2 p = 0$	$\mathbf{curl}\,\mathbf{curl}\,\mathbf{e} - \kappa^2 \mathbf{e} = 0$
Boundary integral operators:	
Dirichlet trace $\gamma p \in H^{\frac{1}{2}}(\Gamma)$	Dirichlet trace $\gamma_{\mathbf{t}}\mathbf{e} \in \boldsymbol{H}_\times^{-\frac{1}{2}}(\mathrm{div}_\Gamma, \Gamma)$
Neumann trace $\frac{1}{\kappa}\gamma_{\mathbf{n}}(\mathbf{grad}\, p) \in H^{-\frac{1}{2}}(\Gamma)$	Neumann trace $\frac{1}{\kappa}\gamma_{\mathbf{t}}(\mathbf{curl}\,\mathbf{e}) \in \boldsymbol{H}_\times^{-\frac{1}{2}}(\mathrm{div}_\Gamma, \Gamma)$
$\mathbb{A}_\kappa = \begin{pmatrix} K_\kappa & V_\kappa \\ D_\kappa & \tilde{K}_\kappa \end{pmatrix}$	$\mathbb{A}_\kappa = \begin{pmatrix} \mathbf{C}_\kappa & \mathbf{S}_\kappa \\ \mathbf{S}_\kappa & \mathbf{C}_\kappa \end{pmatrix}$
The issue of coercivity	
$\mathbb{A}_\kappa = \begin{pmatrix} K_\kappa & V_0 \\ D_0 & K_\kappa^* \end{pmatrix} + \text{compact pert.}$	$\mathbb{A}_\kappa = \begin{pmatrix} \mathbf{C}_\kappa & \mathbf{S}_0 \\ \mathbf{S}_0 & \mathbf{C}_\kappa \end{pmatrix} + \text{compact pert.}$
Ellipticity on trace spaces:	No ellipticity, because $\mathbf{S}_0$ indefinite:
$\langle \varphi, V_0\overline{\varphi}\rangle_{\frac{1}{2},\Gamma} \geq C \left\|\varphi\right\|^2_{H^{-\frac{1}{2}}(\Gamma)}$,	$\mathbf{S}_0 = \mathbf{V}_0 + \frac{1}{\kappa}\mathbf{curl}_\Gamma \circ V_0 \circ \mathrm{div}_\Gamma$.
$\langle D_0\varphi, \overline{\varphi}\rangle_{\frac{1}{2},\Gamma} \geq C \left\|\varphi\right\|^2_{H^{\frac{1}{2}}(\Gamma)/\mathbb{C}}$.	Yet, individual terms are (semi)-definite.

Proof. Since $\mathbf{V}_\kappa : \boldsymbol{H}_\times^{-\frac{1}{2}}(\Gamma) \mapsto \boldsymbol{H}_\times^{\frac{1}{2}}(\Gamma)$ is continuous according to Theorem 3, the compact embedding $\boldsymbol{\mathcal{X}}(\Gamma) \hookrightarrow \boldsymbol{H}_\times^{-\frac{1}{2}}(\Gamma)$ (Corollary 1) gives the result. □

We can even establish a generalised Gårding inequality for $\mathbf{S}_0$ on Lipschitz boundaries: looking at the formula (31) and, in particular, the bilinear form

$$\langle \mathbf{S}_0\boldsymbol{\mu}, \boldsymbol{\xi}\rangle_{\boldsymbol{\tau},\Gamma} = \frac{1}{\kappa}\langle \mathrm{div}_\Gamma\boldsymbol{\mu}, V_0 \mathrm{div}_\Gamma\boldsymbol{\mu}\rangle_{\frac{1}{2},\Gamma} - \kappa\langle \boldsymbol{\mu}, \mathbf{V}_0\boldsymbol{\xi}\rangle_{\boldsymbol{\tau},\Gamma} , \tag{37}$$

we realise a striking similarity to the bilinear form of the Maxwell source problem (15). Thus, it is natural to employ the splitting idea of Sect. 3 based on (21) and the isomorphism

$$\mathsf{X}_\Gamma = \mathsf{R}^\Gamma - \mathsf{Z}^\Gamma : \boldsymbol{H}_\times^{-\frac{1}{2}}(\mathrm{div}_\Gamma, \Gamma) \mapsto \boldsymbol{H}_\times^{-\frac{1}{2}}(\mathrm{div}_\Gamma, \Gamma) . \tag{38}$$

Lemma 10 (Generalised Gårding inequality for $\mathbf{S}_\kappa$). *There is a compact bilinear form* $c_\Gamma : \boldsymbol{H}_\times^{-\frac{1}{2}}(\mathrm{div}_\Gamma, \Gamma) \times \boldsymbol{H}_\times^{-\frac{1}{2}}(\mathrm{div}_\Gamma, \Gamma) \mapsto \mathbb{C}$ *and a constant* $C_G > 0$ *such that*

$$|\,(\mathbf{S}_\kappa\boldsymbol{\mu},\mathsf{X}_\Gamma\overline{\boldsymbol{\mu}})_{\boldsymbol{\tau}}+c_\Gamma(\boldsymbol{\mu},\overline{\boldsymbol{\mu}})|\geq C_G\,\|\boldsymbol{\mu}\|^2_{\boldsymbol{H}_\times^{-\frac{1}{2}}(\mathrm{div}_\Gamma,\Gamma)}\qquad\forall\boldsymbol{\mu}\in\boldsymbol{H}_\times^{-\frac{1}{2}}(\mathrm{div}_\Gamma,\Gamma)\,.$$

Proof. We set

$$c_\Gamma(\boldsymbol{\mu},\boldsymbol{\xi}):=-\left\langle\mathbf{V}_\kappa\mathsf{R}^\Gamma\boldsymbol{\mu},\mathsf{R}^\Gamma\boldsymbol{\xi}\right\rangle_{\boldsymbol{\tau},\Gamma}+\left\langle\mathbf{V}_\kappa\mathsf{R}^\Gamma\boldsymbol{\mu},\mathsf{Z}^\Gamma\boldsymbol{\xi}\right\rangle_{\boldsymbol{\tau},\Gamma}-\left\langle\mathbf{V}_\kappa\mathsf{Z}^\Gamma\boldsymbol{\mu},\mathsf{R}^\Gamma\boldsymbol{\xi}\right\rangle_{\boldsymbol{\tau},\Gamma},$$

which is compact by Lemma 9. Noting that

$$\left\langle\mathbf{S}_0\boldsymbol{\mu},\mathsf{X}_\Gamma\overline{\boldsymbol{\xi}}\right\rangle_{\boldsymbol{\tau},\Gamma}=\frac{1}{\kappa}\left\langle V_0\mathrm{div}_\Gamma\mathsf{R}^\Gamma\boldsymbol{\mu},\mathrm{div}_\Gamma\mathsf{R}^\Gamma\overline{\boldsymbol{\xi}}\right\rangle_{0;\Gamma}+\kappa\left\langle\mathsf{Z}^\Gamma\boldsymbol{\mu},\mathbf{V}_0\mathsf{Z}^\Gamma\overline{\boldsymbol{\xi}}\right\rangle_{\boldsymbol{\tau},\Gamma}-c_\Gamma(\boldsymbol{\mu},\overline{\boldsymbol{\xi}}),$$

we invoke Lemma 8 and the stability of the decomposition to finish the proof. □

In the case of smooth domains this result is sufficient to obtain coercivity of $\mathbb{A}_\kappa$, because for smooth boundaries the singularity of the kernel of $\mathbf{C}_\kappa$ partly cancels. This is a well-known effect in the case of double layer potentials for second order elliptic operators. For $\mathbf{C}_\kappa$ the observation was made by Nédélec [53, Sect. 5.5].

Lemma 11. *If Γ is smooth, that is, of class C^∞, then $\mathbf{C}_\kappa$ is continuous as an operator $\mathbf{C}_\kappa:\boldsymbol{H}^s_{\mathbf{t}}(\Gamma)\mapsto\boldsymbol{H}^{s+1}_{\mathbf{t}}(\Gamma)$ and $\mathbf{C}_\kappa:\boldsymbol{TH}^{s-\frac{1}{2}}(\mathrm{div}_\Gamma;\Gamma)\mapsto\boldsymbol{TH}^{s+\frac{1}{2}}(\mathrm{div}_\Gamma;\Gamma)$ for all $s\in\mathbb{R}$.*

Proof. The first part of the proof boils down to manipulations of (33) using the product rule for $\mathbf{curl}_{\mathbf{x}}$ and the identity $(\mathbf{b}\times\mathbf{c})\times\mathbf{a}=\mathbf{c}(\mathbf{a}\cdot\mathbf{b})-\mathbf{b}(\mathbf{a}\cdot\mathbf{c})$. For $\boldsymbol{\mu},\boldsymbol{\xi}\in\mathbf{L}^\infty(\Gamma)\cap\boldsymbol{TH}(\mathrm{div}_\Gamma;\Gamma)$ we end up with

$$\begin{aligned}
&\langle\mathbf{C}_\kappa\boldsymbol{\mu},\boldsymbol{\xi}\rangle_{\boldsymbol{\tau},\Gamma}=\\
&=-\int_\Gamma\int_\Gamma((\boldsymbol{\mu}(\mathbf{y})\times\mathbf{grad}_{\mathbf{x}}\,E_\kappa(\mathbf{x}-\mathbf{y}))\times\mathbf{n}(\mathbf{x}))\cdot(\boldsymbol{\xi}(\mathbf{x})\times\mathbf{n}(\mathbf{x}))\,\mathrm{d}S(\mathbf{y},\mathbf{x})\\
&=-\int_\Gamma\int_\Gamma\boldsymbol{\mu}(\mathbf{y})(\mathbf{grad}_{\mathbf{x}}\,E_\kappa(\mathbf{x}-\mathbf{y})\cdot\mathbf{n}(\mathbf{x}))\cdot(\boldsymbol{\xi}(\mathbf{x})\times\mathbf{n}(\mathbf{x}))\,\mathrm{d}S(\mathbf{y},\mathbf{x})+\\
&\quad+\int_\Gamma\int_\Gamma\mathbf{grad}_{\mathbf{x}}\,E_\kappa(\mathbf{x}-\mathbf{y})\,(\boldsymbol{\mu}(\mathbf{y})\cdot(\mathbf{n}(\mathbf{x})-\mathbf{n}(\mathbf{y})))\cdot(\boldsymbol{\xi}(\mathbf{x})\times\mathbf{n}(\mathbf{x}))\,\mathrm{d}S(\mathbf{y},\mathbf{x})\,.
\end{aligned}$$

According to [25, Sect. 6.4] we have $|\mathbf{n}(\mathbf{x})-\mathbf{n}(\mathbf{y})|=O(|\mathbf{x}-\mathbf{y}|)$ for smooth surfaces. Thus, a closer scrutiny of the formulae shows that

$$\mathbf{grad}_{\mathbf{x}}\,E_\kappa(\mathbf{x}-\mathbf{y})\cdot\mathbf{n}(\mathbf{x})\simeq\mathbf{grad}_{\mathbf{x}}\,E_\kappa(\mathbf{x}-\mathbf{y})(\mathbf{n}(\mathbf{x})-\mathbf{n}(\mathbf{y}))^T\simeq O(|\mathbf{x}-\mathbf{y}|^{-1})\,,$$

for $\mathbf{x}\to\mathbf{y}$. Both kernels are weakly singular, as is the kernel of $\mathbf{S}_\kappa$, so the theory of pseudo-differential operators [25, Chap. 4.4] shows that $\mathbf{C}_\kappa$ is continuous as an operator from $\boldsymbol{H}^s_{\mathbf{t}}(\Gamma)\mapsto\boldsymbol{H}^{s+1}_{\mathbf{t}}(\Gamma)$, $s\in\mathbb{R}$ (Note that on a smooth boundary the infinite scale of Sobolev spaces is available).

Next, pick a smooth tangential vector-field $\boldsymbol{\mu}$, use Lemma 5 and apply simple manipulations based on vector identities

$$\operatorname{div}_\Gamma \mathbf{C}_\kappa(\boldsymbol{\mu})(\mathbf{x}) = \frac{1}{\kappa}\,\mathbf{curl}\,\mathbf{curl}\int_\Gamma E_\kappa(\mathbf{x}-\mathbf{y})\boldsymbol{\mu}(\mathbf{y})\,\mathrm{d}S\cdot\mathbf{n}(\mathbf{x})$$
$$= \frac{1}{\kappa}\int_\Gamma \frac{\partial}{\partial\mathbf{n}(\mathbf{x})}E_\kappa(\mathbf{x}-\mathbf{y})\operatorname{div}_\Gamma\boldsymbol{\mu}(\mathbf{y})\,\mathrm{d}S + \kappa\int_\Gamma E_\kappa(\mathbf{x}-\mathbf{y})\boldsymbol{\mu}(\mathbf{y})\,\mathrm{d}S\cdot\mathbf{n}(\mathbf{x})\,.$$

By density, we conclude that $\operatorname{div}_\Gamma\circ\mathbf{C}_\kappa : \boldsymbol{TH}^{s-\frac{1}{2}}(\operatorname{div}_\Gamma;\Gamma)\mapsto H^{s+\frac{1}{2}}(\Gamma)$ is continuous. This can be combined with the previous results and confirms the second assertion of the theorem. □

The crucial message sent by this lemma and Lemma 7 is that *on smooth boundaries* the operator $\mathbb{A}_\kappa : \boldsymbol{H}_\times^{-\frac{1}{2}}(\operatorname{div}_\Gamma,\Gamma)^2 \mapsto \boldsymbol{H}_\times^{-\frac{1}{2}}(\operatorname{div}_\Gamma,\Gamma)^2$ can be converted into

$$\mathbb{A}_\kappa \simeq \begin{pmatrix} 0 & \mathbf{S}_0 \\ \mathbf{S}_0 & 0 \end{pmatrix}\,.$$

by dropping "compact perturbations"[5]. In other words, Dirichlet and Neumann traces are coupled by compact terms only. On smooth boundaries we merely have to examine $\mathbf{S}_\kappa$, if we are interested in coercivity.

Unfortunately, the coupling terms $\mathbf{C}_\kappa$ in $\mathbb{A}_\kappa$ cannot be discarded in the case of non-smooth boundaries for want of a result like Lemma 11: in general, we have to deal with the two different traces $\gamma_\mathbf{t}\mathbf{e}$ and $\gamma_N\mathbf{e}$ together. This is a completely new aspect of boundary integral operators that we have not encountered in the case of the Maxwell source problem. Thus, splitting alone is not enough, but has to be accompanied by an appropriate grouping of the components. This is where the "physical meaning" of the splitting (21) that we discussed in Table 2 offers an important hint: it suggests that we distinguish between trace components of electric and magnetic nature. The ultimate justification for this idea is the profound result that it is merely compact terms, by which electric and magnetic components are coupled in the boundary integral operator $\mathbb{A}_\kappa$. The next lemma rigorously expresses this insight, *cf.* Prop. 3.13 in [21].

Lemma 12. *The bilinear form $\langle\mathbf{C}_\kappa\cdot,\cdot\rangle_{\tau,\Gamma}$ is compact both on $\boldsymbol{\mathcal{N}}(\Gamma)\times\boldsymbol{\mathcal{N}}(\Gamma)$ and $\boldsymbol{\mathcal{X}}(\Gamma)\times\boldsymbol{\mathcal{X}}(\Gamma)$.*

Proof. We restrict ourselves to the proof of the second assertion. We choose some $\boldsymbol{\zeta},\boldsymbol{\mu}\in\boldsymbol{\mathcal{X}}(\Gamma)$ and recall the definition of $\mathbf{C}_\kappa$ along with the jump relations. It is important to note that, by virtue of the definition of $\boldsymbol{\mathcal{X}}(\Gamma)$, $\boldsymbol{\mu}$ can be extended by $\mathbf{v} := \mathsf{J}(\operatorname{div}_\Gamma\boldsymbol{\mu})\in\boldsymbol{H}^1(\Omega_s)$, J defined in the proof of Lemma 2, such that $\gamma_\mathbf{t}\mathbf{v}=\boldsymbol{\mu}$ and $\|\mathbf{v}\|_{\boldsymbol{H}^1(\Omega_s)}\leq C\,\|\boldsymbol{\mu}\|_{\boldsymbol{H}_\times^{-\frac{1}{2}}(\operatorname{div}_\Gamma,\Gamma)}$. Also, exploiting $\operatorname{div}\mathbf{v}=0$, we get

$$\langle\mathbf{C}_\kappa\boldsymbol{\zeta},\boldsymbol{\mu}\rangle_{\tau,\Gamma} = \langle\gamma_N^-\boldsymbol{\Psi}_\mathbf{V}^\kappa(\boldsymbol{\zeta}),\boldsymbol{\mu}\rangle_{\tau,\Gamma} - \tfrac{1}{2}\langle\boldsymbol{\zeta},\boldsymbol{\mu}\rangle_{\tau,\Gamma}\,.$$

[5] Here and below we use the symbol $\simeq$ to express equality of operators and bilinear forms up to addition of compact terms

Using the identity $\mathbf{curl}\,\mathbf{curl}\,\boldsymbol{\Psi}_{\mathbf{V}}^{\kappa}(\boldsymbol{\zeta}) = \mathbf{grad}\,\Psi_V^{\kappa}(\mathrm{div}_\Gamma\boldsymbol{\zeta}) + \kappa^2\boldsymbol{\Psi}_{\mathbf{V}}^{\kappa}(\boldsymbol{\zeta})$ and the integration by parts (5), we obtain:

$$\langle\gamma_N^-\boldsymbol{\Psi}_{\mathbf{V}}^{\kappa}(\boldsymbol{\zeta}),\boldsymbol{\mu}\rangle_{\boldsymbol{\tau},\Gamma} = -\int_{\Omega_s} \mathbf{curl}\,\boldsymbol{\Psi}_{\mathbf{V}}^{\kappa}\cdot\mathbf{curl}\,\mathbf{v} + \kappa^2\boldsymbol{\Psi}_{\mathbf{V}}^{\kappa}(\boldsymbol{\zeta})\cdot\mathbf{v}\,d\mathbf{x} + \\ + \langle\gamma^-\Psi_V^{\kappa}(\mathrm{div}_\Gamma\mu),\gamma_{\mathbf{n}}^-\mathbf{v}\rangle_{\frac{1}{2},\Gamma}\,.$$

This means that

$$\begin{aligned}|\,(\mathbf{C}_\kappa(\boldsymbol{\zeta}),\boldsymbol{\mu})_{\boldsymbol{\tau}}\,| \leq\ & |\boldsymbol{\Psi}_{\mathbf{V}}^{\kappa}(\boldsymbol{\zeta})|_{\boldsymbol{H}^1(\Omega_s)}\,\|\mathbf{curl}\,\mathbf{v}\|_{\boldsymbol{L}^2(\Omega_s)} + \kappa^2\,\|\boldsymbol{\Psi}_{\mathbf{V}}^{\kappa}(\boldsymbol{\zeta})\|_{\boldsymbol{L}^2(\Omega_s)}\,\|\mathbf{v}\|_{\boldsymbol{L}^2(\Omega_s)} \\ & + \|V_\kappa(\mathrm{div}_\Gamma\boldsymbol{\zeta})\|_{L^2(\Gamma)}\,\|\gamma_{\mathbf{n}}^-\mathbf{v}\|_{L^2(\Gamma)} + \|\boldsymbol{\zeta}\|_{\boldsymbol{L}_{\mathbf{t}}^2(\Gamma)}\|\gamma_{\mathbf{t}}^-\mathbf{v}\|_{\boldsymbol{L}_{\mathbf{t}}^2(\Gamma)} \\ \leq\ & C(\|\boldsymbol{\zeta}\|_{\boldsymbol{H}_\times^{-\frac{1}{2}}(\Gamma)} + \|V_\kappa(\mathrm{div}_\Gamma\boldsymbol{\zeta})\|_{L^2(\Gamma)} + \|\boldsymbol{\zeta}\|_{\boldsymbol{L}_{\mathbf{t}}^2(\Gamma)})\,\|\mathbf{v}\|_{\boldsymbol{H}^1(\Omega_s)}\,,\end{aligned}$$

with some $C = C(\Omega_s) > 0$. It goes without saying that the operator $V_\kappa : H^{-\frac{1}{2}}(\Gamma) \mapsto L^2(\Gamma)$ is compact. Then, the compact embedding of $\boldsymbol{\mathcal{X}}(\Gamma)$ in $\boldsymbol{L}_{\mathbf{t}}^2(\Gamma)$ according to Corollary 1 finishes the proof. □

To understand the meaning of these results, we consider the combined boundary integral operator $\mathbb{A}_\kappa : \boldsymbol{H}_\times^{-\frac{1}{2}}(\mathrm{div}_\Gamma,\Gamma)^2 \mapsto \boldsymbol{H}_\times^{-\frac{1}{2}}(\mathrm{div}_\Gamma,\Gamma)^2$ with respect to the splitting (21). As usual, we adopt a variational perspective and study the bilinear form associated with $\mathbb{A}_\kappa$. It will be based on the following anti-symmetric pairing on the product space $\boldsymbol{H}_\times^{-\frac{1}{2}}(\mathrm{div}_\Gamma,\Gamma)\times\boldsymbol{H}_\times^{-\frac{1}{2}}(\mathrm{div}_\Gamma,\Gamma)$,

$$\left\langle\begin{pmatrix}\boldsymbol{\zeta}\\ \boldsymbol{\mu}\end{pmatrix},\begin{pmatrix}\boldsymbol{\xi}\\ \boldsymbol{\lambda}\end{pmatrix}\right\rangle_{\boldsymbol{\tau}\times\boldsymbol{\tau}} := \langle\boldsymbol{\zeta},\boldsymbol{\lambda}\rangle_{\boldsymbol{\tau},\Gamma} + \langle\boldsymbol{\mu},\boldsymbol{\xi}\rangle_{\boldsymbol{\tau},\Gamma}\,.$$

Now, pick $\boldsymbol{\zeta},\boldsymbol{\mu},\boldsymbol{\xi},\boldsymbol{\lambda}\in\boldsymbol{H}_\times^{-\frac{1}{2}}(\mathrm{div}_\Gamma,\Gamma)$ and use superscripts $\perp$ and 0 to tag their components in $\boldsymbol{\mathcal{X}}(\Gamma)$ and $\boldsymbol{\mathcal{N}}(\Gamma)$, respectively.

$$\begin{aligned}\left\langle\mathbb{A}_\kappa\begin{pmatrix}\boldsymbol{\zeta}\\ \boldsymbol{\mu}\end{pmatrix},\begin{pmatrix}\boldsymbol{\xi}\\ \boldsymbol{\lambda}\end{pmatrix}\right\rangle_{\boldsymbol{\tau}\times\boldsymbol{\tau}} =& \left\langle\mathbb{A}_\kappa\begin{pmatrix}\boldsymbol{\zeta}^\perp\\ \boldsymbol{\mu}^0\end{pmatrix},\begin{pmatrix}\boldsymbol{\xi}^\perp\\ \boldsymbol{\lambda}^0\end{pmatrix}\right\rangle_{\boldsymbol{\tau}\times\boldsymbol{\tau}} + \left\langle\mathbb{A}_\kappa\begin{pmatrix}\boldsymbol{\zeta}^0\\ \boldsymbol{\mu}^\perp\end{pmatrix},\begin{pmatrix}\boldsymbol{\xi}^0\\ \boldsymbol{\lambda}^\perp\end{pmatrix}\right\rangle_{\boldsymbol{\tau}\times\boldsymbol{\tau}} + \\ &+ \left\langle\mathbb{A}_\kappa\begin{pmatrix}\boldsymbol{\zeta}^\perp\\ \boldsymbol{\mu}^0\end{pmatrix},\begin{pmatrix}\boldsymbol{\xi}^0\\ \boldsymbol{\lambda}^\perp\end{pmatrix}\right\rangle_{\boldsymbol{\tau}\times\boldsymbol{\tau}} + \left\langle\mathbb{A}_\kappa\begin{pmatrix}\boldsymbol{\zeta}^0\\ \boldsymbol{\mu}^\perp\end{pmatrix},\begin{pmatrix}\boldsymbol{\xi}^\perp\\ \boldsymbol{\lambda}^0\end{pmatrix}\right\rangle_{\boldsymbol{\tau}\times\boldsymbol{\tau}}.\end{aligned} \tag{39}$$

Let us take a look at the bilinear forms in the second line:

$$\left\langle\mathbb{A}_\kappa\begin{pmatrix}\boldsymbol{\zeta}^\perp\\ \boldsymbol{\mu}^0\end{pmatrix},\begin{pmatrix}\boldsymbol{\xi}^0\\ \boldsymbol{\lambda}^\perp\end{pmatrix}\right\rangle_{\boldsymbol{\tau}\times\boldsymbol{\tau}} = \begin{cases}\left\langle\mathbf{C}_\kappa\boldsymbol{\zeta}^\perp,\lambda^\perp\right\rangle_{\boldsymbol{\tau},\Gamma} + \kappa\left\langle\mathbf{V}_\kappa\boldsymbol{\mu}^0,\boldsymbol{\lambda}^\perp\right\rangle_{\boldsymbol{\tau},\Gamma} + \\ +\kappa\left\langle\mathbf{V}_\kappa\boldsymbol{\zeta}^\perp,\boldsymbol{\xi}^0\right\rangle_{\boldsymbol{\tau},\Gamma} + \langle\mathbf{C}_\kappa\boldsymbol{\mu}^0,\boldsymbol{\xi}^0\rangle_{\boldsymbol{\tau},\Gamma}\,.\end{cases}$$

Lemmas 9 and 12 show that this is a compact bilinear form! The same applies to the other term in the second line of (39). Harking back to the discussion in Sect. 3, we emphasise that both Dirichlet and Neumann traces involve electric and magnetic components, which are isolated by the splitting:

$$\text{Electric components: } \boldsymbol{\zeta}^0, \boldsymbol{\mu}^\perp \quad \longleftrightarrow \quad \text{Magnetic components: } \boldsymbol{\zeta}^\perp, \boldsymbol{\mu}^0 \,.$$

The bottom line is that up to compact terms electric and magnetic components of the traces are decoupled in $\mathbb{A}_\kappa$. It has turned out that the decoupling observed in the case of smooth boundaries does not reflect the "physics of the fields".

Using the appropriate splitting and decoupling, we can proceed as in the case of $\mathbf{S}_0$: we introduce the isomorphism $\mathbb{X}_\Gamma : \boldsymbol{H}_\times^{-\frac{1}{2}}(\mathrm{div}_\Gamma, \Gamma)^2 \mapsto \boldsymbol{H}_\times^{-\frac{1}{2}}(\mathrm{div}_\Gamma, \Gamma)^2$ by

$$\mathbb{X}_\Gamma \begin{pmatrix} \boldsymbol{\zeta} \\ \boldsymbol{\mu} \end{pmatrix} := \begin{pmatrix} \mathsf{X}_\Gamma \boldsymbol{\zeta} \\ \mathsf{X}_\Gamma \boldsymbol{\mu} \end{pmatrix}, \quad \boldsymbol{\zeta}, \boldsymbol{\mu} \in \boldsymbol{H}_\times^{-\frac{1}{2}}(\mathrm{div}_\Gamma, \Gamma) \,. \tag{40}$$

Then we get the following generalisation of Lemma 10.

Theorem 9 (Generalised Gårding inequality for $\mathbb{A}_\kappa$). *There is a constant $C_G > 0$ and a compact bilinear form $\mathbf{c}_\Gamma$ on $\boldsymbol{H}_\times^{-\frac{1}{2}}(\mathrm{div}_\Gamma, \Gamma) \times \boldsymbol{H}_\times^{-\frac{1}{2}}(\mathrm{div}_\Gamma, \Gamma)$ such that*

$$\left| \left\langle \mathbb{A}_\kappa \begin{pmatrix} \boldsymbol{\zeta} \\ \boldsymbol{\mu} \end{pmatrix}, \mathbb{X}_\Gamma \begin{pmatrix} \overline{\boldsymbol{\zeta}} \\ \overline{\boldsymbol{\mu}} \end{pmatrix} \right\rangle_{\mathcal{T}\times\mathcal{T}} - \mathbf{c}_\Gamma\left(\begin{pmatrix} \boldsymbol{\zeta} \\ \boldsymbol{\mu} \end{pmatrix}, \begin{pmatrix} \overline{\boldsymbol{\zeta}} \\ \overline{\boldsymbol{\mu}} \end{pmatrix} \right) \right| \geq C_G \left\| \begin{pmatrix} \boldsymbol{\zeta} \\ \boldsymbol{\mu} \end{pmatrix} \right\|^2_{\boldsymbol{H}_\times^{-\frac{1}{2}}(\mathrm{div}_\Gamma, \Gamma)}$$

for all $\boldsymbol{\zeta}, \boldsymbol{\mu} \in \boldsymbol{H}_\times^{-\frac{1}{2}}(\mathrm{div}_\Gamma, \Gamma)$.

Proof. We have already found that up to compact perturbations

$$\left\langle \mathbb{A}_\kappa \begin{pmatrix} \boldsymbol{\zeta} \\ \boldsymbol{\mu} \end{pmatrix}, \begin{pmatrix} \boldsymbol{\xi} \\ \boldsymbol{\lambda} \end{pmatrix} \right\rangle_{\mathcal{T},\Gamma} \simeq \left\langle \mathbb{A}_\kappa \begin{pmatrix} \boldsymbol{\zeta}^\perp \\ \boldsymbol{\mu}^0 \end{pmatrix}, \begin{pmatrix} \boldsymbol{\xi}^\perp \\ \boldsymbol{\lambda}^0 \end{pmatrix} \right\rangle_{\mathcal{T},\Gamma} + \left\langle \mathbb{A}_\kappa \begin{pmatrix} \boldsymbol{\zeta}^0 \\ \boldsymbol{\mu}^\perp \end{pmatrix}, \begin{pmatrix} \boldsymbol{\xi}^0 \\ \boldsymbol{\lambda}^\perp \end{pmatrix} \right\rangle_{\mathcal{T},\Gamma} .$$

What comes next amounts to reusing arguments from the proof of Lemma 10. We inspect the first summand and find, using Lemmas 6 and 9,

$$\begin{aligned} \left\langle \mathbb{A}_\kappa \begin{pmatrix} \boldsymbol{\zeta}^\perp \\ \boldsymbol{\mu}^0 \end{pmatrix}, \begin{pmatrix} \overline{\boldsymbol{\zeta}}^\perp \\ -\overline{\boldsymbol{\mu}}^0 \end{pmatrix} \right\rangle_{\mathcal{T}\times\mathcal{T}} &= -\left\langle \mathbf{C}_\kappa \boldsymbol{\zeta}^\perp, \overline{\boldsymbol{\mu}}^0 \right\rangle_{\mathcal{T},\Gamma} - \kappa \left\langle \mathbf{V}_\kappa \boldsymbol{\mu}^0, \overline{\boldsymbol{\mu}^0} \right\rangle_{\mathcal{T},\Gamma} + \\ &\quad + \left\langle \mathbf{S}_\kappa \boldsymbol{\zeta}^\perp, \overline{\boldsymbol{\zeta}}^\perp \right\rangle_{\mathcal{T},\Gamma} + \left\langle \mathbf{C}_\kappa \boldsymbol{\mu}^0, \overline{\boldsymbol{\zeta}}^\perp \right\rangle_{\mathcal{T},\Gamma} \\ &\simeq -2i \,\mathrm{Im} \left\{ \left\langle \mathbf{C}_\kappa \boldsymbol{\zeta}^\perp, \overline{\boldsymbol{\mu}}^0 \right\rangle_{\mathcal{T},\Gamma} \right\} - \kappa \left\langle \mathbf{V}_0 \boldsymbol{\mu}^0, \overline{\boldsymbol{\mu}^0} \right\rangle_{\mathcal{T},\Gamma} + \frac{1}{\kappa} \left\langle V_0(\mathrm{div}_\Gamma \boldsymbol{\zeta}, \mathrm{div}_\Gamma \overline{\boldsymbol{\zeta}} \right\rangle_{\mathcal{T},\Gamma} . \end{aligned}$$

Appealing to Lemmas 2 and 8, we conclude that

$$\left| \left\langle \mathbb{A}_\kappa \begin{pmatrix} \boldsymbol{\zeta}^\perp \\ \boldsymbol{\mu}^0 \end{pmatrix}, \begin{pmatrix} \overline{\boldsymbol{\zeta}}^\perp \\ -\overline{\boldsymbol{\mu}}^0 \end{pmatrix} \right\rangle_{\mathcal{T}\times\mathcal{T}} + \text{ comp.} \right| \geq C \left(\left\| \boldsymbol{\mu}^0 \right\|^2_{\boldsymbol{H}_\times^{-\frac{1}{2}}(\Gamma)} + \left\| \boldsymbol{\zeta}^\perp \right\|^2_{\boldsymbol{H}_\times^{-\frac{1}{2}}(\mathrm{div}_\Gamma, \Gamma)} \right) .$$

The same manipulations can be carried out for the second summand. Together with the stability of (21) this gives the assertion. □

7 Boundary Integral Equations

Boundary integral equations (BIE) can be obtained in two ways, either by the *direct method* or the *indirect method*. The distinct feature of the direct method is that traces of the solution of the transmission problem/boundary value problem occur as unknowns in the formulation. Its integral equations immediately arise from the Calderon projectors $\mathbb{P}_\kappa^-$ and $\mathbb{P}_\kappa^+$ via Theorem 8. Conversely, the unknowns of the indirect methods are jumps of traces across Γ. It can be motivated by the fact that the potentials $\boldsymbol{\Psi}_{SL}^\kappa$ and $\boldsymbol{\Psi}_{DL}^\kappa$ already provide solutions to the homogeneous equations, *cf.* (29). An excellent presentation of the main ideas of indirect methods is given in [36, Vol. IV, Chap. XI].

7.1 The Direct Method

We start with the discussion of direct methods for scattering at a perfect conductor, that is, the exterior Dirichlet problem for the homogeneous electric wave equation

$$\mathbf{curl}\,\mathbf{curl}\,\mathbf{e} - \kappa^2\mathbf{e} = 0 \quad \text{in } \Omega' \quad , \quad \gamma_{\mathbf{t}}^+\mathbf{e} = \gamma_{\mathbf{t}}^+\mathbf{e}_i \; , \tag{41}$$

plus Silver–Müller radiation conditions. We know that we can always find a unique solution of (41) [31, Thm. 6.10]. However, it is a bewildering feature of many boundary integral equations connected with (41) that they fail to have unique solutions, if κ coincides with "forbidden wave numbers" [26, 38]. Those are related to interior eigenvalues of the operator of (41).

Definition 4. $\lambda \in \mathbb{R}$ *is called an* interior electric/magnetic Maxwell eigenvalue, *if there is a non-zero* $\mathbf{e} \in \boldsymbol{H}_0(\mathbf{curl};\Omega_s)$ *or* $\mathbf{e} \in \boldsymbol{H}(\mathbf{curl};\Omega_s)$, *respectively, such that*

$$(\mathbf{curl}\,\mathbf{e}, \mathbf{curl}\,\mathbf{v})_{0;\Omega_s} = \lambda\,(\mathbf{e},\mathbf{v})_{0;\Omega_s} \quad \forall \mathbf{v} \in \boldsymbol{H}_0(\mathbf{curl};\Omega_s) \text{ or } \mathbf{v} \in \boldsymbol{H}(\mathbf{curl};\Omega_s) \, .$$

Note that these eigenvalues form a discrete sequence accumulating at ∞.

The first direct method relies on Theorem 8, which tells us that $(\gamma_{\mathbf{t}}^+\mathbf{e}_i, \boldsymbol{\lambda})$ are exterior Cauchy data according to Definition 3, if $\mathbb{P}_\kappa^-(\gamma_{\mathbf{t}}^+\mathbf{e}_i,\boldsymbol{\lambda}) = 0$. From the first row of this equation we obtain the integral equation of the first kind $\mathbf{S}_\kappa\boldsymbol{\lambda} = -(\frac{1}{2}\,\mathrm{Id} + \mathbf{C}_\kappa)(\gamma_{\mathbf{t}}^+\mathbf{e}_i)$ for the unknown Neumann data $\boldsymbol{\lambda} := \gamma_N^+\mathbf{e}$ of a solution $\mathbf{e}$ of (41). In weak form it reads: seek $\boldsymbol{\lambda} \in \boldsymbol{H}_\times^{-\frac{1}{2}}(\mathrm{div}_\Gamma,\Gamma)$ such that

$$\langle \mathbf{S}_\kappa\boldsymbol{\lambda}, \boldsymbol{\mu}\rangle_{\boldsymbol{\tau},\Gamma} = -\left\langle (\tfrac{1}{2}\,\mathrm{Id} + \mathbf{C}_\kappa)(\gamma_{\mathbf{t}}^+\mathbf{e}_i), \boldsymbol{\mu}\right\rangle_{\boldsymbol{\tau},\Gamma} \quad \forall \boldsymbol{\mu} \in \boldsymbol{H}_\times^{-\frac{1}{2}}(\mathrm{div}_\Gamma,\Gamma) \, . \tag{42}$$

Conversely, if $(\gamma_{\mathbf{t}}^+\mathbf{e}_i, \boldsymbol{\lambda})$ satisfies (42), we find

$$\mathbb{P}_\kappa^- \begin{pmatrix} \gamma_{\mathbf{t}}^+\mathbf{e}_i \\ \boldsymbol{\lambda} \end{pmatrix} = \begin{pmatrix} 0 \\ \boldsymbol{\xi} \end{pmatrix} \quad \text{for some } \boldsymbol{\xi} \in \boldsymbol{H}_\times^{-\frac{1}{2}}(\mathrm{div}_\Gamma,\Gamma) \, .$$

Hence, by Theorem 8, $\boldsymbol{\xi}$ is the Neumann trace of an electric eigenmode of Ω_s. If κ^2 does not coincide with an interior electric eigenvalue, this eigenmode can only be trivial, which means $\boldsymbol{\xi} = 0$. The next lemma summarises our findings.

Lemma 13. *If κ^2 is not an interior electric eigenvalue, then $\boldsymbol{\lambda} \in \boldsymbol{H}_{\times}^{-\frac{1}{2}}(\mathrm{div}_\Gamma, \Gamma)$ is a solution of* (42) *if and only if $(\gamma_{\mathbf{t}}^{+}\mathbf{e}_i, \boldsymbol{\lambda})$ are Cauchy data for* (41).

Remark 1. If κ^2 is an interior electric eigenvalue, then $\boldsymbol{\lambda}$ is unique up to Neumann traces $\boldsymbol{\xi}$ of the corresponding eigenmodes. Thanks to the representation formula (26), we find that $\boldsymbol{\Psi}_{SL}^{\kappa}(\boldsymbol{\xi})$ vanishes in Ω'. In other words, the representation

$$\mathbf{e} = -\boldsymbol{\Psi}_{DL}^{\kappa}(\gamma_{\mathbf{t}}^{+}\mathbf{e}_i) - \boldsymbol{\Psi}_{SL}^{\kappa}(\boldsymbol{\lambda}) \tag{43}$$

will produce the unique field solution in Ω'.

Now, a standard Fredholm alternative argument can be applied:

Theorem 10. *If κ satisfies the assumptions of Lemma 13, then there exists a unique solution of* (42) *for any $\mathbf{e}_i$.*

Using the second row of $\mathbb{P}_{\kappa}^{-}$ we obtain the B.I.E. $(\frac{1}{2}\,\mathrm{Id} + \mathbf{C}_\kappa)\boldsymbol{\lambda} = -\mathbf{S}_\kappa(\gamma_{\mathbf{t}}^{+}\mathbf{e}_i)$, whose associated variational problem can be stated as: seek $\boldsymbol{\lambda} \in \boldsymbol{H}_{\times}^{-\frac{1}{2}}(\mathrm{div}_\Gamma, \Gamma)$ such that

$$\left\langle (\tfrac{1}{2}\,\mathrm{Id} + \mathbf{C}_\kappa)\boldsymbol{\lambda}, \boldsymbol{\mu}\right\rangle_{\tau,\Gamma} = -\left\langle \mathbf{S}_\kappa(\gamma_{\mathbf{t}}^{+}\mathbf{e}_i), \boldsymbol{\mu}\right\rangle_{\tau,\Gamma} \quad \forall \boldsymbol{\mu} \in \boldsymbol{H}_{\times}^{-\frac{1}{2}}(\mathrm{div}_\Gamma, \Gamma)\,. \tag{44}$$

In contrast to (42), in order to show unique solvability of (44) we not only need to avoid "forbidden wave numbers", we have to assume smooth boundaries too.

Theorem 11. *If κ^2 is not an interior magnetic eigenvalue and if Γ is C^∞-smooth, then* (44) *has a unique solution $\boldsymbol{\lambda} \in \boldsymbol{TH}^{-\frac{1}{2}}(\mathrm{div}_\Gamma; \Gamma)$ and the pair $(\gamma_{\mathbf{t}}^{+}\mathbf{e}^i, \boldsymbol{\lambda})$ supplies Cauchy data for the electric wave equation in Ω'.*

Proof. As in the justification of Lemma 13 it turns out that

$$\mathbb{P}_{\kappa}^{-}\begin{pmatrix}\gamma_{\mathbf{t}}^{+}\mathbf{e}_i\\ \boldsymbol{\lambda}\end{pmatrix} = \begin{pmatrix}\boldsymbol{\xi}\\ 0\end{pmatrix} \quad \text{for some } \boldsymbol{\xi} \in \boldsymbol{TH}^{-\frac{1}{2}}(\mathrm{div}_\Gamma; \Gamma)\,,$$

if $\boldsymbol{\lambda}$ satisfies (44). By the assumption on κ we have $\boldsymbol{\xi} = 0$ and, by Theorem 8, $(\gamma_{\mathbf{t}}^{+}\mathbf{e}_i, \boldsymbol{\lambda})$ is identified as valid Maxwell Cauchy data for the exterior problem. Recalling the uniqueness result for (41), this means that solutions of (44) are unique. Next, use Lemma 11, which asserts the compactness of $\mathbf{C}_\kappa : \boldsymbol{TH}^{-\frac{1}{2}}(\mathrm{div}_\Gamma; \Gamma) \to \boldsymbol{TH}^{-\frac{1}{2}}(\mathrm{div}_\Gamma; \Gamma)$. This confirms that the operator in (44) is Fredholm of index zero. □

If $\Gamma \in C^\infty$ and $\gamma_{\mathbf{t}}^{+}\mathbf{e}_i \in \boldsymbol{H}_{\mathbf{t}}^{1}(\Gamma)$ (which, e.g., is fulfilled for exciting plane waves), the lifting properties of the operators $\mathbf{C}_\kappa$ according to Lemma 11 and the fact that $\mathrm{div}_\Gamma \circ \mathbf{S}_\kappa = \kappa\,\mathrm{div}_\Gamma \circ \mathbf{V}_\kappa$ bear out that the solution of (44) will belong to $\boldsymbol{TH}(\mathrm{div}_\Gamma, \Gamma)$. Hence, a *completely equivalent* variational formulation in $\boldsymbol{TH}(\mathrm{div}_\Gamma, \Gamma)$ is possible: find $\boldsymbol{\lambda} \in \boldsymbol{TH}(\mathrm{div}_\Gamma, \Gamma)$ such that $\forall \boldsymbol{\mu} \in \boldsymbol{TH}(\mathrm{div}_\Gamma, \Gamma)$

$$\left((\tfrac{1}{2}\,\mathrm{Id} + \mathbf{C}_\kappa)\boldsymbol{\lambda}, \boldsymbol{\mu}\right)_{\boldsymbol{TH}(\mathrm{div}_\Gamma,\Gamma)} = -\left(\mathbf{S}_\kappa(\gamma_{\mathbf{t}}^{+}\mathbf{e}_i), \boldsymbol{\mu}\right)_{\boldsymbol{TH}(\mathrm{div}_\Gamma,\Gamma)}\,. \tag{45}$$

Given a sufficiently smooth $\gamma_{\mathbf{t}}^{+}\mathbf{e}_i$, the right hand side is a continuous functional on $\boldsymbol{TH}(\mathrm{div}_\Gamma, \Gamma)$. In addition, Lemma 11 shows that $\mathbf{C}_\kappa : \boldsymbol{TH}(\mathrm{div}_\Gamma, \Gamma) \mapsto \boldsymbol{TH}^1(\mathrm{div}_\Gamma, \Gamma)$ and, hence, the sesqui-linear form in (45) is $\boldsymbol{TH}(\mathrm{div}_\Gamma, \Gamma)$-coercive. Thus, Theorem 11 will remain valid for (45). The real rational behind the lifting of (44) into $\boldsymbol{TH}(\mathrm{div}_\Gamma, \Gamma)$ will be elaborated in Sect. 9.

Next, we tackle scattering at an isotropic, homogeneous dielectric object occupying Ω_s. Inside Ω_s material parameters $\epsilon^- > 0$ and $\mu^- > 0$ prevail, leading to a wave number $\kappa^- := \omega\sqrt{\epsilon^-\mu^-}$. Outside we face ϵ_0, μ_0 and wave number κ^+. These wave numbers underlie the definition of γ_N^- and γ_N^+. The transmission conditions from (1) become

$$\gamma_{\mathbf{t}}^{-}\mathbf{e} = \gamma_{\mathbf{t}}^{+}\mathbf{e} \quad , \quad \tfrac{\kappa^-}{\mu^-}\gamma_N^-\mathbf{e} = \tfrac{\kappa^+}{\mu_0}\gamma_N^+\mathbf{e} \,.$$

Taking our cue from the approach to acoustic scattering in [61], we introduce scaled boundary integral operators

$$\widehat{\mathbb{A}}_{\kappa^-} = \begin{pmatrix} \mathrm{Id} & 0 \\ 0 & \frac{\kappa^-}{\mu^-} \end{pmatrix} \mathbb{A}_{\kappa^-} \begin{pmatrix} \mathrm{Id} & 0 \\ 0 & \frac{\mu^-}{\kappa^-} \end{pmatrix} \quad , \quad \widehat{\mathbb{A}}_{\kappa^+} = \begin{pmatrix} \mathrm{Id} & 0 \\ 0 & \frac{\kappa^+}{\mu_0} \end{pmatrix} \mathbb{A}_{\kappa^+} \begin{pmatrix} \mathrm{Id} & 0 \\ 0 & \frac{\mu_0}{\kappa^+} \end{pmatrix} .$$

The following scaled traces match the scaled operators

$$(\boldsymbol{\zeta}^+, \boldsymbol{\lambda}^+) = (\gamma_{\mathbf{t}}^{+}\mathbf{e}, \tfrac{\kappa^+}{\mu_0}\gamma_N^+\mathbf{e}) \quad , \quad (\boldsymbol{\zeta}^-, \boldsymbol{\lambda}^-) = (\gamma_{\mathbf{t}}^{-}\mathbf{e}, \tfrac{\kappa^-}{\mu^-}\gamma_N^-\mathbf{e}) \,.$$

For them the transmission condition takes the simple form

$$\begin{pmatrix} \boldsymbol{\zeta}^- \\ \boldsymbol{\lambda}^- \end{pmatrix} - \begin{pmatrix} \boldsymbol{\zeta}^+ \\ \boldsymbol{\lambda}^+ \end{pmatrix} = \begin{pmatrix} \gamma_{\mathbf{t}}^{+}\mathbf{e}_i \\ \gamma_{\mathbf{t}}^{+}\mathbf{h}_i \end{pmatrix} . \tag{46}$$

A scaled version of Theorem 8 bears out that $(\boldsymbol{\zeta}^-, \boldsymbol{\lambda}^-)$ and $(\boldsymbol{\zeta}^+, \boldsymbol{\lambda}^+)$ are interior/exterior Cauchy data for the electric wave equation with wave numbers κ^- and κ^+, respectively, if and only if

$$(\tfrac{1}{2}\,\mathrm{Id} - \widehat{\mathbb{A}}_{\kappa^-})\begin{pmatrix} \boldsymbol{\zeta}^- \\ \boldsymbol{\lambda}^- \end{pmatrix} = 0 \quad , \quad (\tfrac{1}{2}\,\mathrm{Id} + \widehat{\mathbb{A}}_{\kappa^+})\begin{pmatrix} \boldsymbol{\zeta}^+ \\ \boldsymbol{\lambda}^+ \end{pmatrix} = 0 \,. \tag{47}$$

Using (46), this immediately implies that

$$\left(\widehat{\mathbb{A}}_{\kappa^-} + \widehat{\mathbb{A}}_{\kappa^+}\right)\begin{pmatrix} \boldsymbol{\zeta}^+ \\ \boldsymbol{\lambda}^+ \end{pmatrix} = (\tfrac{1}{2}\,\mathrm{Id} - \widehat{\mathbb{A}}_{\kappa^-})\begin{pmatrix} \gamma_{\mathbf{t}}^{+}\mathbf{e}_i \\ \gamma_{\mathbf{t}}^{+}\mathbf{h}_i \end{pmatrix} . \tag{48}$$

These are the boundary integral equations of the direct method for the transmission problem. Conversely, if $(\boldsymbol{\zeta}^+, \boldsymbol{\lambda}^+)$ is a solution of (48), set $\binom{\boldsymbol{\zeta}^-}{\boldsymbol{\lambda}^-} = \binom{\boldsymbol{\zeta}^+}{\boldsymbol{\lambda}^+} + \binom{\gamma_{\mathbf{t}}^{+}\mathbf{e}_i}{\gamma_{\mathbf{t}}^{+}\mathbf{h}_i}$, and consider

$$\begin{pmatrix} \widetilde{\boldsymbol{\zeta}}^- \\ \widetilde{\boldsymbol{\lambda}}^- \end{pmatrix} := (\tfrac{1}{2}\,\mathrm{Id} - \widehat{\mathbb{A}}_{\kappa^-})\begin{pmatrix} \boldsymbol{\zeta}^- \\ \boldsymbol{\lambda}^- \end{pmatrix} \quad , \quad \begin{pmatrix} \widetilde{\boldsymbol{\zeta}}^+ \\ \widetilde{\boldsymbol{\lambda}}^+ \end{pmatrix} := (\tfrac{1}{2}\,\mathrm{Id} + \widehat{\mathbb{A}}_{\kappa^+})\begin{pmatrix} \boldsymbol{\zeta}^+ \\ \boldsymbol{\lambda}^+ \end{pmatrix} .$$

Owing to Theorem 8, the pairs $(\widetilde{\boldsymbol{\zeta}}^-, \widetilde{\boldsymbol{\lambda}}^-)$ and $(\widetilde{\boldsymbol{\zeta}}^+, \widetilde{\boldsymbol{\lambda}}^+)$ are Maxwell Cauchy data for Ω' and Ω_s (and κ^-, κ^+), respectively. From equation (48) we infer that $(\widetilde{\boldsymbol{\zeta}}^-, \widetilde{\boldsymbol{\lambda}}^-) = (\widetilde{\boldsymbol{\zeta}}^+, \widetilde{\boldsymbol{\lambda}}^+)$. Thus, the interior and exterior Dirichlet and Neumann traces of the related Maxwell solutions agree. A combination of these Maxwell solutions solves the homogeneous electric wave equation (with κ^+ inside Ω_s and κ^- outside) in $\mathbb{R}^3$ and satisfies the Silver–Müller radiation conditions. Thanks to the uniqueness of solutions of the exterior Maxwell problem, it has to vanish. This implies $(\widetilde{\boldsymbol{\zeta}}^-, \widetilde{\boldsymbol{\lambda}}^-) = 0$ and $(\widetilde{\boldsymbol{\zeta}}^+, \widetilde{\boldsymbol{\lambda}}^+) = 0$, so that we recover (48). This confirms the following result.

Lemma 14. *Any solution* $(\boldsymbol{\zeta}^+, \boldsymbol{\lambda}^+)$ *of* (48) *provides (scaled) exterior Cauchy data for the transmission problem with excitation by an incident wave* $(\mathbf{e}_i, \mathbf{h}_i)$.

Using the pairing $\langle\cdot,\cdot\rangle_{\boldsymbol{\tau}\times\boldsymbol{\tau}}$, the variational formulation of (48) in $\boldsymbol{H}_\times^{-\frac{1}{2}}(\mathrm{div}_\Gamma, \Gamma) \times \boldsymbol{H}_\times^{-\frac{1}{2}}(\mathrm{div}_\Gamma, \Gamma)$ is straightforward. So is the next theorem that arises from Theorem 9, the previous Lemma, and a Fredholm argument.

Theorem 12. *The boundary integral equation* (48) *has a unique solution for any excitation.*

7.2 The Indirect Method

We will only discuss the exterior Dirichlet problem for the electric wave equation. Let $\mathbf{e}^+$ denote the unique solution of the exterior Dirichlet boundary value problem, satisfying $\gamma_\mathbf{t}^+\mathbf{e}^+ = \gamma_\mathbf{t}^+\mathbf{e}_i$ and the Silver–Müller radiation conditions at ∞. Write $\mathbf{e}^-$ for the solution of an interior Dirichlet problem for the electric wave equation, such that $\gamma_\mathbf{t}^-\mathbf{e}^- = \gamma_\mathbf{t}^+\mathbf{e}^+$. Again it is crucial to stay away from "forbidden wave numbers": let us *assume* that κ^2 does not coincide with an interior electric eigenvalue. Therefore, such an $\mathbf{e}^-$ exists and is unique. Call $\mathbf{e}$ the Maxwell solution in $\Omega_s \cup \Omega'$ that emerges by combining $\mathbf{e}^+$ and $\mathbf{e}^-$. As $[\gamma_\mathbf{t}]_\Gamma(\mathbf{e}) = 0$, the representation formula (26) becomes

$$\mathbf{e} = -\boldsymbol{\Psi}^\kappa_{SL}([\gamma_N]_\Gamma(\mathbf{e})) \quad \text{in } \boldsymbol{H}_{\mathrm{loc}}(\mathbf{curl}^2, \Omega_s \cup \Omega') .$$

Applying the exterior Dirichlet trace $\gamma_\mathbf{t}^+$ gives us the final integral equation in weak form: seek the unknown jump $\boldsymbol{\lambda} := [\gamma_N]_\Gamma(\mathbf{e}) \in \boldsymbol{H}_\times^{-\frac{1}{2}}(\mathrm{div}_\Gamma, \Gamma)$, which satisfies

$$\langle \mathbf{S}_\kappa \boldsymbol{\lambda}, \boldsymbol{\mu}\rangle_{\boldsymbol{\tau},\Gamma} = -\langle \gamma_\mathbf{t}^+\mathbf{e}_i, \boldsymbol{\mu}\rangle_{\boldsymbol{\tau},\Gamma} \quad \forall \boldsymbol{\mu} \in \boldsymbol{H}_\times^{-\frac{1}{2}}(\mathrm{div}_\Gamma, \Gamma) . \tag{49}$$

This integral equation is also known as *electric field integral equation* (EFIE) or *Rumsey's principle*. Theorem 10 applies, because (42) and (49) feature the same bilinear form.

Parallel to the case of direct methods for the exterior Dirichlet problem, we have a second option also in the case of the indirect approach. We *assume* that κ^2 is not an interior magnetic eigenvalue. Then, we may choose $\mathbf{e}^-$ as a Neumann extension of $\mathbf{e}^+$, that is, $\mathbf{e}^-$ is the solution of the interior Neumann problem for the electric wave equation with Neumann data $\gamma_N^-\mathbf{e}^- = \gamma_N^+\mathbf{e}^+$. Combining $\mathbf{e}^+$ and $\mathbf{e}^-$ to form $\mathbf{e}$, we conclude from (26) that $\mathbf{e} = -\boldsymbol{\Psi}_{DL}^\kappa([\gamma_\mathbf{t}]_\Gamma(\mathbf{e}))$ in $\boldsymbol{H}_{\mathrm{loc}}(\mathbf{curl}^2, \Omega_s \cup \Omega')$. Applying the exterior Dirichlet trace to this equation, we get the so-called *magnetic field integral equation* (MFIE), an integral equation of the second kind: find $\boldsymbol{\zeta} \in \boldsymbol{H}_\times^{-\frac{1}{2}}(\mathrm{div}_\Gamma, \Gamma)$ with

$$\left\langle (\tfrac{1}{2}\,\mathrm{Id} - \mathbf{C}_\kappa)\boldsymbol{\zeta}, \boldsymbol{\mu} \right\rangle_{\boldsymbol{\tau},\Gamma} = \left\langle \gamma_\mathbf{t}^+ \mathbf{e}_i, \boldsymbol{\mu} \right\rangle_{\boldsymbol{\tau},\Gamma} \quad \forall \boldsymbol{\mu} \in \boldsymbol{H}_\times^{-\frac{1}{2}}(\mathrm{div}_\Gamma, \Gamma) \,. \tag{50}$$

Its theoretical analysis on smooth surfaces is already covered by Theorem 11.

A serious drawback of the integral equations stated so far is their vulnerability to the presence of forbidden wave numbers, though the related boundary value problem always possesses a unique solution. Only one class of indirect BIE, the so-called *combined field integral equations* (CFIE), enjoys immunity. They owe their name to the fact that both $\boldsymbol{\Psi}_{DL}^\kappa$ and $\boldsymbol{\Psi}_{SL}^\kappa$ enter the trial expression for $\mathbf{e}$. A crucial prerequisite is a *compact* "smoothing operator" $\mathsf{M} : \boldsymbol{H}_\times^{-\frac{1}{2}}(\mathrm{div}_\Gamma, \Gamma) \mapsto \boldsymbol{H}_\times^{-\frac{1}{2}}(\mathrm{div}_\Gamma, \Gamma)$ that satisfies

$$\boldsymbol{\mu} \in \boldsymbol{H}_\times^{-\frac{1}{2}}(\mathrm{div}_\Gamma, \Gamma): \quad \langle \mathsf{M}\boldsymbol{\mu}, \overline{\boldsymbol{\mu}} \rangle_{\boldsymbol{\tau},\Gamma} > 0 \quad \Leftrightarrow \quad \boldsymbol{\mu} \neq 0 \,.$$

It is an important building block of the trial representation formula

$$\mathbf{e} = -i\eta \boldsymbol{\Psi}_{SL}^\kappa(\boldsymbol{\zeta}) - \boldsymbol{\Psi}_{DL}^\kappa(\mathsf{M}\boldsymbol{\zeta}) \,, \tag{51}$$

where $\boldsymbol{\zeta} \in \boldsymbol{H}_\times^{-\frac{1}{2}}(\mathrm{div}_\Gamma, \Gamma)$, $\eta > 0$. By (29), this field is a Maxwell solution in $\Omega_s \cup \Omega'$. The exterior Dirichlet trace applied to (51) results in the combined field integral equation: find $\boldsymbol{\zeta} \in \boldsymbol{H}_\times^{-\frac{1}{2}}(\mathrm{div}_\Gamma, \Gamma)$ such that $\forall \boldsymbol{\mu} \in \boldsymbol{H}_\times^{-\frac{1}{2}}(\mathrm{div}_\Gamma, \Gamma)$

$$-i \left\langle \eta \mathbf{S}_\kappa(\boldsymbol{\zeta}), \boldsymbol{\mu} \right\rangle_{\boldsymbol{\tau},\Gamma} + \left\langle (\tfrac{1}{2}\,\mathrm{Id} - \mathbf{C}_\kappa)(\mathsf{M}\boldsymbol{\zeta}), \boldsymbol{\mu} \right\rangle_{\boldsymbol{\tau},\Gamma} = \left\langle \gamma_\mathbf{t}^+ \mathbf{e}_i, \boldsymbol{\mu} \right\rangle_{\boldsymbol{\tau},\Gamma} \,. \tag{52}$$

The idea to use a regularising operator to state a combined field integral equation is due to Kress [48].

Theorem 13. *The boundary integral equation* (52) *has a unique solution* $\boldsymbol{\zeta} \in \boldsymbol{H}_\times^{-\frac{1}{2}}(\mathrm{div}_\Gamma, \Gamma)$ *for all* $\eta > 0, \kappa > 0$.

Proof. To demonstrate uniqueness, we assume that $\boldsymbol{\zeta} \in \boldsymbol{H}_\times^{-\frac{1}{2}}(\mathrm{div}_\Gamma, \Gamma)$ solves

$$-i\eta \mathbf{S}_\kappa(\boldsymbol{\zeta}) + (\tfrac{1}{2}\,\mathrm{Id} - \mathbf{C}_\kappa)(\mathsf{M}\boldsymbol{\zeta}) = 0 \,. \tag{53}$$

It is immediate from the jump relations that $\mathbf{e}$ given by (51) is an exterior Maxwell solution with $\gamma_\mathbf{t}^+\mathbf{e} = 0$. By uniqueness we infer that $\mathbf{e} = 0$ in Ω'. Appealing to the jump relations from Theorem 7 once more, we find

$$\gamma_\mathbf{t}^-\mathbf{e} = -\mathsf{M}\boldsymbol{\zeta} \quad , \quad \gamma_N^-\mathbf{e} = -i\eta\boldsymbol{\zeta} \,.$$

Next, we use (11) and see that

$$i\eta \left\langle \zeta, \overline{\mathsf{M}\zeta} \right\rangle_{\boldsymbol{\tau},\Gamma} = \left\langle \gamma_N^- \mathbf{e}, \overline{\gamma_\mathbf{t}^- \mathbf{e}} \right\rangle_{\boldsymbol{\tau},\Gamma} = \int_{\Omega_s} \frac{1}{\kappa} |\,\mathbf{curl}\,\mathbf{e}|^2 \,\mathrm{d}\mathbf{x} - \kappa |\mathbf{e}|^2 \,\mathrm{d}\mathbf{x} \in \mathbb{R}\,.$$

Necessarily, $\left(\zeta, \overline{\mathsf{M}\zeta}\right)_{\boldsymbol{\tau}} = 0$, so that the requirements on M imply $\zeta = 0$.

Knowing that M is compact, we conclude from Lemma 10 that the bilinear form of (52) satisfies a generalised Gårding inequality. Thus, Theorem 4 gives existence from uniqueness. □

A possible candidate for M can be introduced through a variational definition: for $\zeta \in \boldsymbol{H}_\times^{-\frac{1}{2}}(\mathrm{div}_\Gamma, \Gamma)$ and all $\mathbf{q} \in \boldsymbol{H}_\times(\mathrm{div}_\Gamma, \Gamma)$, $\mathsf{M}\zeta \in \boldsymbol{H}_\times(\mathrm{div}_\Gamma, \Gamma)$ is to satisfy

$$\langle \mathsf{M}\zeta, \mathbf{q} \rangle_{0;\Gamma} + \langle \mathrm{div}_\Gamma \mathsf{M}\zeta, \mathrm{div}_\Gamma \mathbf{q} \rangle_{0;\Gamma} = \langle \mathbf{q}, \zeta \rangle_{\boldsymbol{\tau},\Gamma} \,. \tag{54}$$

Obviously, $\mathsf{M} : \boldsymbol{H}_\times^{-\frac{1}{2}}(\mathrm{div}_\Gamma, \Gamma) \mapsto \boldsymbol{H}_\times(\mathrm{div}_\Gamma, \Gamma)$ is a continuous linear operator. By density of $\boldsymbol{H}_\times(\mathrm{div}_\Gamma, \Gamma)$ in $\boldsymbol{H}_\times^{-\frac{1}{2}}(\mathrm{div}_\Gamma, \Gamma)$, M must be injective, which also means

$$\left\langle \mathsf{M}\zeta, \overline{\zeta} \right\rangle_{\boldsymbol{\tau},\Gamma} = \|\mathsf{M}\zeta\|^2_{\boldsymbol{H}_\times(\mathrm{div}_\Gamma,\Gamma)} > 0 \quad \Leftrightarrow \quad \zeta \neq 0\,.$$

It is easy to see that M inherits compactness from the embedding $\boldsymbol{H}_\times(\mathrm{div}_\Gamma, \Gamma) \hookrightarrow \boldsymbol{H}_\times^{-\frac{1}{2}}(\mathrm{div}_\Gamma, \Gamma)$.

8 Boundary Element Spaces

We equip the piecewise smooth compact two-dimensional surface Γ with an oriented triangulation Γ_h. This means that all its edges are endowed with a direction. We assume a perfect resolution of Γ, that is $\Gamma = \bar{K}_1 \cup \ldots \cup \bar{K}_N$, where $\mathcal{K}_h := \{K_1, \ldots, K_N\}$ is the set of mutually disjoint open cells of Γ_h. Moreover, no cell may straddle boundaries of the smooth faces Γ^j of Γ. We will admit triangular and quadrilateral cells only: for each $K \in \mathcal{K}_h$ there is a diffeomorphism $\boldsymbol{\Phi}_K : \widehat{K} \mapsto K$, where $\widehat{K}$ is the "unit triangle" or unit square in $\mathbb{R}^2$, depending on the shape of K [27, Sect. 5].

This paves the way for a parametric construction of boundary elements: to begin with, choose finite-dimensional local spaces $\mathcal{W}(\widehat{K}) \subset (C^\infty(\widehat{K}))^2$ of polynomial vector fields together with a dual basis of so-called local degrees of freedom (d.o.f.). Possible choices for $\mathcal{W}(\widehat{K})$ and related d.o.f. abound: we may use the classical triangular Raviart-Thomas (RT_p) elements of polynomial order $p \in \mathbb{N}_0$ [56],

$$\mathcal{W}(\widehat{K}) := \{\mathbf{x} \mapsto \mathbf{p}_1(\mathbf{x}) + p_2(\mathbf{x}) \cdot \mathbf{x},\, \mathbf{x} \in \widehat{K},\, \mathbf{p}_1 \in (\mathcal{P}_p(\widehat{K}))^2,\, p_2 \in \mathcal{P}_p(\widehat{K})\}\,,$$

where $\mathcal{P}_p(\widehat{K})$ is the space of two-variable polynomials of total degree $\leq p$. An alternative are the triangular BDM_p elements of degree p [12], $p \in \mathbb{N}_0$, which rely on $\mathcal{W}(\widehat{K}) := (\mathcal{P}_{p+1}(\widehat{K}))^2$. In both cases, the usual d.o.f. involve certain polynomial moments of normal components on edges, together with interior vectorial moments for $p > 0$. For instance, in the case of RT_0, edge fluxes are the appropriate degrees of freedom:

$$\boldsymbol{\mu}_h \in \boldsymbol{\mathcal{W}}(\widehat{K}) \mapsto \int_{\widehat{e}} \boldsymbol{\mu}_h \cdot \widehat{\mathbf{n}}\, \mathrm{d}S \;, \quad \widehat{e} \text{ edge of } \widehat{K} \;.$$

Similar local spaces and degrees of freedom are available for the unit square.

Using the pull-back of 1-forms the local spaces can be lifted to the cells of Γ_h. In terms of vector fields this is equivalent to the *Piola transformation*

$$(\mathfrak{F}_K \boldsymbol{\mu})(\mathbf{x}) := \sqrt{\det(\mathrm{G})}\, \mathrm{G}^{-1}\, D\boldsymbol{\Phi}_K^T(\widehat{\mathbf{x}}) \boldsymbol{\mu}(\widehat{\mathbf{x}}) \;, \tag{55}$$

where $\mathrm{G} := D\boldsymbol{\Phi}(\widehat{\mathbf{x}})^T D\boldsymbol{\Phi}(\widehat{\mathbf{x}})$, $\mathbf{x} = \boldsymbol{\Phi}_K(\widehat{\mathbf{x}})$, $\widehat{\mathbf{x}} \in \widehat{K}$. Thus, we can introduce the global boundary element space

$$\boldsymbol{\mathcal{W}}_h := \{\boldsymbol{\mu} \in \boldsymbol{H}_\times(\mathrm{div}_\Gamma, \Gamma) : \boldsymbol{\mu}_{|K} \in \mathfrak{F}_K(\boldsymbol{\mathcal{W}}(\widehat{K}))\, \forall K \in \mathcal{K}_h\} \;. \tag{56}$$

In practice, $\boldsymbol{\mathcal{W}}_h \subset \boldsymbol{H}_\times(\mathrm{div}_\Gamma, \Gamma)$ is ensured by a suitable choice of d.o.f. Remember that d.o.f. have to be associated with individual edges of $\widehat{K}$ or the interior of $\widehat{K}$. It is crucial that the normal component of any $\widehat{\boldsymbol{\mu}}_h \in \boldsymbol{\mathcal{W}}(\widehat{K})$ on any edge $\widehat{e}$ of $\widehat{K}$ vanishes if and only if $\widehat{\boldsymbol{\mu}}_h$ belongs to the kernel of all local d.o.f. associated with $\widehat{e}$. In light of (12), this ensures $\boldsymbol{\mathcal{W}} \subset \boldsymbol{H}_\times(\mathrm{div}_\Gamma, \Gamma)$. In the sequel $\boldsymbol{\mathcal{W}}_h$ will designate a generic $\boldsymbol{H}_\times(\mathrm{div}_\Gamma, \Gamma)$-conforming boundary element space. It may arise from the RT_p family of elements, $p \in \mathbb{N}_0$, the BDM_p family, or a combination of both.

Based on the degrees of freedom we can introduce *local interpolation operators* $\Pi_h : \mathrm{Dom}(\Pi_h) \mapsto \boldsymbol{\mathcal{W}}_h$. It is a projector onto $\boldsymbol{\mathcal{W}}_h$ and enjoys the fundamental *commuting diagram property*

$$\mathrm{div}_\Gamma \circ \Pi_h = \mathrm{Q}_h \circ \mathrm{div}_\Gamma \quad \text{on } \boldsymbol{H}_\times(\mathrm{div}_\Gamma, \Gamma) \cap \mathrm{Dom}(\Pi_h) \;. \tag{57}$$

Here, Q_h is the $L^2(\Gamma)$-orthogonal projection onto a suitable space Q_h of Γ_h-piecewise polynomial discontinuous functions. It must be emphasised that the interpolation operators Π_h fail to be bounded on $\boldsymbol{H}_\times(\mathrm{div}_\Gamma, \Gamma)$; slightly more regularity of tangential vector fields in $\mathrm{Dom}(\Pi_h)$ is required [7, Lemma 4.7].

Next, we turn our attention to asymptotic properties of the boundary element spaces, in particular to estimates of interpolation errors and best approximation errors. We restrict ourselves to the h-version of boundary elements, which relies on shape-regular families $\{\Gamma_h\}_{h \in \mathbb{H}}$ of triangulations of Γ [30, Chap. 3,§ 3.1]. Here, $\mathbb{H}$ stands for a decreasing sequence of meshwidths, and $\mathbb{H}$ is assumed to converge to zero.

By means of transformation to reference elements, the commuting diagram property, and Bramble-Hilbert arguments, interpolation error estimates can easily be obtained [13, III.3.3].

Lemma 15 (Interpolation error estimate). *For $0 < s \leq p + 1$ we find constants $C > 0$, depending only on the shape regularity of the meshes and s, such that for all $\boldsymbol{\mu} \in \boldsymbol{H}_\times^s(\Gamma) \cap \boldsymbol{H}_\times(\mathrm{div}_\Gamma, \Gamma)$, $h \in \mathbb{H}$,*

$$\|\boldsymbol{\mu} - \Pi_h \boldsymbol{\mu}\|_{\boldsymbol{L}^2(\Gamma)} \leq C h^s \left(\|\boldsymbol{\mu}\|_{\boldsymbol{H}^s_\times(\Gamma)} + \|\mathrm{div}_\Gamma \boldsymbol{\mu}\|_{L^2(\Gamma)} \right) ,$$

and such that for all $\boldsymbol{\mu} \in \boldsymbol{H}_\times(\mathrm{div}_\Gamma, \Gamma)$, $\mathrm{div}_\Gamma \boldsymbol{\mu} \in H^s_-(\Gamma)$ *and*

$$\|\mathrm{div}_\Gamma(\boldsymbol{\mu} - \Pi_h \boldsymbol{\mu})\|_{L^2(\Gamma)} \leq C h^s \|\mathrm{div}_\Gamma \boldsymbol{\mu}\|_{H^s_-(\Gamma)} .$$

Corollary 5. *The union of all boundary element spaces* $\mathcal{W}_h$, $h \in \mathbb{H}$, *is dense in* $\boldsymbol{H}_\times^{-\frac{1}{2}}(\mathrm{div}_\Gamma, \Gamma)$.

A particular variant of the above interpolation error estimate addresses vector fields with discrete surface divergence:

Lemma 16. *If* $\boldsymbol{\mu} \in \boldsymbol{H}^s_\times(\Gamma)$, $0 < s \leq 1$, *and* $\mathrm{div}_\Gamma \boldsymbol{\mu} \in Q_h$, *then there is a constant* $C > 0$, *depending only on the shape-regularity of the meshes, such that*

$$\|\boldsymbol{\mu} - \Pi_h \boldsymbol{\mu}\|_{\boldsymbol{L}^2_\mathbf{t}(\Gamma)} \leq C h^s \|\boldsymbol{\mu}\|_{\boldsymbol{H}^s_\times(\Gamma)} .$$

From the interpolation error estimates we instantly get best approximation estimates in terms of the $\boldsymbol{H}_\times(\mathrm{div}_\Gamma, \Gamma)$-norm. Yet, what we actually need is a result about approximation in the "energy norm" $\|\cdot\|_{\boldsymbol{H}_\times^{-\frac{1}{2}}(\mathrm{div}_\Gamma, \Gamma)}$ of the form

$$\inf_{\boldsymbol{\xi}_h} \|\boldsymbol{\mu}_h - \boldsymbol{\xi}_h\|_{\boldsymbol{H}_\times^{-\frac{1}{2}}(\mathrm{div}_\Gamma, \Gamma)} \leq C h^{s+\frac{1}{2}} \|\boldsymbol{\mu}\|_{\boldsymbol{H}^s_\times(\mathrm{div}_\Gamma, \Gamma)} . \tag{58}$$

The estimate in $\boldsymbol{H}_\times(\mathrm{div}_\Gamma, \Gamma)$ does not directly provide (58). Even worse, standard duality arguments cannot be applied. Recall their main idea: we set out from a Hilbertian triple $V \subset H \subset V'$, have a finite dimensional subspace of H, say V_h, and we want to estimate the best approximation error in V'. Then it is crucial that we know how to use the difference in regularity between H and V through an estimate of the type $\exists\, v_h \in V_h \;:\; \|u - v_h\|_H \leq C(h)\|u\|_V$, with $C(h)$ optimal in a suitable sense.

Here, we have an estimate between $\boldsymbol{H}_\times(\mathrm{div}_\Gamma, \Gamma)$ and $\boldsymbol{H}^s_\times(\mathrm{div}_\Gamma, \Gamma)$, $s > 0$. Thus, we should use $\boldsymbol{H}_\times(\mathrm{div}_\Gamma, \Gamma)$ as self dual space, i.e. the standard inner product in $\boldsymbol{H}_\times(\mathrm{div}_\Gamma, \Gamma)$. But, in order to conclude, we should be able to prove that $\boldsymbol{H}^{-s}_\times(\mathrm{div}_\Gamma, \Gamma)$ is dual of $\boldsymbol{H}^s_\times(\mathrm{div}_\Gamma, \Gamma)$ for $0 < s \leq \frac{1}{2}$ with respect to the $\boldsymbol{H}_\times(\mathrm{div}_\Gamma, \Gamma)$ inner product. Unfortunately this is the case for regular surfaces but not for non-regular ones [29].

The question of obtaining (58) has been addressed in [16] and the idea is to use the duality argument face by face (which are seen as regular open manifolds), exploiting continuity of the normal components of vector-fields in $\boldsymbol{H}_\times(\mathrm{div}_\Gamma, \Gamma)$. At the end of a technical procedure we obtain the following result:

Theorem 14. *Let* $\mathcal{P}_h : \boldsymbol{H}_\times^{-\frac{1}{2}}(\mathrm{div}_\Gamma, \Gamma) \to \mathcal{W}_h$ *be the orthogonal projection with respect to the* $\boldsymbol{H}_\times^{-\frac{1}{2}}(\mathrm{div}_\Gamma, \Gamma)$ *inner product. Then, for any* $-\frac{1}{2} \leq s \leq p+1$ *we have*

$$\|\boldsymbol{\mu} - \mathcal{P}_h \boldsymbol{\mu}\|_{\boldsymbol{H}_\times^{-\frac{1}{2}}(\mathrm{div}_\Gamma, \Gamma)} \leq C h^{s+\frac{1}{2}} \|\boldsymbol{\mu}\|_{\boldsymbol{H}^s_\times(\mathrm{div}_\Gamma, \Gamma)} \quad \forall \boldsymbol{\mu} \in \boldsymbol{H}^s_\times(\mathrm{div}_\Gamma, \Gamma) . \tag{59}$$

This theorem tells us that we can expect good approximation properties, but these cannot be obtained using local interpolation operators.

9 Galerkin Discretisation

The Galerkin approach consists of replacing the Hilbert spaces $\boldsymbol{H}_\times^{-\frac{1}{2}}(\mathrm{div}_\Gamma, \Gamma)$ and $\boldsymbol{H}_\times(\mathrm{div}_\Gamma, \Gamma)$ in the variational formulations by finite dimensional subspaces $\boldsymbol{\mathcal{W}}_h$.

9.1 Integral Equations of the First Kind

First, we study the simplest BIE of the first kind, namely the electric field integral equations (42) and (49), that is, we examine variational problems like: seek $\boldsymbol{\lambda} \in \boldsymbol{H}_\times^{-\frac{1}{2}}(\mathrm{div}_\Gamma, \Gamma)$ such that

$$a(\boldsymbol{\lambda}, \boldsymbol{\mu}) := \langle \mathbf{S}_\kappa \boldsymbol{\lambda}, \boldsymbol{\mu} \rangle_{\boldsymbol{\tau},\Gamma} = \text{r.h.s.}(\boldsymbol{\mu}) \quad \forall \boldsymbol{\mu} \in \boldsymbol{H}_\times^{-\frac{1}{2}}(\mathrm{div}_\Gamma, \Gamma) , \tag{60}$$

for a suitable continuous functional on the right hand side. If κ stays away from interior electric Maxwell eigenvalues, we saw that the operator $\mathbf{S}_\kappa : \boldsymbol{H}_\times^{-\frac{1}{2}}(\mathrm{div}_\Gamma, \Gamma) \mapsto \boldsymbol{H}_\times^{-\frac{1}{2}}(\mathrm{div}_\Gamma, \Gamma)$ defines an isomorphism. This is equivalent to the existence of a constant $C_S > 0$ such that the following *continuous inf-sup condition* holds true:

$$\sup_{\boldsymbol{\eta} \in \boldsymbol{H}_\times^{-\frac{1}{2}}(\mathrm{div}_\Gamma, \Gamma)} \frac{|a(\boldsymbol{\mu}, \boldsymbol{\eta})|}{\|\boldsymbol{\eta}\|_{\boldsymbol{H}_\times^{-\frac{1}{2}}(\mathrm{div}_\Gamma, \Gamma)}} \geq C_S \|\boldsymbol{\mu}\|_{\boldsymbol{H}_\times^{-\frac{1}{2}}(\mathrm{div}_\Gamma, \Gamma)} \quad \forall \boldsymbol{\mu} \in \boldsymbol{H}_\times^{-\frac{1}{2}}(\mathrm{div}_\Gamma, \Gamma) . \tag{61}$$

We aim at establishing a *uniform discrete inf-sup-condition* of the form: there exists $C_D > 0$ such that $\forall \boldsymbol{\mu}_h \in \boldsymbol{\mathcal{W}}_h$,

$$\sup_{\boldsymbol{\eta}_h \in \boldsymbol{\mathcal{W}}_h} \frac{|a(\boldsymbol{\mu}_h, \boldsymbol{\eta}_h)|}{\|\boldsymbol{\eta}_h\|_{\boldsymbol{H}_\times^{-\frac{1}{2}}(\mathrm{div}_\Gamma, \Gamma)}} \geq C_D \|\boldsymbol{\mu}_h\|_{\boldsymbol{H}_\times^{-\frac{1}{2}}(\mathrm{div}_\Gamma, \Gamma)} , \quad h \in \mathbb{H} . \tag{62}$$

According to Babuška's theory [8] refined in [63] this guarantees existence of discrete solutions $\boldsymbol{\lambda}_h \in \boldsymbol{\mathcal{W}}_h$ and translates into their quasi-optimal behaviour:

$$\|\boldsymbol{\lambda} - \boldsymbol{\lambda}_h\|_{\boldsymbol{H}_\times^{-\frac{1}{2}}(\mathrm{div}_\Gamma, \Gamma)} \leq C_D^{-1} C_A \inf_{\boldsymbol{\eta}_h \in \boldsymbol{\mathcal{W}}} \|\boldsymbol{\lambda} - \boldsymbol{\eta}_h\|_{\boldsymbol{H}_\times^{-\frac{1}{2}}(\mathrm{div}_\Gamma, \Gamma)} \quad \forall h \in \mathbb{H} , \tag{63}$$

where $C_A > 0$ is the operator norm of $a(\cdot, \cdot)$. As a first step towards a discrete inf-sup condition, we have to find a suitable candidate for $\boldsymbol{\eta}$ in (61). To that end, introduce the operator $\mathsf{T} : \boldsymbol{H}_\times^{-\frac{1}{2}}(\mathrm{div}_\Gamma, \Gamma) \mapsto \boldsymbol{H}_\times^{-\frac{1}{2}}(\mathrm{div}_\Gamma, \Gamma)$ through

$$a(\boldsymbol{\eta}, \mathsf{T}\boldsymbol{\mu}) = c_\Gamma(\boldsymbol{\mu}, \boldsymbol{\eta}) \quad \forall \boldsymbol{\eta} \in \boldsymbol{H}_\times^{-\frac{1}{2}}(\mathrm{div}_\Gamma, \Gamma), \ \boldsymbol{\mu} \in \boldsymbol{H}_\times^{-\frac{1}{2}}(\mathrm{div}_\Gamma, \Gamma) ,$$

where c_Γ is the compact bilinear form of Lemma 10. Owing to (61) this is a valid definition of a compact operator T. It is immediate from (61) and Lemma 10 that

$$|a(\boldsymbol{\mu}, (\mathsf{X}_\Gamma + \mathsf{T})\overline{\boldsymbol{\mu}})| = |a(\boldsymbol{\mu}, \mathsf{X}_\Gamma \overline{\boldsymbol{\mu}}) + c_\Gamma(\boldsymbol{\mu}, \overline{\boldsymbol{\mu}})| \geq C_G \|\boldsymbol{\mu}\|^2_{\boldsymbol{H}_\times^{-\frac{1}{2}}(\mathrm{div}_\Gamma, \Gamma)} \tag{64}$$

for all $\boldsymbol{\mu} \in \boldsymbol{H}_{\times}^{-\frac{1}{2}}(\mathrm{div}_\Gamma, \Gamma)$. The choice $\boldsymbol{\eta} := (\mathsf{X}_\Gamma + \mathsf{T})\boldsymbol{\mu}$ will make (61) hold with $C_S = C_G$. The challenge is that $(\mathsf{X}_\Gamma + \mathsf{T})\boldsymbol{\mu}_h$ will not be a boundary element function even for $\boldsymbol{\mu}_h \in \boldsymbol{\mathcal{W}}_h$. This is clear because neither X_Γ nor T may leave the boundary element spaces invariant. It will be necessary to project $\mathsf{X}_\Gamma\boldsymbol{\mu}_h$ and $\mathsf{T}\boldsymbol{\mu}_h$ back to $\boldsymbol{\mathcal{W}}_h$. This can be achieved by applying suitable continuous projection operators $\mathsf{P}_h^X : \mathsf{X}_\Gamma(\boldsymbol{\mathcal{W}}_h) \mapsto \boldsymbol{\mathcal{W}}_h$, $\mathsf{P}_h^T : \boldsymbol{H}_{\times}^{-\frac{1}{2}}(\mathrm{div}_\Gamma, \Gamma) \mapsto \boldsymbol{\mathcal{W}}_h$. Then, for an arbitrary $\boldsymbol{\mu}_h \in \boldsymbol{\mathcal{W}}_h$ we can hope that $\boldsymbol{\eta}_h := (\mathsf{P}_h^X \circ \mathsf{X}_\Gamma + \mathsf{P}_h^T \circ \mathsf{T})\boldsymbol{\mu}_h$ is an appropriate choice for $\boldsymbol{\eta}_h$ in (62). Making use of (64) we see that

$$|a(\boldsymbol{\mu}_h, \boldsymbol{\eta}_h)| = |a(\boldsymbol{\mu}_h, (\mathsf{X}_\Gamma + \mathsf{T})\boldsymbol{\mu}_h) - a(\boldsymbol{\mu}_h, ((\mathrm{Id} - \mathsf{P}_h^X)\mathsf{X}_\Gamma + (\mathrm{Id} - \mathsf{P}_h^T)\mathsf{T})\boldsymbol{\mu}_h)|. \quad (65)$$

We know that $|a(\boldsymbol{\mu}_h, (\mathsf{X}_\Gamma + \mathsf{T})\boldsymbol{\mu}_h)| \geq C_G \|\boldsymbol{\mu}_h\|^2_{\boldsymbol{H}(\mathbf{curl};\Omega)}$ and we need to estimate the second term in the left hand side by the triangle inequality. Obviously, the projectors $\mathsf{P}_h^X, \mathsf{P}_h^T$ have to guarantee *uniform* convergence $(\mathrm{Id} - \mathsf{P}_h^T) \circ \mathsf{T}_{|\boldsymbol{\mathcal{W}}_h} \to 0$ and $(\mathrm{Id} - \mathsf{P}_h^X) \circ \mathsf{X}_{\Gamma|\boldsymbol{\mathcal{W}}_h} \to 0$ in $\boldsymbol{H}_{\times}^{-\frac{1}{2}}(\mathrm{div}_\Gamma, \Gamma)$ as $h \to 0$. For P_h^T this is easy: we choose P_h^T as the $\boldsymbol{H}_{\times}^{-\frac{1}{2}}(\mathrm{div}_\Gamma, \Gamma)$-orthogonal projection. Due to the compactness of T, we know [49, Corollary 10.4] that there exists a decreasing function $\epsilon = \epsilon(h)$ such that $\lim_{h\to 0} \epsilon(h) = 0$ and, for all $\boldsymbol{\mu}_h \in \boldsymbol{H}_{\times}^{-\frac{1}{2}}(\mathrm{div}_\Gamma, \Gamma)$,

$$\left\|(\mathrm{Id} - \mathsf{P}_h^T) \circ \mathsf{T}\boldsymbol{\mu}_h\right\|_{\boldsymbol{H}_{\times}^{-\frac{1}{2}}(\mathrm{div}_\Gamma,\Gamma)} < \epsilon(h) \left\|\boldsymbol{\mu}_h\right\|_{\boldsymbol{H}_{\times}^{-\frac{1}{2}}(\mathrm{div}_\Gamma,\Gamma)} . \quad (66)$$

As regards P_h^X, a crucial hint lies in the observation that P_h^X acts on functions in $\mathsf{X}_\Gamma(\boldsymbol{\mathcal{W}}_h)$. From $\mathrm{div}_\Gamma \mathsf{R}^\Gamma \boldsymbol{\mu} = \mathrm{div}_\Gamma \boldsymbol{\mu}$ we conclude that $\mathrm{div}_\Gamma(\mathsf{X}_\Gamma(\boldsymbol{\mathcal{W}}_h)) \subset Q_h$. We see that P_h^X has to be applied to functions with discrete div_Γ only. We recall Lemma 16, which bears out that $\mathsf{X}_\Gamma(\boldsymbol{\mathcal{W}}_h)$ is contained in the domain of the local interpolation operators Π_h. We discover that a perfectly valid candidate for P_h^X is the local interpolation operator: $\mathsf{P}_h^X := \Pi_h$. Then, Lemma 16 is the key to uniform convergence $(\mathrm{Id} - \mathsf{P}_h^X)\mathsf{X}_{\Gamma|\boldsymbol{\mathcal{W}}_h} \to 0$.

Lemma 17. *There is a $C_* = C_*(\Omega, p, \text{shape regularity}) > 0$ such that for all $\boldsymbol{\mu}_h \in \boldsymbol{\mathcal{W}}_h$*

$$\left\|(\mathrm{Id} - \mathsf{P}_h^X)\mathsf{X}_\Gamma\boldsymbol{\mu}_h\right\|_{\boldsymbol{H}_\times(\mathrm{div}_\Gamma,\Gamma)} \leq C_* h^{1/2} \left\|\mathrm{div}_\Gamma\boldsymbol{\mu}_h\right\|_{H^{-\frac{1}{2}}(\Gamma)} . \quad (67)$$

Proof. Note that $(\mathrm{Id} - \Pi_h)\mathsf{X}_\Gamma\boldsymbol{\mu}_h = (\mathrm{Id} - \Pi_h)(2\mathsf{R}^\Gamma - \mathrm{Id})\boldsymbol{\mu}_h = 2(\mathrm{Id} - \Pi_h)\mathsf{R}^\Gamma\boldsymbol{\mu}_h$, and that $\mathrm{div}_\Gamma\mathsf{R}^\Gamma\boldsymbol{\mu}_h = \mathrm{div}_\Gamma\boldsymbol{\mu}_h$. Thus, for the estimate of the $\mathbf{L}^2_t$ norm, we need only combine Lemma 16 (applied to $\mathsf{R}^\Gamma\boldsymbol{\mu}_h$) with (20). The observation, based on the commuting diagram property (57), that

$$\mathrm{div}_\Gamma((\mathrm{Id} - \Pi_h)\mathsf{X}_\Gamma\boldsymbol{\mu}_h) = (\mathrm{Id} - Q_h)\mathrm{div}_\Gamma(\mathsf{X}_\Gamma\boldsymbol{\mu}_h) = (\mathrm{Id} - Q_h)\mathrm{div}_\Gamma\boldsymbol{\mu}_h = 0$$

finishes the proof. □

Using (66) and (67) in (65), we obtain:

$$|a(\boldsymbol{\mu}_h, \boldsymbol{\eta}_h)| \geq (C_G - C_A(\epsilon(h) + C_* h^{\frac{1}{2}})) \left\|\boldsymbol{\mu}_h\right\|^2_{\boldsymbol{H}_\times^{-\frac{1}{2}}(\mathrm{div}_\Gamma, \Gamma)} .$$

This means that for h small enough to ensure $1 - C_A(\epsilon(h) + C_* h^{\frac{1}{2}})/C_G > \frac{1}{2}$ we have the discrete inf-sup condition (62). This yields the main result:

Theorem 15. *If κ^2 is not an interior electric eigenvalue, then there exists $h^* > 0$, depending on the parameters of the continuous problem and the shape-regularity of the triangulation, such that a unique solution $\boldsymbol{\lambda}_h \in \boldsymbol{\mathcal{W}}_h$ of the discretised problem (60) exists, provided that $h < h_*$. It supplies an asymptotically optimal approximation to the continuous solution $\boldsymbol{\lambda}$ of* (60) *in the sense of* (63).

Exactly the same arguments apply to (48) and give us an analogue of Theorem 15 for the Galerkin BEM discretisation in the case of the transmission problem.

Remark 2. Asymptotic quasi-optimality alone does not provide information about the actual speed of convergence as $h \to 0$, unless we have information about the smoothness of $\boldsymbol{\lambda}$. To assess the regularity of $\boldsymbol{\lambda}$ it is necessary to recall its meaning as a boundary value or the jump of a trace of Maxwell solutions. Then the results on the regularity of Maxwell solutions given in [34] can be used. Ultimately we will always have $\boldsymbol{\lambda} \in \boldsymbol{H}_\times^s(\mathrm{div}_\Gamma, \Gamma)$ for some $s > 0$ depending on the excitation and the geometry of Γ. In combination with Theorem 14 we can predict asymptotic rates of convergence for the h-version of the Galerkin boundary element schemes.

Remark 3. A striking difference between (42) and (49) is the choice of unknowns. In the indirect method $\boldsymbol{\lambda}$ is a jump. Hence, when solving the boundary integral equations on a polyhedron, the unknown of the indirect method will be affected by the corner and edge singularities of both interior and exterior Maxwell solutions [34]. As any edge is re-entrant when seen from either Ω_s or Ω', the jump $[\gamma_N]_\Gamma(\mathbf{e})$ will invariably possess a very low regularity. As a consequence, it might be much harder to approximate by boundary elements than the unknown of the direct method.

In terms of Galerkin discretisation the CFIE from Sect. 7.2 poses an extra difficulty, because of the composition of the integral operator $\mathbf{C}_\kappa$ and the smoothing operator M. The usual trick to avoid such operator products is to switch to a mixed formulation introducing the new unknown $\mathbf{p} := \mathsf{M}\boldsymbol{\zeta}$. If we use the particular smoothing operator from (54), we get $\mathbf{p} \in \boldsymbol{H}_\times(\mathrm{div}_\Gamma, \Gamma)$ and may simply incorporate (54) into the eventual mixed variational problem: find $\boldsymbol{\zeta} \in \boldsymbol{H}_\times^{-\frac{1}{2}}(\mathrm{div}_\Gamma, \Gamma)$, $\mathbf{p} \in \boldsymbol{H}_\times(\mathrm{div}_\Gamma, \Gamma)$ such that for all $\boldsymbol{\mu} \in \boldsymbol{H}_\times^{-\frac{1}{2}}(\mathrm{div}_\Gamma, \Gamma)$, $\mathbf{q} \in \boldsymbol{H}_\times(\mathrm{div}_\Gamma, \Gamma)$,

$$\begin{aligned} -i\eta \langle \mathbf{S}_\kappa \boldsymbol{\zeta}, \boldsymbol{\mu} \rangle_{\boldsymbol{\tau},\Gamma} + \quad & \left\langle (\tfrac{1}{2}\,\mathrm{Id} - \mathbf{C}_\kappa)\mathbf{p}, \boldsymbol{\mu} \right\rangle_{\boldsymbol{\tau},\Gamma} && = \left\langle \gamma_{\mathbf{t}}^+ \mathbf{e}_i, \boldsymbol{\mu} \right\rangle_{\boldsymbol{\tau},\Gamma} , \\ \langle \mathbf{q}, \boldsymbol{\zeta} \rangle_{\boldsymbol{\tau},\Gamma} \quad & - \langle \mathbf{p}, \mathbf{q} \rangle_{0;\Gamma} - \langle \mathrm{div}_\Gamma \mathbf{p}, \mathrm{div}_\Gamma \mathbf{q} \rangle_{0;\Gamma} && = 0 . \end{aligned} \tag{68}$$

Thanks to the compact embedding $\boldsymbol{H}_\times(\mathrm{div}_\Gamma, \Gamma) \hookrightarrow \boldsymbol{H}_\times^{-\frac{1}{2}}(\mathrm{div}_\Gamma, \Gamma)$ the off-diagonal terms in (68) are compact. Thus, a generalised Gårding inequality is immediate from Lemma 10. As far as the analysis of the Galerkin discretisation in

$\mathcal{W}_h \times \mathcal{W}_h$ is concerned, we only need to deal with the diagonal terms in (68): using exactly the same arguments as above we conclude the quasi-optimality of Galerkin solutions on sufficiently fine meshes. Please note that the estimates now employ the norm of the product space $\boldsymbol{H}_\times^{-\frac{1}{2}}(\mathrm{div}_\Gamma, \Gamma) \times \boldsymbol{H}_\times(\mathrm{div}_\Gamma, \Gamma)$. Thus, asymptotic rates of convergence will depend on the smoothness of both $\boldsymbol{\zeta}$ and $\mathbf{p}$.

9.2 Integral Equations of the Second Kind

Pitfalls have to be avoided when performing a Galerkin boundary element discretisation of the Fredholm integral equation of the second kind (44). A straightforward Galerkin discretisation would lead to: seek $\boldsymbol{\lambda}_h \in \mathcal{W}_h$ such that

$$\left\langle (\tfrac{1}{2}\,\mathrm{Id} + \mathbf{C}_\kappa)\boldsymbol{\lambda}_h, \boldsymbol{\mu}_h \right\rangle_{\boldsymbol{\tau},\Gamma} = \text{r.h.s.}(\boldsymbol{\mu}_h) \quad \forall \boldsymbol{\mu}_h \in \mathcal{W}_h\,. \tag{69}$$

If $\Gamma \in C^\infty$, then existence and uniqueness of solutions of the continuous variational problem are clear from Theorem 11. However, this does not necessarily remain true for (69). The cause of the difficulties is the failure of Theorem 2 to hold in the discrete setting. In other words, $\mathcal{W}_h$ may not be dual to itself with respect to the pairing $\langle\cdot,\cdot\rangle_{\boldsymbol{\tau},\Gamma}$. More precisely, in [29, Sect. 3.1] it has been shown by means of Hodge decompositions that for RT_0 boundary elements and quasiuniform families of surface meshes Γ_h there is $\alpha > 0$ and spaces $\mathcal{K}_h \subset \mathcal{W}_h$ such that, for all $h \in \mathbb{H}$, $\dim \mathcal{K}_h \geq \alpha \dim \mathcal{W}_h$ and

$$\forall \boldsymbol{\mu}_h \in \mathcal{K}_h \qquad \sup_{\boldsymbol{\xi}_h \in \mathcal{W}_h} \frac{|\langle \boldsymbol{\mu}_h, \boldsymbol{\xi}_h \rangle_{\boldsymbol{\tau},\Gamma}|}{\|\boldsymbol{\xi}_h\|_{\boldsymbol{H}_\times^{-\frac{1}{2}}(\mathrm{div}_\Gamma,\Gamma)}} \leq C h^{\frac{1}{2}} \|\boldsymbol{\mu}_h\|_{\boldsymbol{H}_\times^{-\frac{1}{2}}(\mathrm{div}_\Gamma,\Gamma)}\,.$$

The discretisation of $\langle\cdot,\cdot\rangle_{\boldsymbol{\tau},\Gamma}$ on $\mathcal{W}_h$ is not stable! This bars us from deriving a discrete inf-sup condition, though the continuous bilinear form satisfies a generalised Gårding inequality.

Remark 4. The instability of $\langle\cdot,\cdot\rangle_{\boldsymbol{\tau},\Gamma}$ in $\mathcal{W}$ also thwarts the straightforward application of an otherwise effective preconditioning strategy for boundary integral equations of the first kind, which is based on Calderón projectors [59]. The gist of the remedy, devised in [28], is to express $\langle\cdot,\cdot\rangle_{\boldsymbol{\tau},\Gamma}$ via an approximate discrete Hodge decomposition of $\mathcal{W}$.

This instability forces us to switch from (44) to (45), before a Galerkin discretisation by means of $\mathcal{W}_h$ becomes feasible: the stability of the $\boldsymbol{TH}(\mathrm{div}_\Gamma, \Gamma)$ inner product in the discrete setting is a moot point. Hence, provided that the assumptions of Theorem 11 hold, the Galerkin discretisation in $\mathcal{W}_h$ will produce asymptotically optimally convergent solutions on sufficiently fine meshes. The proof follows the standard approach to coercive variational problems [62]. However, note that all estimates will be based on the $\boldsymbol{H}_\times(\mathrm{div}_\Gamma, \Gamma)$-norm, that is

$$\|\boldsymbol{\lambda} - \boldsymbol{\lambda}_h\|_{\boldsymbol{H}_\times(\mathrm{div}_\Gamma,\Gamma)} \leq C \inf_{\boldsymbol{\eta} \in \mathcal{W}_h} \|\boldsymbol{\lambda} - \boldsymbol{\eta}_h\|_{\boldsymbol{H}_\times(\mathrm{div}_\Gamma,\Gamma)}\,.$$

We point out that the bilinear expressions $(\boldsymbol{\lambda}_h, \boldsymbol{\mu}_h) \mapsto \langle \operatorname{div}_\Gamma \mathbf{C}_\kappa \boldsymbol{\lambda}_h, \operatorname{div}_\Gamma \boldsymbol{\mu}_h \rangle_{0;\Gamma}$ that have to be evaluated for basis functions of $\boldsymbol{\mathcal{W}}_h$ in order to get the system matrix can be converted into sums of two integrals over $\Gamma \times \Gamma$ featuring weakly singular kernels. Details can be found in the proof of Lemma 11.

10 Coupling of Finite Elements and Boundary Elements

The solution of the transmission problem of electromagnetic scattering by means of direct boundary integral equations is confined to the case of homogeneous scatterers, because the simple representation formula (26) for Maxwell solutions cannot accommodate variable material coefficients $\underline{\epsilon} = \underline{\epsilon}(\mathbf{x})$, $\underline{\mu} = \underline{\mu}(\mathbf{x})$, $\mathbf{x} \in \Omega_s$. This situation poses no problems for a Galerkin finite element discretisation of the spatial variational problem inside Ω_s. On the other hand, the field problem in the air region Ω' is not amenable to a treatment by classical finite elements, but offers a perfect setting for the boundary element methods discussed in the previous sections. Thus, it is natural to tackle scattering at an inhomogeneous body by a combined Galerkin discretisation involving both finite elements and boundary elements. In this section the focus will be on a method based on the Calderon projector $\mathbb{P}_\kappa^+$ from (34).

Using $a_M(\cdot,\cdot)$ defined in (15), the electric field in Ω_s satisfies

$$a_M(\mathbf{e}, \mathbf{v}) - \left\langle \underline{\mu}^{-1} \gamma_{\mathbf{t}}^- \operatorname{\mathbf{curl}} \mathbf{e}, \gamma_{\mathbf{t}}^- \mathbf{v} \right\rangle_{\tau,\Gamma} = 0 \tag{70}$$

for all $\mathbf{v} \in \boldsymbol{H}(\operatorname{\mathbf{curl}}; \Omega_s)$. The gist of coupling is to employ an operator representation of the *Dirichlet-to-Neumann* map $\mathsf{DtN}_\kappa^+ : \boldsymbol{H}_\times^{-\frac{1}{2}}(\operatorname{div}_\Gamma, \Gamma) \mapsto \boldsymbol{H}_\times^{-\frac{1}{2}}(\operatorname{div}_\Gamma, \Gamma)$, which is a linear operator returning $\gamma_N^+ \mathbf{e}$ for a Maxwell solution $\mathbf{e}$ in Ω' if $\gamma_{\mathbf{t}}^+ \mathbf{e}$ is prescribed. If this was available, we could use the transmission conditions

$$\gamma_{\mathbf{t}}^- \mathbf{e} = \gamma_{\mathbf{t}}^+ \mathbf{e} + \gamma_{\mathbf{t}}^+ \mathbf{e}_i \quad , \quad \underline{\mu}^{-1} \gamma_{\mathbf{t}}^- \operatorname{\mathbf{curl}} \mathbf{e} = \tfrac{\kappa}{\mu_0} \gamma_N^+ \mathbf{e} + \gamma_{\mathbf{t}}^+ \mathbf{h}_i \ , \tag{71}$$

to cast the scattering problem in the variational form: seek $\mathbf{e} \in \boldsymbol{H}(\operatorname{\mathbf{curl}}; \Omega_s)$ such that for all $\mathbf{v} \in \boldsymbol{H}(\operatorname{\mathbf{curl}}; \Omega_s)$

$$a_M(\mathbf{e}, \mathbf{v}) - \kappa \mu_0^{-1} \left\langle \mathsf{DtN}_\kappa^+ \gamma_{\mathbf{t}}^- \mathbf{e}, \gamma_{\mathbf{t}}^- \mathbf{v} \right\rangle_{\tau,\Gamma} = \text{r.h.s}(\mathbf{v}) \ .$$

By Theorem 8 either row of the interior Calderon projector $\mathbb{P}_\kappa^-$ immediately supplies a realisation of DtN_κ^+:

$$\mathsf{DtN}_\kappa^+ = -(\tfrac{1}{2} \operatorname{Id} + \mathbf{C}_\kappa)^{-1} \mathbf{S}_\kappa \quad , \quad \mathsf{DtN}_\kappa^+ = -\mathbf{S}_\kappa^{-1} (\tfrac{1}{2} \operatorname{Id} + \mathbf{C}_\kappa) \ . \tag{72}$$

Both formulae describe the same operator, but appear vastly different. The reason is that they both break the inherent symmetry of magnetic and electric fields. Symmetry can be preserved by combining both rows of $\mathbb{P}_\kappa^+$ in a clever manner:

$$\mathsf{DtN}_\kappa^+ = -\mathbf{S}_\kappa - (\tfrac{1}{2} \operatorname{Id} - \mathbf{C}_\kappa) \mathbf{S}_\kappa^{-1} (\tfrac{1}{2} \operatorname{Id} + \mathbf{C}_\kappa) \ . \tag{73}$$

This discovery was first presented in [32] and is the foundation for the so-called symmetric approach to marrying finite elements and boundary elements. It has been applied to a wide range of transmission problems, see, for instance [22, 23, 50]. In the case of electromagnetism the idea was examined theoretically in [4–6], and in [9] for a related problem involving impedance boundary conditions.

Of course, a variational formulation suited for Galerkin discretisation has to dispense with the explicit inverse $\mathbf{S}_\kappa^{-1}$. Instead another equation is added, which leads to: seek $\mathbf{e} \in \boldsymbol{H}(\mathbf{curl}; \Omega_s)$, $\boldsymbol{\lambda} \in \boldsymbol{H}_\times^{-\frac{1}{2}}(\mathrm{div}_\Gamma, \Gamma)$ with

$$\begin{aligned} a_M(\mathbf{e}, \mathbf{e}') + \left\langle \tfrac{\kappa}{\mu_0} \mathbf{S}_\kappa \gamma_{\mathbf{t}}^- \mathbf{e}, \gamma_{\mathbf{t}}^- \mathbf{e}' \right\rangle_{\boldsymbol{\tau},\Gamma} - \left\langle \tfrac{\kappa}{\mu_0} (\tfrac{1}{2}\,\mathrm{Id} - \mathbf{C}_\kappa)\boldsymbol{\lambda}, \gamma_{\mathbf{t}}^- \mathbf{e}' \right\rangle_{\boldsymbol{\tau},\Gamma} &= \ldots, \\ \langle (\tfrac{1}{2}\,\mathrm{Id} + \mathbf{C}_\kappa)\gamma_{\mathbf{t}}^- \mathbf{e}, \boldsymbol{\lambda}' \rangle_{\boldsymbol{\tau},\Gamma} \quad + \quad \langle \mathbf{S}_\kappa \boldsymbol{\lambda}, \boldsymbol{\lambda}' \rangle_{\boldsymbol{\tau},\Gamma} &= \ldots, \end{aligned} \tag{74}$$

for all $\mathbf{e}' \in \boldsymbol{H}(\mathbf{curl}; \Omega_s)$, $\boldsymbol{\lambda}' \in \boldsymbol{H}_\times^{-\frac{1}{2}}(\mathrm{div}_\Gamma, \Gamma)$. The new unknown $\boldsymbol{\lambda}$ will provide the exterior Neumann trace $\gamma_N^+ \mathbf{e}$.

Note that the symmetric version of DtN_κ^+ involves the inverse of $\mathbf{S}_\kappa$. This suggests that "forbidden wave numbers" will also haunt the coupled formulations, *cf.* Sect. 7. Similar to Lemma 14 one proves the following theorem, see [44].

Theorem 16. *If κ^2 is not an interior electric eigenvalue, then a solution $(\mathbf{e}, \boldsymbol{\lambda})$ of (74) provides a solution of the transmission problem (1) by retaining $\mathbf{e}$ in Ω_s and using the exterior Stratton-Chu representation formula (26) with data $(\gamma_{\mathbf{t}}^- \mathbf{e} - \gamma_{\mathbf{t}}^+ \mathbf{e}_i, \boldsymbol{\lambda})$.*

Corollary 6. *If κ^2 is not an interior electric eigenvalue, the solution $(\mathbf{e}, \boldsymbol{\lambda})$ of (74) is unique.*

We point out that even if κ violates the assumption of the theorem, the solution for $\mathbf{e}$ will remain unique. This will no longer be true for $\boldsymbol{\lambda}$, which is unique only up to Neumann traces of interior electric eigenmodes. This can be seen by refining the arguments in the proof of Theorem 16.

We denote by $\mathbf{d}_\kappa$ the bilinear form on $\boldsymbol{H}(\mathbf{curl}; \Omega_s) \times \boldsymbol{H}_\times^{-\frac{1}{2}}(\mathrm{div}_\Gamma, \Gamma)$ that is associated with the variational problem (74). Pursuing the same policy as in Sect. 3 and 7, we aim to establish a generalised Gå rding inequality for $\mathbf{d}_\kappa$. Of course, the splitting idea will pave the way. More precisely, the crucial "sign flipping isomorphism" $\mathbb{X}_{\mathcal{V}}$ will involve both splittings (18) and (21) employed in Sects. 3 and 7. Writing, $\mathcal{V} := \boldsymbol{H}(\mathbf{curl}; \Omega_s) \times \boldsymbol{H}_\times^{-\frac{1}{2}}(\mathrm{div}_\Gamma, \Gamma)$, it reads

$$\mathbb{X}_{\mathcal{V}} \begin{pmatrix} \mathbf{u} \\ \boldsymbol{\xi} \end{pmatrix} := \begin{pmatrix} (\mathsf{R} - \mathsf{Z})\mathbf{u} \\ (\mathsf{R}^\Gamma - \mathsf{Z}^\Gamma)\boldsymbol{\xi} \end{pmatrix} : \mathcal{V} \mapsto \mathcal{V} .$$

We make the important observation that the trace $\gamma_{\mathbf{t}}^-$ maps $\mathbf{curl}$-free vector fields into $\mathcal{N}(\Gamma)$. In addition we can use the symmetry of $\mathbf{C}_\kappa$ stated in lemma 6 and proceed as in the proof of Theorem 9. This will give us the desired strengthened Gå rding inequality:

Theorem 17. *There exists a compact bilinear form* $\mathsf{c} : \mathcal{V} \times \mathcal{V} \mapsto \mathbb{C}$ *and a constant* $C_G > 0$ *such that*

$$\left| \mathsf{d}_\kappa \left(\begin{pmatrix} \mathbf{u} \\ \boldsymbol{\mu} \end{pmatrix}, \mathbb{X}_{\mathcal{V}} \begin{pmatrix} \overline{\mathbf{u}} \\ \overline{\boldsymbol{\mu}} \end{pmatrix} \right) - c \left(\begin{pmatrix} \mathbf{u} \\ \boldsymbol{\mu} \end{pmatrix}, \begin{pmatrix} \overline{\mathbf{u}} \\ \overline{\boldsymbol{\mu}} \end{pmatrix} \right) \right| \geq$$
$$\geq C_G \left(\|\mathbf{u}\|^2_{\boldsymbol{H}(\mathbf{curl};\Omega_s)} + \|\boldsymbol{\mu}\|^2_{\boldsymbol{H}_\times^{-\frac{1}{2}}(\mathrm{div}_\Gamma,\Gamma)} \right)$$

holds for all $\mathbf{u} \in \boldsymbol{H}(\mathbf{curl};\Omega_s)$, $\boldsymbol{\mu} \in \boldsymbol{H}_\times^{-\frac{1}{2}}(\mathrm{div}_\Gamma,\Gamma)$.

Hence, in conjunction with Cor. 6, a Fredholm alternative argument confirms the existence of solutions of the variational problem (74).

Besides Γ_h the Galerkin discretisation of (74) requires a triangulation Ω_h of Ω_s. In principle, both can be independent of each other, but implementation is greatly facilitated if $\Gamma_h = \Omega_{h|\Gamma}$. Then, we can rely on the $\boldsymbol{H}_\times(\mathrm{div}_\Gamma,\Gamma)$-conforming boundary element spaces $\mathcal{W}_h$ to approximate $\boldsymbol{\lambda}$, and special $\boldsymbol{H}(\mathbf{curl};\Omega_s)$-conforming finite elements for $\mathbf{e}$. The latter are thoroughly discussed in [45, Chap. 3]. They enjoy all the properties that permit us to prove a discrete inf-sup-condition as in Sect. 9.1. Thus we can get asymptotic quasi-optimality of discrete solutions obtained by the symmetric coupling of finite elements and boundary elements for the electromagnetic scattering problem.

References

1. R. Adams, *Sobolev Spaces*, Academic Press, New York, 1975.
2. A. Alonso and A. Valli, *Some remarks on the characterization of the space of tangential traces of* $H(\mathrm{rot};\Omega)$ *and the construction of an extension operator*, Manuscripta mathematica, 89 (1996), pp. 159–178.
3. H. Ammari and S. He, *Effective impedance boundary conditions for an inhomogeneous thin layer on a curved metallic surface*, IEEE Trans. Antennas and Propagation, 46 (1998), pp. 710–715.
4. H. Ammari and J. Nédélec, *Couplage éléments finis-équations intégrales pour la résolution des équations de Maxwell en milieu hétérogene*, in Equations aux derivees partielles et applications. Articles dedies a Jacques-Louis Lions, Gauthier-Villars, Paris, 1998, pp. 19–33.
5. H. Ammari and J.-C. Nédélec, *Coupling of finite and boundary element methods for the time-harmonic Maxwell equations. II: A symmetric formulation*, in The Maz'ya anniversary collection. Vol. 2, J. Rossmann, ed., vol. 110 of Oper. Theory, Adv. Appl., Birkhäuser, Basel, 1999, pp. 23–32.
6. H. Ammari and J.-C. Nédélec, *Coupling integral equations method and finite volume elements for the resolution of the Leontovich boundary value problem for the time-harmonic Maxwell equations in three dimensional heterogeneous media*, in Mathematical aspects of boundary element methods. Minisymposium during the IABEM 98 conference, dedicated to Vladimir Maz'ya on the occasion of his 60th birthday on 31st December 1997, M. Bonnet, ed., vol. 41 of CRC Research Notes in Mathematics, CRC Press, Boca Raton, FL, 2000, pp. 11–22.

7. C. Amrouche, C. Bernardi, M. Dauge, and V. Girault, *Vector potentials in three-dimensional nonsmooth domains*, Math. Meth. Appl. Sci., 21 (1998), pp. 823–864.
8. I. Babuška, *Error bounds for the finite element method*, Numer. Math., 16 (1971), pp. 322–333.
9. A. Bendali, *Boundary element solution of scattering problems relative to a generalized impedance boundary condition*, in Partial differential equations: Theory and numerical solution. Proceedings of the ICM'98 satellite conference, Prague, Czech Republic, August 10-16, 1998., W. Jäger, ed., vol. 406 of CRC Res. Notes Math., Boca Raton, FL, 2000, Chapman & Hall/CRC, pp. 10–24.
10. A. Bendali and L. Vernhet, *The Leontovich boundary value problem and its boundary integral equations solution*, Preprint, CNRS-UPS-INSA, Department de Genie Mathematique, Toulouse, France, 2001.
11. H. Brezis, *Analyse Fonctionnelle. Théorie et Applications*, Masson, Paris, 1983.
12. F. Brezzi, J. Douglas, and D. Marini, *Two families of mixed finite elements for 2nd order elliptic problems*, Numer. Math., 47 (1985), pp. 217–235.
13. F. Brezzi and M. Fortin, *Mixed and Hybrid Finite Element Methods*, Springer, 1991.
14. A. Buffa, *Hodge decompositions on the boundary of a polyhedron: The multiconnected case*, Math. Mod. Meth. Appl. Sci., 11 (2001), pp. 1491–1504.
15. A. Buffa, *Traces theorems for functional spaces related to Maxwell equations: An overwiew*. To appear in Proceedings of the GAMM Workshop on Computational Electromagnetics, Kiel, January 26th - 28th, 2001.
16. A. Buffa and S. Christiansen, *The electric field integral equation on Lipschitz screens: Definition and numerical approximation*, Numer. Mathem. (2002), DOI 10.1007/s00211-002-0422-0.
17. A. Buffa and P. Ciarlet, Jr., *On traces for functional spaces related to Maxwell's equations. Part I: An integration by parts formula in Lipschitz polyhedra.*, Math. Meth. Appl. Sci., 24 (2001), pp. 9–30.
18. A. Buffa and P. Ciarlet, Jr., *On traces for functional spaces related to Maxwell's equations. Part II: Hodge decompositions on the boundary of Lipschitz polyhedra and applications*, Math. Meth. Appl. Sci., 21 (2001), pp. 31–48.
19. A. Buffa, M. Costabel, and C. Schwab, *Boundary element methods for Maxwell's equations on non-smooth domains*, Numer. Mathem. 92 (2002) 4, pp. 679-710.
20. A. Buffa, M. Costabel, and D. Sheen, *On traces for* $\mathbf{H}(\mathbf{curl}, \Omega)$ *in Lipschitz domains*, J. Math. Anal. Appl., 276/2 (2002), pp. 845-876.
21. A. Buffa, R. Hiptmair, T. von Petersdorff, and C. Schwab, *Boundary element methods for Maxwell equations on Lipschitz domains*, Numer. Math., (2002). To appear.
22. C. Carstensen, *A posteriori error estimate for the symmetric coupling of finite elements and boundary elements*, Computing, 57 (1996), pp. 301–322.
23. C. Carstensen and P. Wriggers, *On the symmetric boundary element method and the symmetric coupling of boundary elements and finite elements*, IMA J. Numer. Anal., 17 (1997), pp. 201–238.
24. M. Cessenat, *Mathematical Methods in Electromagnetism*, vol. 41 of Advances in Mathematics for Applied Sciences, World Scientific, Singapore, 1996.
25. G. Chen and J. Zhou, *Boundary Element Methods*, Academic Press, New York, 1992.
26. S. Christiansen, *Discrete Fredholm properties and convergence estimates for the EFIE*, Technical Report 453, CMAP, Ecole Polytechique, Paris, France, 2000.
27. S. Christiansen, *Mixed boundary element method for eddy current problems*, Research Report 2002-16, SAM, ETH Zürich, Zürich, Switzerland, 2002.

28. S. Christiansen and J.-C. Nédélec, *Des prćonditionneurs pour la résolution numérique des équations intégrales de frontiére de l'electromagnétisme*, C.R. Acad. Sci. Paris, Ser. I Math, 31 (2000), pp. 617–622.
29. S. Christiansen and J.-C. Nédélec, *A preconditioner for the electric field integral equation based on Calderón formulae*. To appear in SIAM J. Numer. Anal.
30. P. Ciarlet, *The Finite Element Method for Elliptic Problems*, vol. 4 of Studies in Mathematics and its Applications, North-Holland, Amsterdam, 1978.
31. D. Colton and R. Kress, *Inverse Acoustic and Electromagnetic Scattering Theory*, vol. 93 of Applied Mathematical Sciences, Springer, Heidelberg, 2nd ed., 1998.
32. M. Costabel, *Symmetric methods for the coupling of finite elements and boundary elements*, in Boundary Elements IX, C. Brebbia, W. Wendland, and G. Kuhn, eds., Springer, Berlin, 1987, pp. 411–420.
33. M. Costabel, *Boundary integral operators on Lipschitz domains: Elementary results*, SIAM J. Math. Anal., 19 (1988), pp. 613–626.
34. M. Costabel and M. Dauge, *Singularities of Maxwell's equations on polyhedral domains*, in Analysis, Numerics and Applications of Differential and Integral Equations, M. Bach, ed., vol. 379 of Longman Pitman Res. Notes Math. Ser., Addison Wesley, Harlow, 1998, pp. 69–76.
35. M. Costabel and M. Dauge, *Maxwell and Lamé eigenvalues on polyhedra*, Math. Methods Appl. Sci., 22 (1999), pp. 243–258.
36. R. Dautray and J.-L. Lions, *Mathematical Analysis and Numerical Methods for Science and Technology*, vol. 4, Springer, Berlin, 1990.
37. A. de La Bourdonnaye, *Some formulations coupling finite element amd integral equation method for Helmholtz equation and electromagnetism*, Numer. Math., 69 (1995), pp. 257–268.
38. L. Demkowicz, *Asymptotic convergence in finite and boundary element methods: Part 1, Theoretical results*, Comput. Math. Appl., 27 (1994), pp. 69–84.
39. V. Girault and P. Raviart, *Finite Element Methods For Navier–Stokes Equations*, Springer, Berlin, 1986.
40. P. Grisvard, *Elliptic Problems in Nonsmooth Domains*, Pitman, Boston, 1985.
41. P. Grisvard, *Singularities in Boundary Value Problems*, vol. 22 of Research Notes in Applied Mathematics, Springer, New York, 1992.
42. W. Hackbusch, *Integral Equations. Theory and Numerical Treatment*, vol. 120 of International Series of Numerical Mathematics, Birkhäuser, Basel, 1995.
43. C. Hazard and M. Lenoir, *On the solution of time-harmonic scattering problems for Maxwell's equations*, SIAM J. Math. Anal., 27 (1996), pp. 1597–1630.
44. R. Hiptmair, *Coupling of finite elements and boundary elements in electromagnetic scattering*, Report 164, SFB 382, Universität Tübingen, Tübingen, Germany, July 2001. Submitted to SIAM J. Numer. Anal.
45. R. Hiptmair, *Finite elements in computational electromagnetism*, Acta Numerica, (2002), pp. 237–339.
46. R. Hiptmair, *Symmetric coupling for eddy current problems*, SIAM J. Numer. Anal., 40 (2002), pp. 41–65.
47. R. Hiptmair and C. Schwab, *Natural boundary element methods for the electric field integral equation on polyhedra*, SIAM J. Numer. Anal., 40 (2002), pp. 66–86.
48. R. Kress, *On the boundary operator in electromagnetic scattering*, Proc. Royal Soc. Edinburgh, 103A (1986), pp. 91–98.
49. R. Kress, *Linear Integral Equations*, vol. 82 of Applied Mathematical Sciences, Springer, Berlin, 1989.

50. M. Kuhn and O. Steinbach, *FEM-BEM coupling for 3d exterior magnetic field problems*, Math. Meth. Appl. Sci., (2002). To appear.
51. R. McCamy and E. Stephan, *Solution procedures for three-dimensional eddy-current problems*, J. Math. Anal. Appl., 101 (1984), pp. 348–379.
52. W. McLean, *Strongly Elliptic Systems and Boundary Integral Equations*, Cambridge University Press, Cambridge, UK, 2000.
53. J.-C. Nédélec, *Acoustic and Electromagnetic Equations: Integral Representations for Harmonic Problems*, vol. 44 of Applied Mathematical Sciences, Springer, Berlin, 2001.
54. A. Nethe, R.Quast, and H. Stahlmann, *Boundary conditions for high frequency eddy current problems*, IEEE Trans. Mag., 34 (1998), pp. 3331–3334.
55. L. Paquet, *Problemes mixtes pour le systeme de Maxwell*, Ann. Fac. Sci. Toulouse, V. Ser., 4 (1982), pp. 103–141.
56. P. A. Raviart and J. M. Thomas, *A Mixed Finite Element Method for Second Order Elliptic Problems*, vol. 606 of Springer Lecture Notes in Mathematics, Springer, Ney York, 1977, pp. 292–315.
57. M. Reissel, *On a transmission boundary-value problem for the time-harmonic Maxwell equations without displacement currents*, SIAM J. Math. Anal., 24 (1993), pp. 1440–1457.
58. A. Schatz, *An observation concerning Ritz-Galerkin methods with indefinite bilinear forms*, Math. Comp., 28 (1974), pp. 959–962.
59. O. Steinbach and W. Wendland, *The construction of some efficient preconditioners in the boundary element method*, Adv. Comput. Math., 9 (1998), pp. 191–216.
60. J. Stratton and L. Chu, *Diffraction theory of electromagnetic waves*, Phys. Rev., 56 (1939), pp. 99–107.
61. T. von Petersdorff, *Boundary integral equations for mixed Dirichlet, Neumann and transmission problems*, Math. Meth. Appl. Sci., 11 (1989), pp. 185–213.
62. W. Wendland, *Boundary element methods for elliptic problems*, in Mathematical Theory of Finite and Boundary Element Methods, A. Schatz, V. Thomée, and W. Wendland, eds., vol. 15 of DMV-Seminar, Birkhäuser, Basel, 1990, pp. 219–276.
63. J. Xu and L. Zikatanov, *Some observations on Babuška and Brezzi theories*, Report AM222, PennState Department of Mathematics, College Park, PA, September 2000. To appear in Numer. Math.

Computation of resonance frequencies for Maxwell equations in non-smooth domains

Martin Costabel and Monique Dauge

IRMAR, Université de Rennes 1, Campus de Beaulieu, 35042 Rennes, France
costabel@univ-rennes1.fr, dauge@univ-rennes1.fr

Summary. We address the computation by finite elements of the non-zero eigenvalues of the $(\mathbf{curl}, \mathbf{curl})$ bilinear form with perfect conductor boundary conditions in a polyhedral cavity. One encounters two main difficulties: (i) The infinite dimensional kernel of this bilinear form (the gradient fields), (ii) The unbounded singularities of the eigen-fields near corners and edges of the cavity. We first list possible variational spaces with their functional properties and provide a short description of the edge and corner singularities. Then we address different formulations using a Galerkin approximation by edge elements or nodal elements.

After a presentation of edge elements, we concentrate on the functional issues connected with the use of nodal elements. In the framework of conforming methods, nodal elements are mandatory if one regularises the bilinear form $(\mathbf{curl}, \mathbf{curl})$ in order to get rid of the gradient fields. A plain regularisation with the $(\mathrm{div}, \mathrm{div})$ bilinear form converges to a wrong solution if the domain has reentrant edges or corners. But remedies do exist. We will present the method of addition of singular functions, and the method of regularisation with weight, where the $(\mathrm{div}, \mathrm{div})$ bilinear form is modified by the introduction of a weight which can be taken as the distance to reentrant edges or corners.

Introduction

Computing Maxwell eigenfrequencies has been an interesting challenge for the numerical analysis community for many years. Besides its many obvious and very important practical applications, ranging from signal processing over heart and brain biology to nuclear fusion, the Maxwell eigenvalue problem has been attractive because of some mathematical features that set it apart from the standard fields of elliptic eigenvalue problems.

There is, on one hand, its simplicity as one of the very basic problems in partial differential equations. On this basic level there is the relation between the Maxwell equations and the de Rham complex and algebraic topology. In the construction and analysis of special families of finite elements, based on Nédélec's edge elements or "Whitney elements", this relation plays an important role. Major progress has been made in the theory of these special elements in recent years and many questions have found satisfactory answers, but some questions concerning the approximation of the eigenvalue problem remain open.

On the other hand, the Maxwell eigenvalue problem has several rather irritating peculiarities:

(i) The gauge invariance allows for many different variational formulations. The simplest of these are non-elliptic, have a non-empty essential spectrum, and the energy space is not compactly embedded into L^2. The effect is that straightforward discretisation do not generally give any useful approximation of the eigenvalues. Some Galerkin schemes may produce convergence of the numerical eigenvalues, but the limits may contain, in addition to the exact eigenvalues, extra spurious numbers. In other cases, the eigenvalues may converge to the exact values, but their multiplicities may be wrong.

(ii) More elaborate variational formulations can be constructed to avoid the problems coming from the infinite-dimensional eigenspace associated with zero frequency, i.e. with electro- or magnetostatic fields. If these formulations involve non-conforming or mixed finite element methods, then the error analysis of the eigenvalue problem is rather difficult, involves conditions whose range of validity is not yet fully understood, and is still incomplete in some important practical situations.

(iii) Another class of variational formulations that recover ellipticity is based on regularisation or penalisation. Whereas the error analysis of Galerkin approximations can be standard and simple in such cases, it is the equivalence between the original Maxwell eigenvalue problem and the regularised variational formulation that can lead to serious problems here. There are always spurious eigenvalues even for the continuous formulation, coming from some auxiliary problem, and some care has to be taken in the approximation procedure to separate the true Maxwell eigenvalues from the spurious eigenvalues. More seriously, on non-smooth domains one can have a situation where the numerical eigenvalues, although converging as the number of degrees of freedom is increased, converge to the spectrum of an entirely different problem. In other cases, it may be possible to prove convergence of the numerical eigenvalues to the correct values, but the observed convergence rate is extremely slow so the method is practically useless.

(iv) Near edges and corners of conducting bodies, electromagnetic fields tend to infinity. This has obvious implications for their numerical approximation, requiring strong mesh refinements, high degree polynomials or special singular trial functions. What is worse is that for some perfectly natural variational formulations and their finite element discretisation, these singularities are not just difficult to approximate, but are *impossible* to approximate. Typically, the eigenmode associated with the lowest non-zero eigenfrequency exhibits these strong singularities, and therefore the approximation of this principal eigenfrequency using a standard regularised formulation discretised by conforming finite elements, can be impossible. Recently a solution to this problem has been found in the weighted regularisation method.

In this article, we will explain and illustrate these interesting phenomena in detail. We give descriptions of the corner and edge singularities of the Maxwell eigenfunctions, based on those of the solutions of Helmholtz boundary value problems. Then we discuss different possibilities for choosing variational formulations and associate function spaces and explain in particular the phenomenon of non-density of smooth functions. The numerical results, even in the simplest case of a square, show very clearly the necessity of choosing the right variational formulation and the right finite-dimensional space of trial functions.

1 Maxwell Resonance Frequencies

Let Ω be a bounded three-dimensional domain filled with a dielectric material of permittivity ε and permeability μ. The electromagnetic resonance frequencies of Ω are the numbers $\omega > 0$ such that there exists an electromagnetic field $(\mathbf{E}, \mathbf{H}) \neq 0$ satisfying the equations

$$\mathbf{curl}\, \varepsilon\mathbf{E} - i\omega\, \mu\mathbf{H} = 0 \qquad \text{and} \qquad \mathbf{curl}\, \mu\mathbf{H} + i\omega\, \varepsilon\mathbf{E} = 0 \quad \text{in} \quad \Omega. \tag{1}$$

As ω is supposed to be non-zero, taking the divergence of these two equations we obtain

$$\operatorname{div} \varepsilon\mathbf{E} = 0 \qquad \text{and} \qquad \operatorname{div} \mu\mathbf{H} = 0. \tag{2}$$

We assume the perfect conductor boundary conditions[1], that is, denoting by $\mathbf{n}$ the outer unit normal on $\partial\Omega$:

$$\mathbf{E} \times \mathbf{n} = 0 \qquad \text{and} \qquad \mathbf{H} \cdot \mathbf{n} = 0 \quad \text{on} \quad \partial\Omega. \tag{3}$$

In this paper we discuss variational methods to solve (1)-(3) with a special emphasis on the situation where Ω *is not smooth*. As a standard model for this, we assume that Ω is a polyhedron, that is any domain whose boundary $\partial\Omega$ is a finite union of plane faces (which are thus polygonal). Concentrating on the problems posed by the singularities of the boundary of Ω, we assume that the dielectric material filling Ω is homogeneous and isotropic, that is, after a possible change of unknowns $\varepsilon = \mu = 1$. We will address generalisations of this model situation in remarks or footnotes.

Recapitulating, we are looking for non-trivial solutions of the system

$$\begin{cases} \mathbf{curl}\, \mathbf{E} - i\omega\, \mathbf{H} = 0 \quad \text{and} \quad \mathbf{curl}\, \mathbf{H} + i\omega\, \mathbf{E} = 0 & \text{in} \quad \Omega \\ \operatorname{div} \mathbf{E} = 0 \quad \text{and} \quad \operatorname{div} \mathbf{H} = 0. & \text{in} \quad \Omega \\ \mathbf{E} \times \mathbf{n} = 0 \quad \text{and} \quad \mathbf{H} \cdot \mathbf{n} = 0 & \text{on} \quad \partial\Omega. \end{cases} \tag{4}$$

Note that, besides all non-zero ω solving (1),(3), a finite-dimensional kernel (solutions for $\omega = 0$) will appear in (4) if (and only if) the topology of Ω is not trivial (i.e. if Ω is not simply connected, or $\partial\Omega$ is not connected).

---

[1] One could also consider impedance boundary conditions

$$\mathbf{n} \times \mathbf{H} - \lambda(\mathbf{n} \times \mathbf{E}) \times \mathbf{n} = 0. \tag{3b}$$

Although the regularity of the eigenmodes would be the same as with perfect conductor boundary conditions, certain approximation properties would be, in principle, better. We will discuss this briefly later.

2 Maxwell Spaces and Density Properties

As usual in papers devoted to Maxwell equations, we will recall or introduce a number of functional spaces. We will not use all of them. But the interesting point is to study the *density* properties of smooth functions. In general everything goes as expected, except for a couple of spaces (the "bad" ones). *In this section only, we assume for simplicity that Ω is a* Lipschitz *polyhedron.*

2.1 A Collection of Spaces

Considering system (4), it is natural to assume that E, H belong to $H(\mathbf{curl}, \Omega)$, the space of $L^2(\Omega)^3$ fields with $\mathbf{curl}$ in $L^2(\Omega)^3$. As their divergence vanishes, the fields E and H also belong to $H(\mathrm{div}, \Omega)$, the space of $L^2(\Omega)^3$ fields with div in $L^2(\Omega)$. Since the following identities hold

$$\forall \mathsf{E} \in H(\mathbf{curl}, \Omega),\ \forall \mathsf{v} \in H^1(\Omega)^3 : \langle \mathsf{E} \times \mathsf{n}, \mathsf{v} \rangle_{\partial\Omega} = \int_\Omega \mathsf{E} \cdot \mathbf{curl}\, \mathsf{v} - \mathbf{curl}\, \mathsf{E} \cdot \mathsf{v},$$

$$\forall \mathsf{E} \in H(\mathrm{div}, \Omega),\ \forall \varphi \in H^1(\Omega) : \langle \mathsf{H} \cdot \mathsf{n}, \varphi \rangle_{\partial\Omega} = \int_\Omega \mathrm{div}\, \mathsf{H}\, \varphi + \mathsf{H} \cdot \mathbf{grad}\, \varphi,$$

the tangential trace $\mathsf{E} \times \mathsf{n}$ makes sense in $H^{-1/2}(\partial\Omega)^3$ and the normal trace in $H^{-1/2}(\partial\Omega)$, whence the possibility of defining $H_0(\mathbf{curl}; \Omega)$ as the subspace of $H(\mathbf{curl}; \Omega)$ with zero tangential traces and $H_0(\mathrm{div}; \Omega)$ as the subspace of $H(\mathrm{div}; \Omega)$ with zero normal traces.

For impedance conditions (3b), the above spaces have to be enlarged to

$$H(\mathbf{curl}, \Omega; TL^2) = \{\mathsf{u} \in H(\mathbf{curl}, \Omega)\ ; \quad \mathsf{u} \times \mathsf{n}|_{\partial\Omega} \in L^2(\partial\Omega)^3\}$$

$$H(\mathrm{div}, \Omega; TL^2) = \{\mathsf{u} \in H(\mathrm{div}, \Omega)\ ; \quad \mathsf{u} \cdot \mathsf{n}|_{\partial\Omega} \in L^2(\partial\Omega)\},$$

where the symbol TL^2 means that the corresponding traces belong to L^2.

Finally we introduce the combined spaces (which are suitable for "regularised" formulations)

$$X_N = H_0(\mathbf{curl}) \cap H(\mathrm{div}) \quad \text{and} \quad X_T = H(\mathbf{curl}) \cap H_0(\mathrm{div})$$

$$W_N = H(\mathbf{curl}; TL^2) \cap H(\mathrm{div}) \quad \text{and} \quad W_T = H(\mathbf{curl}) \cap H(\mathrm{div}; TL^2)$$

2.2 Density of Smooth Functions

The following result describes the "good" spaces where density holds:

Theorem 1.
$C^\infty(\overline{\Omega})^3$ *is dense in* $H(\mathbf{curl}, \Omega)$ *and* $C_0^\infty(\Omega)^3$ *is dense in* $H_0(\mathbf{curl}, \Omega)$, [45].
$C^\infty(\overline{\Omega})^3$ *is dense in* $H(\mathbf{curl}, \Omega; TL^2)$, [7].

$C^\infty(\overline{\Omega})^3$ *is dense in* $H(\mathrm{div}, \Omega)$ *and* $C_0^\infty(\Omega)^3$ *is dense in* $H_0(\mathrm{div}, \Omega)$, [45].

$C^\infty(\overline{\Omega})^3$ *is dense in* $H(\mathbf{curl}, \Omega; TL^2)$, [24].

$H_0(\mathbf{curl}) \cap H_0(\mathrm{div}) = H_0^1(\Omega)^3$.

The spaces W_N *and* W_T *coincide and* $C^\infty(\overline{\Omega})^3$ *is dense in* $W_N = W_T$, [22, 25].

However, there are certain "bad" spaces where the closure of smooth functions stops somewhere in between. Let the spaces of smooth fields satisfying the zero tangential and normal trace condition be denoted by $C_N^\infty(\Omega)$ and $C_T^\infty(\Omega)$ respectively, then:

Theorem 2. [19, 23, 26] *The closure of* C_N^∞ *in* X_N *is* $H_N := H^1(\Omega)^3 \cap X_N$.
The closure of C_T^∞ *in* X_T *is* $H_T := H^1(\Omega)^3 \cap X_T$.

The possible difference between X_N and H_N, or X_T and H_T, is fully spanned by gradients: we note that $\mathbf{grad}\,\varphi \in X_N$ if and only if φ belongs to the domain of the Dirichlet Laplacian

$$D(\Delta^{\mathrm{Dir}}) = \{\varphi \in H_0^1(\Omega)\,; \quad \Delta\varphi \in L^2(\Omega)\}$$

and $\mathbf{grad}\,\varphi \in X_T$ if and only if $\varphi \in D(\Delta^{\mathrm{Neu}})$, the domain of the Neumann Laplacian. In [9], it is shown that:

$$X_N = H_N + \mathbf{grad}\,\big(D(\Delta^{\mathrm{Dir}})\big) \quad \text{and} \quad X_T = H_T + \mathbf{grad}\,\big(D(\Delta^{\mathrm{Neu}})\big). \tag{5}$$

If Ω is convex, then it is known [46] that the domains $D(\Delta^{\mathrm{Dir}})$ and $D(\Delta^{\mathrm{Neu}})$ are contained in $H^2(\Omega)$, but if not, then neither $D(\Delta^{\mathrm{Dir}})$ nor $D(\Delta^{\mathrm{Neu}})$ are contained in $H^2(\Omega)$.

Decompositions of the type (5) have been studied in [9, 10, 42] for various non-smooth domains. For the decomposition of X_N, Ω can be any Lipschitz domain and it can have cuts or "screen" parts. For the decomposition of X_T, more regularity is needed: Piecewise C^α with $\alpha > \frac{3}{2}$ will do, but in [42] is given an example of a $C^{3/2}$ domain (without any edges and corners) where no such decomposition exists.

3 Singular Functions of the Dirichlet Laplace Operator

In order to investigate the decompositions (5) for non-convex polyhedra and to prepare for a description of the non-regular parts in the solutions of (4), we give some information on the structure of the singular functions of the Dirichlet problem Δ^{Dir} for the Laplacian on Ω:

$$\Delta\varphi = g \quad \text{in} \quad \Omega, \quad \text{and} \quad g \in H_0^1(\Omega). \tag{6}$$

Let us denote the set of the edges $\mathbf{e}$ by $\mathcal{E}$ of Ω and the set of its corners $\mathbf{c}$ by $\mathcal{C}$.

Associated with each edge or corner is: *(i)* a local system of coordinates, *(ii)* a countable set of singular functions, and *(iii)* corresponding templates for singular parts – here we will only mention the singular parts which are not contained in $H^2(\Omega)$.

3.1 For an Edge $\mathbf{e} \in \mathcal{E}$

(i) In a neighbourhood of **e**, Ω coincides with a dihedron of the form $\Gamma_\mathbf{e} \times \mathbb{R}$ where $\Gamma_\mathbf{e}$ is a plane sector with angle $\omega_\mathbf{e}$. A local system of cylindrical coordinates (r, θ, z) depending on **e** is introduced, where r denotes the distance to the edge, $\theta \in (0, \omega_\mathbf{e})$ and z denotes a Cartesian coordinate along the edge.

(ii) The singular functions $\Phi_\mathbf{e}^{\ell,\mathrm{Dir}}$ are indexed by the positive integers $\ell \in \mathbb{N}$, and given explicitly by

$$\Phi_\mathbf{e}^{\ell,\mathrm{Dir}} = r^{\ell\pi/\omega_\mathbf{e}} \sin \frac{\ell\pi\theta}{\omega_\mathbf{e}}. \tag{7}$$

The degree of homogeneity $\ell\pi/\omega_\mathbf{e}$ is the *singular exponent*. It is called "singular" because it is a measure of the lack of regularity of $\Phi_\mathbf{e}^{\ell,\mathrm{Dir}}$: The latter belongs to $H^{1+\tau}(\Omega)$ only if $\tau < \ell\pi/\omega_\mathbf{e}$. This limit is sharp, except if $\ell\pi/\omega_\mathbf{e}$ is an integer.

(iii) Non-H^2 singular parts along **e** only occur if $\omega_\mathbf{e} > \pi$ and $\ell = 1$: The non-H^2 part of a solution of problem (6) with smooth right hand side g has the form $\gamma_\mathbf{e}(z)\,\Phi_\mathbf{e}^{1,\mathrm{Dir}}(r,\theta)$ where the *edge coefficient* function $\gamma_\mathbf{e}$, defined along the edge **e**, only depends on g.

3.2 For a Corner $\mathbf{c} \in \mathcal{C}$

(i) In a neighbourhood of **c**, Ω coincides with a cone $\Gamma_\mathbf{c}$. Let (ρ, ϑ), $\rho > 0$, $\vartheta \in \mathbb{S}^2$, be spherical coordinates with origin at the vertex **c**. The cone $\Gamma_\mathbf{c}$ is characterised by its spherical section $G_\mathbf{c} := \Gamma_\mathbf{c} \cap \mathbb{S}^2$.

(ii) The singular functions $\Phi_\mathbf{c}^{\ell,\mathrm{Dir}}$ at the corner **c** are given by:

$$\Phi_\mathbf{c}^{\ell,\mathrm{Dir}} = \rho^\lambda \phi_\mathbf{c}^{\ell,\mathrm{Dir}}(\vartheta), \quad \text{with} \quad \lambda = \lambda^{\ell,\mathrm{Dir}} = -\tfrac{1}{2} + \sqrt{\mu^{\ell,\mathrm{Dir}} + \tfrac{1}{4}} \tag{8}$$

where the $\mu^{\ell,\mathrm{Dir}}$ are the eigenvalues of the Laplace-Beltrami Dirichlet problem on the spherical polygon $G_\mathbf{c}$ and $\phi_\mathbf{c}^{\ell,\mathrm{Dir}}$ are its eigenvectors. The singular exponent $\lambda^{\ell,\mathrm{Dir}}$ also measures the regularity of $\Phi_\mathbf{c}^{\ell,\mathrm{Dir}}$ in the following sense: Let $\mathcal{V}_\mathbf{c}$ be any cone with vertex **c** which does not intersect with the edges[2], then $\Phi_\mathbf{c}^{\ell,\mathrm{Dir}}$ belongs to $H^{1+\tau}(\mathcal{V}_\mathbf{c})$ for $\tau < \lambda^{\ell,\mathrm{Dir}} + \frac{1}{2}$.

(iii) Non-H^2 corner contributions (as distinct from the edge contributions) appear for $\lambda^{\ell,\mathrm{Dir}} \le \frac{1}{2}$. In fact, a necessary (but not sufficient) condition for this is "*$\Gamma_\mathbf{c}$ non convex and $\ell = 1$*". Then the non-H^2 corner part has the form $\gamma_\mathbf{c}\,\Phi_\mathbf{c}^{1,\mathrm{Dir}}(\rho, \vartheta)$ where the *corner coefficient* $\gamma_\mathbf{c}$ is a real number only depending on g.

3.3 Regularity of Dirichlet Solutions

The smallest corner exponent is $\lambda_\mathbf{c}^{1,\mathrm{Dir}}$ while the smallest edge exponent is $\pi/\omega_\mathbf{e}$. If g is smooth enough, then

[2] We have introduced these cones $\mathcal{V}_\mathbf{c}$ because the spherical functions $\phi_\mathbf{c}^{\ell,\mathrm{Dir}}$ have themselves singular parts at the vertices $v_\mathbf{e}$ of $G_\mathbf{c}$ – with $v_\mathbf{e} \in \mathbf{e}$ for each edge containing **c**.

$$\varphi \in H^{1+\tau}(\Omega), \qquad \forall \tau < \min\left\{\tau_{\mathcal{E}}, \tau_{\mathcal{C}}^{\mathrm{Dir}} + \tfrac{1}{2}\right\}, \tag{9}$$

$$\text{where} \quad \tau_{\mathcal{E}} = \min_{\mathbf{e}\in\mathcal{E}} \frac{\pi}{\omega_{\mathbf{e}}} \quad \text{and} \quad \tau_{\mathcal{C}}^{\mathrm{Dir}} = \min_{\mathbf{c}\in\mathcal{C}} \lambda_{\mathbf{c}}^{1,\mathrm{Dir}}. \tag{10}$$

As a consequence of the results given in §3.1-3.2, we can see that the implication $g \in L^2(\Omega) \Rightarrow u \in H^2(\Omega)$ holds if and only if Ω is convex. Thus when Ω is not convex, the first decomposition in equation (5) is not trivial.

3.4 Origin of Singularities

A common feature of the singular functions $\Phi_{\mathbf{e}}^{\ell,\mathrm{Dir}}$ and $\Phi_{\mathbf{c}}^{\ell,\mathrm{Dir}}$ is that they are both harmonic in $\Gamma_{\mathbf{e}}$ and $\Gamma_{\mathbf{c}}$ and satisfy zero Dirichlet conditions on $\partial\Gamma_{\mathbf{e}}$ and $\partial\Gamma_{\mathbf{c}}$ respectively. Moreover they have H^1-regularity in any bounded neighbourhood of the origin. Therefore, multiplied by a suitable cut-off function, each of them yields a singular[3] solution of (6) with smooth right hand side.

The full theory including a splitting into regular and singular parts goes back to [52] for conical points and to [35] for polyhedra. See also [34, 37, 55].

3.5 Neumann Singularities

The edge and corner singular functions of the Laplace Neumann problem Δ^{Neu} are determined in the same way. The Neumann singularities are thus denoted by $\Phi_{\mathbf{e}}^{\ell,\mathrm{Neu}}$ and $\Phi_{\mathbf{c}}^{\ell,\mathrm{Neu}}$. The edge singular function has the form (7) with sin replaced with cos and the corner singular function has the form (8) with exponents $\lambda_{\mathbf{c}}^{\ell,\mathrm{Neu}}$ corresponding to the non-zero Neumann eigenvalues of the Laplace-Beltrami operator on $G_{\mathbf{c}}$.

4 Maxwell Singular Functions

Eliminating $\mathbf{H}$ or $\mathbf{E}$ from (4), we obtain the uncoupled system

$$\mathbf{curl}\,\mathbf{curl}\,\mathbf{E} - \omega^2\mathbf{E} = 0, \quad \operatorname{div}\mathbf{E} = 0, \ \text{in } \Omega, \quad \mathbf{E}\times\mathbf{n} = 0 \ \text{on } \partial\Omega \tag{11}$$

$$\mathbf{curl}\,\mathbf{curl}\,\mathbf{H} - \omega^2\mathbf{H} = 0, \quad \operatorname{div}\mathbf{H} = 0, \ \text{in } \Omega, \quad \mathbf{H}\cdot\mathbf{n} = 0 \ \text{on } \partial\Omega. \tag{12}$$

4.1 Standard Singularities

Following [27], we obtain the electric singular functions $\mathbf{E}_{\mathbf{e}}(r,\theta)$ and $\mathbf{E}_{\mathbf{c}}(\rho,\vartheta)$ of the system (11) by solving two-dimensional and three-dimensional versions of the system

$$\mathbf{curl}\,\mathbf{curl}\,\mathbf{E} = 0, \quad \operatorname{div}\mathbf{E} = 0, \quad \text{and} \quad \mathbf{E}\times\mathbf{n} = 0 \ \text{on the boundary} \tag{13}$$

[3] Non-smooth in Cartesian coordinates for non-integer singular exponents, and always non-smooth in polar or spherical coordinates.

on $\Gamma_{\mathbf{e}} \times \mathbb{R}$ and $\Gamma_{\mathbf{c}}$ respectively, and similarly for the magnetic singular functions $\mathbf{H}_{\mathbf{e}}$ and $\mathbf{H}_{\mathbf{c}}$. In each case, we find two[4] types **1** and **2** of electric singular functions: $\mathbf{E}_{\mathbf{e}}^{\ell,1}$, $\mathbf{E}_{\mathbf{c}}^{\ell,1}$ and $\mathbf{E}_{\mathbf{e}}^{\ell,2}$, $\mathbf{E}_{\mathbf{c}}^{\ell,2}$, and their magnetic counterparts.

Type **1** contains the gradients of the singular functions of Δ:

$$\mathbf{E}_{\mathbf{e}}^{\ell,1} = \big(\mathbf{grad}_{x,y}\, \Phi_{\mathbf{e}}^{\ell,\mathrm{Dir}}, 0\big) \quad \text{and} \quad \mathbf{E}_{\mathbf{c}}^{\ell,1} = \mathbf{grad}_{x,y,z}\, \Phi_{\mathbf{c}}^{\ell,\mathrm{Dir}} \tag{14}$$

(for the magnetic field, Dir is replaced with Neu) and type **2** is defined as

$$\mathbf{E}_{\mathbf{e}}^{\ell,2} = \big(0, 0, \Phi_{\mathbf{e}}^{\ell,\mathrm{Dir}}\big) \quad \text{and} \quad \mathbf{E}_{\mathbf{c}}^{\ell,2} = \mathbf{grad}_{x,y,z}\, \Phi_{\mathbf{c}}^{\ell,\mathrm{Neu}} \times \mathbf{x}. \tag{15}$$

Here $\mathbf{x} = (x, y, z)$ always refers to local Cartesian coordinates (centred in $\mathbf{c}$, or z along $\mathbf{e}$ and (x, y) transverse to $\mathbf{e}$) and the components of the field $\mathbf{E}$ are written in the same system. The corresponding results for the magnetic fields are obtained by interchanging Dir and Neu.

It is obvious that the singular functions of type **1** satisfy (13). For type **2**, this is a consequence of the relation, valid for any scalar function Φ homogeneous of degree λ in the cone $\Gamma_{\mathbf{c}}$ [27, Lemma 6.2]:

$$\mathbf{curl}\big(\mathbf{grad}\,\Phi \times \mathbf{x}\big) = (\lambda + 1)\,\mathbf{grad}\,\Phi. \tag{16}$$

The corresponding singular exponents are given in Table 1. Besides that, corner types **1** and **2** exchange between the electric and magnetic fields $\mathbf{E}$ and $\mathbf{H}$ solutions of (4): the coefficient of $\mathbf{E}_{\mathbf{c}}^{\ell,1}$ in $\mathbf{E}$ is the same as the coefficient of $-ik\mathbf{H}_{\mathbf{c}}^{\ell,2}$ in $\mathbf{H}$ with $k = \omega(\lambda_{\mathbf{c}}^{\ell,\mathrm{Dir}} + 1)^{-1}$ and $\mathbf{H}_{\mathbf{c}}^{\ell,1}$ corresponds to $ik\mathbf{E}_{\mathbf{c}}^{\ell,2}$.

$\mathbf{E}_{\mathbf{e}}^{\ell,1}$	$\mathbf{E}_{\mathbf{e}}^{\ell,2}$	$\mathbf{E}_{\mathbf{c}}^{\ell,1}$	$\mathbf{E}_{\mathbf{c}}^{\ell,2}$
$\dfrac{\ell\pi}{\omega_{\mathbf{e}}} - 1$	$\dfrac{\ell\pi}{\omega_{\mathbf{e}}}$	$\lambda_{\mathbf{c}}^{\ell,\mathrm{Dir}} - 1$	$\lambda_{\mathbf{c}}^{\ell,\mathrm{Neu}}$

Table 1. Electric singular exponents of types **1** and **2**

4.2 Topological Singularities

The above collection of corner singular functions may not be complete in the situation when the spherical section $G_{\mathbf{c}}$ of the cone $\Gamma_{\mathbf{c}}$ has a non-trivial topology, namely, when its boundary is multiply connected. Let us denote the Laplace-Beltrami operator on $G_{\mathbf{c}}$ by $\Delta_{\mathbf{c}}$ and the distinct connected components of $\partial G_{\mathbf{c}}$ by $\partial_j G_{\mathbf{c}}$, $j = 0, \ldots, J$. Then the space $P_{\mathbf{c}}^{\mathrm{Dir}}$ of functions $\phi \in H^1(G_{\mathbf{c}})$ such that $\Delta_{\mathbf{c}}\phi = 0$ and with constant traces c_j on each $\partial_j G_{\mathbf{c}}$ has the dimension $J + 1$. Then scalar functions Φ, homogeneous of degree 0, defined by

[4] In [27] three types are described. But for divergence free solutions the third type is absent.

$$\Phi(\rho, \vartheta) = \phi(\vartheta) \quad \text{in} \quad \Gamma_{\mathbf{c}}$$

satisfy $\Delta\Phi = 0$ in $\Gamma_{\mathbf{c}}$ and have constant traces on the connected components of $\partial\Gamma_{\mathbf{c}}$. Therefore the singular fields defined as

$$\mathbf{E_c}(\rho, \vartheta) = \mathbf{grad}\,\Phi \quad \text{in} \quad \Gamma_{\mathbf{c}}$$

are **curl** and div free in $\Gamma_{\mathbf{c}}$ and their tangential component $\mathbf{E_c} \times \mathbf{n}$ is zero on $\partial\Gamma_{\mathbf{c}}$. This space of dimension J, together with its type **2** counterpart (space of the $\mathbf{E_c} \times \mathbf{x}$), definitely completes[5] the set of corner singular fields for the electric part of eigenfields. For the full story, see [27, §6.4].

4.3 Regularity of Maxwell Eigenfields

For the regularity statement of the Dirichlet Laplacian, we have introduced $\tau_{\mathcal{E}}$ and $\tau_{\mathcal{C}}^{\mathrm{Dir}}$ in (10). We similarly define $\tau_{\mathcal{C}}^{\mathrm{Neu}}$ as the smallest of the non-zero Neumann exponents $\lambda_{\mathbf{c}}^{\mathrm{Neu}}$ over all corners. In order to take the topological singularities into account we define

$$\tilde{\tau}_{\mathcal{C}}^{\mathrm{Dir}} = \begin{cases} \tau_{\mathcal{C}}^{\mathrm{Dir}} & \text{if all } G_{\mathbf{c}} \text{ have a connected boundary} \\ 0 & \text{if at least one } G_{\mathbf{c}} \text{ has a multiply connected boundary} \end{cases}$$

and similarly for $\tilde{\tau}_{\mathcal{C}}^{\mathrm{Dir}}$.

Theorem 3. *Let* $(\mathbf{E}, \mathbf{H})$ *be solution of* (4). *Then*

$$\mathbf{E} \in H^{\tau}(\Omega)^3, \qquad \forall \tau < \min\left\{\tau_{\mathcal{E}}, \tilde{\tau}_{\mathcal{C}}^{\mathrm{Dir}} + \tfrac{1}{2}, \tilde{\tau}_{\mathcal{C}}^{\mathrm{Neu}} + \tfrac{3}{2}\right\},$$

$$\textit{and} \quad \mathbf{H} \in H^{\tau}(\Omega)^3, \qquad \forall \tau < \min\left\{\tau_{\mathcal{E}}, \tilde{\tau}_{\mathcal{C}}^{\mathrm{Neu}} + \tfrac{1}{2}, \tilde{\tau}_{\mathcal{C}}^{\mathrm{Dir}} + \tfrac{3}{2}\right\}.$$

Thus we see that the condition $\tau_{\mathcal{E}} > 1$ (i.e. all $\omega_{\mathbf{e}} < \pi$) is necessary to have $H^1(\Omega)$-regularity. As we assume that Ω is a polyhedron, this condition implies that Ω is *convex*. Therefore all cones $\Gamma_{\mathbf{c}}$ are convex and their spherical part $G_{\mathbf{c}}$ are convex subsets of the sphere. The minimal exponents satisfy then, see [36, §4]:

$$\Omega \text{ convex} \Longrightarrow \quad \tau_{\mathcal{C}}^{\mathrm{Dir}} \geq \tau_{\mathcal{E}} \quad \text{and} \quad \tau_{\mathcal{C}}^{\mathrm{Neu}} \geq \frac{\sqrt{5}-1}{2}.$$

Therefore, when the polyhedron Ω is convex, $\mathbf{E}$ belongs to $H^{\tau}(\Omega)^3$ for all $\tau < \min\{\tau_{\mathcal{E}}, 1 + \sqrt{5}/2\}$.

In general, as $\tau_{\mathcal{E}} \geq \frac{1}{2}$ and $\tilde{\tau}_{\mathcal{C}}^{\mathrm{Dir}}$, $\tilde{\tau}_{\mathcal{C}}^{\mathrm{Neu}} \geq 0$, we see that in any case $\mathbf{E}$, $\mathbf{H} \in H^{\tau}(\Omega)^3$ for all $\tau < \frac{1}{2}$. Moreover, if Ω is Lipschitz, all $\omega_{\mathbf{e}}$ are less than 2π and the topological singularities are absent. Therefore $\tau_{\mathcal{E}} > \frac{1}{2}$ and $\tilde{\tau}_{\mathcal{C}}^{\mathrm{Dir}}$, $\tilde{\tau}_{\mathcal{C}}^{\mathrm{Neu}} > 0$, which implies that $\mathbf{E}$, $\mathbf{H} \in H^{\frac{1}{2}+\delta}(\Omega)^3$ for all $\delta < \delta_{\Omega}$ for some $\delta_{\Omega} > 0$.

[5] Since we have no source term, we do not have to consider polynomial right hand sides for the determination of the singular function, as was done in [27].

4.4 A Decomposition of Electric Maxwell Eigenfields

The splitting (5) is non-trivial if and only if Ω is non-convex. So, let us consider a non-convex Lipschitz[6] polyhedron Ω. As the electric part of eigenvectors $\mathbf{E}$ belongs to X_N, we have

$$\mathbf{E} = \mathbf{E}_0 + \mathbf{grad}\,\Phi \quad \text{with} \quad \mathbf{E}_0 \in H_N \;\text{ and }\; \Phi \in H^1_0(\Omega) \text{ s.t. } \Delta\Phi \in L^2(\Omega). \tag{17}$$

Noting that singular fields of type **2** have positive exponents, it is possible to prove the following decomposition of the electric part $\mathbf{E}$ of any solution of (4):

$$\mathbf{E} = \mathbf{E}_0 + \mathbf{grad}\,\Phi : \begin{cases} \mathbf{E}_0 \in H^{1+\tau}(\Omega)^3, \;\forall \tau < \min\left\{\tau_{\boldsymbol{\mathcal{E}}}, \tau_{\boldsymbol{C}}^{\mathrm{Neu}} + \frac12, \tau_{\boldsymbol{C}}^{\mathrm{Dir}} + \frac12\right\} \\ \varphi \in H^{1+\tau}(\Omega), \quad \forall \tau < \min\left\{\tau_{\boldsymbol{\mathcal{E}}}, \tau_{\boldsymbol{C}}^{\mathrm{Dir}} + \frac12\right\}. \end{cases} \tag{18}$$

Similar results hold for the magnetic field $\mathbf{H}$.

Let us emphasise that the decomposition (17)–(18) does not coincide with the Hodge decomposition of $\mathbf{E}$: $\mathbf{E}_0$ and $\mathbf{grad}\,\Phi$ are not orthogonal and there even exists an alternative to (17) as a decomposition with a singular part of the form $\mathbf{curl}\,\mathbf{A}$ with a singular field $\mathbf{A} \in H_T$, see [27, Remark 4.10].

4.5 Non-homogeneous Materials

When the physical parameters ε and μ are piecewise constant on a polyhedral partition (Ω_j) of Ω, a similar theory applies, see [31]. Now the set $\boldsymbol{C}$ includes the corners of all sub-domains Ω_j and similarly for edges. We then still have singularities of type **1** and **2** for both $\mathbf{E}$ and $\mathbf{H}$: but instead of being generated by the singular functions of the plain Laplace problems Δ^{Dir} and Δ^{Neu}, their scalar densities are the singular functions of the operators $\Delta^{\mathrm{Dir}}_\varepsilon$ and $\Delta^{\mathrm{Neu}}_\mu$ (corresponding to the bilinear forms $\int_\Omega \varepsilon\, \mathbf{grad}\, u \cdot \mathbf{grad}\, v$ and $\int_\Omega \mu\, \mathbf{grad}\, u \cdot \mathbf{grad}\, v$ respectively).

The outcome can be a very low regularity for $\mathbf{E}$ or $\mathbf{H}$: for any $\delta > 0$, there exists ε such that the generic regularity of $\mathbf{E}$ is less than $H^\delta(\Omega)$.

5 A Question and Two Unsatisfactory Answers

5.1 Typical Spectral Problems

Hoping that Maxwell problems share the desirable properties of Laplace problems, we first look for a Galerkin formulation of problem (11), i.e. we seek a space $\mathfrak{X}$ and a bilinear form a such that non-trivial solutions $(\mathbf{u}, \omega)$ of the variational problem

$$(\mathfrak{P}) \qquad \mathbf{u} \in \mathfrak{X}, \quad \forall \mathbf{v} \in \mathfrak{X}, \qquad a(\mathbf{u}, \mathbf{v}) = \omega^2 \int_\Omega \mathbf{u} \cdot \mathbf{v}\, \mathrm{d}x$$

[6] The decomposition (17) holds more generally when there are no topological singularities, and (18) holds also in the presence of topological singularities if τ is replaced by $\widetilde{\tau}$.

coincide with non-trivial solutions $(\mathbf{E}, \omega) \in H(\mathbf{curl}, \Omega) \times \mathbb{R}$ of problem (11). Moreover, if we discretise the above problem using finite element subspaces $\mathfrak{X}_h$ of $\mathfrak{X}$

$$(\mathfrak{P}_h) \qquad \mathbf{u}_h \in \mathfrak{X}_h, \quad \forall \mathbf{v}_h \in \mathfrak{X}_h, \qquad a(\mathbf{u}_h, \mathbf{v}_h) = \omega_h^2 \int_\Omega \mathbf{u}_h \cdot \mathbf{v}_h \, \mathrm{d}\mathbf{x}$$

then we require that $\omega_h \to \omega$ in the sense that the k-th non-zero eigenfrequency $\omega_{h,k}$ of $(\mathfrak{P}_h)$ converges to the k-th non-zero eigenfrequency ω_k of $(\mathfrak{P})$.

5.2 The Minimal Space is a Bad Choice

The minimal space $\mathfrak{X}$ which one could take is $H_0(\mathbf{curl}) \cap H(\mathrm{div}; 0)$ i.e. the space of divergence free fields in $H_0(\mathbf{curl})$. The bilinear form a is then taken as the curl bilinear form a_0:

$$a_0(\mathbf{u}, \mathbf{v}) = \int_\Omega \mathbf{curl}\,\mathbf{u} \cdot \mathbf{curl}\,\mathbf{v} \, \mathrm{d}\mathbf{x}\,. \tag{19}$$

With this choice of $\mathfrak{X}$ and a, the continuous Problem $(\mathfrak{P})$ has exactly the Maxwell electric eigenmodes as solutions.

However any finite element space $\mathfrak{X}_h$ (if it exists) contained in the space $\mathfrak{X}$ would be **curl** and div conforming. Therefore the tangential and normal jumps of any $\mathbf{u}_h \in \mathfrak{X}_h$ across neighbouring elements of the mesh will be zero, which implies that the piecewise polynomial $\mathbf{u}_h$ is continuous on $\overline{\Omega}$, thus **grad** conforming. As a consequence any $\mathbf{u}_h$ is contained in H_N, the subspace of H^1 fields in $X_N = H_0(\mathbf{curl}) \cap H(\mathrm{div})$.

Although this seems harmless[7] for convex domains for which $H_N = X_N$, this fact is a real *obstruction* to the convergence $\omega_{h,k} \to \omega_k$ for a general non-convex polyhedra, since H_N is closed in X_N, (cf. Theorem 2), and $H_N \neq X_N$, (see (5), (9)).

5.3 The Maximal Space is a Bad Choice

The maximal space one could take is $\mathfrak{X} = H_0(\mathbf{curl})$. The form a is still the curl bilinear form a_0 (19). The non-zero eigenfrequencies of the continuous problem $(\mathfrak{P})$ are then exactly the Maxwell eigenfrequencies.

However now $(\mathfrak{P})$ has also the infinite dimensional kernel

$$K = \{u \in H_0(\mathbf{curl})\,, \quad \mathbf{curl}\,u = 0\}, \tag{20}$$

and we note that

$$K \subset \{u = \mathbf{grad}\,p\,, \quad p \in H_0^1(\Omega)\} \tag{21}$$

(equality holds if Ω is simply connected).

[7] For edge angles $\omega_\mathbf{e}$ close to π, the extra regularity beyond $H^1(\Omega)$ is very low, compare with Theorem 3.

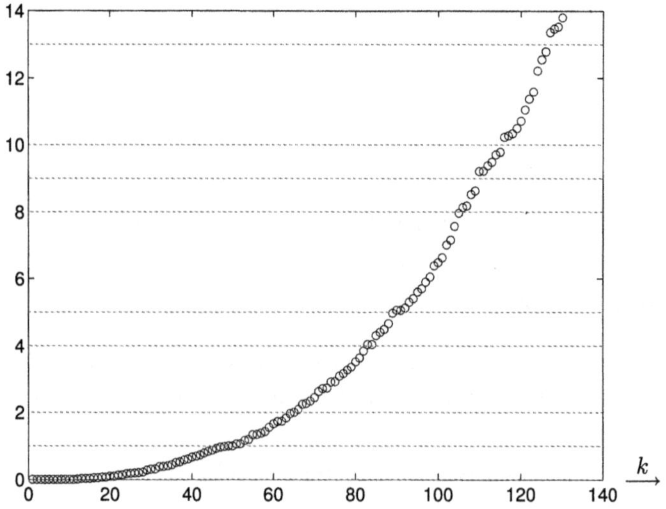

Fig. 1. Nodal triangles (15 nodes per side, $\mathbb{P}_1$) [18]. *Eigenvalue* ω_k *vs rank* k

The interesting spectrum lies between the two "points" 0 and $+\infty$ of the essential spectrum[8] of $(\mathfrak{P})$. The general results [39, 40] for the approximation of the discrete spectrum in presence of essential spectrum do not apply here. This fact has been investigated from the theoretical and computational point of view in [14, 18].

Here we give a simple illustration. The domain taken to be the two-dimensional[9] square $[0,\pi]^2$. We compute the eigenvalues of the bilinear form a_0 on different discretisation of the space $H_0(\mathbf{curl})$ and we plot them in the following way: The eigenvalues are sorted by increasing order, the abscissa is the rank of the eigenvalue ω^2, the ordinate is its value, which is marked by a circle. The horizontal lines indicate the exact values of the non-zero spectrum, which is, repeated according to multiplicity in the region $\omega^2 < 14$, given by:

$$[1\ \ 1\ \ 2\ \ 4\ \ 4\ \ 5\ \ 5\ \ 8\ \ 9\ \ 9\ \ 10\ \ 10\ \ 13\ \ 13].$$

In Fig.1, we plot the results from [18] of a computation made with an unstructured triangular mesh containing 15 nodes per side of the square and Lagrange $\mathbb{P}_1$ polynomials for each of the two components of the field. The total number of unknowns is 440. The distribution of eigenvalues is very similar to that obtained for the Cosserat problem in [62] with a p-version code.

We have also tried the computation in our square with the p-version of finite elements. In Fig.2, we use only one element with tensor $\mathbb{Q}_p$ elements for each of

[8] Both points are accumulation points of the spectrum.

[9] In principle the computation in the 2D case has no "physical" interest since the electric Maxwell eigenvalues coincide with the Neumann eigenvalues of the Laplacian, the **curl** of the Laplace eigenvectors being the electric Maxwell eigenvectors. Nevertheless, a priori knowledge of eigenpairs is very valuable to test any numerical method.

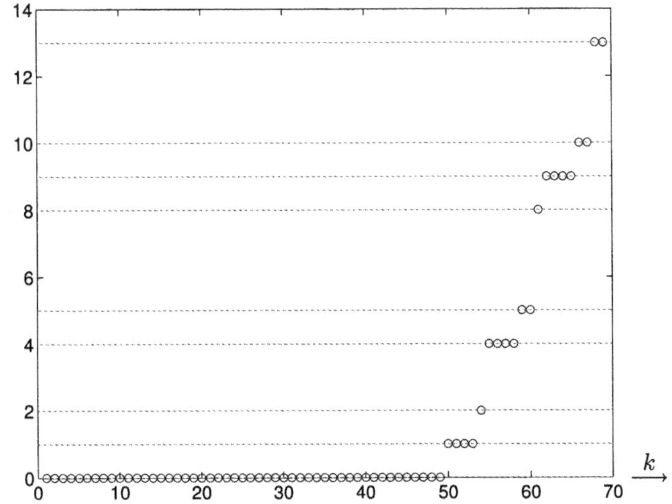

Fig. 2. One square element ($\mathbb{Q}_8$). *Eigenvalue* ω_k *vs rank k*

the two components and the number of unknowns is $2(p+1)^2$. The result looks much better: we get a large kernel (computed values between −5E-14 and 5E-14) of dimension $(p-1)^2$ and the (first) correct values for the next eigenvalues, *except* that the multiplicity of the pure squares $(1, 4, 9, ...)$ is 4 instead of 2 as it should be (the numerical multiplicity corresponds to values that coincide very accurately with 14 common digits). The corresponding effect in a cube has been noticed by Wang & Monk in earlier computations with $\mathbb{Q}_4$ elements.

The explanation of this curiosity is the following[10]: If we denote the eigenpairs of the discrete 1D Dirichlet problem in $\mathbb{P}_p(0,\pi)$ by $\big(v_j^{(p)}, \lambda_j^{(p)}\big)$, and the Legendre polynomial of degree p by L_p, we find the 4 eigenvectors

$$\big(v_j^{(p)}(y), 0\big), \qquad \big(0, v_j^{(p)}(x)\big), \tag{22}$$

$$\big(L_p(\tfrac{2}{\pi}x - 1)v_j^{(p)}(y), 0\big), \qquad \big(0, v_j^{(p)}(x)L_p(\tfrac{2}{\pi}y - 1)\big), \tag{23}$$

associated with the eigenvalue $\lambda_j^{(p)}$. We can see immediately that the divergence of the first two is zero, whereas the divergence of the latter ones blows up with p (the ratio of the L^2 norm of the divergence against the $H(\mathbf{curl})$ norm of the eigenvector behaves as $p^{3/2}$).

In Fig.3, we now use 4 elements (each side is split into two equal parts) with tensor $\mathbb{Q}_p$ elements, and the number of unknowns is $2(2p+1)^2$. We can see the results starting to deteriorate: the kernel has dimension $4(p-1)^2$, the eigenvalues are correctly computed (with the same extra multiplicity) but pairs of *spurious* values appear also. Compare with the results obtained on a criss-cross mesh [17, §III.B].

[10] In the case of the square, although we expect that a similar explanation holds for the cube.

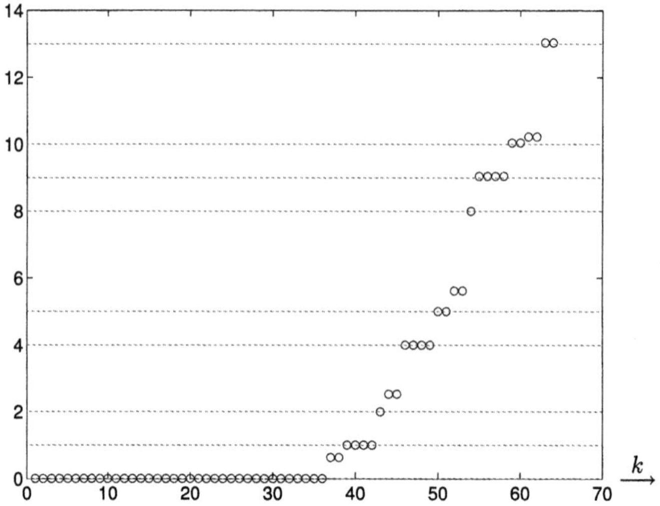

Fig. 3. Four square elements ($\mathbb{Q}_4$). *Eigenvalue* ω_k *vs rank k*

6 Mimicking the Kernel: Edge Elements

6.1 The Principle

Let us assume for simplicity that $\partial\Omega$ is connected. The principle[11] is to use compatible finite element spaces and projection operators for 0-forms (potential p) and 1-forms (electric field $\mathbf{u}$) according to

p	space $\mathsf{P}_h \subset H_0^1$	$\mathbf{grad}$ conforming	projector π_h
$\mathbf{u}$	space $\mathbf{V}_h \subset H_0(\mathbf{curl})$	$\mathbf{curl}$ conforming	projector r_h

with the commuting diagram property

$$\boxed{r_h(\mathbf{grad}\, p) = \mathbf{grad}(\pi_h p)} \tag{24}$$

This leads to the following alteration of the idea of the minimal space (§5.2). Let us denote the scalar product in $L^2(\Omega)$ by $\langle \cdot, \cdot \rangle$. We go back to the minimal $\mathfrak{X}$:

$$\mathfrak{X} = \mathfrak{X}^{\mathsf{min}} = \big\{\mathbf{u} \in H_0(\mathbf{curl})\,, \quad \forall \mathbf{v} \in K, \ \ \langle \mathbf{u}, \mathbf{v} \rangle = 0\big\} \tag{25}$$

where K is the kernel (20). But now we propose as discrete space:

$$\mathfrak{X}_h = \mathfrak{X}_h^{\mathsf{min}} = \big\{\mathbf{u}_h \in \mathbf{V}_h\,, \quad \forall \mathbf{v}_h \in K \cap \mathbf{V}_h, \ \ \langle \mathbf{u}_h, \mathbf{v}_h \rangle = 0\big\}. \tag{26}$$

Since $\partial\Omega$ is assumed to be connected, K coincides with $\mathbf{grad}(H_0^1)$.

[11] For further details on edge elements, see the survey [49].

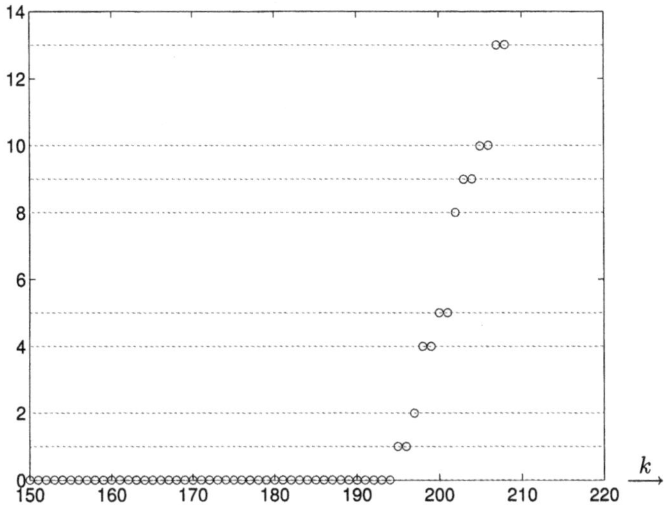

Fig. 4. Triangular edge elements (15 nodes per side, $\mathbf{P}_1$) [18]. *Eigenvalue* ω_k *vs rank* k

Lemma 1. $K \cap \mathbf{V}_h = \mathbf{grad}(\mathsf{P}_h)$.

Proof. *(i)* As P_h is contained in $H_0^1(\Omega)$, $\mathbf{grad}(\mathsf{P}_h)$ is contained in K. On the other hand, for any $p_h \in \mathsf{P}_h$, (24) gives that $\mathbf{grad}(p_h) = \mathbf{grad}(\pi_h p_h) = r_h(\mathbf{grad}\, p_h)$ which belongs to $\mathbf{V}_h$.
(ii) Let $\mathbf{u}_h$ belong to $K \cap \mathbf{V}_h$. As $\mathbf{u}_h \in K$, there exists $p \in H_0^1(\Omega)$ such that $\mathbf{u}_h = \mathbf{grad}\, p$. We have, thanks to (24) again, $\mathbf{u}_h = r_h(\mathbf{u}_h) = r_h(\mathbf{grad}\, p) = \mathbf{grad}(\pi_h p)$. Hence $\mathbf{u}_h \in \mathbf{grad}(\mathsf{P}_h)$. □

In fact, the minimal version (26) of the space $\mathfrak{X}_h$ will never be explicitly employed. But if we go back to the maximal space idea, §5.3, we take now $\mathfrak{X}_h$ as the whole space $\mathbf{V}_h$ and we are looking for the eigenpairs $(\mathbf{u}_h, \omega_h)$ satisfying

$$\mathbf{u}_h \in \mathbf{V}_h, \quad \forall \mathbf{v}_h \in \mathbf{V}_h, \qquad \int_\Omega \mathbf{curl}\, \mathbf{u}_h, \mathbf{curl}\, \mathbf{v}_h \, \mathrm{d}\mathbf{x} = \omega_h^2 \int_\Omega \mathbf{u}_h \cdot \mathbf{v}_h \, \mathrm{d}\mathbf{x}.$$

Then the kernel (the $\mathbf{u}_h$ corresponding to $\omega_h = 0$) coincides with $\mathbf{grad}(\mathsf{P}_h)$ and the remainder of the spectrum is associated with the eigenvectors in the space $\mathfrak{X}_h^{\mathsf{min}}$. Therefore, if $\mathfrak{X}_h^{\mathsf{min}}$ is a good approximation of the space $\mathfrak{X}^{\mathsf{min}}$, we expect (the reality is more involved, see §6.4) a good approximation of the Maxwell eigenvalues.

This is indeed illustrated, see Fig.4, by results of computations from [18] where the edge elements of order 1 on triangles are used (same square domain and same mesh as in Fig.1). The number of degrees of freedom is equal to 635. The dimension of the kernel is 194, which is the dimension of P_h. We show the ranks k of eigenvalues in the transition region $k \in [150, 220]$ between the kernel and the remainder of the spectrum.

Note 1. The space $\mathfrak{X}^{\mathsf{min}}$ is contained in the space $H(\mathrm{div};0)$ of divergence free fields. But, for any of the families $(\mathbf{V}_h, \mathsf{P}_h)$, there holds $\mathfrak{X}_h^{\mathsf{min}} \not\subset H(\mathrm{div};0)$, which means that, from the minimal space point of view, $\mathfrak{X}_h^{\mathsf{min}}$ is a non-conforming approximation. In fact $\mathfrak{X}_h^{\mathsf{min}}$ is the *discrete divergence free* space and a certain control of the divergence of its elements is a necessary condition to the convergence of eigenvalues as $h \to 0$.

6.2 Mixed Formulations

As we will see later, a much larger number of degrees of freedom is necessary in the presence of non-convex corners to obtain a decent approximation. The maximal space method obliges one to compute a large number of eigenvalues (here, nearly one third of the total spectrum in 2D) and we expect that the computational effort and the quality of approximation deteriorate for such a large number. Nevertheless see [63] for the special issue of the computation of the eigenvalues of the Laplacian by spectral methods.

To get rid of this embarrassing kernel, a *mixed formulation* in $\mathbf{V}_h \times \mathsf{P}_h$ is preferred, [2, 45, 50, 56]:

$$(\mathbf{u}_h, p_h) \in \mathbf{V}_h \times \mathsf{P}_h, \quad \forall (\mathbf{v}_h, q_h) \in \mathbf{V}_h \times \mathsf{P}_h,$$

$$(\mathfrak{M}_h) \qquad \left\{ \begin{aligned} \int_\Omega \mathbf{curl}\, \mathbf{u}_h \cdot \mathbf{curl}\, \mathbf{v}_h + \mathbf{grad}\, p_h \cdot \mathbf{v}_h &= \omega_h^2 \int_\Omega \mathbf{u}_h \cdot \mathbf{v}_h \\ \int_\Omega \mathbf{u}_h\, \mathbf{grad}\, q_h &= 0. \end{aligned} \right.$$

We note that the continuous version of this formulation is posed in $H_0(\mathbf{curl}, \Omega) \times H_0^1(\Omega)$, that it satisfies the inf-sup condition and that its non-zero solutions provide the Maxwell eigenvalues. The analysis of the convergence of this mixed formulation is done in [13, 14].

In order to have a non-degenerate saddle point problem, one can also use a regularised formulation instead, see [8], for example for given $s > 0$:

$$(\mathbf{u}_h, p_h) \in \mathbf{V}_h \times \mathsf{P}_h, \quad \forall (\mathbf{v}_h, q_h) \in \mathbf{V}_h \times \mathsf{P}_h,$$

$$(\mathfrak{M}_h^s) \qquad \left\{ \begin{aligned} \int_\Omega \mathbf{curl}\, \mathbf{u}_h \cdot \mathbf{curl}\, \mathbf{v}_h + \mathbf{grad}\, p_h \cdot \mathbf{v}_h &= \omega_h^2 \int_\Omega \mathbf{u}_h \cdot \mathbf{v}_h \\ \int_\Omega \mathbf{u}_h\, \mathbf{grad}\, q_h - \frac{1}{s} \int_\Omega p_h q_h &= 0. \end{aligned} \right.$$

The continuous version of this formulation is also posed in $H_0(\mathbf{curl}, \Omega) \times H_0^1(\Omega)$. It has *spurious* eigenfrequencies which are exactly the values of ω such that $\omega^2 = s\,\nu$, with ν any eigenvalue of the Laplace operator on $H_0^1(\Omega)$. The other eigenvalues are the Maxwell eigenvalues.

6.3 Finite Element Spaces

The requirement (24) naturally provides subspaces of finite elements (of type $\mathfrak{X}_h^{\mathsf{min}}$) which are *almost divergence-free*. This is why such elements were first introduced for the discretisation of Stokes and Navier-Stokes equations, see [45, 59]. Their application to Maxwell equations and their development is due to Nédélec [56, 57]. Let us very briefly describe the basic families of reference elements.

For P_h one takes as reference element the polynomial space $\mathbb{P}_p$: polynomials of degree $\leq p$ in 2 variables on triangles or 3 variables on tetrahedra, or $\mathbb{Q}_p$: polynomials of partial degrees $\leq p$ on the unit square $(0,1)^2$ or the unit cube $(0,1)^3$.

The corresponding reference elements for $\mathbf{V}_h$ are $\mathbf{P}_p$ and $\mathbf{Q}_p$ respectively, defined as follows (in 3D), with the space of homogeneous polynomials of degree p denoted by $\overline{\mathbb{P}}_p$,

$$\mathbf{P}_p = \{\mathbf{u}+\mathbf{v} \mid \quad \mathbf{u} \in (\mathbb{P}_{p-1})^3, \quad \mathbf{v} \in (\overline{\mathbb{P}}_p)^3 \text{ with } \mathbf{v}\cdot\mathbf{x} = 0\}$$

and, with the tensor product $\mathbb{P}_{p_1} \otimes \mathbb{P}_{p_2} \otimes \mathbb{P}_{p_3}$ denoted by $\mathbb{Q}_{p_1,p_2,p_3}$,

$$\mathbf{Q}_p = \{\mathbf{u} = (u_1, u_2, u_3) \mid \quad u_1 \in \mathbb{Q}_{p-1,p,p}, \quad u_2 \in \mathbb{Q}_{p,p-1,p}, \quad u_3 \in \mathbb{Q}_{p,p,p-1}\}.$$

It is very easy to describe the projectors π_h and r_h for first degree elements. The projector π_h is the Lagrange interpolant at the nodes (vertices $\mathbf{a}$ of the elements), i.e. $(\pi_h p)(\mathbf{a}) = p(\mathbf{a})$, whereas r_h interpolates the tangential *moments* along the edges $(\mathbf{a}, \mathbf{a}')$ of the elements: $\int_{\mathbf{a}}^{\mathbf{a}'} (r_h \mathbf{u}) \cdot \boldsymbol{\tau}\, \mathrm{d}\boldsymbol{\tau} = \int_{\mathbf{a}}^{\mathbf{a}'} \mathbf{u} \cdot \boldsymbol{\tau}\, \mathrm{d}\boldsymbol{\tau}$. The chain of identities for any scalar p (with enough regularity)

$$\begin{aligned}\int_{\mathbf{a}}^{\mathbf{a}'} (r_h\, \mathbf{grad}\, p - \mathbf{grad}\, \pi_h p)\cdot \boldsymbol{\tau}\, \mathrm{d}\boldsymbol{\tau} &= \int_{\mathbf{a}}^{\mathbf{a}'} \mathbf{grad}(p - \pi_h p)\cdot \boldsymbol{\tau}\, \mathrm{d}\boldsymbol{\tau}\\ &= (p-\pi_h p)(\mathbf{a}') - (p - \pi_h p)(\mathbf{a}) = 0.\end{aligned}$$

proves the commuting diagram property (24).

The description of the general case can be carried out in the general framework of Whitney forms, see [49, §3].

Remark 1. Examining the expressions (22)-(23) for the discrete eigenvectors in the standard space of $\mathbb{Q}_p^2$ vector fields on the square, we see that the two first (good) ones also belong to the edge reference space $\mathbf{Q}_p$, whereas the last (bad) ones do not.

6.4 Convergence Analysis

For the abstract framework, we follow [20, 21] which provides a necessary and sufficient criterion for a family of discrete spaces $(\mathfrak{X}_h)_{h\in\mathfrak{h}}$ to provide a "spurious-free" approximation. Here $\mathfrak{h}$ is a subset of positive numbers with the only accumulation point at zero and for any $h \in \mathfrak{h}$, $\mathfrak{X}_h$ is a subspace of $H_0(\mathbf{curl}, \Omega)$.

We recall that K is the subspace (20) of the curl-free fields in $H_0(\mathbf{curl}, \Omega)$ and that the subspace of the fields with zero discrete divergence is, cf. (26):

$$\mathfrak{X}_h^{\mathsf{min}} = \big\{\mathbf{u}_h \in \mathfrak{X}_h\,, \quad \forall \mathbf{v}_h \in K \cap \mathfrak{X}_h,\ \langle \mathbf{u}_h, \mathbf{v}_h \rangle = 0\big\}.$$

Finally we recall the natural norm of $H(\mathbf{curl}, \Omega)$:

$$\|\mathbf{v}\|_{H(\mathbf{curl},\Omega)} = \big(\|\,\mathbf{curl}\,\mathbf{v}\|^2_{L^2(\Omega)^3} + \|\mathbf{v}\|^2_{L^2(\Omega)^3}\big)^{1/2}.$$

The characteristic criteria for a spurious-free approximation are the three following conditions:

(CAS) *Completeness of the Approximating Subspace*

$$\forall \mathbf{v} \in H_0(\mathbf{curl}, \Omega), \qquad \lim_{h\to 0}\ \inf_{\mathbf{v}_h \in \mathfrak{X}_h} \|\mathbf{v} - \mathbf{v}_h\|_{H(\mathbf{curl},\Omega)} = 0.$$

(CDK) *Completeness of the Discrete Kernel*

$$\forall \mathbf{v} \in H_0(\mathbf{curl}, \Omega) \cap K, \qquad \lim_{h\to 0}\ \inf_{\mathbf{v}_h \in \mathfrak{X}_h \cap K} \|\mathbf{v} - \mathbf{v}_h\|_{H(\mathbf{curl},\Omega)} = 0.$$

(DCP) *Discrete Compactness Property*
For any sequence $\{\mathbf{v}_h\}_{h\in\mathfrak{h}}$ of discrete divergence free fields $\mathbf{v}_h \in \mathfrak{X}_h^{\mathsf{min}}$ bounded in $H(\mathbf{curl}, \Omega)$, there exists a subsequence $\{\mathbf{v}_h\}_{h\in\mathfrak{h}'}$ and $\mathbf{w} \in L^2(\Omega)^3$ such that

$$\lim_{h\in\mathfrak{h}',\ h\to 0} \|\mathbf{v}_h - \mathbf{w}\|_{L^2(\Omega)^3} = 0.$$

Let us comment on these three conditions in the framework of general edge elements according to §6.1.

(i) Condition (CAS) is satisfied in a natural way if $\mathbf{V}_h$ enjoys good classical local approximation properties.

(ii) The commuting diagram property (24) shows that in this situation (CDK) is equivalent to the classical approximation property in $H^1(\Omega)$ for the nodal space family (P_h)

$$\forall \varphi \in H^1_0(\Omega), \qquad \lim_{h\to 0}\ \inf_{\varphi_h \in \mathsf{P}_h} \|\varphi - \varphi_h\|_{H^1(\Omega)} = 0.$$

(iii) The most original of these conditions is the Discrete Compactness Property, which was introduced by Kikuchi [51]. It implies a certain control of the true divergence when the discrete divergence is zero. For example, (DCP) is satisfied if any sequence $\mathbf{v}_h \in \mathfrak{X}_h^{\mathsf{min}}$ bounded in $H(\mathbf{curl}, \Omega)$ satisfies

$$\lim_{h\to 0} \|\operatorname{div} \mathbf{v}_h\|_{H^{-1}(\Omega)} = 0.$$

The *spurious-free* approximation of the spectrum is defined in [20] as the conjunction of five different properties, which, when restricted to the non-zero part of the spectrum, is equivalent to the following:

Definition 1. *Let the increasing sequence of* non-zero *eigenfrequencies of* $(\mathfrak{P})$ *be denoted by* $(\omega_k)_{k\geq 1}$ *and for any* $h \in \mathfrak{h}$ *denote the increasing sequence of* non-zero *eigenfrequencies of* $(\mathfrak{P}_h)$ *by* $(\omega_{h,k})_{k\geq 1}$. *In each case the eigenfrequencies are* *repeated according to their multiplicities. For $h \in \mathfrak{h}$ denote the associated orthonormal systems of eigenvectors by $(\mathbf{u}_{h,k})_{k\geq 1}$.*

The problems $(\mathfrak{P}_h)_{h\to 0}$ are a positively spurious-free *approximation of $(\mathfrak{P})$ if*
(i) *For all $k \geq 1$,*

$$\lim_{h\to 0} \omega_{h,k} = \omega_k,$$

(ii) *For all $k \geq 1$ and for all $h \in \mathfrak{h}$, there exists an eigenvector $\mathbf{u}_k^h$ of $(\mathfrak{P})$ associated with the eigenfrequency ω_k so that*

$$\lim_{h\to 0} \left\|\mathbf{u}_k^h - \mathbf{u}_{h,k}\right\|_{H(\mathbf{curl},\Omega)} = 0.$$

For our Maxwell problem, only the positive eigenvalues are meaningful. This is the reason for the introduction of a "positively spurious-free" approximation. The spurious-free approximation in [20] also requires that the discrete kernels are an approximation of the kernel of the continuous problem $(\mathfrak{P})$.

Theorem 4. [20] *The problems $(\mathfrak{P}_h)_{h\to 0}$ are a positively spurious-free approximation of $(\mathfrak{P})$ if conditions* (CAS), (CDK) *and* (DCP) *are satisfied. These three conditions are also necessary if a spurious-free approximation is required.*

Within the h-version, it is proved, see [11, 12, 21, 38], that the Nédélec families satisfy the above three conditions. But within the p- and hp-version, no definitive result seems to be known, see [15].

Remark 2. In [20], a counter-example is given showing that conditions (CAS), (CDK) do not imply (DCP). Our family of tensor product spaces $(\mathbb{Q}_p^2)_{p\to\infty}$ on the square provides another counter-example: as $h = 1/p \to 0$, conditions (CAS) and (CDK) are obviously satisfied. However the growth estimate on the divergence of the eigenvectors (23) shows that (DCP) does not hold for this family. On the other hand, the extra multiplicity for the eigenvalues close to 1, 4,... invalidates condition *(i)* for the convergence of eigenvalues. A corresponding phenomenon was observed in the case of a mixed $(\mathbb{Q}_1, \mathbb{P}_0)$ formulation in [16].

7 Back to Nodal Elements: The Plain Regularisation

Although edge elements have their attractions, one may nevertheless prefer to try to use nodal elements for various reasons:
- Few free source codes providing edge elements are available;
- Modifying an existing FEM code to incorporate edge elements is hard work;
- Coupling with heat or elasticity is easier when the same elements are used for each field;
- If applicable, dealing with a coercive bilinear form has advantages over a mixed formulation: monotonicity of eigenvalues, optimal convergence analysis, etc. cf. [5];
- Or one may simply like to develop alternative methods...

7.1 The Principle

The idea goes back to Leis [53] and consists in blowing up the kernel K by transforming the eigenvalue zero into a family of non-zero spurious eigenvalues. This is easily done, at the continuous level, by the introduction of a parameter $s > 0$ and a new bilinear form – note that a_0 coincides with the **curl** form (19),

$$a_s(\mathbf{u}, \mathbf{v}) = \int_\Omega \mathbf{curl}\,\mathbf{u} \cdot \mathbf{curl}\,\mathbf{v}\; \mathrm{d}\mathbf{x} + s \int_\Omega \operatorname{div} \mathbf{u} \operatorname{div} \mathbf{v}\; \mathrm{d}\mathbf{x}\,. \tag{27}$$

The variational space naturally associated with this form is $X_N = H_0(\mathbf{curl}) \cap H(\operatorname{div})$. So we have defined a new version of the generic problem ($\mathfrak{P}$) of §5.1 with $\mathfrak{X} = X_N$ and $a = a_s$. For any $s > 0$, the form a_s is *coercive* over X_N.

Theorem 5. *There exists a basis* $(\mathbf{u}_k)$ *of eigenvectors of problem* ($\mathfrak{P}$) *with* $\mathfrak{X} = X_N$ *and* $a = a_s$*, associated with the increasing sequence of eigenvalues* $\omega_k^2[s]$*, for which the following alternative holds: For each* $k > 0$ *we have either* (i) *or* (ii),

(i) *The pair* $(\mathbf{u}_k, \omega_k^2[s])$ *is a Maxwell eigenpair solution of* (11).
(ii) *There exists an eigenpair* (φ, ν) *of the Dirichlet Laplacian* Δ^{Dir} *such that* $\mathbf{u}_k = \mathbf{grad}\,\varphi$ *and* $\omega_k^2 = s\,\nu$.

Conversely any non-trivial solution $(\mathbf{E}, \omega)$ *of* (11) *is a combination of pairs of type* (i) *with* $\omega = \omega_k$.

This result is proved in [26]. As a consequence, we see that the eigenvectors do not depend on s (only their ranks do) and can be organised in two families: the Maxwell eigenvectors for which the divergence part of the energy $a_s(\mathbf{u}, \mathbf{u})$ is zero, and the "spurious" eigenvectors for which the curl part of the energy is zero.

In the literature, this method is often referred to as a "penalty method", as opposed to the "mixed methods" based on edge elements, see §6.1. We prefer the term of "regularisation", because the exact determination of a Maxwell eigenvalue does not require that s tends to infinity.

When discretised using finite element subspaces $\mathfrak{X}_h$, the form a_s is expected to provide two families of eigenvectors $\mathbf{u}_{h,k}$ which can be identified by the value of the ratio

$$\tau(\mathbf{u}_h) = \frac{\|\,\mathbf{curl}\,\mathbf{u}_h\|^2_{L^2(\Omega)}}{s\|\operatorname{div}\mathbf{u}_h\|^2_{L^2(\Omega)}}\,. \tag{28}$$

We expect large values of τ for the approximation of Maxwell eigenvectors and small ones for the approximation of spurious eigenvectors.

7.2 In the Square

We apply this procedure on the same square domain as before (cf. Fig.s 1-4): We compute the eigenpairs with the one-element $\mathbb{Q}_8$ approximation and a small value of

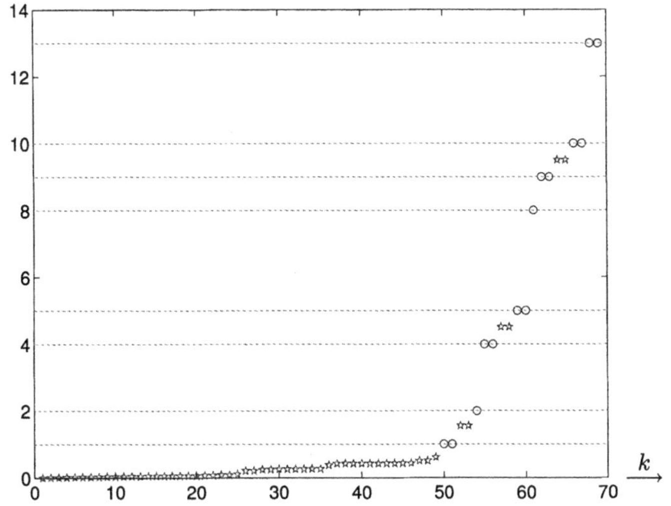

Fig. 5. One square element ($\mathbb{Q}_8$) and $s = 0.002$. *Eigenvalue* ω_k *vs rank* k

$s = 0.002$. For each eigenpair, we compute τ and mark the corresponding eigenvalue by a circle if $\tau \geq \tau_0$, by a "pentagram" if $\tau \leq \tau_0^{-1}$ and by a triangle if $\tau_0^{-1} < \tau < \tau_0$.

No triangles are observed when we choose $\tau_0 = 50$. If we compare with Fig.2, we see that the 49 zero eigenvalues rise to positive values, and that the 2 extra ones on 1, 4, 9,... rise also! This is due to the large value of the divergences of the related eigenvectors, cf. (23).

We represent the eigenvalues of a_s in a different way now. We let s *vary* and try to follow the representatives of the two families. We do that by post-processing the results where we list for each value of s, the first 66 eigenvalues with the curl and divergence energies of their eigenvectors, which allows us to compute τ.

We can see that the two families are easily distinguishable, the Maxwell ones are constant with respect to s and provide a good approximation of the true ones (still represented by horizontal solid lines) and the spurious ones are perfectly linear with respect to s (we have only plotted the first 15 lines for the sake of clarity of the figure – you see only 9 due to multiple eigenvalues). Similar computations are shown in [1, Fig. 5.2] (without postprocessing of the eigenvalues).

So, everything well in the best possible world ?

7.3 In the L-Shaped Domain

Let us consider the L-shaped domain $(-1,1)^2 \setminus (-1,0)^2$. We use a 9 element mesh: 3 small squares of size $1/16$ around $(0,0)$ and the remaining "annulus" split into 6 identical trapezia. We take $\mathbb{Q}_{10}$ polynomials on the reference element. There are 1902 degrees of freedom. We have a precise evaluation of the true solution by a computation of the eigenvalues of the scalar Neumann problem on the same domain: These values are represented as horizontal (gray) lines.

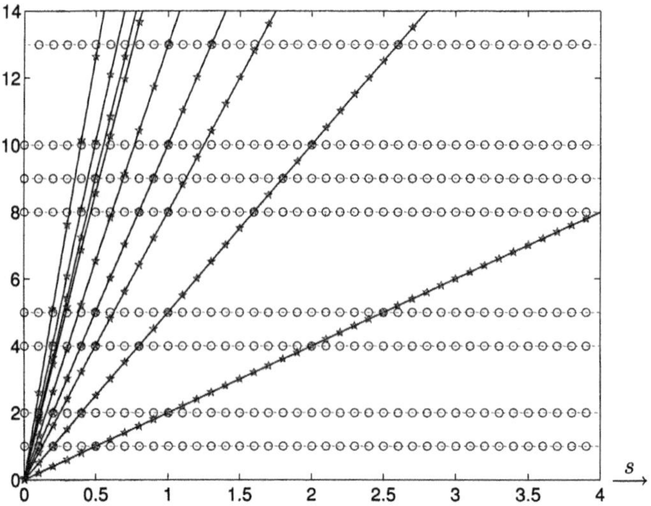

Fig. 6. One square element ($\mathbb{Q}_8$) and $s = 0.002 \to 4$. *Eigenvalues vs parameter s*

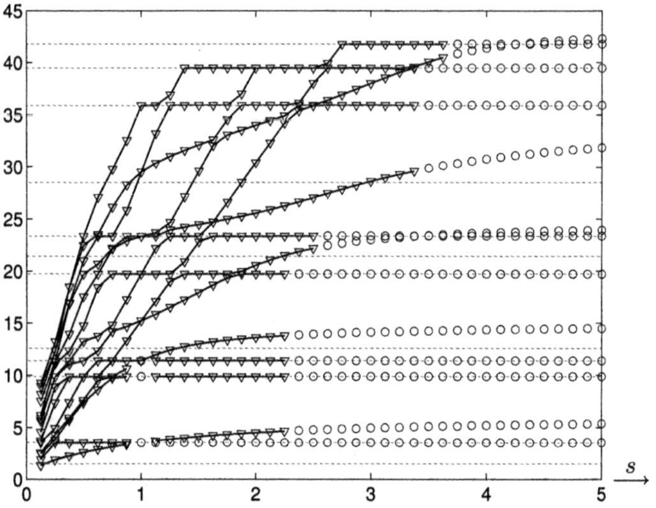

Fig. 7. L-shape (9 elements $\mathbb{Q}_{10}$) and $s = 0.125 \to 5$. *Eigenvalues vs parameter s*

On Fig.7, we can see that the sorting criterion does not work everywhere, and that certain families of eigenvalues prefer to organise themselves on *curved* lines. Moreover, while certain true eigenvalues are correctly approximated for any value of s (# 2, 3, 4, 6, 8, 10, 11, 12, – disregarding the multiplicity) others are completely left apart (# 1, 5, 7, 9). The reason for this is that the value of the coefficient γ_c of the first singularity $\mathbf{grad}(r^{2/3} \sin 2\pi\theta/3)$: When γ_c is non-zero, the eigenvector does not belong to H^1 and cannot therefore be approximated.

The reason behind these odd phenomena is not a poor approximation: More refined meshes with more elements would give almost the same results. The reason is that each of these Galerkin spaces is contained in H_N which is closed in the continuous variational space X_N, and nevertheless does not coincide with it since Ω is not convex, see Theorem 2 and (5).

In the two next sections we present two main strategies to overcome this major problem: The singular function method (SFM) and the weighted regularisation method (WRM). Finally we also mention from [17] two more methods based on regularisation.

8 Plain Regularisation With a Singular Function Method

The methods based on this idea are described in the survey note [47] which also provides many references. Here we will briefly mention them, referring the reader to [47] for more details. The common idea of these methods amounts to replacing the standard finite element spaces with spaces augmented by these singular functions which cannot be approximated by the (nodal) finite element spaces.

To our knowledge, these methods have not been applied to eigenvalue problems, but insofar as they rely on a Galerkin method, they can be applied to the computation of eigenvalues as well.

They require an explicit knowledge of the singularities and, until now, they have been successfully applied only to two-dimensional problems or axisymmetric geometries (where the angular Fourier decomposition transforms the three-dimensional problem into a series of problems in a meridian plane).

8.1 Singularity as a Gradient

In a 2D polygonal domain, one augments nodal finite element spaces by the fields

$$\mathbf{S}_{\mathbf{c}}^{\mathrm{Dir}} := \mathbf{grad}\,\big(r^{\pi/\omega_{\mathbf{c}}} \sin \frac{\pi\theta}{\omega_{\mathbf{c}}}\big), \quad \forall \mathbf{c} \text{ reentrant corner.}$$

In order to implement the essential boundary conditions, the usual cut-off is not used (because this would generate large errors) and a discrete correction of $\mathbf{S}_{\mathbf{c}}^{\mathrm{Dir}}$ is preferred. Instead of $\mathbf{S}_{\mathbf{c}}^{\mathrm{Dir}}$, one considers the corrected singularities $\mathbf{E}_{\mathbf{c},h} := \mathbf{S}_{\mathbf{c}}^{\mathrm{Dir}} - \mathbf{T}_{\mathbf{c},h}$ where $\mathbf{T}_{\mathbf{c},h} \in X_h$ is the solution of the discrete problem

$$\forall \mathbf{v}_h \in X_{h,0}, \quad a_1(\mathbf{T}_{\mathbf{c},h}, \mathbf{v}_h) = 0 \quad \text{and} \quad \mathbf{T}_{\mathbf{c},h} \times \mathbf{n} = \mathbf{S}_{c}^{\mathrm{Dir}} \times \mathbf{n} \text{ on } \partial\Omega.$$

Here X_h is the full nodal vector field space and $X_{h,0}$ is the subspace of the fields $\mathbf{v}$ which satisfy the boundary condition $\mathbf{v} \times \mathbf{n} = 0$. The bilinear form a_1 is defined in (27) for $s = 1$. For further details, see [19, 48].

A similar technique may be applied at an axisymmetric 3D conical point [43].

8.2 Divergence-free Singularities

In a 2D polygonal domain, the main singularity to be considered is

$$\mathbf{S}_{\mathbf{c}}^{\mathrm{Neu}} := \mathbf{curl}\,\Big(r^{\pi/\omega_{\mathbf{c}}} \cos\frac{\pi\theta}{\omega_{\mathbf{c}}}\Big), \quad \forall \mathbf{c} \text{ reentrant corner},$$

which in fact coincides with $\mathbf{S}_{\mathbf{c}}^{\mathrm{Dir}}$! The distinction lies in the correction which is adopted for $\mathbf{S}_{\mathbf{c}}^{\mathrm{Neu}}$. Let $\Phi_{\mathbf{c}}^{\mathrm{Neu}}$ be the Neumann singular function $r^{\pi/\omega_{\mathbf{c}}}\cos(\pi\theta/\omega_{\mathbf{c}})$ and let us denote by $H^2_{\mathrm{Neu}}(\Omega)$ the subspace of functions with zero normal derivative on $\partial\Omega$. There exists a unique $\Psi_{\mathbf{c}} \in H^2_{\mathrm{Neu}}(\Omega)$ such that

$$\forall \Psi \in H^2_{\mathrm{Neu}}(\Omega), \quad \int_\Omega \Delta(\Phi_{\mathbf{c}}^{\mathrm{Neu}} - \Psi_{\mathbf{c}})\, \Delta\Psi = 0.$$

Then the singularity which is added in the method is $\mathbf{S}_{\mathbf{c}}^{\mathrm{Neu}} - \mathbf{curl}(\Phi_{\mathbf{c}}^{\mathrm{Neu}})$. This singularity is divergence-free and orthogonal to $H_N \cap H(\mathrm{div}; 0)$. Further details are given in [4].

9 The Weighted Regularisation Method

The singular function method appears to be difficult to analyse and implement in the case of a three-dimensional polyhedron, because of the complex structure of the infinite dimensional space of non-H^1 singular functions. The alternative to the SFM is to use the regularisation together with a relaxation of the topology of the variational space. In this section we focus on the weighted regularisation where the divergence part of the energy is lowered by the adjunction of a weight which tends to zero near non convex corners and edges. This idea can be compared to the technique in [58] where the contribution to the divergence integral of the elements near non convex corners is simply ignored.

9.1 The Principle

The idea and the analysis of the weighted regularisation is presented in [28]. It consists of generalising the basic regularised form (27) by the introduction of a weight w in the divergence term, so that a_s is now replaced by $a_{s,w}$ defined as

$$a_{s,w}(\mathbf{u},\mathbf{v}) := a_0(\mathbf{u},\mathbf{v}) + s\int_\Omega w(\mathbf{x})^2 \,\mathrm{div}\,\mathbf{u}\,\mathrm{div}\,\mathbf{v}\,\mathrm{d}\mathbf{x},$$

where $w > 0$ is a weight. The associated variational space is now

$$X_N^w := H_0(\mathbf{curl}) \cap \{\mathbf{v} \in L^2(\Omega)^3\,,\ w\,\mathrm{div}\,\mathbf{v} \in L^2(\Omega)\}.$$

Taking $a = a_{s,w}$ and $\mathfrak{X} = X_N^w$ yields a new version of our problem $(\mathfrak{P})$, §5.1:

$$(\mathfrak{P}_{s,w}) \qquad \mathbf{u} \in X_N^w, \quad \forall \mathbf{v} \in X_N^w, \qquad a_{s,w}(\mathbf{u},\mathbf{v}) = \omega^2 \int_\Omega \mathbf{u}\cdot\mathbf{v}\,\mathrm{d}\mathbf{x}\,.$$

The following result generalises Theorem 5 on the natural splitting of the spectrum of problem $(\mathfrak{P}_{s,w})$.

Theorem 6. *Suppose that the weight w ensures the following embeddings hold:*

$$L^2(\Omega) \subset \big\{p \in L^2_{\mathsf{loc}}(\Omega)\,;\ wp \in L^2(\Omega)\big\} \subset H^{-1}(\Omega). \tag{29}$$

Then there exists a basis $(\mathbf{u}_k)$ of eigenvectors of problem $(\mathfrak{P}_{s,w})$ associated with the increasing sequence of eigenvalues $\omega_k^2[s,w]$, for which the following alternative holds: For each $k > 0$ we have either (i) *or* (ii)*,*

(i) *The pair $(\mathbf{u}_k, \omega_k^2[s,w])$ is a Maxwell eigenpair solution of* (11).
(ii) *There exists an eigenpair (φ,ν) of the Dirichlet weighted Laplacian $w\Delta(w\,\cdot)$ so that $\mathbf{u}_k = \mathbf{grad}\,\varphi$ and $\omega_k^2 = s\,\nu$.*

Conversely any non-trivial solution $(\mathbf{E},\omega)$ of (11) *is a combination of pairs of type* (i) *with $\omega = \omega_k$.*

9.2 The Class of Weights

We choose the weight w in the form

$$w(\mathbf{x}) = d_0(\mathbf{x})^\gamma \quad \text{with} \quad d_0(\mathbf{x}) = \operatorname{dist}(\mathbf{x}, \mathfrak{S}_0) \tag{30}$$

where $\mathfrak{S}_0$ is the set of non-convex corners if Ω is a polygon, or the set of non-convex edges if Ω is a polyhedron.

Note 2. Of course we can replace d_0 in (30) by any function $\tilde{d}_0$ equivalent to d_0 on $\overline{\Omega}$. In polygons, the set $\mathfrak{S}_0$ is the union of a finite number of corner points and it is very easy to define a function $\tilde{d}_0$: we can take for example the product of the distances to each corner. In polyhedra, one has to be more careful. If several non-convex edges $\mathbf{e}$ have a common end point $\mathbf{c} \in \mathcal{C}$, then we *cannot* take $\tilde{d}_0$ as the product of distances to each $\mathbf{e}$. We will again comment on this choice when giving numerical results. □

We have:

Lemma 2. *Let $L^2_\gamma(\Omega)$ be the weighted space $\big\{p \in L^2_{\mathsf{loc}}(\Omega)\,;\ d_0^\gamma p \in L^2(\Omega)\big\}$. The embeddings* (29)*: $L^2(\Omega) \subset L^2_\gamma(\Omega) \subset H^{-1}(\Omega)$ hold if and only if $0 \le \gamma \le 1$.*

It is possible to prove a direct analogue of the splitting (5) in the setting of weighted spaces (Theorem 7 below). The standard Dirichlet Laplacian Δ^{Dir} has to be replaced by its weighted counterpart $\Delta^{\mathrm{Dir}}_\gamma$ defined as

$$\begin{aligned} \Delta^{\mathrm{Dir}}_\gamma : D(\Delta^{\mathrm{Dir}}_\gamma) := \Big\{\varphi \in H^1_0(\Omega)\,;\ \Delta\varphi \in L^2_\gamma(\Omega)\Big\} &\longrightarrow L^2_\gamma(\Omega) \\ \varphi \qquad\qquad &\longmapsto \Delta\varphi. \end{aligned}$$

The variational space adapted to the regularisation using the weight $w = d_0^\gamma$ is given by

$$X_N^\gamma := H_0(\mathbf{curl}) \cap \{\mathbf{v} \in L^2(\Omega)^3 \ , \ \operatorname{div} \mathbf{v} \in L^2_\gamma(\Omega)\}$$

with norm

$$\|\mathbf{v}\|_{X_N^\gamma} = \left(\|\mathbf{v}\|^2_{L^2(\Omega)} + \|\,\mathbf{curl}\,\mathbf{v}\,\|^2_{L^2(\Omega)} + \|\operatorname{div}\mathbf{v}\|^2_{L^2_\gamma(\Omega)}\right)^{1/2}.$$

Theorem 7. [28, Th.1.2 & 1.4] (i) *Any element* $\mathbf{v} \in X_N^\gamma$ *can be decomposed into the sum*

$$\mathbf{v} = \mathbf{w} + \mathbf{grad}\,\varphi, \quad \textit{with} \quad \mathbf{w} \in H_N \;\; \textit{and} \;\; \varphi \in D(\Delta_\gamma^{\mathrm{Dir}})$$

such that the following estimate holds

$$\|\mathbf{w}\|_{H^1(\Omega)^3} + \|\varphi\|_{H^1(\Omega)} + \|\Delta\varphi\|_{L^2_\gamma(\Omega)} \le C\|\mathbf{v}\|_{X_N^\gamma}\,.$$

(ii) *If* $H^2 \cap H^1_0(\Omega)$ *is dense in* $D(\Delta_\gamma^{\mathrm{Dir}})$*, then* H_N *is dense in* X_N^γ*.*

9.3 Choosing the Weight

With the previous theorem, we do not yet know whether the introduction of a weight helps for the density of smooth functions. The answer is provided in the next statement.

Theorem 8. [28, Th.4.1] *For any polygonal or polyhedral domain, there exists* $\tau(\Omega) > 0$ *such that*

$$\forall\gamma,\;\; 1 - \tau(\Omega) < \gamma \le 1, \quad H_N \;\; \textit{is dense in} \;\; X_N^\gamma. \tag{31}$$

Let $\tau_{\mathcal{E}} = \min_{\mathbf{e}\in\mathcal{E}}(\pi/\omega_{\mathbf{e}})$ *and let* $\tau_{\mathcal{C}}^{\mathrm{Dir}}$ *be defined in* (10)*, then* $\tau(\Omega)$ *is given by:*

$$\tau(\Omega) = \tau_{\mathcal{E}} \;\; \textit{in 2D} \qquad \textit{and} \qquad \tau(\Omega) = \min\{\tau_{\mathcal{E}},\, \tau_{\mathcal{C}}^{\mathrm{Dir}} + \tfrac{1}{2}\} \;\; \textit{in 3D}.$$

The proof of this theorem relies on *(ii)* of Theorem 7 together with a characterisation of the domain of $\Delta_\gamma^{\mathrm{Dir}}$: when (31) holds, the domain $D(\Delta_\gamma^{\mathrm{Dir}})$ coincides with the following weighted space:

$$V^2_\gamma(\Omega) = \{p \in L^2_{\mathrm{loc}}(\Omega)\,;\; d_0^\gamma d^{|\alpha|-2}\partial^\alpha p \in L^2(\Omega),\; |\alpha| \le 2\}$$

where d is the distance to the set $\mathfrak{S}$ of *all* corners (in 2D) and edges (in 3D). In fact, the singularities described in §3 belong to $V^2_\gamma(\Omega)$ for all γ satisfying condition (31).

9.4 Convergence of the Spectrum

Let L_γ^{Dir} denote the weighted Laplacian $d^\gamma\Delta(d^\gamma.)$ which is responsible for the "spurious" part of the spectrum, see *(ii)* in Theorem 6. Its spectrum $\sigma(L_\gamma^{\mathrm{Dir}})$ has the following properties:

Lemma 3. (i) $\sigma(L_\gamma^{\mathrm{Dir}})$ *decreases as* γ *increases.*
(ii) *There exists* $\sigma_0 > 0$ *such that:*

$$\forall\gamma \le 1, \qquad 0 < \sigma_0 \le \sigma(L_\gamma^{\mathrm{Dir}})\,. \tag{32}$$

(iii) *For all* $\gamma < 1$, L_γ^{Dir} *has a purely discrete spectrum with accumulation at* $+\infty$.
(iv) *For* $\gamma = 1$, $\sigma(L_\gamma^{\mathrm{Dir}})$ *contains a full interval* $[\sigma_1, +\infty)$ *of essential spectrum.*

Let $(\mathfrak{X}_h)_h$ be a conforming finite element approximation of X_N^γ. We can apply the result of [5] to our case. Let ω be an eigenfrequency of problem $(\mathfrak{P})$ with multiplicity q, $\omega = \omega_\ell = \ldots = \omega_{\ell+q-1}$, and let us denote the corresponding approximate eigenpairs by $(\omega_{h,k}, \mathbf{u}_{h,k})$, $k = \ell, \ldots, \ell+q-1$. We have the estimate:

$$|(\omega_{h,k} - \omega)/\omega| \le C \inf_{\mathbf{v}_h\in\mathfrak{X}_h} \|E_\omega \mathbf{u}_{h,k} - \mathbf{v}_h\|^2_{X_N^\gamma}, \qquad k = \ell, \ldots, \ell+q-1. \tag{33}$$

Here E_ω is the orthogonal projection of X_N^γ onto the exact eigenspace of problem $(\mathfrak{P})$ corresponding to ω. Combining this with Theorem 6 and Lemma 3, we obtain:

Corollary 1. *Let* ω_k *be the* k*-th Maxwell eigenfrequency, cf. problem* (11). *Let* $\gamma \in [0,1]$ *and* s *be such that* $\omega_k < s \min\sigma(L_\gamma^{\mathrm{Dir}})$. *Let* $\omega_{h,k}$ *be the* k*-th eigenfrequency of the discrete problem* $(\mathfrak{P}_h)$: *Find* $\mathbf{u}_h \in \mathfrak{X}_h$

$$\forall \mathbf{v}_h \in \mathfrak{X}_h, \quad a_0(\mathbf{u}_h, \mathbf{v}_h) + s\int_\Omega d_0(\mathbf{x})^{2\gamma} \operatorname{div}\mathbf{u}_h \operatorname{div}\mathbf{v}_h \,\mathrm{d}\mathbf{x} = \omega_h^2 \int_\Omega \mathbf{u}_h\cdot\mathbf{v}_h\,\mathrm{d}\mathbf{x}.$$

If $S[\omega_k]$ *denotes the unit ball of the Maxwell eigenspace associated with the eigenvalue* ω_k, *then the estimate holds*

$$|\omega_{h,k} - \omega_k| \le C \sup_{\mathbf{u}\in S[\omega_k]} \inf_{\mathbf{v}_h\in\mathfrak{X}_h} \|\mathbf{u} - \mathbf{v}_h\|^2_{X_N^\gamma}\,. \tag{34}$$

Given suitable weight according to condition (31), we obtain from [28, Th.7.4] convergence rates for eigenfrequencies for the h-version of the finite element method with nodal elements via the weighted regularisation method.

If an *hp*-version of finite elements is used (geometrical refinement of the mesh and high degree polynomials [61]), it is possible to prove that the weighted regularisation method inherits the exponential convergence rates obtained for the Laplace operator, see [32] for 2D domains.

9.5 Experiments on the L-shaped Domain

We return to the example given in §7.3 and illustrated in Fig.7. We can show that for such a domain the lower bound σ_0 (32) of the spurious Laplacian is $\frac{4}{9}$. We use the same 9 element mesh and the space $\mathbb{Q}_{10}$. We only change the bilinear form, using $a_{s,w}$ with $w = r^{1/2}$ in Fig.8 and $w = r$ in Fig.9 instead of the plain a_s (in this case we have $d_0(\mathbf{x}) = r$). We recall that the "exact" values are computed as the Neumann eigenvalues of the Laplace operator and are represented as horizontal lines

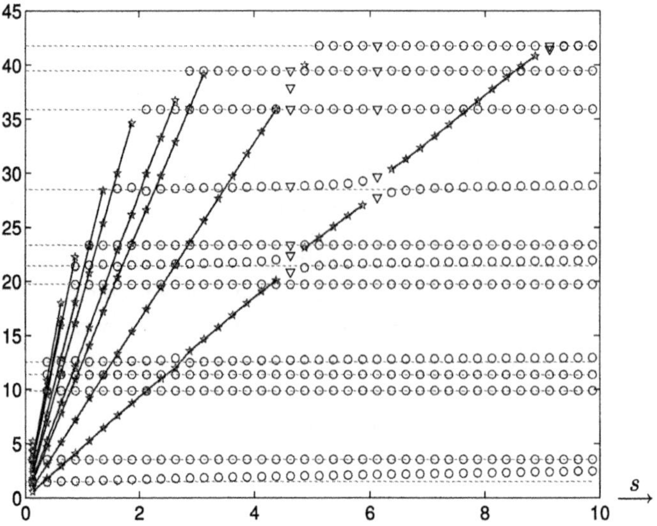

Fig. 8. L-shape, WRM with $\gamma = \frac{1}{2}$ and $s : 0.125 \rightarrow 10$. *Eigenvalues vs parameter s*

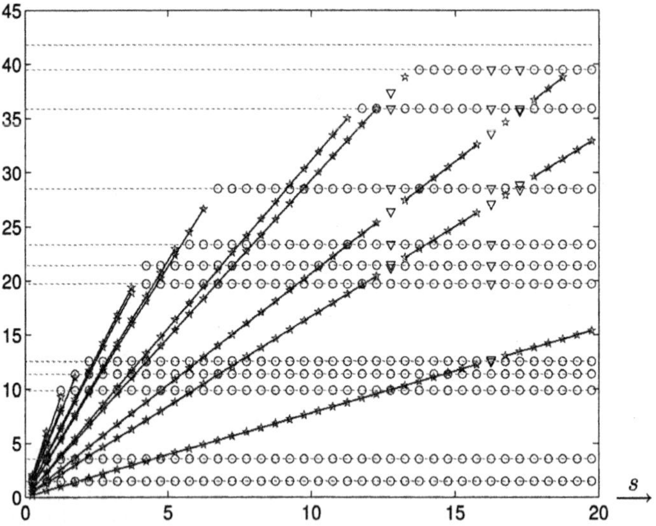

Fig. 9. L-shape, WRM with $\gamma = 1$ and $s : 0.25 \rightarrow 20$. *Eigenvalues vs parameter s*

in both figures. The representation of computed eigenvalues as circles, triangles or pentagrams follow the same criteria described in §7.2.

The improvement of the approximation when going from the plain regularisation to the weighted regularisation is visible (recall that the same number of unknowns (1902) is employed for the results of Fig.7 to 9): For $\gamma = \frac{1}{2}$ (recall that the infimum of admissible values for γ is $\frac{1}{3}$), we see that the eigenvalues # 1, 5, 7, 9 which were "forgotten" by the plain regularisation, are now approximated – albeit less accurately than the others.

The results with $\gamma = 1$ have a similar appearance to those of the square in Fig.6: Each Maxwell eigenvalue is approximated by an eigenvalue of the discrete problem (almost constant with respect to s) and conversely the spurious eigenvalues depend linearly on s and escape when s increases – the smallest slope is indeed larger than $\frac{4}{9}$. We note that larger values of s have to be preferred when γ increases.

For the h-version of finite element method, combining:

a. The general convergence estimate of eigenvalues (34),
b. The best approximation rates of the WRM proved in [28, Th.7.4],
c. The regularity of the first two Maxwell eigenvectors $\mathbf{u}_1$ and $\mathbf{u}_2$ on this symmetric L-shape (i.e. $\mathbf{u}_1 \in H^{\frac{2}{3}-\varepsilon}(\Omega)$, $\mathbf{u}_2 \in H^{\frac{4}{3}-\varepsilon}(\Omega)$ for all $\varepsilon > 0$),

we obtain the theoretical convergence rates *with respect to the number of unknowns* $N \simeq h^{-2}$:

$$|\lambda_{h,1} - \lambda_1| \leq CN^{-(\gamma-\frac{1}{3})+\varepsilon} \quad \text{and} \quad |\lambda_{h,2} - \lambda_2| \leq CN^{-(\gamma+\frac{1}{3})+\varepsilon}, \tag{35}$$

for example using rectangular elements of degree $p \geq 3$. We stress that the exponent $\gamma - \frac{1}{3}$ is optimal for the first eigenvector and thus the condition $\gamma > \frac{1}{3}$ is necessary to obtain convergence. The second eigenvector is more regular and can be approximated without WRM as was visible on Fig.7, but the convergence rate is improved with the use of the WRM.

	Neumann		WRM $\gamma = 0$		WRM $\gamma = \frac{1}{2}$		WRM $\gamma = 1$	
N	$\|\lambda_1 - \lambda_{h,1}\|$	τ	$\|\lambda_1 - \lambda_{h,1}\|$	τ	$\|\lambda_1 - \lambda_{h,1}\|$	τ	$\|\lambda_1 - \lambda_{h,1}\|$	τ
42	0.02232261		4.1209226		3.6789139		1.0963354	
130	0.00861570	0.843	3.3219792	0.191	2.8324970	0.231	0.4075287	0.876
450	0.00339645	0.750	3.1616433	0.040	2.5032137	0.100	0.1554590	0.776
1666	0.00134663	0.707	3.1077517	0.013	2.2467554	0.083	0.0597067	0.731
N	$\|\lambda_2 - \lambda_{h,2}\|$	τ	$\|\lambda_2 - \lambda_{h,2}\|$	τ	$\|\lambda_2 - \lambda_{h,2}\|$	τ	$\|\lambda_2 - \lambda_{h,2}\|$	τ
42	0.01491466		0.3439764		0.2690502		0.0846120	
130	0.00113782	2.277	0.1058913	1.043	0.0782876	1.093	0.0130193	1.657
450	0.00010186	1.944	0.0392901	0.798	0.0271478	0.853	0.0021616	1.446
1666	0.00001117	1.689	0.0152519	0.723	0.0097170	0.785	0.0003549	1.380

Table 2. L-shape, uniform grids, $p = 2$. Errors and convergence rates

We illustrate the error estimates (35) by choosing $\mathbb{Q}_p$ elements on a sequence of uniform grids, starting with a very simple grid containing only 3 squares which are

	Neumann		WRM $\gamma=0$		WRM $\gamma=\frac{1}{2}$		WRM $\gamma=1$	
N	$\|\lambda_1-\lambda_{h,1}\|$	τ	$\|\lambda_1-\lambda_{h,1}\|$	τ	$\|\lambda_1-\lambda_{h,1}\|$	τ	$\|\lambda_1-\lambda_{h,1}\|$	τ
450	0.00084498		3.1789407		2.1954444		0.0374398	
1666	0.00033527	0.706	3.1149073	0.016	1.9323080	0.098	0.0147321	0.713
6402	0.00013226	0.691	3.0899818	0.006	1.6883397	0.100	0.0057792	0.695
25090	0.00005169	0.688	3.0801659	0.002	1.4581948	0.107	0.0022747	0.683
N	$\|\lambda_2-\lambda_{h,2}\|$	τ	$\|\lambda_2-\lambda_{h,2}\|$	τ	$\|\lambda_2-\lambda_{h,2}\|$	τ	$\|\lambda_2-\lambda_{h,2}\|$	τ
450	0.00000215		0.0478032		0.0176105		0.0001714	
1666	0.00000034	1.408	0.0185193	0.724	0.0063407	0.780	0.0000305	1.318
6402	0.00000005	1.372	0.0072806	0.694	0.0021963	0.788	0.0000050	1.335
25090	0.00000001	1.345	0.0028786	0.679	0.0007433	0.793	0.0000008	1.338

Table 3. L-shape, uniform grids, $p = 8$. Errors and convergence rates

then repeatedly sub-divided into four. In Tables 2 and 3, we show errors and convergence rates for the computation of the first two Maxwell eigenvalues. We choose $s = 100$ to avoid the spurious eigenvalues. With weight exponent $\gamma = 1$, we see the expected convergence rates $\tau = 2/3$ for the first eigenvalue and $\tau = 4/3$ for the second eigenvalue.

Other numerical results are provided in [30], both for the above L-shaped domain and for another L-shaped domain with curved sides. We show in particular how the error decreases when a geometrically refined mesh is used together with high degree p: With the degree p equal to the number of layers of the mesh, an exponential rate of convergence is obtained, see also [32].

Another situation where the WRM turns out to be very efficient is the problem with an impedance boundary condition. In [29] we showed numerical results for a boundary penalisation method where the perfect conductor boundary condition on an L-shaped domain is replaced by an impedance-like condition. According to the density result in Theorem 1 for W_N, the standard regularisation method leads to a convergent Galerkin approximation even in the presence of non-H^1 singularities. In the experiments in [29], one sees convergence, but the convergence rate is extremely low, so that the method is not practical. By using the WRM with $\gamma = 1$ on the L-shape domain, one observes the same good convergence properties as reported here for the perfect conductor boundary condition. The boundary penalisation, when combined with the weighted regularisation, can therefore be employed efficiently as an alternative to the use of essential boundary conditions.

9.6 3D Computations

To test our code (the finite element library MÉLINA [54]) and our weighted regularisation method for three-dimensional domains, we first performed experiments with the thick L-shaped domain formed by taking the tensor product of the 2D L-shape with the interval $(0,1)$ and obtained the expected behaviour of computed eigenvalues. These experiments will be reported elsewhere.

Instead, we shall report on computations using the archetypical three-dimensional corner:

$$\Omega = (-1,1)^3 \setminus (-1,0)^3.$$

We refer to this as the *Fichera corner*, in honour of the fact that Fichera was the first to estimate the lowest Laplace-Dirichlet corner exponent at the corner $\mathbf{c} = 0$: in [41] the estimate $0.4335 < \lambda_{\mathbf{c}}^{1,\mathrm{Dir}} < 0.4645$ is given. In [60], a computation by a boundary element method yields the approximate value $\lambda_{\mathbf{c}}^{1,\mathrm{Dir}} \simeq 0.45418$.

Let us recall, cf. Table 1, that the set of Maxwell corner exponents is the union of the Dirichlet-Laplace exponents minus 1 and the Neumann-Laplace. Thomas Apel provided us with the following numerical approximation of the first Laplace exponents for the Fichera corner (computed with the method of [3]), see Table 4. Thus

Dirichlet	Neumann
0.45418	0.00000
1.23088	0.84001
1.23090	0.84002
1.78432	1.20637
2.11766	1.80618
2.11768	1.80619

Table 4. Fichera corner. Laplacian corner exponents

the first Maxwell exponents are -0.5459, 0.2309, 0.7843, 0.8400, etc...

The domain Ω has three non-convex edges $\mathbf{e}_1$, $\mathbf{e}_2$ and $\mathbf{e}_3$ which coincide with the interval $(-1,0)$ in each coordinate axis. As distance function d_0, we can choose the minimum of the distance functions d_j to each edge $\mathbf{e}_j$, or take the product $d_1 d_2 d_3$ *divided by* ρ^2 where ρ is the distance to 0.

We performed computations with a tensor mesh (each side of the 7 cubes which form Ω is divided into 3 subintervals according to the ratios $0, \frac{1}{64}, \frac{1}{2}, 1$) and $\mathbb{Q}_4$ elements are used (which produces 41691 degrees of freedom for the Maxwell system). In Fig.10 we plot the 8 lowest computed eigenvalues obtained with the WRM and $\gamma = 1$ (circles, pentagrams and triangles according to the conventions in §7.2), and also the results obtained without weight ($\gamma = 0$), represented *on the same graph* as dots $^\bullet$. Both families (Maxwell and spurious) are connected by lines. Reference values for the exact Maxwell eigenvalues are not available.

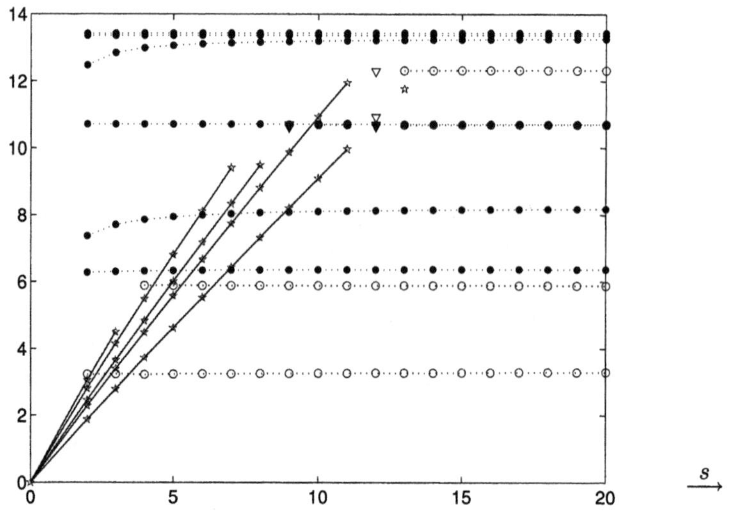

Fig. 10. Fichera corner, WRM with $\gamma = 1$ and $s : 1 \to 20$. *Parameter s vs eigenvalues*

10 Comparisons

10.1 Miscellaneous Penalised Formulations

Two more methods for plane domains using "penalised" formulations (i.e. in our words "regularisation") are described in [17] together with numerical experiments. One of them consists of the use of the piecewise linear nonconforming elements of Crouzeix-Raviart [33] in the plain regularised bilinear form (27).

The other method can be expressed as a three field method involving the electric field and its (scalar) curl and divergence. It is similar to that of [6] and has been partially analysed in [44]. This method is equivalent to a plain regularised formulation with *projection*. Given a mesh $\mathcal{T}_h$, let V_h, M_h and Q_h be the following three spaces:

$$V_h = \{\mathbf{v}_h \in X_N, \quad \mathbf{v}_h \text{ piecewise } \mathbb{Q}_2\} \tag{36}$$
$$M_h = \{\mu_h \in L^2_0(\Omega), \quad \mu_h \text{ piecewise } \mathbb{P}_1\} \tag{37}$$
$$Q_h = \{q_h \in L^2(\Omega), \quad q_h \text{ piecewise } \mathbb{P}_1\} \tag{38}$$

Here $L^2_0(\Omega)$ denotes the space of L^2 functions with zero mean value over Ω. Let P_1 and $P_{1,0}$ denote the L^2 projections over Q_h and M_h, respectively. Then the three field method is equivalent to solve

$$\mathbf{u}_h \in \mathfrak{X}_h, \quad \forall \mathbf{v}_h \in \mathfrak{X}_h, \qquad a(\mathbf{u}_h, \mathbf{v}_h) = \omega_h^2 \int_\Omega \mathbf{u}_h \cdot \mathbf{v}_h \, \mathrm{d}\mathbf{x}$$

with the space $\mathfrak{X}_h = V_h$ and the bilinear form $a = a^h_s$

$$a^h_s(\mathbf{u}, \mathbf{v}) = \int_\Omega P_{1,0}(\operatorname{curl} \mathbf{u})\, P_{1,0}(\operatorname{curl} \mathbf{v}) \, \mathrm{d}\mathbf{x} + s \int_\Omega P_1(\operatorname{div} \mathbf{u})\, P_1(\operatorname{div} \mathbf{v}) \, \mathrm{d}\mathbf{x} \,.$$

10.2 Numerical Results on the L-Shaped Domain

it is interesting to compare the results in [17] with what can be obtained by the weighted regularisation. The L-shape is now $(0,\pi)^2 \setminus (0,\pi/2)^2$. We add to the results from [17, Table II] reference values (the first twelve eigenvalues of the Neumann scalar problem) and two series of computations made by the WRM with $\mathbb{Q}_2$ and $\mathbb{Q}_4$ elements.

Neumann 1581 d.o.f.	Edge $\mathbf{P}_1$ 910 d.o.f.	$\mathbb{P}_1^{CR}$ 966 d.o.f.	$\mathbb{Q}_2^P$ 962 d.o.f.	WRM $\mathbb{Q}_2$ 1050 d.o.f.	WRM $\mathbb{Q}_4$ 1050 d.o.f.
0.59805	0.59179	0.61595	0.60510	0.61216	0.60166
1.43229	1.43232	1.45070	1.43314	1.43265	1.43232
4.00000	4.00554	3.93313	4.00047	4.00180	4.00000
4.00000	4.00554	4.01422	4.00049	4.00182	4.00000
4.61598	4.61320	4.60196	4.62364	4.62132	4.61600
5.09540	5.06733	5.07813	5.13854	5.10554	5.09633
8.00000	7.95513	7.90668	8.08411	8.03128	8.00008
8.68312	8.64737	8.71656	8.76070	8.70613	8.67862
9.46112	9.48166	9.57140	9.48655	9.49236	9.46145
11.54689	11.4261	11.5640	11.8268	11.64367	11.54926
14.54106	14.4486	14.5241	14.8680	14.72710	14.54433
16.00000	16.0862	15.6111	16.0254	16.10483	16.00022
Number of computed zeros					
1	267	305	214	0	0

Table 5. L-shape. First 12 eigenvalues by different methods

In Table 5, columns 2, 3 and 4 are taken from [17]: $\mathbb{P}_1^{CR}$ refers to the Crouzeix-Raviart elements, whereas $\mathbb{Q}_2^P$ refers to the nodal biquadratic elements with projection. The value of s is 100 for both, and 12.5 for the WRM. In the WRM series we have removed the spurious values which were present: 10.4818282 for the degree 2 and 8.9545298 for the degree 4. Such a spurious value is easily recognizee by its small ratio $\tau(\mathbf{u}_h)$, see (28). Another way to trace them is to compute for two different value of s.

We can note that in the series $\mathbb{P}_1^{CR}$ and $\mathbb{Q}_2^P$ it often happens that a computed eigenvalue is below its exact counterpart. For the WRM, this never happens *below* $s\sigma_0$, *with* σ_0 *the lower limit of the spurious Laplacian* L_γ^{Dir}, cf. Cor.1, (here $\gamma = 1$, $\sigma_0 = \frac{4}{9}$ and $s\sigma_0 \simeq 5.5555$). It may (seldom) happen above (the value 8.67862 in the degree 4 series).

10.3 The Benchmark Page

As a step towards further comparison of the performance of different numerical methods for the Maxwell eigenvalue problem, a web page has been created containing the results of the 2D and 3D computations presented in this section. At

`http://www.maths.univ-rennes1.fr/~dauge/benchmax.html`

one can find detailed descriptions of the geometries of the model boundary value and eigenvalue problems as well as the numerical results of computations using the weighted regularisation method. In addition to the L-shaped domain and the Fichera corner, there are results in 2D for a square with a crack, a curved rectangle, a curved L and a transmission problem on a square composed of four subdomains with different permittivities, and in 3D for the "thick L".

Contributions to this web page are invited in the form of numerical results for the same benchmark problems. In particular, it will be very interesting to compare results obtained with the weighted regularisation method and nodal finite elements to results obtained with various other variational formulations and edge elements.

Acknowledgements

The authors are pleased to thank Thomas Apel for performing the computation of three-dimensional singularity exponents, Daniele Boffi for providing the numerical results with edge elements and for his advice, and Peter Monk for discussions and proofreading.

References

1. S. Adam, P. Arbenz, R. Geus. Eigenvalue solvers for electromagnetic fields in cavities. Technical Report 275, Institute of Scientific Computing, ETH Zürich 1997.
2. C. Amrouche, C. Bernardi, M. Dauge, V. Girault. Vector potentials in three-dimensional nonsmooth domains. *Math. Meth. Appl. Sci.* **21** (1998) 823–864.
3. T. Apel, V. Mehrmann, D. Watkins. Structured eigenvalue methods for the computation of corner singularities in 3D anisotropic elastic structures. *Comput. Methods Appl. Mech. Engrg.* **191** (2002) 4459–4473.
4. F. Assous, P. Ciarlet, E. Sonnendrücker. Résolution des équations de Maxwell dans un domaine avec un coin rentrant. *C. R. Acad. Sc. Paris, Série I* **323** (1996) 203–208.
5. I. Babuška, J. E. Osborn. Finite element-Galerkin approximation of the eigenvalues and eigenvectors of selfadjoint problems. *Math. Comp.* **52**(186) (1989) 275–297.
6. K.-J. Bathe, C. Nitikitpaiboon, X. Wang. A mixed displacement-based finite element formulation for acoustic fluid-structure interaction. *Comput. & Structures* **56**(2-3) (1995) 225–237.
7. F. Ben Belgacem, C. Bernardi, M. Costabel, M. Dauge. Un résultat de densité pour les équations de Maxwell. *C. R. Acad. Sci. Paris Sér. I Math.* **324**(6) (1997) 731–736.
8. A. N. Bespalov. Finite element method for the eigenmode problem of a RF cavity resonator. *Soviet J. Numer. Anal. Math. Modelling* **3**(3) (1988) 163–178.

9. M. Birman, M. Solomyak. L^2-theory of the Maxwell operator in arbitrary domains. *Russ. Math. Surv.* **42 (6)** (1987) 75–96.
10. M. Birman, M. Solomyak. On the main singularities of the electric component of the electro-magnetic field in regions with screens. *St. Petersbg. Math. J.* **5 (1)** (1993) 125–139.
11. D. Boffi. Fortin operator and discrete compactness for edge elements. *Numer. Math.* **87**(2) (2000) 229–246.
12. D. Boffi. A note on the de Rham complex and a discrete compactness property. *Appl. Math. Lett.* **14**(1) (2001) 33–38.
13. D. Boffi, F. Brezzi, L. Gastaldi. On the convergence of eigenvalues for mixed formulations. *Ann. Scuola Norm. Sup. Pisa Cl. Sci. (4)* **25**(1-2) (1997) 131–154 (1998). Dedicated to Ennio De Giorgi.
14. D. Boffi, F. Brezzi, L. Gastaldi. On the problem of spurious eigenvalues in the approximation of linear elliptic problems in mixed form. *Math. Comp.* **69**(229) (2000) 121–140.
15. D. Boffi, L. Demkowicz, M. Costabel. Discrete compctness for p and hp 2d edge finite elements. TICAM Report 02-21, Université de Bordeaux 1, 2002.
16. D. Boffi, R. G. Duran, L. Gastaldi. A remark on spurious eigenvalues in a square. *Appl. Math. Lett.* **12**(3) (1999) 107–114.
17. D. Boffi, M. Farina, L. Gastaldi. On the approximation of Maxwell's eigenproblem in general 2D domains. *Comput. & Structures* **79** (2001) 1089–1096.
18. D. Boffi, P. Fernandes, L. Gastaldi, I. Perugia. Computational models of electromagnetic resonators: analysis of edge element approximation. *SIAM J. Numer. Anal.* **36** (1999) 1264–1290.
19. A.-S. Bonnet-Ben Dhia, C. Hazard, S. Lohrengel. A singular field method for the solution of Maxwell's equations in polyhedral domains. *SIAM J. Appl. Math.* **59**(6) (1999) 2028–2044 (electronic).
20. S. Caorsi, P. Fernandes, M. Raffetto. On the convergence of Galerkin finite element approximations of electromagnetic eigenproblems. *SIAM J. Numer. Anal.* **38**(2) (2000) 580–607 (electronic).
21. S. Caorsi, P. Fernandes, M. Raffetto. Spurious-free approximations of electromagnetic eigenproblems by means of Nedelec-type elements. *M2AN Math. Model. Numer. Anal.* **35**(2) (2001) 331–354.
22. P. Ciarlet, Jr., C. Hazard, S. Lohrengel. Les équations de Maxwell dans un polyèdre : un résultat de densité. *C. R. Acad. Sc. Paris, Série I Math.* **326**(11) (1998) 1305–1310.
23. M. Costabel. A coercive bilinear form for Maxwell's equations. *J. Math. Anal. Appl.* **157 (2)** (1991) 527–541.
24. M. Costabel, M. Dauge. Espaces fonctionnels Maxwell: Les gentils, les méchants et les singularités. On line publication (Dec. 1998): `http://www.maths.univ-rennes1.fr/~dauge`.
25. M. Costabel, M. Dauge. Un résultat de densité pour les équations de Maxwell régularisées dans un domaine lipschitzien. *C. R. Acad. Sc. Paris, Série I* **327** (1998) 849–854.
26. M. Costabel, M. Dauge. Maxwell and Lamé eigenvalues on polyhedra. *Math. Meth. Appl. Sci.* **22** (1999) 243–258.
27. M. Costabel, M. Dauge. Singularities of electromagnetic fields in polyhedral domains. *Arch. Rational Mech. Anal.* **151**(3) (2000) 221–276.
28. M. Costabel, M. Dauge. Weighted regularization of Maxwell equations in polyhedral domains. *Numer. Math.* **93 (2)** (2002) 239–277.
29. M. Costabel, M. Dauge, D. Martin. Numerical investigation of a boundary penalization method for Maxwell equations. In P. Neittaanmäki, T. Tiihonen, P. Tarvainen, editors,

Proceedings of the 3rd European Conference on Numerical Mathematics and Advanced Applications, pp. 214–221. World Scientific, Singapore 2000.

30. M. Costabel, M. Dauge, D. Martin, G. Vial. Weighted regularization of Maxwell equations – computations in curvilinear polygons. In *Proceedings of the 4th European Conference on Numerical Mathematics and Advanced Applications*. Springer 2002.
31. M. Costabel, M. Dauge, S. Nicaise. Singularities of Maxwell interface problems. *M2AN Math. Model. Numer. Anal.* **33**(3) (1999) 627–649.
32. M. Costabel, M. Dauge, C. Schwab. Exponential convergence of the *hp*-FEM for the weighted regularization of Maxwell equations in polygonal domains. In preparation.
33. M. Crouzeix, P.-A. Raviart. Conforming and nonconforming finite element methods for solving the stationary Stokes equations. I. *Rev. Française Automat. Informat. Recherche Opérationnelle Sér. Rouge* **7**(R-3) (1973) 33–75.
34. M. Dauge. "Simple" corner-edge asymptotics. On line publication (Dec. 2000): `http://www.maths.univ-rennes1.fr/~dauge`.
35. M. Dauge. *Elliptic Boundary Value Problems in Corner Domains – Smoothness and Asymptotics of Solutions.* Lecture Notes in Mathematics, Vol. 1341. Springer-Verlag, Berlin 1988.
36. M. Dauge. Neumann and mixed problems on curvilinear polyhedra. *Integral Equations Oper. Theory.* **15** (1992) 227–261.
37. M. Dauge. Singularities of corner problems and problems of corner singularities. In *Actes du 30ème Congrès d'Analyse Numérique: CANum '98 (Arles, 1998)*, pp. 19–40 (electronic). Soc. Math. Appl. Indust., Paris 1999.
38. L. Demkowicz, P. Monk. Discrete compactness and the approximation of Maxwell's equations in $\mathbb{R}^3$. *Math. Comp.* **70** (2001) 507–523.
39. J. Descloux. Essential numerical range of an operator with respect to a coercive form and the approximation of its spectrum by the Galerkin method. *SIAM J. Numer. Anal.* **18**(6) (1981) 1128–1133.
40. J. Descloux, N. Nassif, J. Rappaz. On spectral approximation. I. The problem of convergence. *RAIRO Anal. Numér.* **12**(2) (1978) 97–112, iii.
41. G. Fichera. Comportamento asintotico del campo elettrico e della densità elettrica in prossimità dei punti singolari della superficie conduttore. *Rend. Sem. Mat. Univ. e Politec. Torino* **32** (1973/74) 111–143.
42. N. Filonov. Système de Maxwell dans des domaines singuliers. Thesis, Université de Bordeaux 1, 1996.
43. E. Garcia. Résolution des équation de Maxwell instationnaires dans des domaine non convexe, la méthode du complément singulier. Thèse, Université Pierre et Marie Curie 2002.
44. L. Gastaldi. Mixed finite element methods in fluid structure systems. *Numer. Math.* **74**(2) (1996) 153–176.
45. V. Girault, P. Raviart. *Finite Element Methods for the Navier–Stokes Equations, Theory and Algorithms.* Springer series in Computational Mathematics, 5. Springer-Verlag, Berlin 1986.
46. P. Grisvard. *Boundary Value Problems in Non-Smooth Domains.* Pitman, London 1985.
47. C. Hazard. Numerical simulation of corner singularities: a paradox in Maxwell-like problems. *C. R. Mecanique* **330** (2002) 57–68.
48. C. Hazard, S. Lohrengel. A singular field method for Maxwell's equations: Numerical aspects in two dimensions. *SIAM J. Numer. Anal.* (2002) To appear.
49. R. Hiptmair. Finite elements in computational electromagnetism. *Acta Numerica* (2002) 237–339.

50. F. Kikuchi. Mixed and penalty formulations for finite element analysis of an eigenvalue problem in electromagnetism. In *Proceedings of the first world congress on computational mechanics (Austin, Tex., 1986)*, Vol. 64, pp. 509–521 1987.
51. F. Kikuchi. On a discrete compactness property for the Nédélec finite elements. *J. Fac. Sci. Univ. Tokyo Sect. IA Math.* **36**(3) (1989) 479–490.
52. V. A. Kondrat'ev. Boundary-value problems for elliptic equations in domains with conical or angular points. *Trans. Moscow Math. Soc.* **16** (1967) 227–313.
53. R. Leis. Zur Theorie elektromagnetischer Schwingungen in anisotropen inhomogenen Medien. *Math. Z.* **106** (1968) 213–224.
54. D. Martin. Mélina. On line documentation: `http://www.maths.univ-rennes1.fr/~dmartin`.
55. S. A. Nazarov, B. A. Plamenevskii. *Elliptic Problems in Domains with Piecewise Smooth Boundaries*. Expositions in Mathematics 13. Walter de Gruyter, Berlin 1994.
56. J.-C. Nédélec. Mixed finite elements in $\mathbb{R}^3$. *Numer. Math.* **35** (1980) 315–341.
57. J.-C. Nédélec. A new family of mixed finite elements in $\mathbb{R}^3$. *Numer. Math.* **50**(1) (1986) 57–81.
58. K. Preiss, O. Biró, I. Ticar. Gauged current vector potential and reentrant corners in the FEM analysis of 3D eddy currents. *IEEE Transactions on Magnetics* **36**(4) (2000) 840–843.
59. P.-A. Raviart, J. M. Thomas. Primal hybrid finite element methods for 2nd order elliptic equations. *Math. Comp.* **31**(138) (1977) 391–413.
60. H. Schmitz, K. Volk, W. Wendland. Three-dimensional singularities of elastic fields near vertices. *Numer. Methods Partial Differential Equations* **9**(3) (1993) 323–337.
61. C. Schwab. *p- and hp-finite element methods. Theory and applications in solid and fluid mechanics*. The Clarendon Press Oxford University Press, New York 1998.
62. M. Suri, C. Xenophontos. Reliability of an *hp* algorithm for buckling analysis. Proceedings of IASS-IACM 2000, Fourth International Colloquium on Computation of Shell and Spatial Structures, 2000 (CD-Rom).
63. H. Vandeven. On the eigenvalues of second-order spectral differentiation operators. *Comput. Methods Appl. Mech. Engrg.* **80**(1-3) (1990) 313–318. Spectral and high order methods for partial differential equations (Como, 1989).

hp-Adaptive Finite Elements for Time-Harmonic Maxwell Equations

Leszek Demkowicz

Texas Institute for Computational and Applied Mathematics, The University of Texas at Austin, Austin, TX 78712, USA
leszek@ticam.utexas.edu

Summary. We review the fundamentals of hp-finite element discretisation of Maxwell equations and their numerical implementation, and describe an automatic hp-adaptive scheme for the time-harmonic Maxwell equations.

1 Introduction

In this article, we shall review the fundamentals of hp-finite element discretisation of Maxwell equations and the numerical implementation, and describe an automatic hp-adaptive scheme for the time-harmonic Maxwell equations. However, we should point out that study in this area is in its infancy, and that the material presented here is still far from being complete, both from the mathematical and computational points of view.

A brief history of hp-edge elements

The approximation of Maxwell equations requires setting a variational problem in the space $H(curl)$. A piecewise regular (e.g. polynomial) vector-valued function belongs to the space $H(curl)$ if and only if the tangential component is continuous across the interelement boundaries. For the lowest order approximation, this leads naturally to degrees-of-freedom related to edges (e.g. edge averages) and hence the name of *edge elements*. Credit for the construction of the lowest order edge tetrahedral elements goes to Whitney [58], and they are frequently called the Whitney elements or Whitney shape functions. Two complete families of edge elements of arbitrary order for both hexahedral (quadrilateral) and tetrahedral (triangular) elements were introduced in two fundamental contributions of Nédélec [37, 38]. Hence, edge elements are also known as Nédélec elements in the mathematics community. Nédélec did not study the possibility of p-convergence, and assumed arbitrary, but uniform, order of approximation throughout the mesh.

A parallel development of edge elements took place in the engineering community with various degree of mathematical sophistication, see [49] for a representative selection of related papers published prior to 1994. Several authors proposed the construction of higher order shape functions but, to my best knowledge, Webb and Forghani [56], and Wang and Ida [53] were among the first ones to introduce *hierarchical* vector-valued (edge) shape functions for orders p=1,2 or 3, (see also [55, 57]), thereby giving the possibility of p-adaptivity.

Monk [35] was the first to analyse p and hp-convergence of Nédélec elements, although the convergence rates obtained were not optimal. Wang, Monk and Szabo [54] were also the first to investigate the p-convergence of Maxwell eigenvalues numerically.

The construction of edge elements of arbitrary, locally varying order, used in conjunction with Kikuchi's mixed variational formulation [31] and the corresponding stability analysis, was presented in [26, 50]. The de Rham diagram for the whole family of hp elements was first presented in [24], extending the original ideas of Bossavit [11]. The construction was later generalised in [18] to include both families of Nédélec elements. Optimal p and hp interpolation estimates for triangular edge elements were proved in [19].

The first proof of convergence for Maxwell eigenvalues and Whitney elements is due to Kikuchi [32]. The importance of the de Rham diagram for the eigenvalue convergence analysis was recognised early by Boffi [8, 9], and Monk and Demkowicz [36]. Convergence of Maxwell eigenvalues in context of p and hp-extensions, especially using adaptive meshes, remains still only partially resolved [10].

An elegant construction of edge elements of arbitrary (but uniform) order in the framework of differential forms was presented by Hiptmair [28, 29].

Hierarchical shape functions of arbitrary order were constructed by Ainsworth and Coyle [2–4], along with a study of the conditioning of the mass and stiffness matrices. and of the pollution (dispersion) properties of edge elements of arbitrary order, following an earlier analysis for Helmholtz equation in one dimension by Ihlenburg and Babuška [30].

To the best of my knowledge, the first hp codes, both in 2D (hybrid meshes consisting of both quadrilaterals and triangles), and in 3D (hexahedra) were developed by Rachowicz and Demkowicz [41, 43–45]. The codes permitted both h and p adaptivity, with the possibility of anisotropic refinement by means of the constrained approximation for one-irregular meshes.

Mathematical foundations for the analysis of hp convergence for Maxwell equations start with the fundamental contributions of Costabel and Dauge [15]. In his fundamental contribution [16], Costabel demonstrated the failure of standard, penalty based, H^1-conforming elements to converge in the case of problems with singular solutions that are not in H^1. In their later work, Costabel and Dauge [17] modified the penalty term to employ a weighted regularisation, and proved and demonstrated the possibility of exponential convergence of the modified method. Experimentally obtained exponential convergence rates were reported by Rachowicz et. al. in [42], and by Ainsworth and Coyle [2].

The possibility of a fully *automatic* hp-adaptive approach for Maxwell equations using the edge elements is discussed in this paper. hp edge elements have been applied to approximate waveguide problems [51, 52] and to scattering problems by coupling hp elements with infinite elements [12, 13, 20, 60], or a perfectly matched layer (PML) [33]. A separate development is taking place in the field of discontinuous Galerkin (DG) methods for Maxwell equations [40].

Time harmonic Maxwell's equations

Our goal is to approximate the time-harmonic Maxwell's equations in both bounded and unbounded domains. In this presentation, we shall confine ourselves to bounded domains only. The techniques presented here can be extended to unbounded domains using, for instance, infinite elements [13]. The first order time-harmonic Maxwell's equations are usually reformulated as a reduced wave equation expressed either in terms of the electric field $\boldsymbol{E}$ or the magnetic field $\boldsymbol{H}$. The choice usually depends upon boundary conditions. As both formulations share the same mathematical properties, we shall focus on the formulation in terms of the electric field only.

Given a *bounded* domain $\Omega \subset \mathbb{R}^3$, with boundary Γ consisting of two disjoint parts Γ_1 and Γ_2. For the sake of simplicity, we shall restrict ourselves to simply connected domains Ω, avoiding the technical issues connected with *cohomology spaces*, see e.g. [14]. We wish to find the electric field $\boldsymbol{E}(\boldsymbol{x}), \boldsymbol{x} \in \bar{\Omega}$, which satisfies, for a given angular frequency ω:

- the reduced wave equation in Ω,

$$\boldsymbol{\nabla} \times \left(\frac{1}{\mu}\boldsymbol{\nabla} \times \boldsymbol{E}\right) - (\omega^2\epsilon - j\omega\sigma)\boldsymbol{E} = -j\omega\boldsymbol{J}^{imp}\,, \tag{1}$$

- Dirichlet (ideal conductor) boundary condition on Γ_1,

$$\boldsymbol{n} \times \boldsymbol{E} = \boldsymbol{0}\,, \tag{2}$$

- Neumann boundary condition on Γ_2,

$$\boldsymbol{n} \times \left(\frac{1}{\mu}\boldsymbol{\nabla} \times \boldsymbol{E}\right) = -j\omega\boldsymbol{J}_S^{imp}\,. \tag{3}$$

Here, ϵ, μ, σ denote dielectric permittivity, magnetic permeability and conductivity of the medium, $\boldsymbol{J}^{imp}$ is a prescribed, impressed (source) current, $\boldsymbol{J}_S^{imp}$ is a prescribed, impressed surface current tangent to the boundary Γ_2, satisfying $\boldsymbol{n} \cdot \boldsymbol{J}_S^{imp} = 0$, with $\boldsymbol{n}$ denoting the normal outward unit vector to Γ, and j is the imaginary unit.

The standard variational formulation is obtained by multiplying (1) by a vector test function $\bar{\boldsymbol{F}}$, integrating over domain Ω, integrating by parts, and using the Neumann boundary condition to obtain the following problem. Find $\boldsymbol{E} \in \boldsymbol{Q}$ such that:

$$\begin{aligned}\int_\Omega \frac{1}{\mu}(\boldsymbol{\nabla} \times \boldsymbol{E}) \cdot (\boldsymbol{\nabla} \times \bar{\boldsymbol{F}})dx - \int_\Omega (\omega^2\epsilon - j\omega\sigma)\boldsymbol{E} \cdot \bar{\boldsymbol{F}}dx = \\ -j\omega\left\{\int_\Omega \boldsymbol{J}^{imp} \cdot \bar{\boldsymbol{F}}dx + \int_{\Gamma_2} \boldsymbol{J}_S^{imp} \cdot \bar{\boldsymbol{F}}dS\right\}\end{aligned} \tag{4}$$

for all $\boldsymbol{F} \in \boldsymbol{Q}$. Here $\boldsymbol{Q}$ is the *space of admissible solutions*,

$$\boldsymbol{Q} := \{\boldsymbol{E} \in \boldsymbol{L}^2(\Omega) \, : \, \boldsymbol{\nabla} \times \boldsymbol{E} \in \boldsymbol{L}^2(\Omega), \, \boldsymbol{n} \times \boldsymbol{E} = \boldsymbol{0} \text{ on } \Gamma_1\} \, . \tag{5}$$

The classical and variational formulations are equivalent to one another whenever a classical solution exists.

Introducing a *space of Lagrange multipliers* (scalar potentials):

$$W := \{q \in H^1(\Omega) \, : \, q = 0 \text{ on } \Gamma_1\}, \tag{6}$$

we employ a special test function $\boldsymbol{F} = \boldsymbol{\nabla} q, q \in W$, to discover that the solution $\boldsymbol{E}$ to (4) must *automatically* satisfy the weak form of the continuity equation,

$$-\int_\Omega (\omega^2 \epsilon - j\omega\sigma)\boldsymbol{E} \cdot \boldsymbol{\nabla}\bar{q} \, dx = -j\omega \left\{ \int_\Omega \boldsymbol{J}^{imp} \cdot \boldsymbol{\nabla}\bar{q} \, dx + \int_{\Gamma_2} \boldsymbol{J}_S^{imp} \cdot \boldsymbol{\nabla}\bar{q} dS \right\} . \tag{7}$$

We also recall the Helmholtz decomposition:

$$\boldsymbol{E} = \boldsymbol{\nabla}\phi + \boldsymbol{E}_0, \quad \text{where } \phi \in V \text{ and } (k^2 \boldsymbol{E}_0, \boldsymbol{\nabla} q) = 0 \, \forall q \in W \tag{8}$$

Here (,) denotes standard L^2 products on domain Ω, and k is the wave number given by $k^2 = \mu(\omega^2\epsilon - j\omega\sigma)$.

It is well-known that the standard variational equation is *not* uniformly stable with respect to the wave number k. As $k \to 0$, we lose control over the gradient term in (8). This corresponds to the fact that, in the limiting case $k = 0$, the problem is ill-posed as the gradient component remains undetermined. One remedy to this problem is to enforce the continuity equation explicitly at the expense of introducing a Lagrange multiplier $p \in W$.

The so-called *regularised variational formulation* takes the following form. Find $\boldsymbol{E} \in \boldsymbol{Q}$ and $p \in W$ such that:

$$\begin{aligned}\int_\Omega \frac{1}{\mu}(\boldsymbol{\nabla} \times \boldsymbol{E}) \cdot (\boldsymbol{\nabla} \times \bar{\boldsymbol{F}}) dx - \int_\Omega (\omega^2\epsilon - j\omega\sigma)\boldsymbol{E} \cdot \bar{\boldsymbol{F}} dx \\ - \int_\Omega (\omega^2\epsilon - j\omega\sigma)\boldsymbol{\nabla} p \cdot \bar{\boldsymbol{F}} dx = -j\omega \left\{ \int_\Omega \boldsymbol{J}^{imp} \cdot \bar{\boldsymbol{F}} dx + \int_{\Gamma_2} \boldsymbol{J}_S^{imp} \cdot \bar{\boldsymbol{F}} dS \right\}\end{aligned} \tag{9}$$

and

$$-\int_\Omega (\omega^2 \epsilon - j\omega\sigma)\boldsymbol{E} \cdot \boldsymbol{\nabla}\bar{q} \, dx = -j\omega \left\{ \int_\Omega \boldsymbol{J}^{imp} \cdot \boldsymbol{\nabla}\bar{q} \, dx + \int_{\Gamma_2} \boldsymbol{J}_S^{imp} \cdot \boldsymbol{\nabla}\bar{q} dS \right\} \tag{10}$$

for all $\boldsymbol{F} \in \boldsymbol{Q}$ and $q \in W$. By repeating the earlier trick and choosing the test function $\boldsymbol{F} = \boldsymbol{\nabla} q$ in the first equation, we discover that the Lagrange multiplier p *vanishes identically*, and for that reason, it is frequently called the *hidden variable*. In contrast to the original formulation, the stability constant for the regularised formulation converges to unity, as $k \to 0$. The regularised formulation works because gradients of the scalar-valued potentials from W form precisely the null space of the curl-curl operator.

The point about the regularised (mixed) formulation is that, whether we use it or not in the actual computations (improved stability is just one of many good reason to do it), the original variational problem is *equivalent* to the mixed problem. This suggests that we cannot avoid the theory of mixed formulations when analysing the problem.

The mixed formulation and a corresponding stability analysis outlined here, led us to the idea of variable order edge elements [18, 24, 26]. We start with a standard, H^1-conforming, variable order element, (triangle, quadrilateral in 2D, tetrahedron, hexahedron, or prism in 3D), identify the corresponding *space of shape functions*, and consider its image under the gradient operator. For instance, in the case of a triangle, if the space of scalar potentials W_p consists of polynomials of order $p+1$ whose restrictions to element edges e are of *lower or equal* order p_e+1, the space of vector-valued functions $\boldsymbol{Q}_p$ used to approximate the $\boldsymbol{E}$-field will consist of vector-valued polynomials whose *tangential components* (but not normal components) on the element boundary will reduce to the lower order p_e. This corresponds to the simple fact that differentiation lowers the degree of polynomials.

2 De Rham Diagram for *hp* Spaces

2.1 Exact Sequence for Polynomial Spaces

For ease of exposition, we shall restrict our presentation to triangular and tetrahedral elements only. Given a master tetrahedral element,

$$T=\{(x_1,x_2,x_3)\ :\ x_1>0,\ x_2>0,\ x_3>0,\ x_1+x_2+x_3<1\}\ ,$$

we introduce the following polynomial spaces:

- $\mathcal{P}^p_{p_e,p_f}$ - space of polynomials of order p, defined on the tetrahedron, whose traces on edges e reduce to (possibly lower) order p_e, $e=1,\dots,6$, and traces on faces f reduce to (possibly lower) order p_f, $f=1,\dots,4$.
- $\boldsymbol{P}^p_{p_e,p_f}$ - space of vector-valued polynomials of order p, defined on tetrahedron T, with traces of their tangential components on edges e of (possibly lower) order p_e, and with traces of their tangential component on faces f of (possibly lower) order p_f.
- $\boldsymbol{P}^p_{p_f}$ - space of vector-valued polynomials of order p, defined on tetrahedron T, with traces of their normal components on faces f of (possibly lower) order p_f.
- $\mathcal{P}^p$ - space of polynomials of order p, defined on tetrahedron T.

We assume that the order p_e for edge e does not exceed the order p_f on faces f adjacent to the edge, and the order p_f is at most p, for all faces f. In particular, $\mathcal{P}^p_{-1,-1}$ denotes the space of polynomials of order p, vanishing on the boundary of the triangle, $\boldsymbol{P}^p_{-1,-1}$ stands for the space of vector-valued polynomials of order p, with the trace of the tangential component on the boundary equal zero, and $\boldsymbol{P}^p_{-1}$ is the space of vector-valued polynomials of order p, with the trace of the normal component on the boundary equal zero. The assumption on the orders p_e, p_f is realised in practice

by implementing the *minimum rule* that sets an edge order p_e and a face order p_f to the minimum of the orders p corresponding to all elements adjacent to the edge or the face, respectively.
Analogous definitions are introduced in the 2D case for the master triangular element,

$$T = \{(x_1, x_2) \, : \, x_1 > 0, \, x_2 > 0, \, x_1 + x_2 < 1\} \; .$$

- $\mathcal{P}^p_{p_e}$ - space of polynomials of order p, defined on the triangle, whose traces on an edge e reduce to (possibly lower) order $p_e, e = 1, 2, 3$.
- $\boldsymbol{P}^p_{p_e}$ - space of vector-valued polynomials of order p, defined on triangle T, with traces of their tangential components on edges e of (possibly lower) order p_e.
- $\mathcal{P}^p$ - space of polynomials of order p, defined on triangle T.

In particular, $\mathcal{P}^p_{-1}$ denotes the space of polynomials of order p, vanishing on the boundary of the triangle, while $\boldsymbol{P}^p_{-1}$ stands for the space of vector-valued polynomials of order p, with the trace of the tangential component on the boundary equal zero.

The polynomial spaces form an exact sequence, analogous to the corresponding exact sequence on the continuous level.

$$\begin{array}{ccccccccc} \mathbb{R} \longrightarrow & H^1 & \xrightarrow{\nabla} & \boldsymbol{H}(\mathrm{curl}) & \xrightarrow{\nabla\times} & \boldsymbol{H}(\mathrm{div}) & \xrightarrow{\nabla\circ} & L^2 & \longrightarrow 0 \\ \mathbb{R} \longrightarrow & \mathcal{P}^{p+1}_{p_e+1,p_f+1} & \xrightarrow{\nabla} & \mathbf{P}^p_{p_e,p_f} & \xrightarrow{\nabla\times} & \mathbf{P}^{p-1}_{p_f-1} & \xrightarrow{\nabla\circ} & \mathcal{P}^{p-2} & \longrightarrow 0 \end{array} \tag{11}$$

The analogous 2D sequence takes the form:

$$\begin{array}{ccccccc} \mathbb{R} \longrightarrow & H^1 & \xrightarrow{\nabla} & \boldsymbol{H}(\mathrm{curl}) & \xrightarrow{\nabla\times} & L^2 & \longrightarrow 0 \\ \mathbb{R} \longrightarrow & \mathcal{P}^{p+1}_{p_e+1} & \xrightarrow{\nabla} & \mathbf{P}^p_{p_e} & \xrightarrow{\nabla\circ} & \mathcal{P}^{p-2} & \longrightarrow 0 \end{array} \tag{12}$$

2.2 H^1-Conforming Projection Based Interpolation

In the case of the p and hp methods, instead of identifying degrees-of-freedom, it is more convenient to introduce directly the corresponding interpolation operators [18].

The idea of projection based interpolation is founded on determining the interpolant by performing local projections, over the element interior, element faces, and element edges. In order to guarantee ϵ-optimal p-interpolation error estimates, these projections must be performed in norms dictated by traces. For $H(curl)$-conforming elements this leads to the use of dual norms on the boundary.

Throughout the paper, $\|u\|_1 \preceq \|u\|_2$ will mean existence of a constant $C > 0$, *independent* of the order p but possibly dependent on ϵ such that,

$$\|u\|_1 \leq C\|u\|_2 \, ,$$

for all functions u of interest.

H^1 interpolation.

Let $u \in W = H^{1/2+r}(T), r > 0$, denote a function to be interpolated. Note that we have assumed a sufficient degree of regularity to guarantee continuity of the function. The idea of projection-based interpolation is based on 'boundary constrained' projection. Let g denote a piecewise polynomial that belongs to the trace space for the element space of shape functions $\mathcal{P}^p_{p_f,p_e}$ introduced above, i.e. g is a continuous function whose restriction to face f reduces to a polynomial of order p_f, and whose restriction to edge e reduces to polynomial of order p_e. The boundary constrained projection u^p is defined as follows:

$$\begin{cases} u^p \in \mathcal{P}^p_{p_f,p_e}, u^p|_{\partial T} = g \\ |u - u^p|_{1,T} \to \min \,, \end{cases}$$

where $|u|_{1,T}$ denotes the first order seminorm over the tetrahedron T. We also introduce an unconstrained boundary projection w^p:

$$\begin{cases} w^p \in \mathcal{P}^p_{p_f,p_e} \\ |u - w^p|_{1,T} \to \min \\ (u - w^p, 1)_{0,T} = 0 \,, \end{cases}$$

where $(u, v)_{0,T}$ denotes the $L^2(T)$ inner product.

The triangle inequality implies that

$$\|u - u^p\|_{1,T} \leq \|u - w^p\|_{1,T} + \|w^p - u^p\|_{1,T} \,. \tag{13}$$

The function $w^p - u^p$ is the minimum seminorm extension of the boundary trace. It follows from the Poincaré inequality that the H^1-norms of the minimum *seminorm* extension and the minimum *norm* extension are equivalent. Indeed, let u_1 and u_2 denote minimum seminorm and minimum norm polynomial extensions of a particular function. Then,

$$\begin{aligned} \|u_1\|^2_{1,T} &= \|u_1\|^2_{0,T} + |u_1|^2_{1,T} \text{ (definition of } H^1 \text{ norm)} \\ &\preceq \|u_1 - u_2\|^2_{0,T} + \|u_2\|^2_{0,T} + |u_1|^2_{1,T} \text{ (triangle inequality)} \\ &\preceq |u_1 - u_2|^2_{1,T} + \|u_2\|^2_{0,T} + |u_1|^2_{1,T} \text{ (Poincaré inequality)} \\ &\preceq |u_1|^2_{1,T} + |u_2|^2_{1,T} + \|u_2\|^2_{0,T} \text{ (triangle inequality)} \\ &\preceq |u_2|^2_{1,T} + |u_2|^2_{1,T} + \|u_2\|^2_{0,T} \text{ (}u_1 \text{ is minimum seminorm extension)} \\ &\preceq \|u_2\|^2_{1,T} \text{ (definition of } H^1 \text{ norm)} \,. \end{aligned}$$

The opposite inequality follows from the fact that u_2 is the minimum norm extension.

At this point we need a polynomial extension theorem.

Lemma 1. *There exists a polynomial extension map,*

$$A : \mathcal{P}^p(\partial T) \rightarrow \mathcal{P}^p(T) \, , (Au)|_{\partial T} = u \, ,$$

such that

$$\|Au\|_{1,T} \leq C\|u\|_{1/2,\partial T} \, ,$$

with constant C independent of p.

The result has been proved for 2D [5, 6] but, to the best of my knowledge, has not been proved in 3D.

By using the polynomial extension result, we can continue from (13) as follows:

$$\begin{aligned}
&\|u - u^p\|_{1,T} \\
&\preceq \|u - w^p\|_{1,T} + \|w^p - u^p\|_{1/2,\partial T} \text{ (polynomial extension)} \\
&\leq \|u - w^p\|_{1,T} + \|g - w^p\|_{1/2,\partial T} + \|g - u^p\|_{1/2,\partial T} \text{ (triangle inequality)} \\
&\preceq \|u - w^p\|_{1,T} + \|u - g\|_{1/2,\partial T} \text{ (Trace Theorem)}
\end{aligned}$$

In conclusion, we have shown that the error of the constrained boundary projection can be bounded by the error of the unconstrained boundary projection plus the error of the boundary approximation, measured in the right norm dictated by the Trace Theorem.

We now localise the approximation on the boundary of the element to individual faces. This is possible because of the following imbedding, valid for $\epsilon > 0$,

$$\|u\|_{1/2,\partial T} \preceq \sum_f \|u\|_{1/2+\epsilon,f} \, ,$$

for all continuous functions u on the boundary ∂T. Consequently, we can reduce the problem of approximating the function u on the boundary to that of approximation over individual faces, provided we can guarantee that the ultimate (projection-based) interpolant is continuous.

At this point, we can replace the element T by a face f, and replace the $H^1(T)$ norm and seminorm with the $H^{1/2+\epsilon}(f)$ norm and seminorm, and repeat *exactly* the same reasoning as for the whole element, arriving at the estimate,

$$\|u - u^p\|_{1/2+\epsilon,f} \preceq \|u - w^p\|_{1/2+\epsilon,f} + \|u - g\|_{\epsilon,\partial f} \, .$$

Here u^p and w^p are the constrained boundary and unconstrained boundary projections, respectively, and g is an approximation of u on the face boundary. In arriving at the estimate, we have used the same tools as were used for the whole element, a polynomial extension result, and Poincaré inequality, both with respect $H^{1/2+\epsilon}$ norm and seminorm. While the Poincaré inequality for fractional spaces follows easily from the classical inequality by a standard interpolation argument, the polynomial extension result has to be assumed.

By repeating the *localisation argument*, we obtain

$$\|u - u^p\|_{\epsilon,f} \le \sum_e \|u - u^p\|_{\epsilon,e} \,.$$

Once again the problem of approximating the function u on the face boundary ∂f has been reduced to approximation over individual edges e. This time, however, we cannot use our previous argument based on traces [1]. Instead, we use the following result [23]:

Lemma 2. *Let $I = (0,1)$ be the unit interval. For given $U \in H^1(I)$, let u^p denote the L^2-projection of U onto the space $\mathcal{P}^p_{-1}$ of polynomials of order p, vanishing at the endpoints of the interval. Then, for every $\epsilon > 0$, there exists a constant C, dependent upon ϵ, but independent of the polynomial order p and the function U, such that*

$$\|U - u^p\|_{0,I} \le \frac{C}{p^{1-\epsilon}} \|U\|_{1,I} \,. \tag{14}$$

We now define $u^p = u_1 + u_2^p$ where u^1 is the linear interpolant of u and u_2^p is the L^2-projection of $U = u - u_1$ onto the space $\mathcal{P}^{p_e}_{-1}$ of polynomials of order p_e, vanishing at the endpoints of the edge. Using a standard interpolation argument we deduce that

$$\|u - u_1 - u_2^p\|_{0,e} \preceq \frac{1}{p^{1/2+r-\epsilon}} \|u - u_1\|_{1/2+r,e} \,,$$

and finally, by a finite dimensionality argument, we have,

$$\|u - u_1\|_{1/2+r,e} \le \|u\|_{1/2+r,e} + \|u_1\|_{1/2+r,e} \preceq \|u\|_{1/2+r,e} \,.$$

Notice that we have replaced the $H^\epsilon(e)$ norm with just the L^2 norm in the edge projections. It follows from standard inverse inequalities and imbedding arguments that this can be done at an expense of losing another factor of 2ϵ in the estimate (see [23] for details). By the same token, the constant ϵ can be removed from the face projections.

We now define the *projection-based* interpolant of a function $u \in H^r(T), r > 1/2$, to be function $\Pi u = u^p \in \mathcal{P}^p_{p_f,p_e}$ which satisfies the conditions,

$$\begin{cases} u^p(v) = u(v) & \text{for each vertex } v \\ |u^p - u|_{0,e} \to \min & \text{for each edge } e \\ |u^p - u|_{1/2,f} \to \min & \text{for each face } f \\ |u^p - u|_{1,T} \to \min & \text{for tetrahedron } T \,. \end{cases}$$

We have demonstrated that the interpolation error can be bounded by the (unconstrained boundary) approximation errors of the function u on the whole element, the element faces, and the interpolation error over the element edges. Since all of these approximation errors behave optimally in terms of polynomial degree p, we arrive at the following final result, subject to the assumption that the conjectured polynomial extensions exist:

[1] This is the essential difference between 2D and 3D cases

Theorem 1. *There exists a constant $C > 0$, dependent upon ϵ such that*

$$\|u - \Pi u\|_{1,T} \leq C \frac{1}{p^{1/2+r-\epsilon}} \|u\|_{1/2+r,T} ,$$

for every function $u \in H^r(T), r > 0$, where $p = \min\{p, p_f, p_e\}$.

Notice that the projection based interpolation operator reproduces exactly the polynomial space. Consequently, we can use a Bramble-Hilbert argument and replace the H^r norm appearing on the right-hand side with the H^r-seminorm.

The interpolant on a physical element is defined in terms of the interpolant on the master element as follows. If $\boldsymbol{x} = \boldsymbol{x}(\boldsymbol{\xi})$ denotes a bijective transformation from the master element $\hat{T}$ onto the physical element T, and $u \in H^r(T)$, then the interpolant Πu is defined by,

$$\widehat{(\Pi u)} = \hat{\Pi} \hat{u} .$$

Here $\hat{u}(\xi) = u(\boldsymbol{x})$ and $\hat{\Pi}$ is the interpolation operator on the master element. In other words, the operation of passing from master to physical element and the interpolations operators, commute (by definition).

After a standard scaling argument, this leads to the ϵ-optimal hp-estimate,

$$\|u - \Pi u\|_{1,T} \leq C \frac{h^{1/2+r}}{p^{1/2+r-\epsilon}} \|u\|_{1/2+r,T} ,$$

for $r > 0, 1/2 + r \leq p$, where $h = h_T$ is the diameter of element T.

2.3 $H(\mathrm{curl})$-Conforming Projection Based Interpolation

In the same spirit, we define the projection based interpolant for $H(\mathrm{curl})$-conforming elements. Given a vector-valued function $\boldsymbol{E} \in \boldsymbol{Q}$, where

$$\boldsymbol{Q} = \{\boldsymbol{E} : \boldsymbol{E} \in \boldsymbol{H}^{1/2+r}(T), \boldsymbol{\nabla} \times \boldsymbol{E} \in \boldsymbol{H}^r(T), r > 0\} ,$$

we define the projection based interpolant $\Pi^{curl} \boldsymbol{E} = \boldsymbol{E}^p \in \boldsymbol{P}^p_{p_f,p_e}$ by the following conditions:

$$\begin{cases} |E^p_t - E_t|_{-1,e} \to \min & \text{for each edge } e \\ |(\boldsymbol{\nabla} \times \boldsymbol{E}^p) \cdot \boldsymbol{n}_f - (\boldsymbol{\nabla} \times \boldsymbol{E}) \cdot \boldsymbol{n}_f|_{-1/2,f} \to \min & \\ (\boldsymbol{E}^p_t - \boldsymbol{E}_t, \boldsymbol{\nabla}_f \phi)_{-1/2,f} = 0 \quad \forall \phi \in \mathcal{P}^{p_f+1}_{-1} & \text{for each face } f \\ |\boldsymbol{\nabla} \times \boldsymbol{E}^p - \boldsymbol{\nabla} \times \boldsymbol{E}|_{0,T} \to \min & \\ (\boldsymbol{E}^p - \boldsymbol{E}, \boldsymbol{\nabla} \phi)_{0,T} = 0 \quad \forall \phi \in \mathcal{P}^{p+1}_{-1,-1} & \text{for tetrahedron } T . \end{cases}$$

Here E^p_t, E_t denote tangential components along edge e of $\boldsymbol{E}^p, \boldsymbol{E}$, respectively, $\boldsymbol{n}_f$ is the outward normal unit vector for face f, $\boldsymbol{\nabla}_f \phi$ stands for the face gradient,

$\boldsymbol{E}_t^p, \boldsymbol{E}_t$ denote tangential components for the particular face, and $(u, v)_{-1/2,f}$ denotes inner product in $H^1/2(f)$ corresponding to the dual seminorm,

$$|u|_{-1/2,f} = \sup_{v \neq 0} \frac{|\langle u, v \rangle|}{|v|_{1/2,f}} .$$

Finally, the projection along the edges is understood in the following sense. We first factor out the average value of E_t, defined by

$$\int_e (c - E_t) = 0 .$$

Next we integrate the resulting difference to obtain a potential $\psi \in H_0^{1/2+r}(e)$, satisfying

$$\frac{\partial \psi}{\partial s} = E_t - c ,$$

and then project it in the L^2-norm onto polynomials $\mathcal{P}_{-1}^{p_e+1}$,

$$\|\psi^p - \psi\|_{0,e} \to \min .$$

Finally, we define the edge interpolant E_t^p to be the sum of the average c and derivative of the projected potential,

$$\boldsymbol{E}_t^p = c + \frac{\partial \psi^p}{\partial s} .$$

Estimating the projection error for the $H(curl)$-interpolant is more technical that for the H^1 case but the overall philosophy is the same. The Poincaré inequality has to be replaced with the discrete Friedrichs inequality, and we again assume the validity of appropriate polynomial extension theorems. The most technical detail deals with the estimation of the interpolation error over faces in the dual norm,

$$\|\boldsymbol{E}_t\|^2_{curl,-1/2+\epsilon,f} = \|\boldsymbol{E}_t\|^2_{-1/2+\epsilon,f} + \|\mathrm{curl}_f E_t\|^2_{-1/2+\epsilon,f} ,$$

where $\mathrm{curl}_f E$ denotes the surface curl introduced earlier. Full details are given in [23], where the following result is proved:

Theorem 2. *Under the conjectures on the polynomial extensions, there exists a constant $C > 0$, dependent upon ϵ such that*

$$\|\boldsymbol{E} - \Pi^{curl} \boldsymbol{E}\|_{curl,0,T} \leq C \frac{1}{p^{1/2+r-\epsilon}} \|\boldsymbol{E}\|_{curl,1/2+r,T} ,$$

for every function $\boldsymbol{E} \in \boldsymbol{H}^{1/2+r}(\mathrm{curl}, T)$, where $p = \min\{p, p_f, p_e\}$.

The space $\boldsymbol{H}^\alpha(\mathrm{curl}, T)$ is defined by

$$\boldsymbol{H}^\alpha(\mathrm{curl}, T) = \{\boldsymbol{E} \,:\, \boldsymbol{E} \in \boldsymbol{H}^\alpha(T), \boldsymbol{\nabla} \times \boldsymbol{E} \in \boldsymbol{H}^\alpha(T)\} .$$

Notice that we do not need the extra $1/2$ regularity in curl in order to define the interpolant but only for the optimality of the error estimate.

Similarly to the H^1-case, we extend the interpolant to physical elements, and derive the corresponding *hp*-interpolation estimate. We emphasise that the interpolation is again performed on the master element.

2.4 $\boldsymbol{H}(\text{div})$-conforming Projection Based Interpolation

For completeness, we also record the projection based interpolation for $H(\text{div})$-conforming elements. Given a vector-valued function $\boldsymbol{H} \in \boldsymbol{V}$ where

$$\boldsymbol{V} = \{\boldsymbol{H} \, : \boldsymbol{H} \in \boldsymbol{L}^2(T), \boldsymbol{\nabla} \circ \boldsymbol{H} \in \boldsymbol{H}^r(T), r > 0\} ,$$

we define the projection based interpolant $\Pi^{div}\boldsymbol{H} := \boldsymbol{H}^p \in \boldsymbol{P}^p_{p_f}$ by the following conditions:

$$\begin{cases} \|\boldsymbol{H}^p \cdot \boldsymbol{n}_f - \boldsymbol{H} \cdot \boldsymbol{n}_f\|_{-1/2,f} \to \min & \text{for each face } f \\ \|\boldsymbol{\nabla} \circ \boldsymbol{H}^p - \boldsymbol{\nabla} \circ \boldsymbol{H}\|_{0,T} \to \min & \\ (\boldsymbol{H}^p - \boldsymbol{H}, \boldsymbol{\nabla} \times \boldsymbol{\phi})_{0,T} = 0 \quad \forall \boldsymbol{\phi} \in \boldsymbol{P}^{p+1}_{-1,-1} & \text{for tetrahedron } T . \end{cases}$$

We again use the discrete Friedrichs inequality and polynomial extension results to arrive at the following result, assuming the relevant polynomial extensions exist:

Theorem 3. *There exists a constant $C > 0$, dependent upon ϵ such that*

$$\|\boldsymbol{F} - \Pi^{div}\boldsymbol{F}\|_{div,0,T} \leq C\frac{1}{p^{r-\epsilon}}\|\boldsymbol{F}\|_{div,r,T} ,$$

for every function $\boldsymbol{F} \in \boldsymbol{H}^r(\text{div}, T)$, where $p = \min\{p, p_f\}$, and

$$\boldsymbol{H}^\alpha(\text{div}, T) = \{\boldsymbol{F} \, : \, \boldsymbol{F} \in \boldsymbol{H}^\alpha(T), \boldsymbol{\nabla} \circ \boldsymbol{F} \in H^\alpha(T)\} .$$

2.5 Commutativity of the Diagram

It follows [24] from the construction of the projection-based interpolation (although the norms used to define the projections have changed from those used in [24], but the algebraic arguments are identical) that the introduced interpolation operators mean the de Rham diagram commutes:

$$\begin{array}{ccccccccccc} \mathbb{R} \longrightarrow & W & \xrightarrow{\boldsymbol{\nabla}} & \boldsymbol{Q} & \xrightarrow{\boldsymbol{\nabla}\times} & \boldsymbol{V} & \xrightarrow{\boldsymbol{\nabla}\circ} & L^2 & \longrightarrow 0 \\ \big\downarrow id & \big\downarrow \Pi & & \big\downarrow \Pi^{\text{curl}} & & \big\downarrow \Pi^{\text{div}} & & \big\downarrow P & \\ \mathbb{R} \longrightarrow & \mathcal{P}^{p+1}_{p_e+1,p_f+1} & \xrightarrow{\boldsymbol{\nabla}} & \mathbf{P}^p_{p_e,p_f} & \xrightarrow{\boldsymbol{\nabla}\times} & \mathbf{P}^{p-1}_{p_f-1} & \xrightarrow{\boldsymbol{\nabla}\circ} & \mathcal{P}^{p-2} & \longrightarrow \mathbf{0} . \end{array} \tag{15}$$

The interpolation operator P denotes the L^2-projection.

In the 2D case, the diagram takes a simpler form,

$$\begin{array}{ccccccc} \mathbb{R} \longrightarrow & H^{1+r} & \xrightarrow{\boldsymbol{\nabla}} & \boldsymbol{H}^r \cap \boldsymbol{H}(\text{curl}) & \xrightarrow{\boldsymbol{\nabla}\times} & L^2 & \longrightarrow \mathbf{0} \\ \big\downarrow id & \big\downarrow \Pi & & \big\downarrow \Pi^{\text{curl}} & & \big\downarrow P & \\ \mathbb{R} \longrightarrow & \mathcal{P}^{p+1}_{p_e+1} & \xrightarrow{\boldsymbol{\nabla}} & \mathbf{P}^p_{p_e} & \xrightarrow{\boldsymbol{\nabla}\times} & \mathcal{P}^{p-1} & \longrightarrow \mathbf{0} , \end{array}$$

with analogous interpolation error estimates [19].

Such diagrams provide a fundamental tool for studying convergence of Maxwell eigenvalues [9, 10, 25, 36] and stability of hp FE discretisation for time-harmonic equations.

Remark 1. In the 3D case, the extra $1/2$ regularity was needed to define the H^1 and $H(\mathrm{curl})$ interpolants. Notice that we do not need it for the $H(\mathrm{div})$ interpolation. The regularity assumptions can be weakened by making use of quasi-local interpolation [46].

Finally, we reiterate the main point of the results: The extra regularity in 3D is necessary to define the local interpolant, but the projection-based interpolation still delivers ϵ-optimal estimates.

3 Coding *hp* Elements–A Programmer's Nightmare

We briefly describe the main design assumptions behind our latest implementation of hp elements [22] in 3D. This is the fourth such implementation, that I have been involved with (and the best one so far...). The code supports anisotropic and isotropic refinements, both in h and p, for hexagonal elements, with the possibility of adding prisms, tetrahedra, and pyramids, and has been written in Fortran 90.

3.1 Main Design Assumptions

Initial mesh.

We assume that the initial mesh is an arbitrary unstructured, but regular mesh consisting of hexahedra. Edges and vertices may be shared by an arbitrary number of elements but no hanging (constrained) nodes are allowed. The entire data structure is reduced to three matrices storing three types of objects:

- *Initial mesh element* comprises typical information for an unstructured grid element: nodal connectivity, orientation for the element edges and faces, neighbours across the element faces, and boundary condition flags for the element faces.
- *Vertex node* definition includes such attributes as: a boundary condition flag, father node, and geometry and solution degrees-of-freedom (d.o.f.).
- *Non vertex node* provides an abstraction for element edges, faces, and element interiors. The corresponding attributes include the nodal type (edge,face or element interior), order of approximation for the node, a boundary condition flag, refinement kind (indicating a possible anisotropic refinement of the node), father node, nodal sons, and finally geometry and solution d.o.f.

Geometry modelling.

The code comes with a 'home grown' *Geometry Modelling Package* [59]. A 3D geometrical domain is represented as a union of curvilinear hexahedra, much in the spirit of an unstructured hexahedral mesh. For each hexahedron in the model, the code provides a parametrisation $\boldsymbol{x}(\boldsymbol{\eta})$, mapping a reference hexahedron onto the particular hexahedron in the model. The parameterisations are compatible on element faces and edges, for all adjacent blocks. Two principal techniques used are the *implicit parameterisations* and standard *trivariate interpolation with linear blending functions*.

Initial mesh generation.

The code also comes with a simple algebraic mesh generator based on the modelling package. Each of the reference hexahedra is covered with a regular $m_1 \times m_2 \times m_3$ mesh of elements with a specified order $p = (p_1, p_2, p_3)$ (the order of approximation may be anisotropic), and the geometry degrees of freedom (d.o.f.) are generated by performing the H^1 projection-based interpolation of the GMP parameterisations. Assumed numbers of subdivisions for each block must be compatible with one another, so the resulting mesh is regular. The minimum rule is used to determine order of approximation for faces and edges shared by several blocks. The same projection is also performed on the Dirichlet boundary condition data, when initiating the corresponding solution d.o.f.

1-irregular mesh h refinements.

All elements, starting with the initial mesh elements, can be refined into eight (isotropic refinement), four or two (anisotropic refinement) element sons. No element can be refined unless *all of its nodes are regular.* If an element with constrained nodes is to be refined, the constrained nodes are eliminated first, by forcing appropriate refinements of the element neighbours. The algorithm eliminates the possibility of multiple constrained nodes. Parent nodes of a constrained node are always unconstrained.

Supporting h refinements.

A minimal amount of information is stored in the data structure to support the elements resulting from the h-refinements. This is the most dramatic change in coding the hp elements, compared with previous implementations. We view the h refinements as a process of breaking nodes, i.e. element interiors, faces or edges. Each broken father node gives birth to a number of node sons—new interiors, faces, edges or vertex nodes, with their number and type depending upon the requested refinement, and the type of father node. The new nodes are initiated in the data structure arrays (an interface with GMP is maintained to initiate geometry d.o.f. for the new nodes), and the tree structure information (sons for the father, father for the sons) is stored. *This is the only extra information generated* to support the new elements. Based on these trees only, special algorithms are constructed to support:

- a natural order of elements, enabling looping over elements in the mesh,
- finding element neighbours for edge and face nodes,
- finding element nodal connectivity for elements resulting from h-refinements.

We refer to [22] for all technical details.

p-Adaptivity.

At any stage, the order of approximation p for any of the non-vertex (edge, face, or interior) nodes can be changed. Due to the minimal information stored in the data structure arrays, the corresponding modifications *only* include modifying the order of approximation for the node, and initiating its degrees of freedom. The geometry d.o.f. are always generated by interpolating the GMP parameterisations. New solution d.o.f. are simply initialised with zeros.

Constrained approximation.

The code enforces continuity of approximation for one-irregular meshes by means of constrained approximation [22]. Breaking of an element is executed in three stages:

- breaking the element interior node,
- breaking the element face nodes,
- breaking the element edge nodes.

We emphasise that, at any stage of the refinement process, the corresponding mesh is fully operational in the sense that we could enforce the continuity, and solve the corresponding variational problem.

Supporting multigrid operations.

The constrained approximation techniques are also utilised to generate the prolongation matrix in the case of nested meshes. The fine mesh d.o.f. are simply represented as linear combinations of coarse mesh d.o.f., recycling the technique used to initiate new d.o.f. during the h-refinements and enforce continuity for constrained nodes.

Edge hp elements.

Following our previous implementations [41, 45], our 3D *hp* code is extended to also support the $H(curl)$-conforming discretisation. A parallel effort for 2D triangular and quadrilateral elements has been used to generate the results reported in the next section. The whole idea of the new data structure was to separate clearly logical operations on nodes from those on the corresponding d.o.f. Solution of edge elements requires the use of different shape functions and, consequently, different coefficients for enforcing $H(curl)$ continuity, but all logical operations on nodes corresponding to h and p refinements are identical for both H^1 and $H(curl)$ discretisation. The new versions of the codes use the C preprocessor in such a way that the codes can be preprocessed to generate a source either for the elliptic or the Maxwell case. Recall that in the Maxwell case, both H^1-conforming (Lagrange multiplier p) and $H(curl)$-conforming (electric field $\boldsymbol{E}$) discretisation must be supported.

We conclude our short discussion with an example of an automatically generated 3D hp mesh shown in Fig. 1. Different colours correspond to different orders of approximation. It will be noticed that anisotropic refinements are also present.

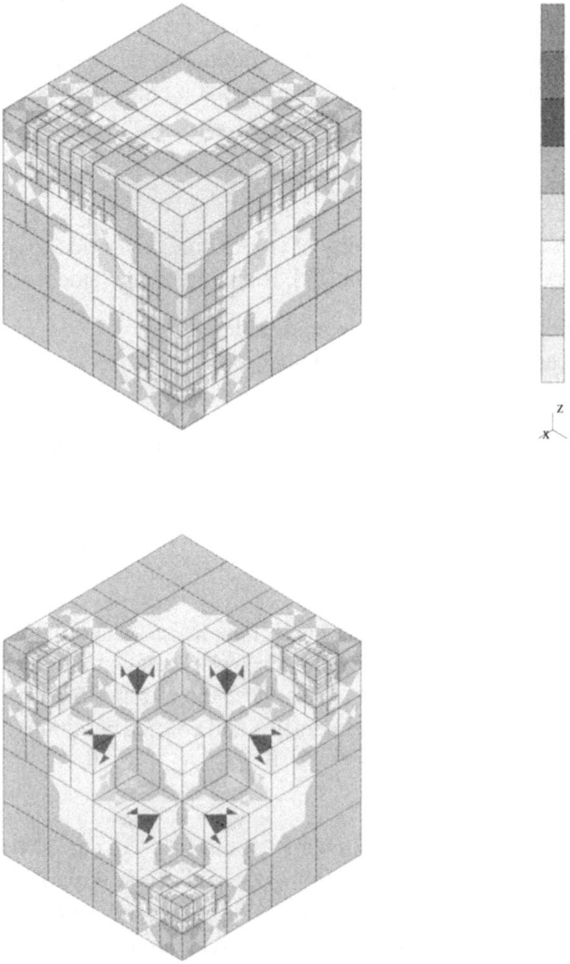

Fig. 1. An automatically generated 3D hp mesh for an elliptic problem. View of the mesh from outside and inside (notice the symmetry)

4 Automatic *hp*-Adaptivity

The projection-based interpolation provides not only a fundamental tool for the mathematical analysis of the *hp* methods but has also turned out to be a basis for generating optimal *hp* meshes. The main idea of using the projection-based interpolation for mesh optimisation [21] is as follows. The algorithm automatically produces a sequence of optimal *hp* meshes that we will call the *coarse* meshes. Given a coarse mesh, we perform a *global hp*-refinement to produce the corresponding *fine mesh*, and determine the corresponding fine mesh solution. We compute the corresponding projection-based interpolation error of the fine mesh solution on the coarse grid. The optimal *hp*- refinements of the coarse grid are then determined by maximising the rate of decrease of the interpolation error versus the number of d.o.f. added,

$$\frac{\|u_{h/2,p+1} - \Pi_{h,p} u_{h/2,p+1}\| - \|u_{h/2,p+1} - \Pi_{h,p}^{next} u_{h/2,p+1}\|}{N_{h,p}^{next} - N_{h,p}} \rightarrow \max ,$$

where $N_{h,p}$ denotes the total number of d.o.f. corresponding to a coarse *hp* mesh. The rationale behind replacing the approximation error with the projection-based interpolation error is twofold:

- contrary to the approximation error, the interpolation error depends upon the mesh attributes *locally* which makes the minimisation problem tractable;
- projection-based interpolation delivers optimal convergence rates, both in p and in h; therefore minimising the projection based error should have the same effect as minimising the approximation error itself.

4.1 The *hp* Algorithm for 2D Elliptic Problems

We shall now discuss in detail the 2D version of the algorithm for the elliptic case.

1. *Storing the current mesh information.* We use a separate data structure to store the information about the current mesh: element vertices, edges, and interiors. If the exact solution is known and the code is run to study the algorithm only, we compute and save the corresponding error. Finally, we output the current mesh data structure.
2. *Global hp-refinement.* We break each element into four element sons, and increase the order of approximation of each node by one. We update the Dirichlet d.o.f. but *do not update* the geometry d.o.f. This is to guarantee that the corresponding coarse and fine meshes are nested. We solve on the fine mesh and save the fine mesh solution in a separate data structure supporting the mesh optimisation. We estimate the error by simply computing the norm of the difference between the fine and coarse mesh solutions. If the exact solution is unknown, we save the error estimate[2]. For quadrilateral elements, when computing the error, we determine simultaneously *element anisotropy flags* indicating whether

[2] We then use it in place of the exact error to report the convergence rates

the element is a candidate for anisotropic h-refinements. For the elliptic case, the computation of the element error, measured in the H_0^1-seminorm, reduces to computing the integral,

$$\int_K \sum_{i=1}^{2}\sum_{j=1}^{2} a_{ij}(\boldsymbol{\xi})\frac{\partial(u-u_{hp})}{\partial\xi_i}\frac{\partial(u-u_{hp})}{\partial\xi_j}\,d\boldsymbol{\xi}\,.$$

Here, and in what follows, $u = u_{h/2,p+1}$ stands for the fine mesh solution. The metric a_{ij} results from the change of variables from the physical to the master element,

$$a_{ij} = \sum_{k=1}^{2}\frac{\partial\xi_i}{\partial x_k}\frac{\partial\xi_j}{\partial x_k}\left|\det\left(\frac{\partial x_m}{\partial\xi_n}\right)\right|\,.$$

The element is declared to be anisotropic provided two conditions are satisfied:

- $$\left|\int_K 2a_{12}\frac{\partial(u-u_{hp})}{\partial\xi_1}\frac{\partial(u-u_{hp})}{\partial\xi_2}\,d\boldsymbol{\xi}\right|$$
$$\leq 0.01\left\{\int_K a_{11}(\frac{\partial(u-u_{hp})}{\partial\xi_1})^2 + a_{22}(\frac{\partial(u-u_{hp})}{\partial\xi_2})^2\,d\boldsymbol{\xi}\right\}$$
- the contribution corresponding to one of the variables is less than one percent of the contribution corresponding to the other variable.

Anisotropic h-refinements are only intended for use in extreme situations when the solution is one-dimensional and aligned with the mesh (as in the case of boundary layers, edge singularities).

3. *Determine the optimal refinements for the edges.* For each of the coarse element edges, we determine,
   - the edge projection error corresponding to the current coarse mesh,
   - the edge projection error corresponding to the p-refined edge,
   - the minimum value of the edge projection error corresponding to a *competitive* h-refinement of the edge.

   If p_e denotes the original order for the edge, and $p_{e,1}, p_{e,2}$ are the orders for the element sons, the refinement is called competitive, if $p_{e,1} + p_{e,2} = p_e$, i.e. the increase in the number of d.o.f. for the edge is the same as for the p-refinement—one new d.o.f. added. Consequently, there are p_e cases to check. The interpolation error for the h-refined element is understood here in a generalised sense. After factoring out the linear interpolant corresponding to the coarse mesh, we project the difference between the fine mesh solution and the linear interpolant onto the space of 'edge bubble functions' corresponding to the appropriate refinement. Thus, for p-refinement, this reduces to standard interpolation, but for an h-refined element, the bubble functions are piecewise polynomials with orders $p_{e,1}, p_{e,2}$, vanishing at the endpoints of the edge. In practice, a weighted H_0^1-seminorm is used in place of the $H_{00}^{\frac{1}{2}}$ norm, see [21] for details. The weight is selected in such a way that the new norm scales with the edge length in the same way as the $H_{00}^{\frac{1}{2}}$ norm. Finally, by comparing the interpolation errors corresponding to the two refinements, we decide whether the edge is to be p- or h-refined.

4. *Deciding which edges to refine.* Having determined *how* to refine each edge, we decide now *which* edges to refine. This is done by evaluating the rate at which the interpolation (projection) error decreases for each edge,

$$\frac{\|u - (\text{old})u^p\|_{1/2,e} - \|u - (\text{new})u^p\|_{1/2,e}}{1},$$

where $(\text{new})u^p$ denotes the interpolant corresponding to the optimal refinement of the edge. The number 1 appearing in the denominator corresponds to the increase in the number of edge d.o.f. We then determine the maximum edge rate and refine only those edges which deliver a rate that is at least of $1/3$ of the maximum rate. The $1/3$ factor is arbitrary.

5. *Applying the edge refinements.* We input the coarse mesh data structure, and enforce the requested edge h-refinements by refining elements adjacent to edges selected for h-refinement. The elements are refined according to their *anisotropy flags*. If the element is declared to be isotropic, it will be refined into four element sons, even if only one of its edges calls for an h-refinement. The coarse mesh edge nodes are assigned the new, optimal orders of approximation, and the corresponding d.o.f. are initiated with those resulting for the edge interpolation. The 1-irregularity rule means that some elements and edges undergo *unwanted* h-refinements. We distinguish between two kinds of unwanted edge refinements.
   - If the edge wanted to be p-refined, and the h-refinement was forced, we determine the order of the edge nodes in such a way that we match the interpolation error for the desired p-refined edge.
   - If the edge did not want to be refined at all, we determine the order of the edge nodes in such a way that we match the interpolation error for the old mesh, within a small (prescribed) tolerance.

6. *Determining the new orders of approximation for middle nodes.* We do not alter the topology of the mesh after the edge refinements. We make an implicit assumption that all 'singularities' are located at vertices or edges, and the necessary h-refinements have already been applied. Thus, we need only determine the order of approximation for element middle nodes. We use the minimum rule to determine the order for new edges interior to old coarse mesh h-refined elements. The minimum rule, and the order of approximation selected for the coarse mesh edges, also provides a starting point for determining the orders for the middle nodes. We consider one (old) coarse mesh element at a time (the procedure is local), and increase the order for the middle node(s) of the new element(s), while monitoring the element projection error decrease rate,

$$\frac{|u - (\text{old})u^p|_{1,T} - |u - (\text{new})u^p|_{1,T}}{N_{new} - N_{old}},$$

where N_{new} and N_{old} denote the number of interior d.o.f. for the new and old coarse meshes, respectively.
This 'investment process' is stopped when the two conditions are satisfied:

- the previous mesh interpolation error is met (within a small tolerance),
- the element rate gets smaller than $1/3$ of an *optimal investment rate.*

In contrast to edges, we cannot afford to determine the optimal rate by first obtaining the rates for all elements, and then taking the maximum. Instead, we determine a *key element* as follows. First, we use the actual element errors (difference between the fine and coarse grid solutions), and limit the candidates for the key element to those for which the element errors are in excess of 1/3 of the largest element contribution. For each of these elements, we determine the projection error corresponding to the previous coarse mesh and pick the element with the largest projection error. This is the *key element.* Finally, we perform the procedure to determine the optimal distribution of order(s) p for the key element, and evaluate the maximum rate of decrease for the element projection error. This rate then serves as the *optimal investment rate* for the global step discussed above.

7. *Enforcing the minimum rule.* At the end of the algorithm, we enforce the minimum rule by revisiting all edge nodes. At this point, we also update all geometry d.o.f.

4.2 Example: Radiation From a Coil Antenna Into a Dispersive Medium

As an illustration, we present a solution to the radiation problem for a single loop, coil antenna wrapped around a metallic cylinder, and radiating into a dispersive medium. The problem is axisymmetric, and can be reduced to a single elliptic equation for the transversal component E of the electric field $\boldsymbol{E}$, in terms of cylindrical coordinates r, z. The variational form of the equation is given by:

$$\begin{cases} E = 0 \text{ on } \Gamma \\ \displaystyle\int_\Omega \left\{ \frac{1}{\mu} \left[\frac{\partial E}{\partial z}\frac{\partial F}{\partial z} + \frac{1}{r^2}\left(E + r\frac{\partial E}{\partial r}\right)\left(F + r\frac{\partial F}{\partial r}\right)\right] - (\omega^2\epsilon - j\omega\sigma)EF \right\} rdrdz \\ \displaystyle = i\omega \int_{\partial D} rJ^{imp} F\, ds, \quad \forall F, F = 0 \text{ on } \Gamma\, . \end{cases}$$

Due the exponential decay of the solution away from the antenna, both the metallic mandrel, and the soil domain can be truncated at a sufficiently large distance, and the equation is approximated in a bounded domain with perfect conductor boundary conditions. The domain is shown, together with the corresponding solution, in Fig 2. For precise geometry and material data, we refer to [48].

In Fig.s 3-6, we zoom in on the antenna (compare the scale on the y-axis) and show the corresponding optimal hp mesh generated automatically by the hp algorithm. Note anisotropic hp-refinements capturing the boundary layer in the mandrel (part of the mandrel is modelled as a medium with high conductivity, not just by a boundary condition), and a characteristic radiation pattern corresponding to a strong reflection from the mandrel.

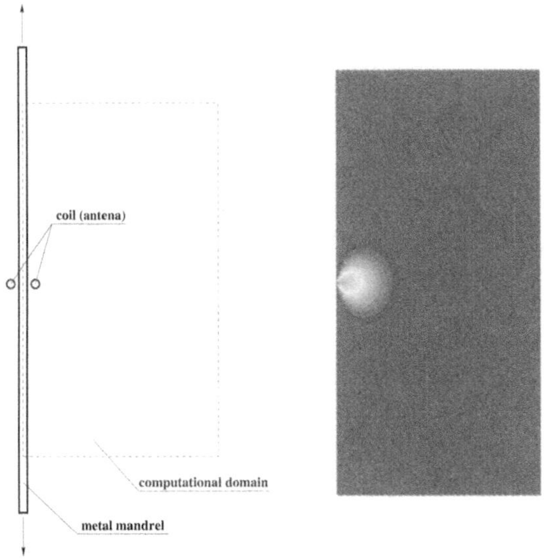

Fig. 2. Radiation from a coil antenna into a dispersive medium. Geometry of the domain and the imaginary part of the solution

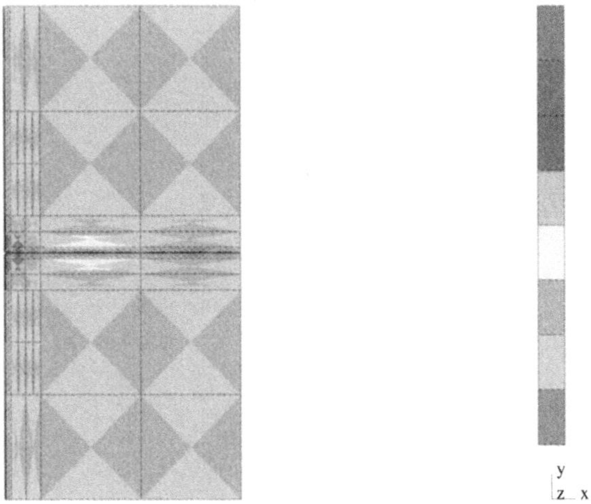

Fig. 3. Radiation from a coil antenna into a dispersive medium. Final mesh

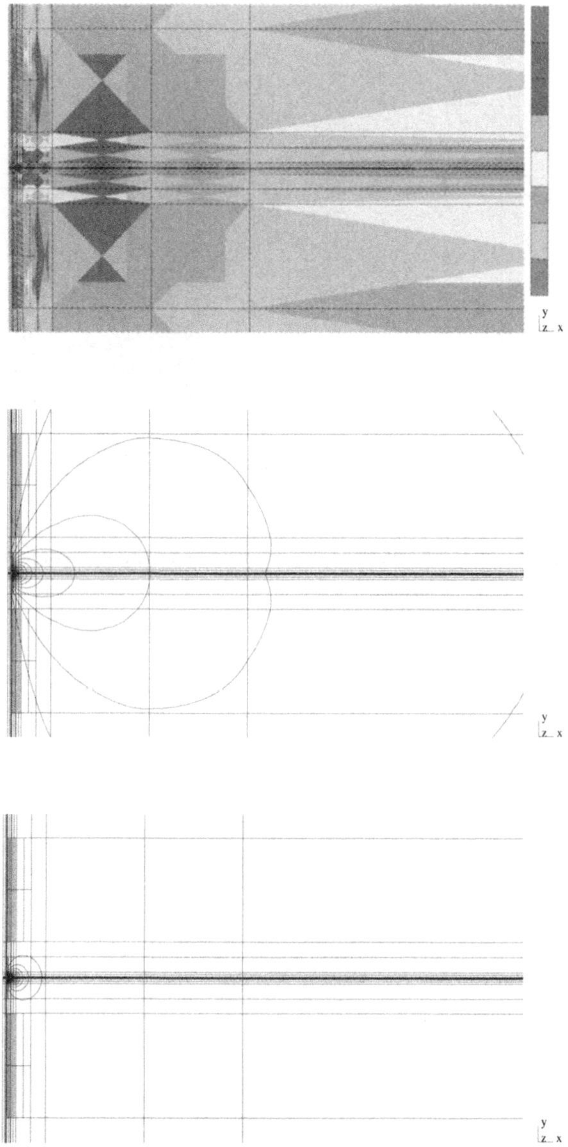

Fig. 4. Radiation from a coil antenna into a dispersive medium. Final mesh with the corresponding real and imaginary components of the solution, zoom with magnification 10 on the antenna

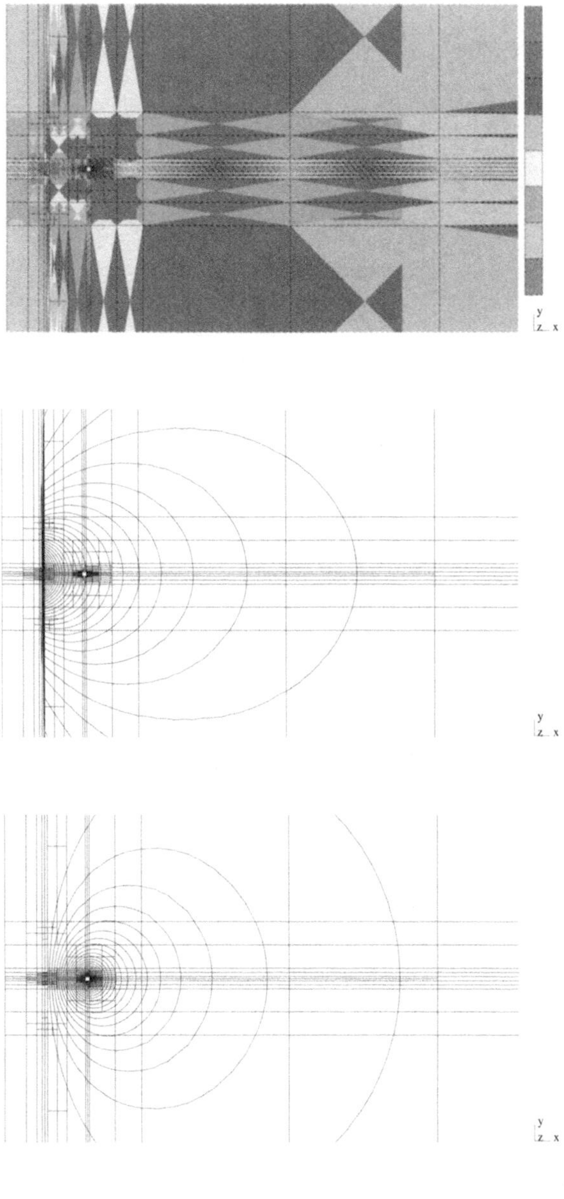

Fig. 5. Radiation from a coil antenna into a dispersive medium. Final mesh with the corresponding real and imaginary components of the solution, zoom with magnification 100 on the antenna

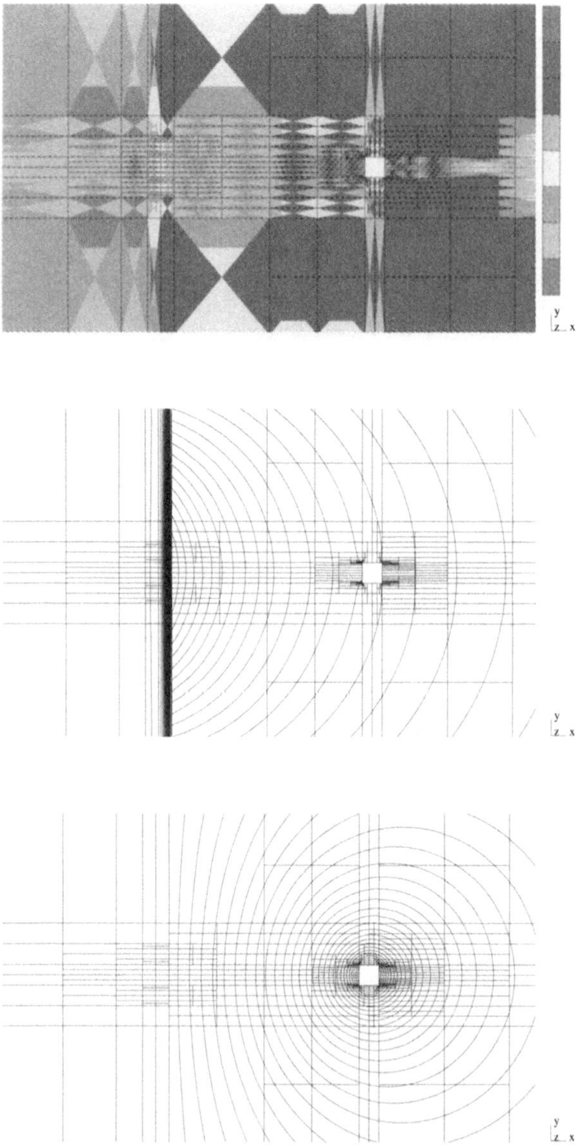

Fig. 6. Radiation from a coil antenna into a dispersive medium. Final mesh with the corresponding real and imaginary components of the solution, zoom with magnification 500 on the antenna

Finally, in Fig. 7, we present the convergence history. The overall straight line indicates the exponential convergence. The irregularity in the graph does not represent a coding error but corresponds to an increasing accuracy of the fine mesh solution. The error reported on the convergence graph is obtained by simply evaluating the norm of the difference between the fine and coarse grid solutions. The monotone behaviour of the error estimate is lost when first h-refinements in the boundary layer take place. The refinements change significantly the fine grid solution, and the error estimate reports an 'increase' of the error.

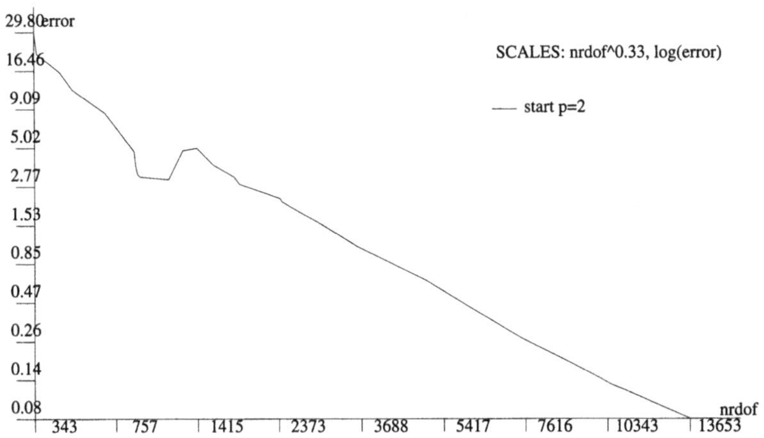

Fig. 7. Radiation from a coil antenna into a dispersive medium. Convergence history on algebraic/logarithmic scales

4.3 The *hp* Algorithm for 2D Time-Harmonic Maxwell Equations

The structure of the algorithm for Maxwell equations is the same as for the elliptic problem, except for the choice of norms. On edges, in place of weighted H^1-seminorm, we use a weighted L^2-norm, scaled to mimic the $H^{-1/2}$ norm,

$$\|E\|^2_{-1/2,e} \approx \int_e |E|^2 \frac{d\xi}{ds}\, ds = \int_0^1 |\hat{E}|^2 \, d\xi \,,$$

where, as usual, $\hat{E}(\xi) = E(\boldsymbol{x}(\xi))$, with $\boldsymbol{x}(\xi)$ denoting the map from the master edge $(0,1)$ onto physical edge e. Similarly, for elements we use the $H(curl)$ norm in place of H^1-norm. As with the definition of the projection-based interpolation in $H(curl)$, projections on edges are performed using the scaled L^2 norm in place of the $H^{-1/2}$ norm, and projections on elements T are replaced with the solution of local mixed problems,

$$\begin{cases} E_t^p \text{ known on boundary } \partial T \\ \|\nabla \times (\boldsymbol{E}^p - \boldsymbol{E})\|_{0,T} \to \min \\ (\boldsymbol{E}^p - \boldsymbol{E}, \nabla \phi)_{0,T}, \quad \text{for every bubble test function } \phi\,. \end{cases}$$

Note that, similarly to the elliptic case, the projection-based interpolation is to be understood in the generalised sense. If an edge or element is h refined, the local projection problems are solved over the original, refined element, and the corresponding spaces are *piecewise* polynomials.

The Lagrange multiplier corresponding to the regularised variational formulation for the Maxwell equations, is equal to zero (up to machine precision), and does not enter the mesh optimisation algorithm.

4.4 Example: Diffraction of a Plane Wave From an Edge

As an illustration, we present an approximation of a 2D diffraction problem illustrated in Fig. 8. A plane wave shines at a 45 degree angle onto a diffracting screen. The exact solution is represented in terms of Fresnel integrals

$$\begin{aligned} &H(r,\theta) \\ &= \tfrac{1}{\sqrt{\pi}} e^{\frac{\pi j}{4} - jkr} \left\{ F\left[(2kr)^{1/2} \sin \tfrac{1}{2}(\theta - \tfrac{\pi}{4})\right] + F\left[(2kr)^{1/2} \sin \tfrac{1}{2}(\theta + \tfrac{\pi}{4})\right] \right\} \end{aligned}$$

where

$$F(u) = 1/2\sqrt{\pi} \left\{ e^{\frac{\pi j}{4}} - \sqrt{2} \left[C\left(\sqrt{\frac{2}{\pi}}u\right) - jS\left(\sqrt{\frac{2}{\pi}}u\right) \right] \right\}$$

and

$$C(z) - jS(z) := \int_0^s e^{-1/2\pi j t^2}\, dt \quad \text{(Fresnel integrals)}\,.$$

The solution represents the most singular solution possible for 2D homogeneous problems. In order to avoid trouble with modelling the open boundary, we cut off a unit square computational domain around the diffracting edge, and use the exact solution to impose Dirichlet boundary conditions. The material data are $\epsilon = \mu = \omega = 1$, and $\sigma = 0$.

Starting with an initial mesh of $4 \times 4 \times 2$ triangular elements of second order, we ask the algorithm to deliver a rather academic relative error of .01 percent. Figures 9 and 10 present eight consecutive zooms on the diffracting edge in the final, optimal mesh.

Figures 11 and 12 present the corresponding zooms on the real and the imaginary part of the first component of the electric field. Only at the level of the smallest

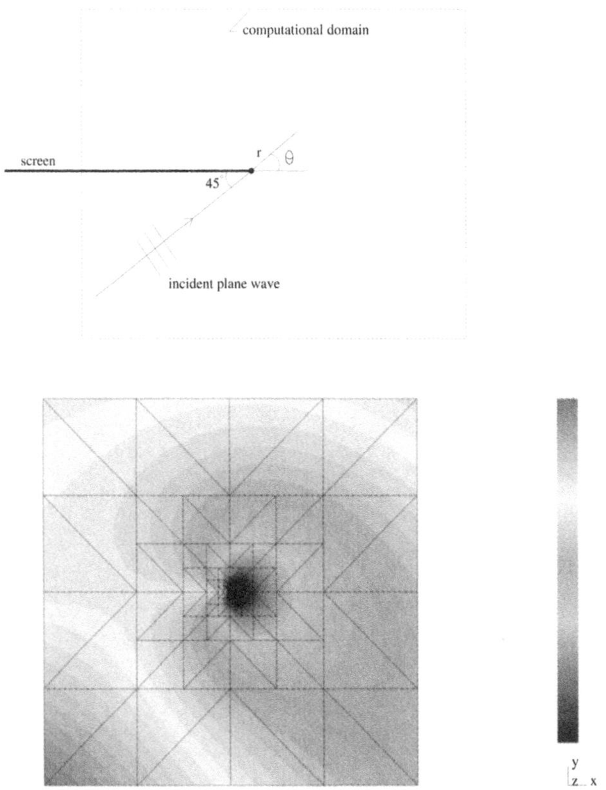

Fig. 8. Diffraction of a plane wave from an edge. Geometry and real part of the first component of the electric field

elements can one see the typical discontinuity in the normal component of the field related to the edge element discretisation.

Finally, Fig. 13 presents the convergence history on a log–log scale. The straight line (disappointingly) indicates that only algebraic rate of convergence is achieved. We attribute it to the loss of accuracy due to round off error. After 26 h-refinements at the tip of the screen, the smallest element size h is of order 10^{-7}. The contribution of the L^2-term to the stiffness matrix scales with h^2, compared with the contribution from the curl-curl term. Thus, with 15 significant digits (double precision), the L^2-term is simply lost.

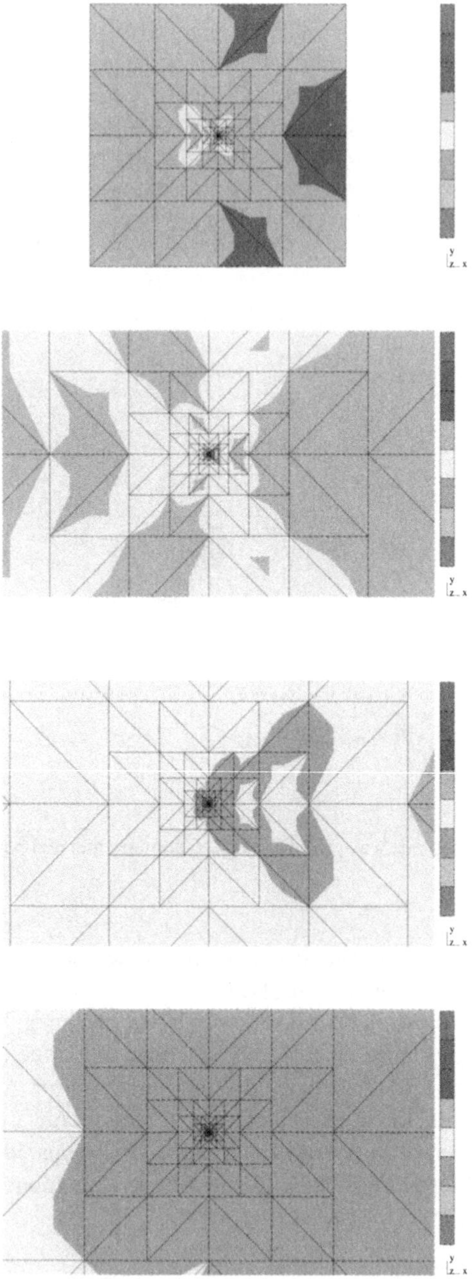

Fig. 9. Diffraction of a plane wave on a screen. Final hp mesh, zooming on the diffracting edge (with factor 10)

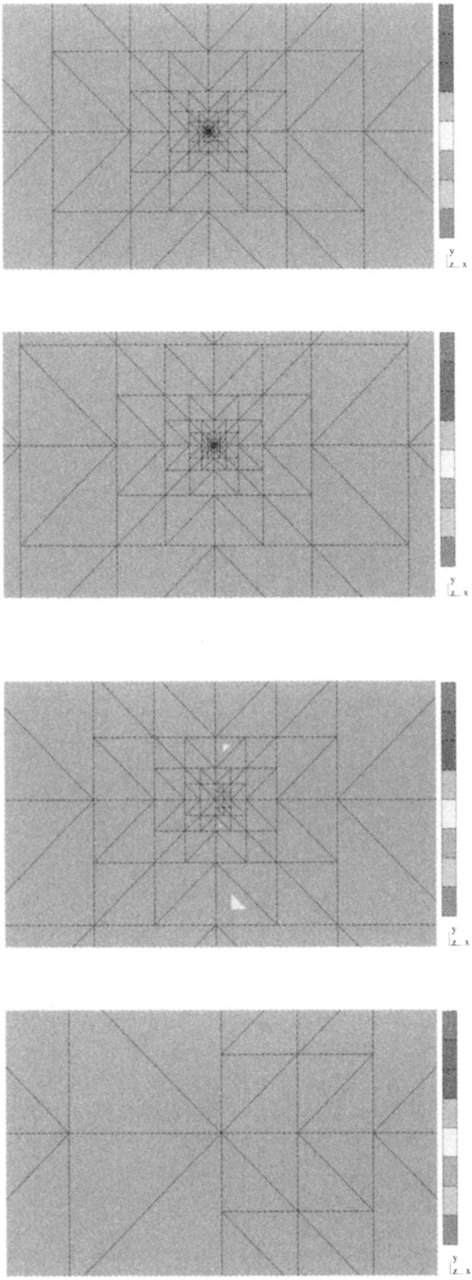

Fig. 10. Diffraction of a plane wave on a screen. Final *hp* mesh, zooming in on the diffracting edge (cont.)

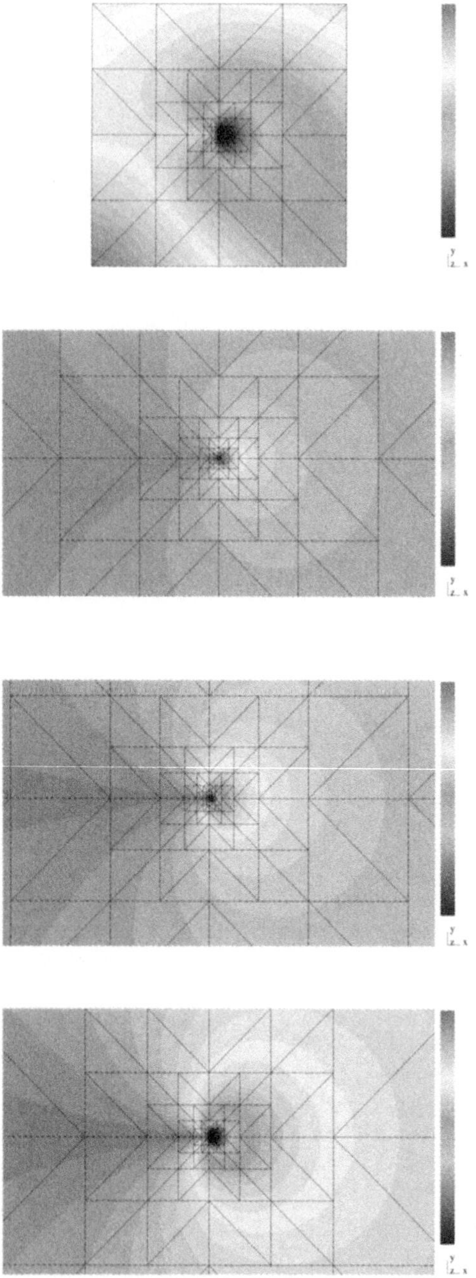

Fig. 11. Diffraction of a plane wave on a screen. Real part of E_1 component, zooming on the diffracting edge (with amplification factor 10)

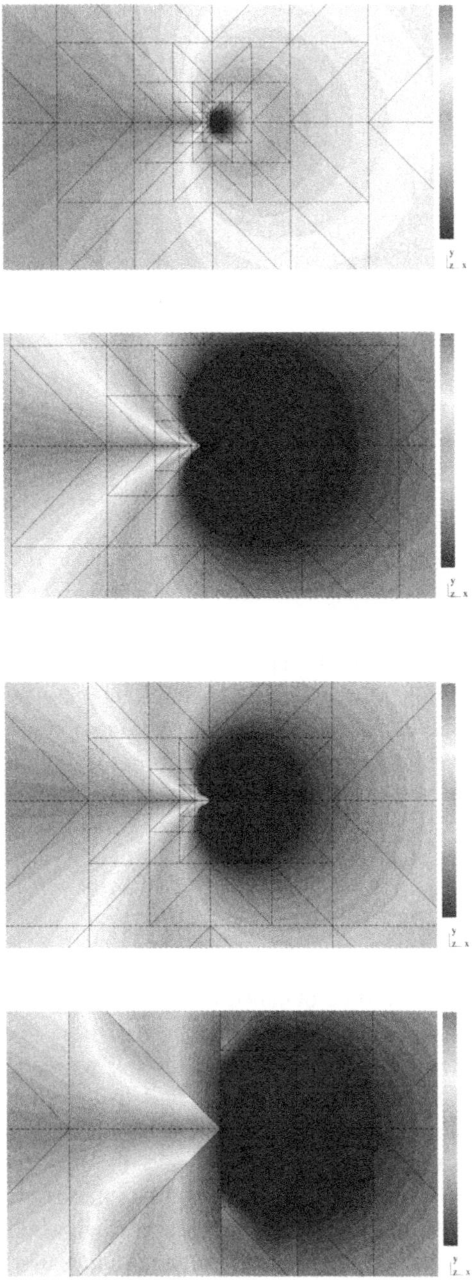

Fig. 12. Diffraction of a plane wave on a screen. Real part of E_1 component, zooming in on the diffracting edge (cont.)

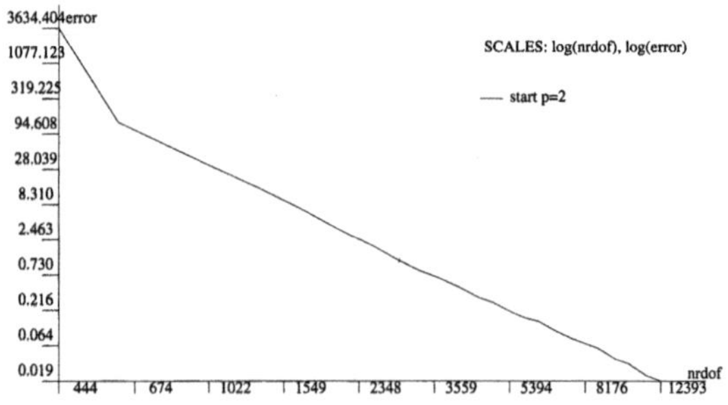

Fig. 13. Diffraction of a planar wave. Convergence history

5 Conclusions

Difficulties

Prohibitive complexity.

The main drawback of the approach presented here is the complexity of the hp algorithm and related coding. Moreover, the 2D algorithm presented here becomes much more technical in 3D. For instance, after determining optimal refinement of edges, one has to store the edge interpolants, and perform the optimisation on faces. Next, the face interpolants have to be stored to determine the optimal distribution of orders of approximation for the old mesh element interiors. Storing and recovering degrees of freedom for edges and faces involves the issue of edge and face orientations and invites possible errors.

Two grid solver.

The fine mesh solution must be obtained using iterative methods. Typically, in 3D, the global hp-refinement increases the problem size by one order of magnitude, and solving it using a direct solver is prohibitive. In [39] we report on the development of a two-grid solver based on a block Jacobi smoother (with an optimal relaxation), and a direct resolution on the coarse mesh. The results are quite encouraging. Convergence properties of the method are practically mesh independent. In the implementation we chose to store the fine mesh stiffness matrix and smoother in a fully assembled form. This limits the size of 3D meshes (on a workstation with 32 bit architecture and 2 GB memory) to less than half a million real-valued d.o.f. This is not particularly large and new implementation strategies have to be worked out. Static condensation and element-by-element techniques suggest themselves. The most encouraging result of our experiments with the two grid solver has been the fact that, for all (elliptic only) problems that have been investigated, it is possibly to guide the optimal hp-refinements using only a partially converged fine mesh solution [39].

Trouble with constraints.

In presence of constrained nodes, in order to yield a globally conforming interpolant, the projection-based interpolation has to incorporate the constraints, which forces it to be constructed in an order proceeding from large edges (or faces) to small edges (or faces). In other words, if we have an edge with a constrained vertex, the edge interpolant can be determined only after the vertex value is known. If the vertex node is constrained by a larger edge (or face), the interpolation over the larger edge (or face) has to be done first. This not only complicates the logic of the implementation but it results in two major deficiencies:

- the interpolation error (defined in this way) does not behave optimally in terms of order p; and to make matters worse,
- the H^1 and $H(curl)$- interpolants do not commute.

The two facts are related to one another, and they are unacceptable from both the theoretical and practical point of view (e.g. the corresponding rates used for the mesh optimisation, do not lead to an optimal selection of new orders of approximation). A simple way out would be to replace the interpolation over a constrained element with the interpolation over its unconstrained element-father. This is consistent with the one-irregularity rule (i.e. no element gets refined unless all its nodes are unconstrained), guarantees commutativity and optimality of the corresponding projection based interpolation. However, it completely spoils the logic of our present implementation of the mesh optimisation routines where we think in terms of edges, faces and elements rather than *groups of elements* ('partially' broken elements). At the end of the day, we have decided to ignore the constraints and perform the interpolation locally while ignoring the constraints. In other words, e.g. in 2D, when we compute element interpolants, we compute the vertex and edge interpolants for all vertices and edges the same way, regardless whether they are constrained or not. This does not seem to affect the overall convergence rates, although the corresponding element interpolants can no longer be 'glued' together to yield a globally conforming function.

Summary

The dispersion analysis [2, 30] clearly shows that it is advantageous to use large elements with high order p. At the same time, complex geometries force us to use small elements with low order p. Only hp-finite elements allow for combining elements of different size and order in the same mesh.

The hp-adaptivity delivers exponential convergence, for solutions with locally varying regularity, including singularities. This is the unique idea introduced by Prof. Babuška and his coworkers [47]. The growing theoretical and numerical evidence demonstrates that the exponential convergence can also be delivered by the hp edge elements.

The hp edge elements should be viewed as a member of the whole family of H^1-$H(curl)$- and $H(div)$- conforming elements forming the de Rham diagram. The theory of the projection-based interpolation applies to the whole family, and provides a consistent basis for automatic *hp*-adaptivity, for all three types of elements.

It is possible to build an *hp* code that will support all three kinds of elements (in 3D) and provide a future basis for solving coupled problems (e.g. electromagnetics coupled with mechanics). I believe that the presented results clearly demonstrate that the automatic *hp*-adaptivity is indeed possible.

Acknowledgement

This work has been supported by Air Force under Contract F49620-98-1-0255. The computations reported in this work were done through the National Science Foundation's National Partnership for Advanced Computational Infrastructure.

The results reported in the paper have been obtained in collaboration with a number of colleagues: Waldek Rachowicz, Ivo Babuška, Peter Monk, Joachim Schöberl, Philippe Devloo, and students: Pavel Solin, David Pardo and Xue Dong.

References

1. H. Adams, *Sobolev Spaces*, Academic Press, New York 1978.
2. M. Ainsworth, and J. Coyle, Hierarchic *hp*-edge Element Families for Maxwell's Equations on Hybrid Quadrilateral/Triangular Meshes, *Computer Methods in Applied Mechanics and Engineering*, **190**, 6709-6733, 2001.
3. M. Ainsworth, and J. Coyle, Conditioning of Hierarchic *p*-Version Nédélec Elements on Meshes of Curvilinear Quadrilaterals and Hexahedra, preprint 2002.
4. M. Ainsworth, and J. Coyle, Hierarchic Finite Element Bases on Unstructured Tetrahedral Meshes, preprint 2002.
5. I. Babuška and M. Suri, The Optimal Convergence Rate of the *p*-Version of the Finite Element Method, *SIAM J. Numer. Anal.*, **24**, 4, 750-776, 1987.
6. I. Babuška, A. Craig, J. Mandel, and J. Pitkaränta, Efficient Preconditioning for the *p*-Version Finite Element Method in Two Dimensions, *SIAM J. Numer. Anal.*, **28**, 3, 624-661, 1991.
7. R. Beck, P. Deuflhard, R. Hiptmair, R.H.W. Hoppe, B. Wohlmuth, Adaptive Multilevel Methods for Edge Element Discretizations of Maxwell's Equations, *Surv. Meth. Ind. (1999), 8: 271-312.*
8. D. Boffi, Discrete Compactness and Fortin Operator for Edge Elements, Preprint, March 1999.
9. D. Boffi, A Note on the Discrete Compactness Property and the de Rham Diagram, *Appl. Math. Lett.*, **14**, 1, 33-38, 2001.
10. D. Boffi, L. Demkowicz, and M Costabel, Discrete Compactness for *p* and *hp* 2D Edge Finite Elements, *TICAM Report 02-21*, May 2002.
11. A. Bossavit, Un noveau point de vue sur les éléments finis mixtes, *Matapli* (bulletin de la Société de Mathématiques Appliquées et Industrielles), 23-35, 1989.
12. W. Cecot, L. Demkowicz, and W. Rachowicz, A Two-Dimensional Infinite Element for Maxwell's Equations, *Computer Methods in Applied Mechanics and Engineering*, **188**, 625-643, 2000.
13. W. Cecot, L. Demkowicz, and W. Rachowicz, An hp-Adaptive Finite Element Method for Electromagnetics. Part 3: A Three-Dimensional Infinite Element for Maxwell's Equations, *International Journal for Numerical Methods in Engineering*, in print.

14. M. Cessenat, *Mathematical Methods in Electromagnetism*, World Scientific, Singapure 1996.
15. M. Costabel and M. Dauge, Singularities of Electromagnetic Fields in Polyhedral Domains, *Arch. Rational Mech. Anal.* **151**, 221-276, 2000.
16. M. Costabel, A Coercive Bilinear Form for Maxwell's Equations, *Math. Methods Appl. Sci.*, **12**, 4, 365-368, 1990.
17. M. Costabel and M. Dauge, Weighted Regularization of Maxwell Equations in Polyhedral Domains, preprint 2001 (http://www.maths.univ-rennes1.fr/ costabel/).
18. L. Demkowicz, Edge Finite Elements of Variable Order for Maxwell's Equations, in *Scientific Computing in Electrical Engineering*, Proceedings of the 3rd International Workshop, August 20-23, Warnemuende, Germany, (Lecture Notes in Computational Science and Engineering **18**), Springer Verlag, Berlin 2000.
19. L. Demkowicz, and I. Babuška, Optimal p Interpolation Error Estimates for Edge Finite Elements of Variable Order in 2D, *TICAM Report 01-11*, accepted to *SIAM Journal on Numerical Analysis*.
20. L. Demkowicz and M. Pal, An Infinite Element for Maxwell's Equations, *Computers Methods in Applied Mechanics and Engineering*, **164**, 77-94, 1998.
21. L. Demkowicz, W. Rachowicz, and Ph. Devloo, A Fully Automatic hp-Adaptivity, *Journal of Scientific Computing*, **17**, 1-3, 127-155, 2002.
22. L. Demkowicz, D. Pardo, and W. Rachowicz, 3D hp-Adaptive Finite Element Package (3Dhp90). Version 2.0, *TICAM Report 02-24*, June 2002.
23. L. Demkowicz, I. Babuška, J. Schöberl, and P. Monk, De Rham Diagram in 3D. Quasi Optimal p-Interpolation Estimates, *TICAM Report*, in preparation.
24. L. Demkowicz, P. Monk, L. Vardapetyan, and W. Rachowicz. De Rham Diagram for hp Finite Element Spaces *Mathematics and Computers with Applications*, **39**, 7-8, 29-38, 2000.
25. L. Demkowicz, P. Monk, Ch. Schwab, and L. Vardapetyan Maxwell Eigenvalues and Discrete Compactness in Two Dimensions, *Mathematics and Computers with Applications*, **40**, 4-5, 598-605, 2000.
26. L. Demkowicz and L. Vardapetyan, Modeling of Electromagnetic Absorption/Scattering Problems Using hp-Adaptive Finite Elements, *Computer Methods in Applied Mechanics and Engineering*, **152**, 1-2, 103-124, 1998.
27. R. Hiptmair, Multigrid Method for Maxwell's Equations., *SIAM J. Numer. Anal. Vol. 36(1): 204-225*,1998.
28. R. Hiptmair, Canonical Construction of Finite Elements, *Mathematics of Computation*, **68**, 1325-1346, 1999.
29. R. Hiptmair, Higher Order Whitney Forms, *Sonderforschungsbereich 382*, Universität Tübingen, Report 156, August 2000.
30. F. Ihlenburg, *Finite Element Analysis of Acoustic Scattering*, Springer-Verlag, New York, 1998
31. F. Kikuchi, Mixed and Penalty Formulations for Finite Element Analysis of an Eigenvalue Problemd in Electromagnetism, *Computer Methods in Applied Mechanics and Engineering*, **64**, 509-521, 1987...
32. F. Kikuchi, On a Discrete Compactness Property for the Nédélec Finite Elements, *J. Fac. Sci. Univ. Tokyo Sct. IA Math*, **36**, 3, 479-490, 1989.
33. P.D. Ledger, J. Peraire, K. Morgan, O. Hassa, and N.P. Weatherill, Efficient, Higly Accurate hp Adaptive Finite Element Computations of the Scattering Width Output of Maxwell's Equations, *International Journal for Numerical Methods in Fluids*, in print.
34. J. L. Lions and E. Magenes, *Non Homogeneous Boundary Value Problems and Applications*, Vol.1, Springer-Verlag, Berlin 1972.

35. P. Monk, On the $p-$ and $hp-$extension of Nédélec's Curl-Conforming Elements, *Journal of Computational and Applied Mathematics*, **53**, 117-137, 1994.
36. P. Monk, L. Demkowicz, Discrete Compactness and the Approximation of Maxwell's Equations in $doubleIR^3$ *Mathematics of Computation*, **70**, 234, 507-523, 2000.
37. J.C. Nédélec, Mixed Finite Elements in $\mathbb{R}^3$, *Numerische Mathematik*, **35**, 315-341, 1980.
38. J.C. Nédélec, A New Family of Mixed Finite Elements in $\mathbb{R}^3$, *Numerische Mathematik*, **50**, 57-81, 1986.
39. D. Pardo and L. Demkowicz, Integration of hp-Adaptivity and Multigrid. I. A Two Grid Solver for hp Finite Elements, *TICAM Report 02-33*.
40. I. Perugia, D. Schötzau, and P. Monk, Stabilized Interior Penalty Methods for the Time-Harmonic Maxwell Equations, *Computer Methods in Applied Mechanics and Engineering*, **191**, 41-42, 4675-4697, 2002.
41. W. Rachowicz and L. Demkowicz, A Two-Dimensional hp-Adaptive Finite Element Package for Electromagnetics (2Dhp90_EM), *TICAM Report 98-15*, July 1998.
42. W. Rachowicz, L. Demkowicz, and L. Vardapetyan, hp-Adaptive FE Modeling for Maxwell's Equations. Evidence of Exponential Convergence ACES' 99, Monterey, CA, March 16-20, 1999.
43. W. Rachowicz and L. Demkowicz, An hp-Adaptive Finite Element Method for Electromagnetics. Part 1: Data Structure and Constrained Approximation, *TICAM Report 98-15 Computer Methods in Applied Mechanics and Engineering*, **187**, 1-2, 307-337, 2000.
44. W. Rachowicz and L. Demkowicz, A Three-Dimensional hp-Adaptive Finite Element Package for Electromagnetics, *TICAM Report 00-04*, February 2000
45. W. Rachowicz and L. Demkowicz, An hp-Adaptive Finite Element Method for Electromagnetics - Part II: A 3D Implementation, *International Journal of Numerical Methods in Engineering*, **53**, 147-180, 2002.
46. Joachim Schoeberl, Commuting Quasi-interpolation Operators for Mixed Finite Elements, Preprint ISC-01-10-MATH, Texas A&M University, 2001.
47. Ch. Schwab, *p and hp-Finite Element Methods*, Clarendon Press, Oxford 1998.
48. P. Solin and L. Demkowicz, Goal-Oriented hp-Adaptivity for Elliptic Problems, *TICAM Report 02-32*, August 2002.
49. P. P. Silvester, and G. Pelosi (eds.), *Finite Elements for Wave Electromagnetics*, IEEE Press, Piscatawy, NJ, 1994.
50. L. Vardapetyan and L. Demkowicz, *hp*-Adaptive Finite Elements in Electromagnetics, *Computers Methods in Applied Mechanics and Engineering*, **169**, 331-344, 1999.
51. L. Vardapetyan, and L. Demkowicz, Full-Wave Analysis of Dielectric Waveguides at a Given Frequency, *Mathematics of Computation*, **72**, 105-129, 2002.
52. L. Vardapetyan, L. Demkowicz, and D. Neikirk, Hp Vector Finite Element Method for Eigenmode Analysis of Waveguides, *Computer Methods in Applied Mechanics and Engineering*, in print.
53. J. Wang, N. Ida, Curvilinear and Higher-Order Edge Elements in Electromagnetic Field Computation, *IEEE Trans. Magn.*, **29**, 1491-1494, 1993.
54. Y. Wang, P. Monk, and B. Szabo, Computing Cavity Modes Using the p Version of the Finite Element Method, *IEEE Transaction on Magnetics*, **32**, 3, 1934-1940, 1996.
55. J. Wang, and J.P. Webb, Hierarchical Vector Boundary Elements and p-Adaption for 3-D Electromagnetic Scattering, *IEEE Transactions on Antennas and Propagation*, **45**, 12, 1997.
56. J. P. Webb, and B. Forghani, Hierarchical Scalar and Vector Tetrahedra, *IEEE Trans. Mag.*, **29**, 2, 1295-1498, 1993.
57. J. P. Webb, Hierarchical Vector Based Funtions of Arbitrary Order for Triangular and Tetrahedral Finite Elements, *IEEE Antennas Propagat. Mag.*, **47**, 8, 1244-1253, 1999.

58. H. Whitney, *Geometric Integration Theory*, Princeton University Press, 1957.
59. Dong Xue and L. Demkowicz, Geometrical Modeling Package. Version 2.0, *TICAM Report 02-30*, August 2002.
60. A. Zdunek, W. Rachowicz, A Three-dimensional *hp*-Adaptive Finite Element Approach to Radar Scattering Problems, *Presented ad Fifth World Congress on Computational Mechanics, Vienna, Austria*, 7-12 July, 2002.

Variational Methods for Time–Dependent Wave Propagation Problems

Patrick Joly

INRIA Rocquencourt, BP105 Le Chesnay, France
patrick.joly@inria.fr

1 Introduction

There is an important need for numerical methods for time dependent wave propagation problems and their many applications, for example in acoustics, electromagnetics and geophysics. Although very old, finite difference time domain methods (FDTD methods in the electromagnetics literature) remain very popular and are widely used in wave propagation simulations, and more generally for the resolution of linear hyperbolic systems, among which Maxwell's system is a typical example. These methods allow us to get discrete equations whose unknowns are generally field values at the points of a regular mesh with spatial step h and time step Δt.

For Maxwell's equations, the Yee scheme [65, 67] is an important and much-used example of such a scheme. In 1D, it concerns the following system:

$$\frac{\partial u}{\partial t} + c\,\frac{\partial v}{\partial x} = 0, \quad \frac{\partial v}{\partial t} + c\,\frac{\partial u}{\partial x} = 0, \qquad x \in \mathbb{R}, \quad t > 0\,. \tag{1}$$

Without any mesh refinement, the equations of the scheme are (with standard notation $u_j^n \approx u(jh, n\Delta t)$)

$$\frac{u_j^{n+1} - u_j^n}{\Delta t} + c\,\frac{v_{j+1/2}^{n+1/2} - v_{j-1/2}^{n+1/2}}{h} = 0, \quad \frac{v_{j+1/2}^{n+1/2} - v_{j+1/2}^{n-1/2}}{\Delta t} + c\,\frac{u_{j+1}^n - u_j^n}{h} = 0, \tag{2}$$

where the discrete unknowns are evaluated on a staggered uniform grid. There are several reasons that explain the success of Yee type schemes, among which their easy implementation, their efficiency, their accuracy and the fact that a lot of properties of continuous Maxwell's equations (energy conservation, divergence free property etc.) are respected at the discrete level. In particular, the good performance of Yee's scheme is related to the following properties:

- A uniform regular grid is used for the space discretisation, so that there is a minimum of information to store and the data to be computed are structured. In other words, one avoids all the complications due to the use of non uniform meshes.

- An explicit time discretisation is applied – no linear system has to be solved at each time step.

This scheme is centred, of order two both in space and time, and is completely explicit. The stability and accuracy properties of such a scheme are well known (at least in a homogeneous medium in which the classical Fourier analysis can be used). As a consequence of its explicit nature, the scheme is stable under the CFL condition

$$\frac{c\Delta t}{h} \leq \frac{\sqrt{d}}{d},$$

where c denotes the propagation speed and d the number of space dimensions. This means that the time step cannot be too large, but it is not restrictive in practice since a small time step is required for reasonable accuracy. On the other hand, the time step must not be too small either because, as is well known, the numerical dispersion – roughly speaking the error committed on the propagation velocity of waves – increases when the ratio $c\Delta t/h$ decreases.

The counterpart of the nice properties of FDTD schemes is a lack of "geometrical flexibility" which makes the use of such schemes not obvious in the case of computational domains of complicated shape (consider here the diffraction of electromagnetic waves by an obstacle as a target problem). It may also be difficult (at least with a theoretical guarantee of stability) to treat boundary conditions and variable coefficients or to be able do local mesh refinement.

To overcomc such difficulties there exist at least two attractive solutions:

- variational methods, particularly the finite element method;
- the finite integration technique [22] and finite volume methods [62].

These are "natural" extensions in the sense that, for instance, Yee's scheme can be interpreted as a particular mixed finite element method or a particular finite volume method on a uniform grid. Moreover, they also provide a systematic way to get a stable extension of FDTD schemes in heterogeneous media, and the convergence theory of such methods is well known.

My first objective in this article will be to provide, in Sect. 2, a brief overview about the construction and the analysis of variational methods. As I said before, these approaches allow the use of general meshes, which leads us to deal with unstructured data and related complications such as the construction of the mesh and the influence of small size elements (in the mesh) on the allowable time step.

My second objective will be to review two recent works that aim to make possible the treatment of complex geometries with FDTD schemes while preserving the nice properties of these methods. This means that:

- the data of the problem remain (mostly) structured;
- the time discretisation remains (essentially) explicit;
- the stability condition is not affected by the geometry of the computational domain.

The first approach, that I will describe in Sect. 3, is the fictitious domain method. Let us consider the model problem of the scattering of an incident electromagnetic wave by a perfectly conducting obstacle. The idea of the method consists of artificially extending the solution inside the obstacle – which makes possible the use of a uniform, regular 3D grid for the electromagnetic field – and to introduce at the same time a conforming surface mesh for the boundary of the obstacle to handle the boundary condition. On this mesh, one computes an auxiliary unknown that can be interpreted as a Lagrange multiplier associated with the boundary condition and coincides, in this case, with the surface electric current. The challenge is then to make these two "independent" meshes communicate in a clever way. This can be done through the use of a mixed variational formulation in which the boundary condition is taken into account in a weak sense. The stability of the method in ensured through a discrete energy conservation. The only additional computational cost (with respect to the standard FDTD scheme) is restricted to the boundary mesh: a sparse positive definite linear system has to be solved at each time step.

An alternative approach (that can be combined with the fictitious domain method) consists of refining the mesh in the neighbourhood of the obstacle. I will present in Sect. 4 some recent research about conservative space-time mesh refinement methods. When one works with regular grids, the transition between a coarse and a fine grid is necessarily "non conforming". Moreover, for efficiency and accuracy considerations, one would like to use a local time step in order to keep the time step/space step ratio constant. Traditional interpolation methods can lead to non standard instability phenomena. Here, we shall propose two alternative methods based on the reformulation of the problem as an artificial domain decomposition. The key issue of these methods is that their stability is guaranteed from the theoretical point of view through the conservation of an appropriate discrete energy. The first method involves the introduction of a Lagrange multiplier on the coarse–fine grid interface (as in the so-called mortar element method), while the second does not. As in the fictitious domain method, both methods require the solution of a small, sparse, positive definite linear system on the interface. We shall also show how spurious numerical phenomena due to a change of grid can be analysed and controlled.

2 Basic Principles of Variational Methods

In this section, we present a brief introduction (a crash course) to variational numerical schemes for time dependent wave propagation models. Almost all the material contained in this section is very classical, but useful for the next sections. We do not pretend here to be exhaustive and completely rigorous. That is why all the assumptions needed for a rigorous development will not always be written and the proofs of a lot of statements will be only sketched or omitted when they are trivial. We only wish to present the main ideas of results with a sufficient degree of abstraction and generality in order to emphasise the interest of the approach.

2.1 Mathematical Models

Most of the basic wave propagation mathematical models can be rewritten in one of the two following forms (abstract wave equations):

- As a first order system in time:

$$\begin{cases} \dfrac{\partial u}{\partial t} - \mathcal{B}v = 0, \\ \dfrac{\partial v}{\partial t} + \mathcal{B}^* u = 0, \end{cases} \tag{3}$$

  where the unknowns u and v are functions from $\Omega \subset \mathbb{R}^N$ (the domain of propagation) into $\mathbb{R}^p$ and $\mathbb{R}^q$ respectively. The two operators $\mathcal{B}$ and $\mathcal{B}^*$ are spatial differential operators which are, at least formally, adjoint for appropriate weighted inner products $(u, \tilde{u})_P$ – with norm $\|u\|_P$ – in $L^2(\Omega)^p$ and $(v, \tilde{v}_D)$ – with norm $\|v\|_D$ – in $L^2(\Omega)^q$. In the applications, verifying such a property amounts in practice to applying an adequate integration by parts formula.
- As a second order system in time, either after elimination of v or u:

$$\frac{\partial^2 u}{\partial t^2} + \mathcal{A}u = 0\,; \tag{4}$$

$$\frac{\partial^2 v}{\partial t^2} + \tilde{\mathcal{A}}v = 0\,. \tag{5}$$

  Both operators $\mathcal{A} = \mathcal{B}\mathcal{B}^*$ and $\tilde{\mathcal{A}} = \mathcal{B}^*\mathcal{B}$ are formally positive self-adjoint (in some weighted L^2 space). By convention, we shall call u the primal variable and v the dual one. Accordingly, we shall refer to problem (4) as the primal problem and to problem (5) as the dual one. In the same way $\|.\|_P$ is the primal L^2 norm and $\|.\|_D$ the dual one.

Of course, the mathematical model has to be completed by initial and boundary conditions (which play a role in the adjointness results etc.). We shall omit them for simplicity in this introductory section.

Let us give several concrete examples for illustration:

- The 1D wave equation:

$$\begin{cases} \mathcal{B} = \dfrac{\partial}{\partial x}, \ \mathcal{B}^* = -\dfrac{\partial}{\partial x}, \ \mathcal{A} = -\dfrac{\partial^2}{\partial x^2}, \\ (u, \tilde{u})_P = \displaystyle\int_\Omega u\, \tilde{u}\, \mathrm{d}x\,, \quad (v, \tilde{v})_D = \int_\Omega v\, \tilde{v}\, \mathrm{d}x\,. \end{cases} \tag{6}$$

- The acoustic wave equation:

$$\begin{cases} u = p \in \mathbb{R}^1 \text{ (the pressure field)}\,, v \in \mathbb{R}^3 \text{ (the velocity field)}\,, \\ \mathcal{B} = K \,\mathrm{div}\,,\ \mathcal{B}^* = -\,\rho^{-1}\nabla,\ \mathcal{A} = -\,K\,\mathrm{div}\,(\rho^{-1}\nabla), \\ (p, \tilde{p})_P = \displaystyle\int_\Omega K^{-1}\, p\, \tilde{p}\, \mathrm{d}x\,, \quad (v, \tilde{v})_D = \int_\Omega \rho^{-1}\, v \cdot \tilde{v}\, \mathrm{d}x\,, \end{cases} \tag{7}$$

where ρ is the density of the fluid and K denotes its Lamé constant.

- Maxwell's equations:

$$\begin{cases} u = e \in \mathbb{R}^3 \text{ (the electric field)}, v = h \in \mathbb{R}^3 \text{ (the magnetic field)}, \\ \mathcal{B} = \varepsilon^{-1} \text{ curl}, \ \mathcal{B}^* = \mu^{-1} \text{ curl}, \ \mathcal{A} = -\ \varepsilon^{-1} \text{ curl } (\mu^{-1}\text{curl}), \\ (e, \tilde{e})_P = \displaystyle\int_\Omega \varepsilon\, e \cdot \tilde{e} \, dx \,, \quad (h, \tilde{h})_D = \displaystyle\int_\Omega \mu\, h \cdot \tilde{h} \, dx \,, \end{cases} \tag{8}$$

where ε and μ are respectively the dielectric permittivity and the magnetic permeability.

- Elastodynamics equations:

$$\begin{cases} u = \mathbf{v} \in \mathbb{R}^3 \text{ (velocity field)}\,, v = \sigma \in \mathbb{R}^6 \text{ ((symmetric) stress tensor)} \\ \mathcal{B} = \rho^{-1} \text{ div}\,, \ \mathcal{B}^* = -\ C\varepsilon(\cdot), \ \mathcal{A} = -\rho^{-1} \text{ div } \ C(\varepsilon(\cdot)), \\ (\mathbf{v}, \tilde{\mathbf{v}})_P = \displaystyle\int_\Omega \rho\, \mathbf{v} \cdot \tilde{\mathbf{v}} \, dx \,, \quad (\sigma, \tilde{\sigma})_D = \displaystyle\int_\Omega C^{-1}\, \sigma \cdot \tilde{\sigma} \, dx \,, \end{cases} \tag{9}$$

where ρ denotes the density, C the operator appearing in Hooke's law and where the (tensor valued) differential operator $\varepsilon(\cdot)$ is defined by:

$$\varepsilon_{ij}(\mathbf{v}) = \frac{1}{2}\left(\frac{\partial \mathbf{v}_i}{\partial x_j} + \frac{\partial \mathbf{v}_j}{\partial x_i}\right).$$

A fundamental property of such models is that they are intrinsically linked to the conservation in time of an energy:

$$E(t) = \frac{1}{2} \left(\|u(t)\|_P^2 + \|v(t)\|_D^2 \right)$$

for model (3) and

$$E(t) = \frac{1}{2} \left(\left\| \frac{\partial u}{\partial t}(t) \right\|_P^2 + (\mathcal{A}u(t), u(t))_P \right)$$

for model (4). These energies make sense once the functional framework has been rigorously prescribed and have a physical meaning in each of the concrete examples evoked above.

Remark 1. The models we have considered correspond to the propagation of waves in the absence of sources. The presence of source terms would imply the addition of right hand sides in the equations and the conservation of the energy would be replaced by an energy identity relating the variation of the energy to the source term.

2.2 Variational Formulations

The variational formulation in space of the above problems is a rigorous reformulation in which the unknown is sought as a function of time with values in spaces of functions of the space variable x. We therefore need to introduce a functional framework. In this way, one separates the role of the space and time variables, which naturally leads to different discretisation procedures in space and time.

Remark 2. In what follows, we shall systematically omit the role of the boundary conditions in such a way that all that follows in this section is rigorously valid only in the case $\Omega = \mathbb{R}^N$. In the other cases, the boundary conditions have a role in the integration by parts and may influence the definition of the functional spaces. However, what is important for our purpose is the abstract form of the problem ((12), (15) or (16)) that we shall obtain. This remains valid for a number of physically relevant boundary conditions, and is only slightly different for some other ones.

Second Order Problems

Let us first present the variational formulation of problem (4). Formally, it is equivalent to take the L^2 scalar product of equation (4) with a test function $\tilde{u}$ that only depends on the space variable (i.e. multiply (4) by $\tilde{u}$ and integrate over Ω). To justify this, we need to introduce a functional framework. Let $H(\mathcal{B}^*)$ be the Hilbert space:

$$H(\mathcal{B}^*) = \{u \in L^2(\Omega)^p \;/\; \mathcal{B}^* u \in L^2(\Omega)^q\}, \tag{10}$$

equipped with the norm

$$\|u\|^2_{H(\mathcal{B}^*)} = \|u\|^2_P + \|\mathcal{B}^* u\|^2_D. \tag{11}$$

Then the variational formulation of (4) can be written as follows:

$$\begin{cases} \text{Find } u(t) : \mathbb{R}^+ \longrightarrow U = H(\mathcal{B}^*) \text{ such that:} \\ \dfrac{\mathrm{d}^2}{\mathrm{d}t^2}(u(t), \tilde{u})_P + a(u(t), \tilde{u}) = 0, \quad \forall\, \tilde{u} \in U, \\ a(u, \tilde{u}) = (\mathcal{B}^* u, \mathcal{B}^* \tilde{u})_D, \quad \forall\, (u, \tilde{u}) \in U \times U. \end{cases} \tag{12}$$

This is also the appropriate formulation for defining the notion of weak solutions to the evolution problem and developing the corresponding existence theory (see for instance [53]). Note that the bilinear form $a(\cdot,\cdot)$ is positive and symmetric (which is intimately related to the energy conservation). As an illustration (again neglecting the boundary conditions etc.):

- For example (7): $U = H^1(\Omega)$ and $a(p,\tilde{p}) = \int_\Omega \rho^{-1}\, \nabla p \cdot \nabla \tilde{p}\, \mathrm{d}x$.
- For example (8) : $U = H(\text{curl}, \Omega)$ and $a(e,\tilde{e}) = \int_\Omega \mu \,\text{curl}\, e \cdot \text{curl}\, \tilde{e}\, \mathrm{d}x$.
- For example (9) : $U = H^1(\Omega)^3$ and $a(\mathbf{v},\tilde{\mathbf{v}}) = \int_\Omega C\, \varepsilon(\mathbf{v}) \cdot \varepsilon(\tilde{\mathbf{v}})\, \mathrm{d}x$.

Indeed, we have an analogous variational formulation for the dual problem (5), which leads to working in the Hilbert space:

$$H(\mathcal{B}) = \{v \in L^2(\Omega)^q \; / \; \mathcal{B}v \in L^2(\Omega)^p\}, \tag{13}$$

equipped with the norm

$$\|v\|^2_{H(\mathcal{B})} = \|v\|^2_D + \|\mathcal{B}v\|^2_P \, . \tag{14}$$

First Order Problems

For these problems, the appropriate variational formulation is the so-called mixed variational formulation (see [16] for stationary problems). Once again the principle is to multiply the two equations of system (3) by test functions and to integrate over Ω, but the key point this time is to apply integration by parts only for one of the two equations.

- **The primal-dual formulation**. This formulation holds in a functional framework in which the regularity is put on the primal variable u. More precisely, u will be sought in the space $U = H(\mathcal{B}^*)$ and v in the space $V = L^2(\Omega)^q$:

$$\begin{cases} \text{Find } (u(t), v(t)) : \mathbb{R}^+ \longrightarrow U \times V \text{ such that:} \\ \dfrac{\mathrm{d}}{\mathrm{d}t}(u(t), \tilde{u})_P - b(\tilde{u}, v(t)) = 0, \quad \forall\, \tilde{u} \in U, \\ \dfrac{\mathrm{d}}{\mathrm{d}t}(v(t), \tilde{v})_D + b(u(t), \tilde{v}) = 0, \quad \forall\, \tilde{v} \in V, \\ b(u, v) = \displaystyle\int_\Omega \mathcal{B}^* u \cdot v \; \mathrm{d}x, \quad \forall\, (u, v) \in U \times V. \end{cases} \tag{15}$$

  In the examples we have
  - (7) : $U = H^1(\Omega), V = L^2(\Omega)^3$ and $b(p, v) = -\int_\Omega \nabla p \cdot v \; \mathrm{d}x$.
  - (8) : $U = H(\text{curl}, \Omega), V = L^2(\Omega)^3$ and $b(e, h) = \int_\Omega \text{curl}\; e \cdot h \; \mathrm{d}x$.
  - (9) : $U = H^1(\Omega)^3, V \equiv L^2(\Omega)^3$ and $b(\mathbf{v}, \sigma) = -\int_\Omega \varepsilon(\mathbf{v}) \cdot \sigma \; \mathrm{d}x$.

- **The dual-primal formulation**. This formulation holds in a functional framework in which the regularity is put on the dual variable v. More precisely, v is sought in the space $V = H(\mathcal{B})$ and u in the space $U = L^2(\Omega)^p$:

$$\begin{cases} \text{Find } (u(t), v(t)) : \mathbb{R}^+ \longrightarrow U \times V \text{ such that:} \\ \dfrac{\mathrm{d}}{\mathrm{d}t}(u(t), \tilde{u})_P - b(\tilde{u}, v(t)) = 0, \quad \forall\, \tilde{u} \in U, \\ \dfrac{\mathrm{d}}{\mathrm{d}t}(v(t), \tilde{v})_D + b(u(t), \tilde{v}) = 0, \quad \forall\, \tilde{v} \in V, \\ b(u, v) = \displaystyle\int_\Omega u \cdot \mathcal{B} v \; \mathrm{d}x, \quad \forall\, (u, v) \in U \times V. \end{cases} \tag{16}$$

For the examples we have

- (7) : $U = L^2(\Omega)^3, V = H(\mathrm{div}, \Omega)$ and $b(p, v) = \int_\Omega p \cdot \mathrm{div}\, v \, \mathrm{d}x$.
- (8) : $U = L^2(\Omega)^3, V = H(\mathrm{curl}, \Omega)$ and $b(e, h) = \int_\Omega e \cdot \mathrm{curl}\, h \, \mathrm{d}x$.
- (9) : $U = L^2(\Omega)^3, V = H^{\mathrm{sym}}(\mathrm{div}, \Omega)$ and $b(\mathbf{v}, \sigma) = \int_\Omega \mathbf{v} \cdot \mathrm{div}\, \sigma \, \mathrm{d}x$ where $H^{\mathrm{sym}}(\mathrm{div}, \Omega)$ is by definition the space of symmetric tensors in $L^2(\Omega)^9$ with divergence in $L^2(\Omega)^3$ (see [11] for instance for details).

2.3 Finite Element Approximation

Contrary to the principle of the finite difference method, which consists of approximating differential operators by difference operators, the philosophy of the finite element methods consists of approximating functional spaces by finite dimensional subspaces and the approximate problem simply amounts to solving the variational problems in these subspaces.

Second Order Problems

Construction.

Let us consider the primal problem (12). We introduce a family $\{U_h, h > 0\}$ of finite dimensional subspaces of U where h is an approximation parameter (designed to tend to zero). In practice, it will be the step-size of a spatial mesh of the computational domain Ω. The approximate problem is simply:

$$\begin{cases} \text{Find } u_h(t) : \mathbb{R}^+ \longrightarrow U_h \text{ such that:} \\ \dfrac{\mathrm{d}^2}{\mathrm{d}t^2}(u_h(t), \tilde{u}_h)_P + a(u_h(t), \tilde{u}_h) = 0, \quad \forall\, \tilde{u}_h \in U_h. \end{cases} \tag{17}$$

Introducing the operator $\mathbf{A}_h \in \mathcal{L}(U_h)$ defined by:

$$(\mathbf{A}_h u_h, \tilde{u}_h)_P = a(u_h(t), \tilde{u}_h), \quad \forall\, (u_h, \tilde{u}_h) \in U_h \times U_h, \tag{18}$$

problem (17) can simply be rewritten as:

$$\frac{\mathrm{d}^2 u_h}{\mathrm{d}t^2} + \mathbf{A}_h u_h = 0. \tag{19}$$

In practice, after expansion of the unknown $u_h(t)$ on a basis (to be chosen) of U_h, the new unknown becomes the vector of the coefficients of this expansion:

$$\mathbf{u}_h(t) \in \mathbb{R}^{\dim U_h},$$

and (17) results into an ordinary differential system:

$$\mathcal{M}_h \frac{\mathrm{d}^2 \mathbf{u}_h}{\mathrm{d}t^2} + \mathcal{A}_h \mathbf{u}_h = 0, \tag{20}$$

where the matrix $\mathcal{M}_h$ (resp. $\mathcal{A}_h$) is symmetric positive definite (resp symmetric and positive). These two matrices are simply defined by:

$$(\mathcal{M}_h \mathbf{u}_h, \tilde{\mathbf{u}}_h) = (u_h(t), \tilde{u}_h)_P, \quad (\mathcal{A}_h \mathbf{u}_h, \tilde{\mathbf{u}}_h) = a(u_h(t), \tilde{u}_h), \tag{21}$$

$\forall (u_h, \tilde{u}_h) \in U_h \times U_h$, where $\mathbf{u}_h$ and $\tilde{\mathbf{u}}_h$ are the vectors associated with u_h and $\tilde{u}_h$ and $(\cdot,\cdot)$ denotes the usual scalar product in $\mathbb{R}^{\dim U_h}$. Note that the matrices represent discrete approximations of (respectively) the "identity" operator and the operator $\mathcal{A}$. Let us also recall that in the applications, the classical choices for the basis of U_h (functions with small support) lead to very sparse matrices.

Convergence and error estimates.

First note that the stability of the finite element method is a direct consequence of the energy identity (that easily derives from the symmetry of $a(\cdot,\cdot)$):

$$\frac{\mathrm{d}}{\mathrm{d}t} E_h(t) = 0, \quad \text{where} \quad E_h(t) = \frac{1}{2}\left(\|\frac{\mathrm{d}u_h}{\mathrm{d}t}(t)\|_P^2 + a(u_h(t), u_h(t)) \right). \tag{22}$$

This stability result is sufficient to prove weak convergence (for instance in the $H^1(0,T;L^2) \cap L^2(0,T;U)$ topology) as soon as the spaces U_h satisfy the standard approximation property:

$$\lim_{h\to 0} \inf_{v_h \in U_h} \|u - v_h\|_U = 0, \quad \forall u \in U. \tag{23}$$

A classical approach to the strong convergence analysis of finite element methods for second order hyperbolic problems consists of combining standard approximation results for elliptic problems with energy estimates (see for instance [34]). Let us give the main ideas of this approach. It uses the notion of elliptic projection defined by:

$$\Pi_h : u \in U \longrightarrow \Pi_h u \in U_h$$

where $\Pi_h u$ is nothing but the orthogonal projection on U_h for the scalar product $(\cdot,\cdot)_U$:

$$a(u - \Pi_h u, \tilde{u}_h) + (u - \Pi_h u, \tilde{u}_h)_P = 0, \quad \forall \tilde{u}_h \in U_h. \tag{24}$$

The approximation assumption for the spaces U_h is simply:

$$\lim_{h\to 0} \|u - \Pi_h u\|_U = 0, \quad \forall u \in U. \tag{25}$$

The idea is then to split the error $e_h = u_h - u$ into two parts:

$$e_h = \eta_h - \varepsilon_h, \quad \eta_h = u_h - \Pi_h u, \quad \varepsilon_h = u - \Pi_h u\,.$$

The convergence of ε_h to 0 results from (23). It remains to look at η_h which satisfies:

$$(\frac{\mathrm{d}^2 \eta_h}{\mathrm{d}t^2}, \tilde{u}_h)_P + a(\eta_h, \tilde{u}_h) = (\frac{\mathrm{d}^2 \varepsilon_h}{\mathrm{d}t^2} + \varepsilon_h, \tilde{u}_h)_P, \quad \forall\, \tilde{u}_h \in U_h.$$

Choosing $\tilde{u}_h = \mathrm{d}\eta_h/\mathrm{d}t$ and setting:

$$\mathcal{E}_h(t) = \frac{1}{2}\{ \, \|\frac{d\eta_h}{\mathrm{d}t}(t)\|_P^2 + a(\eta_h(t), \eta_h(t)) \, \},$$

we get the identity:

$$\frac{\mathrm{d}}{\mathrm{d}t}\mathcal{E}_h = (\frac{\mathrm{d}^2\varepsilon_h}{\mathrm{d}t^2} + \varepsilon_h, \frac{\mathrm{d}\eta_h}{\mathrm{d}t})_P \leq \sqrt{2}\, \|\frac{\mathrm{d}^2\varepsilon_h}{\mathrm{d}t^2} + \varepsilon_h\|_P \, \times \, \mathcal{E}_h^{1/2}.$$

After integration in time, we obtain the estimate:

$$\mathcal{E}_h(t)^{1/2} \leq \mathcal{E}_h(0)^{1/2} + \int_0^t \|\frac{\mathrm{d}^2\varepsilon_h}{\mathrm{d}t^2}(s) + \varepsilon_h(s)\|_P \, \mathrm{d}t\,,$$

where $\mathcal{E}_h(0)$ refers to the approximation of the initial data and the integral term can be shown to tend to 0 as h tends to 0, thanks to (23), given some regularity assumptions (in time) on the exact solution u. This estimate is sufficient to prove the convergence of u_h in $L^\infty(0, T; U)$. In concrete examples and with standard choices of the finite element spaces, this also leads to error estimates in powers of h.

Mass lumping.

One of the practical problems posed by the finite element approach is the presence of the mass matrix $\mathcal{M}_h$: even with an explicit time discretisation (see Sect. 2.4), this matrix has to be inverted at each time step. With finite difference or finite volume methods for instance, this matrix is diagonal by construction (or at least block diagonal, the dimension of the blocks – typically 2 or 3 for anisotropic vectorial problems – is independent of h) which leads to "really" explicit schemes. This is consistent with the completely local nature of the continuous operator which is approximated by the mass matrix (typically the identity operator). With the finite element method, due to the fact that it is in general impossible to construct a basis of functions with disjoint supports, this matrix is no longer diagonal (although it is very sparse). One will generally use an iterative method to solve the linear system, since, in practice, direct methods are prohibited for reasons of size. Typically the conjugate gradient method is used, and it converges in very few iterations (typically between 10 and 50, depending on the complexity of the problem and the desired accuracy, which should a priori increase with the order of the finite element method, see the remark below). Nevertheless, this has a significant cost since one step of the iterative algorithm corresponds to one step of explicit scheme with a diagonal mass matrix.

Remark 3. The iterative methods converge quickly due to the good conditioning of the mass matrix. Note however that this condition number increases with the space dimension, with the size and the quality of the mesh and in the case of very heterogeneous media. It is also greater for the Maxwell system than for the scalar wave equation.

An alternative approach is to apply the so-called *mass lumping* procedure. This consists of replacing the exact mass matrix by an approximation, the lumped mass matrix, which is diagonal. This should be done without losing any (order of) accuracy. The technical justification is the use of a quadrature approximation for the evaluation of the integrals that define the entries of the mass matrix.

Let us give more detail in the special case of P_k (Q_k) Lagrange elements for the approximation of the scalar wave equation on a triangular (quadrilateral) or tetrahedral (hexahedral) mesh. In the case $k = 1$, the technique is well known and corresponds to considering the diagonal matrix in which each diagonal element is the sum of all the elements of the same line in the original mass matrix! This works only for $k = 1$ and corresponds to use the quadrature formula on each triangle (quadrilateral) or tetrahedron (hexahedron) consisting of taking the mean value of the function f over the vertices (which are in this case the location of the degrees of freedom) multiplied by the measure of the element.

The generalisation to higher order without any loss of accuracy is more delicate. For Lagrange elements, the degrees of freedom are values of the unknown function at given points. The idea is to use, in each element K of the mesh, a quadrature formula of the form:

$$\int_K f \, \mathrm{d}x = \text{meas}\, K \sum_l \omega_l \, f(M_l)$$

where the M_l are the quadrature nodes and the ω_l are the corresponding quadrature weights. The success of the procedure is subject to the following criteria:

- The quadrature nodes coincide with the locations of the degrees of freedom: this will provide a diagonal mass matrix.
- The quadrature weights are strictly positive: this is necessary to ensure the invertibility of the lumped mass matrix and the stability of the resulting scheme.
- The quadrature formula must be sufficiently precise in order to preserve the accuracy of the finite element method (appropriate criteria can be found in [21, 40, 64] and the corresponding error analysis for second order hyperbolic problems in [6]).

In the case of quadrilateral or hexahedral elements, these criteria can easily be fulfilled with the standard Q_k approximation spaces. Exploiting the tensor product nature of the finite element, it suffices to treat the 1D situation. The solution simply consists of displacing the degrees of freedom that are strictly interior to the elements at the nodes determined by the Gauss-Lobatto quadrature formulas. Such elements are also called spectral finite elements. They give very good results in theory and practice. We refer the reader to [25] for an error analysis for the wave equation, and to [24] for more details (it is in particular emphasised that it is also of interest to use the same quadrature formulas for the evaluation of the stiffness matrix) and numerical results.

In the case of triangles or tetrahedra, the solution is more complicated. For stability reasons, as soon as $k \geq 2$, it is necessary to enrich the finite element space and to replace P_k by a new space of polynomials $\widetilde{P}_k$ with:

$$P_k \subset \widetilde{P}_k \subset P_{k'}, \quad k' \geq k,$$

to determine the new locations for the degrees of freedom as well as appropriate quadrature formulas, according to the above criteria. In dimension 2, a solution has been found for $k = 2, 3, 4, 5$ and in dimension 3 for $k = 2, 3$ (see [20, 26]). In 2D, various numerical tests show that the method is very efficient. In 3D, the conclusions are less optimistic. It appears that contrary to what happens in two dimensions, the dimension of the space $\widetilde{P}_k$ is much higher than the dimension of P_k. Therefore, one has to deal with many more degrees of freedom than with the standard element and, consequently, the computational time increases significantly, this being amplified by the deterioration of the CFL stability condition (see [47] for more details).

For models whose unknowns are vector fields, such as Maxwell's equations, the problem of mass lumping is even more delicate. As an illustration, let us briefly describe the state of the art of the construction of 3D edge elements (i.e. finite element spaces for the approximation of the space $H(\mathrm{curl}, \Omega)$ as in Nédélec [59, 60]) that are compatible with mass lumping. In the case of hexahedra, the situation is rather simple ([27, 28]): thanks to the tensor product structure, it suffices to adapt the solution of the scalar case. The main change is that all the components of the vector field are not treated in the same way (remember for instance that only the tangential component is continuous from one element to the other) and one uses both Gauss-Lobatto and Gauss-Legendre quadrature points and weights (see the book [24] for more details). In the case of hexahedra, even for the lowest edge element space, the situation is more complicated. As in the scalar case, it is necessary to enrich the finite element space. The key point is that one must be able to evaluate completely the value of the vector field at each quadrature point from the degrees of freedom associated with this point. In practice, this implies for instance to adding the normal components of the field at the points located on the boundary of the element (while one uses only tangential components with the usual edge elements) without enforcing their continuity from one element to the other. We refer the reader to [36, 37] for a description of the elements in 2 and 3 space dimensions and to [35] for numerical computations in 2D (that illustrate the interest of such elements). In 3D, the second order element described in [48] seems to be particularly attractive due to the small number of additional degrees of freedom needed.

First Order Problems

All of what follows is applicable to both the primal-dual and dual-primal formulations of Sect. 2.2, since they have the same abstract form.

Construction.

As for second order problems, the method will be completely determined by two families $\{U_h, h > 0\}$ and $\{V_h, h > 0\}$ of finite dimensional subspaces of U and V. The approximate problem is simply:

$$\begin{cases} \text{Find } (u_h(t), v_h(t)) : \mathbb{R}^+ \longrightarrow U_h \times V_h \text{ such that:} \\ \dfrac{\mathrm{d}}{\mathrm{d}t}(u_h(t), \tilde{u}_h)_P - b(\tilde{u}_h, v_h(t)) = 0, \quad \forall\, \tilde{u}_h \in U_h, \\ \dfrac{\mathrm{d}}{\mathrm{d}t}(v_h(t), \tilde{v}_h)_D + b(u_h(t), \tilde{v}_h) = 0, \quad \forall\, \tilde{v}_h \in V_h\,, \end{cases} \tag{26}$$

or equivalently, in an operator form:

$$\begin{cases} \dfrac{\partial u_h}{\partial t} - \mathbf{B}_h v_h = 0, \\ \dfrac{\partial v_h}{\partial t} + \mathbf{B}_h^* u_h = 0, \end{cases} \tag{27}$$

where $(\mathbf{B}_h, \mathbf{B}_h^*) \in \mathcal{L}(U_h) \times \mathcal{L}(V_h)$ are defined by:

$$\forall\, (u_h, v_h) \in U_h \times V_h, \quad (\mathbf{B}_h^* u_h, v_h)_D = (u_h, \mathbf{B}_h v_h)_P = b(u_h, v_h). \tag{28}$$

In practice, this corresponds, as in Sect. 2.3, to a first order ordinary differential system that can be written as (we omit here the definition of the various matrices which is quite natural):

$$\begin{cases} \mathcal{M}_h^P \dfrac{\mathrm{d}\mathbf{u}_h}{\mathrm{d}t} - \mathcal{B}_h \mathbf{v}_h = 0, \\ \mathcal{M}_h^D \dfrac{\mathrm{d}\mathbf{v}_h}{\mathrm{d}t} + \mathcal{B}_h^* \mathbf{u}_h = 0\,. \end{cases} \tag{29}$$

In (29), $\mathcal{M}_h^P$ and $\mathcal{M}_h^D$ are two (symmetric positive definite) mass matrices. The matrix $\mathcal{B}_h^*$ is the adjoint (in the usual sense of matrices) of $\mathcal{B}_h$ and:

- $(\mathcal{M}_h^P)^{-1}\, \mathcal{B}_h$ represents a discrete approximation of the operator $\mathcal{B}$,
- $(\mathcal{M}_h^D)^{-1}\, \mathcal{B}_h^*$ represents a discrete approximation of the operator $\mathcal{B}^*$.

In the applications considered in Sect. 2.1 :

- (7) : $(\mathcal{M}_h^P)^{-1}\, \mathcal{B}_h$ is a discrete gradient operator while $(\mathcal{M}_h^D)^{-1}\, \mathcal{B}_h^*$ represents a discrete divergence.
- (8) : $(\mathcal{M}_h^P)^{-1}\, \mathcal{B}_h$ and $(\mathcal{M}_h^D)^{-1}\, \mathcal{B}_h^*$ both represent discrete curl operators.
- (9) : $(\mathcal{M}_h^P)^{-1}\, \mathcal{B}_h$ is a discrete deformation operator while $(\mathcal{M}_h^D)^{-1}\, \mathcal{B}_h^*$ represents a discrete divergence.

Remark 4. With two mass matrices, the question of mass lumping is naturally posed and can be solved as explained in Sect. 2.3. Note however, that the question only affects one of the two matrices. With the primal-dual formulation for instance, the "regularity" is put on the primal variable as indicated by the definition (10) of the space U. A consequence is that, in practice, the space U_h will be made of functions which are regular (typically polynomial) inside each element and satisfy certain continuity from one element to the other: the corresponding mass matrix is not (even block) diagonal. In contrast, as V is simply an L^2 space, one will use completely discontinuous finite elements to construct V_h so that the corresponding mass matrix is (block) diagonal by construction.

Error estimates.

Note that one has the discrete energy conservation identity:

$$\frac{\mathrm{d}}{\mathrm{d}t}E_h(t) = 0, \quad \text{where} \quad E_h(t) = \frac{1}{2}\left(\|u_h(t)\|_P^2 + \|v_h(t)\|_D^2 \right). \tag{30}$$

This is an L^2 stability result that is valid independently of the choice of the spaces U_h and V_h. However, it is not sufficient to deduce the weak convergence to the true solution, even if the spaces U_h and V_h satisfy the approximation property (23). Indeed, this estimate does not provide either of the following estimates that would be needed in order to pass to the limit in the weak formulation:

- an estimate of $u_h(t)$ in the space $U = H(\mathcal{B}^*)$ in the case of the primal-dual formulation,
- an estimate of $v_h(t)$ in the space $V = H(\mathcal{B})$ in the case of the dual-primal formulation.

As a matter of fact, the convergence of the mixed finite element method requires some compatibility between the two spaces U_h and V_h. Particular examples of compatibility condition (considered for instance in [56, 58]) are:

$$\begin{cases} \mathcal{B}^* U_h \subset V_h & \text{in the primal-dual case } (i) \\ \mathcal{B} V_h \subset U_h & \text{in the dual-primal case } (ii)\,. \end{cases} \tag{31}$$

Indeed, in such a case, the mixed formulation coincides with the standard (primal or dual) variational formulation of Sect. 2.2. Let us consider for instance (26) rewritten as:

$$\begin{cases} \dfrac{\mathrm{d}}{\mathrm{d}t}(u_h(t), \tilde{u}_h)_P - (\mathcal{B}^* \tilde{u}_h, v_h(t))_D = 0, & \forall\, \tilde{u}_h \in U_h, \\ \dfrac{\mathrm{d}}{\mathrm{d}t}(v_h(t), \tilde{v}_h)_D + (\mathcal{B}^* u_h(t), \tilde{v}_h)_D = 0, & \forall\, \tilde{v}_h \in V_h. \end{cases} \tag{32}$$

If (31) (i) holds, we can take $\tilde{v}_h = \mathcal{B}^* \tilde{u}_h$ in the second equation of (32) to obtain:

$$\frac{\mathrm{d}}{\mathrm{d}t}(v_h(t), \mathcal{B}^* \tilde{u}_h)_D + (\mathcal{B}^* u_h(t), \mathcal{B}^* \tilde{u}_h)_D = 0, \quad \forall\, \tilde{v}_h \in V_h.$$

Using this identity in the first equation of (32) (differentiated in time) allows us to eliminate $v_h(t)$ and to show that $u_h(t)$ is solution of:

$$\frac{\mathrm{d}^2}{\mathrm{d}t^2}(u_h(t), \tilde{u}_h)_P + (\mathcal{B}^* u_h(t), \mathcal{B}^* \tilde{u}_h)_D = 0, \quad \forall\, \tilde{u}_h \in U_h. \tag{33}$$

Reciprocally, it is easy to see that if u_h is a solution of (33), and v_h given by

$$\frac{dv_h}{\mathrm{d}t} = \mathcal{B}^* u_h \quad (+\text{ the appropriate initial condition}),$$

then (u_h, v_h) is a solution of (32). In such a case, the mixed approach actually produces a factorisation of the operator $\mathbf{A}_h$ (cf. 18) as:

$$\mathbf{A}_h = \mathbf{B}_h^* \, \mathbf{B}_h \,.$$

Remark 5. Such a factorisation property can be interesting for the practical implementation of primal finite element methods in a mixed form (see [24]).

A more general compatibility condition can be expressed in terms of the existence of an appropriate elliptic projection. More precisely, it can be shown by energy techniques (see [43, 54]) that convergence is guaranteed provided that it is possible to construct a linear operator:

$$\left| \begin{array}{l} \Pi_h \;:\; U \times V \rightarrow U_h \times V_h \\[2ex] \qquad (u, v) \;\rightarrow (\Pi_h u, \Pi_h v) \end{array} \right. \tag{34}$$

such that

$$\begin{cases} (u - \Pi_h u, \widetilde{u}_h)_P + b(\widetilde{u}_h, v - \Pi_h v) = 0, \; \forall \widetilde{u}_h \in U_h \\[1ex] b(u - \Pi_h u, \widetilde{v}_h) = 0, \qquad\qquad\qquad \forall \widetilde{v}_h \in V_h \, , \end{cases} \tag{35}$$

and moreover

$$\forall \, (u, v) \in U \times V, \quad \lim_{h \rightarrow 0} \; (\, \|u - \Pi_h u\|_U + \|v - \Pi_h v\|_V \,) = 0. \tag{36}$$

The elliptic theory of mixed finite element methods provides sufficient conditions for the existence of such an operator, namely:

$$\begin{cases} \exists \, \beta > 0 \, / \, \forall \, u_h \in U_h, \; \exists \, v_h \in V_h \text{ s.t. } \; b(u_h, v_h) \geq \beta \; \|u_h\|_U \; \|v_h\|_V, \\[1ex] \exists \, \alpha > 0 \, / \, \|u_h\|_P^2 \geq \alpha \; \|u_h\|_U^2, \quad \forall \, u_h \in U_h \text{ s.t. } \forall \, v_h \in V_h, b(u_h, v_h) = 0. \end{cases} \tag{37}$$

Remark 6. Conditions of the type (37) are not necessary for the convergence of the method. See [9] or [11] for examples where the convergence holds under weaker conditions.

Remark 7. Of course, one obtains a set of sufficient conditions for convergence by exchanging the roles of the two variables u and v in (35) or (37).

2.4 Time Discretisation

Centred Schemes

The conservative nature (cf. the conservation of the energy) of the abstract wave equation (4) or (5) can be seen as a consequence of the time reversibility of the equation. That is why we shall prefer centred finite difference schemes which preserve this property at the discrete level.

Second order problems.

Let us consider a time step $\Delta t > 0$ and denote by u_h^n an approximation of $u_h(t^n)$. The simplest finite difference scheme for the approximation of (17) is the so-called leap-frog scheme

$$\left(\frac{u_h^{n+1} - 2u_h^n + u_h^{n-1}}{\Delta t^2}, \tilde{u}_h\right)_P + a(u_h^n, \tilde{u}_h) = 0, \quad \forall\, \tilde{u} \in U_h\,, \tag{38}$$

or equivalently

$$\frac{u_h^{n+1} - 2u_h^n + u_h^{n-1}}{\Delta t^2} + \mathbf{A}_h u_h^n = 0\,. \tag{39}$$

By construction, (39) is explicit (and is "really" explicit when the evaluation of $\mathbf{A}_h u_h^n$ corresponds to a simple matrix-vector product – which is the case with mass lumping) and allows us to compute u_h^{n+1} from the two previous time steps (u_h^n and u_h^{n-1}):

$$u_h^{n+1} = 2u_h^n - u_h^{n-1} - \Delta t^2 \mathbf{A}_h u_h^n.$$

Of course, (39) (or (38)) must be completed by a start-up procedure (that we shall omit here for simplicity) using the initial conditions to compute u_h^0 and u_h^1.

This scheme is second order accurate in time. It is possible to construct higher order schemes (which is a priori natural with high order finite elements in space) which remain explicit and centred. A classical approach is the so-called *modified equation* approach. For instance, to construct a fourth order scheme, we start by looking at the truncation error of (39):

$$\frac{u_h(t^{n+1}) - 2u_h(t^n) + u_h(t^{n-1})}{\Delta t^2} = \frac{\mathrm{d}^2 u_h}{\mathrm{d}t^2}(t^n) + \frac{\Delta t^2}{12}\frac{\mathrm{d}^4 u_h}{\mathrm{d}t^4}(t^n) + O(\Delta t^4).$$

Using the equation satisfied by u_h, we get the identity:

$$\frac{u_h(t^{n+1}) - 2u_h(t^n) + u_h(t^{n-1})}{\Delta t^2} = \frac{\mathrm{d}^2 u_h}{\mathrm{d}t^2}(t^n) + \frac{\Delta t^2}{12}\mathbf{A}_h^2 u_h(t^n) + O(\Delta t^4)\,,$$

which leads naturally to the following fourth order scheme:

$$\frac{u_h^{n+1} - 2u_h^n + u_h^{n-1}}{\Delta t^2} + \mathbf{A}_h u_h^n - \frac{\Delta t^2}{12}\mathbf{A}_h^2 u_h^n = 0\,. \tag{40}$$

This can be implemented in such a way that each time step involves only 2 applications of the operator $\mathbf{A}_h$ (the CPU time for each time step with (40) is about twice that with (39)):

$$u_h^{n+1} = 2u_h^n - u_h^{n-1} - \Delta t^2 \mathbf{A}_h \left(I - \frac{\Delta t^2}{12}\mathbf{A}_h\right) u_h^n.$$

More generally, an explicit centred scheme of order $2m$ is given by:

$$\frac{u_h^{n+1} - 2u_h^n + u_h^{n-1}}{\Delta t^2} + \mathbf{A}_h(\Delta t)\, u_h^n = 0, \quad \mathbf{A}_h(\Delta t) = P_m(\mathbf{A}_h \Delta t), \tag{41}$$

where the polynomial $P_m(x)$ is defined by:

$$P_m(x) = 1 + 2\sum_{l=1}^{m-1}(-1)^l \frac{x^l}{(2l+2)!}. \tag{42}$$

First order problems.

To construct the equivalent of the leap-frog scheme for first order systems, we must use centred finite difference approximations in time. This naturally leads to the use of staggered grid approximations. More precisely:

- the unknown u_h will be computed at times $t^n = n\Delta t$; u_h^n.
- the unknown v_h will be computed at times $t^{n+1/2} = n\Delta t$; $v_h^{n+1/2}$.

The fully discrete scheme is simply:

$$\begin{cases} (\dfrac{u_h^{n+1} - u_h^n}{\Delta t}, \tilde{u}_h)_P - b(\tilde{u}_h, v_h^{n+1/2}) = 0, \quad \forall\, \tilde{u}_h \in U_h, \\ (\dfrac{v_h^{n+1/2} - v_h^{n-1/2}}{\Delta t}, \tilde{v}_h)_D + b(u_h^n, \tilde{v}_h) = 0, \quad \forall\, \tilde{v}_h \in V_h, \end{cases} \tag{43}$$

or equivalently in the operator form:

$$\begin{cases} \dfrac{u_h^{n+1} - u_h^n}{\Delta t} - \mathbf{B}_h^* v_h^{n+1/2} = 0, \\ \dfrac{v_h^{n+1/2} - v_h^{n-1/2}}{\Delta t} + \mathbf{B}_h u_h^n = 0. \end{cases} \tag{44}$$

Remark 8. In practice, keeping the notation of Sect. 2.3, one solves the following problem:

$$\begin{cases} \mathcal{M}_h^P \dfrac{\mathbf{u}_h^{n+1} - \mathbf{u}_h^n}{\Delta t} - \mathcal{B}_h^* \mathbf{v}_h^{n+1/2} = 0, \\ \mathcal{M}_h^D \dfrac{\mathbf{v}_h^{n+1/2} - \mathbf{v}_h^{n-1/2}}{\Delta t} + \mathcal{B}_h u_h^n = 0. \end{cases} \tag{45}$$

This shows that one has a fully explicit scheme if the mass lumping procedure is applied.

Note that if one eliminates the unknown v_h, one sees that u_h satisfies the scheme:

$$\frac{u_h^{n+1} - 2u_h^n + u_h^{n-1}}{\Delta t^2} + \mathbf{B}_h^* \mathbf{B}_h u_h^n = 0. \tag{46}$$

which establishes an obvious link with the second order formulation discussed previously.

Stability and Error Analysis

Second order problems.

We present below the energy technique for analysing the stability of (38) or (39). The idea is first to determine a discrete equivalent of the energy conservation property (22). The principle consists of taking for the test function $\tilde{u}_h$ in (38) a discrete equivalent of the time derivative of $u_h(t)$ at time t^n, namely:

$$\tilde{u}_h = \frac{u_h^{n+1} - u_h^{n-1}}{2\Delta t}.$$

Using this $\tilde{u}_h$ and the symmetry of $a(\cdot,\cdot)$, we observe that:

$$\begin{cases} \left(\dfrac{u_h^{n+1} - 2u_h^n + u_h^{n-1}}{\Delta t^2}, \tilde{u}_h\right) = \dfrac{1}{2\Delta t}\left\{\|\dfrac{u_h^{n+1} - u_h^n}{\Delta t}\|_P^2 - \|\dfrac{u_h^n - u_h^{n-1}}{\Delta t}\|_P^2\right\}, \\ a\left(u_h^n, \tilde{u}_h\right) = \dfrac{1}{2\Delta t}\left\{a(u_h^{n+1}, u_h^n) - a(u_h^n, u_h^{n-1})\right\}. \end{cases}$$

After summation, these two equalities lead to the discrete conservation property:

$$E_h^{n+1/2} = E_h^{n-1/2}, \quad E_h^{n+1/2} \stackrel{\text{def}}{=} \frac{1}{2}\,\|\frac{u_h^{n+1} - u_h^n}{\Delta t}\|_P^2 + \frac{1}{2}\,a(u_h^{n+1}, u_h^n). \tag{47}$$

To get a stability result, it is necessary to prove that $E_h^{n+1/2}$ is a positive energy, which is not obvious since the second term in (47) has a priori no sign. However one can expect that, if Δt is small enough, u_h^{n+1} will be close to u_h^n and $a(u_h^{n+1}, u_h^n)$ will be "almost positive". More precisely:

$$E_h^{n+1/2} = \frac{1}{2}\,\|\frac{u_h^{n+1} - u_h^n}{\Delta t}\|_P^2 + \frac{1}{2}\,a(u_h^{n+1/2}, u_h^{n+1/2}) - \frac{\Delta t^2}{8}\,a\Big(\frac{u_h^{n+1} - u_h^n}{\Delta t}, \frac{u_h^{n+1} - u_h^n}{\Delta t}\Big). \tag{48}$$

Introducing the norm of the operator $\mathbf{A}_h$:

$$\|\mathbf{A}_h\| = \sup_{u_h \in U_h, u_h \neq 0} \frac{a(u_h, u_h)}{\|u_h\|^2}, \tag{49}$$

we get a lower bound for $E_h^{n+1/2}$ (we define $u_h^{n+1/2} = \frac{1}{2}\,(u_h^{n+1} + u_h^n)$):

$$E_h^{n+1/2} \geq \frac{1}{2}\left(1 - \frac{\Delta t^2}{4}\,\|\mathbf{A}_h\|\right)\,\|\frac{u_h^{n+1} - u_h^n}{\Delta t}\|_P^2 + \frac{1}{2}\,a(u_h^{n+1/2}, u_h^{n+1/2}). \tag{50}$$

This is the basic estimate for proving the following stability result:

Theorem 1. *A sufficient condition for the stability of the scheme (38) is:*

$$\frac{\Delta t^2}{4}\,\|\mathbf{A}_h\| \leq 1. \tag{51}$$

Remark 9. This stability result requires some comments:

- By stability, we mean that we are able to obtain uniform estimates (with respect to h and Δt) of the solution, typically of the form $\|u_h^n\|_U \leq C$.
- From (50), it is easy to deduce this type of estimate under the stronger condition:

$$\frac{\Delta t^2}{4}\,\|\mathbf{A}_h\| \leq \alpha, \quad \text{for } \alpha < 1.$$

  Proving the stability result for $\alpha = 1$ requires some additional effort (cf. remark 10).
- The condition (51) is a priori a sufficient stability condition. However, in this simple case, due to the fact that $\mathbf{A}_h$ can be diagonalised, the Von Neumann analysis can be applied to prove that this condition is also necessary. It suffices to look at solutions of the form

$$u_h^n = a^n\,\mathbf{w}_h, \quad a^n \in \mathbb{R},$$

  where $\mathbf{w}_h$ is the eigenvector of $\mathbf{A}_h$ associated with its greatest eigenvalue $\lambda = \|\mathbf{A}_h\|$.
- The condition (51) appears as an abstract CFL condition. In the applications, when $\mathbf{A}$ is a second order differential operator in space, it is possible to get a bound for $\|\mathbf{A}_h\|$ of the form:

$$\|\mathbf{A}_h\| \leq \frac{4c_+^2}{h^2},$$

  where c_+ is a positive constant which is consistent with a velocity and only depends on the continuous problem. Therefore, a (weaker) sufficient stability condition takes the form:

$$\frac{c_+\,\Delta t}{h} \leq 1.$$

  Under an assumption of uniform regularity (see [21] for a definition) of the computational mesh, it is also possible to show that (for $c_- \leq c_+$):

$$\|\mathbf{A}_h\| \geq \frac{4c_-^2}{h^2},$$

  so that a necessary stability condition is:

$$\frac{c_-\,\Delta t}{h} \leq 1.$$

  In the uniform mesh case one even has

$$\|\mathbf{A}_h\| \sim \frac{4c^2}{h^2}, \quad (h \to 0).$$

- The above stability results also apply to the higher order scheme (41) but it is complicated by the fact that one must verify that the operator $\mathbf{A}_h(\Delta t)$ is positive, which also imposes an upper bound on Δt.

Convergence analysis.

Classical convergence theory relies on energy estimates. We only give a flavour of the proof (a rigorous proof would need tedious details) in the case where we assume that the exact solution u is smooth enough in time. Let us introduce the error:

$$e_h^n = u_h^n - u_h(t^n), \qquad \text{(where } u_h(t) \text{ is the exact solution of (19))}. \tag{52}$$

We have immediately:

$$\frac{e_h^{n+1} - 2e_h^n + e_h^{n-1}}{\Delta t^2} + \mathbf{A}_h e_h^n = \varepsilon_h^n, \tag{53}$$

where the truncation error ε_h^n defined by:

$$\varepsilon_h^n = \frac{u_h^{n+1} - 2u_h^n + u_h^{n-1}}{\Delta t^2} + \mathbf{A}_h u_h^n, \tag{54}$$

tends to 0 with Δt. A typical estimate is:

$$\sup_{0 \le t^n \le T} \|\varepsilon_h^n\|_P \le C\,\Delta t^2 \sup_{0 \le t \le T} \|\frac{\mathrm{d}^4 u_h(t^n)}{\mathrm{d}t^4}(t)\| \qquad (\le C(u,T)\,\Delta t^2)\,. \tag{55}$$

Let us introduce the energy of the error:

$$\mathcal{E}_h^{n+1/2} = \frac{1}{2}\,\|\frac{e_h^{n+1} - e_h^n}{\Delta t}\|_P^2 + \frac{1}{2}\,a(e_h^{n+1}, e_h^n)\,. \tag{56}$$

From (53), we easily deduce the identity:

$$\frac{\mathcal{E}_h^{n+1/2} - \mathcal{E}_h^{n-1/2}}{\Delta t} = (\varepsilon_h^n, \frac{e_h^{n+1} - e_h^{n-1}}{2\Delta t})_P\,. \tag{57}$$

Assume that:

$$\frac{\Delta t^2}{4}\,\|\mathbf{A}_h\| \le \alpha^2 < 1\,.$$

From (50), we deduce in particular that:

$$\|\frac{e_h^{n+1} - e_h^n}{\Delta t}\|_P \le (1-\alpha^2)^{-1/2}\sqrt{2\mathcal{E}_h^{n+1/2}},$$

and therefore that:

$$\|\frac{e_h^{n+1} - e_h^{n-1}}{2\Delta t}\|_P \le (1-\alpha^2)^{-1/2}\left(\sqrt{2\mathcal{E}_h^{n-1/2}} + \sqrt{2\mathcal{E}_h^{n+1/2}}\right).$$

Using this inequality in (57) leads to

$$\frac{\sqrt{\mathcal{E}_h^{n+1/2}} - \sqrt{\mathcal{E}_h^{n-1/2}}}{\Delta t} \le \sqrt{2}\,(1-\alpha^2)^{-1/2}\|\varepsilon_h^n\|_P.$$

After summation over n, we finally get an error estimate in terms of the energy of the error:

$$\sqrt{\mathcal{E}_h^{n+1/2}} \leq \sqrt{\mathcal{E}_h^{1/2}} + \sqrt{2}\,(1-\alpha^2)^{-1/2} \sum_{n=0}^{T/\Delta t} \|\varepsilon_h^n\|_P \,\Delta t, \tag{58}$$

where,

- $\sqrt{\mathcal{E}_h^{1/2}}$ represents the error due to the approximation of the initial conditions,
- and the term $\sum_{n=0}^{T/\Delta t} \|\varepsilon_h^n\|_P \;\Delta t$ is a discrete $L^1(0,T;L^2)$ norm of the truncation error, which is $O(\Delta t^2)$.

Remark 10. The estimate (58) blows up when α tends to 1. This is not an optimal result: in many practical applications, the scheme is more accurate when α approaches 1. In fact (58) can be improved as follows:

- Due to the fact that the scheme is "time invariant", it is easy to see (take the half sum of two successive equations in (39)) that the sequence of "intermediate values":

$$e_h^{n+1/2} = \frac{e_h^{n+1} + e_h^n}{2} \tag{59}$$

  satisfies the same scheme (53) as e_h^n, except that n is replaced by $n+1/2$ and the right hand side is:

$$\varepsilon_h^{n+1/2} = \frac{\varepsilon_h^{n+1} + \varepsilon_h^n}{2} \qquad \text{(which is still } O(\Delta t^2)\text{)}.$$

- As a consequence we have the identity:

$$\begin{aligned} \frac{\mathcal{E}_h^{n+1} - \mathcal{E}_h^n}{\Delta t} &= (\varepsilon_h^{n+1/2}, \frac{e_h^{n+3/2} - e_h^{n-1/2}}{2\Delta t})_P \\ &\leq \|\varepsilon_h^{n+1/2}\|_P \left(\|\frac{e_h^{n+2} - e_h^n}{2\Delta t}\|_P + \|\frac{e_h^{n+1} - e_h^{n-1}}{2\Delta t}\|_P \right), \end{aligned} \tag{60}$$

  where the energy $\mathcal{E}_h^n$ is that associated with $e_h^{n+1/2}$:

$$\begin{aligned} \mathcal{E}_h^n &= \frac{1}{2}\,\|\frac{e_h^{n+1/2} - e_h^{n-1/2}}{\Delta t}\|_P^2 + \frac{1}{2}\,a(e_h^{n+1/2}, e_h^{n-1/2}) \\ &= \frac{1}{2}\,\|\frac{e_h^{n+1} - e_h^{n-1}}{2\Delta t}\|_P^2 + \frac{1}{2}\,a(e_h^{n+1/2}, e_h^{n-1/2}). \end{aligned}$$

- Thanks to identity (50), we know that for $\alpha \leq 1$:

$$\mathcal{E}_h^{n+1/2} \geq \frac{1}{2}\,a(e_h^{n+1/2}, e_h^{n+1/2}), \quad \mathcal{E}_h^{n-1/2} \geq \frac{1}{2}\,a(e_h^{n-1/2}, e_h^{n-1/2}).$$

Therefore, if we introduce the new energy:

$$\widetilde{\mathcal{E}}_h^n = \mathcal{E}_h^n + \mathcal{E}_h^{n+1/2} + \mathcal{E}_h^{n-1/2}, \tag{61}$$

we get the inequality:

$$\widetilde{\mathcal{E}}_h^n \geq \frac{1}{2}\,\|\frac{e_h^{n+1} - e_h^{n-1}}{2\Delta t}\|_P^2 + \frac{1}{2}\,a(e_h^{n+1/2}, e_h^{n+1/2})$$
$$+\frac{1}{2}\,a(e_h^{n-1/2}, e_h^{n-1/2}) + \frac{1}{2}\,a(e_h^{n+1/2}, e_h^{n-1/2}),$$

which, thanks to the positivity of $a(\cdot,\cdot)$, leads to the lower bound:

$$\widetilde{\mathcal{E}}_h^n \geq \frac{1}{2}\,\|\frac{e_h^{n+1} - e_h^{n-1}}{2\Delta t}\|_P^2 + \frac{1}{4}\,a(e_h^{n+1/2}, e_h^{n+1/2}) + \frac{1}{4}\,a(e_h^{n-1/2}, e_h^{n-1/2}). \tag{62}$$

Using this inequality in (60) leads to:

$$\frac{\mathcal{E}_h^{n+1} - \mathcal{E}_h^n}{\Delta t} \leq \sqrt{2}\,\|\varepsilon_h^{n+1/2}\|_P \left(\sqrt{\widetilde{\mathcal{E}}_h^{n+1}} + \sqrt{\widetilde{\mathcal{E}}_h^n}\right), \tag{63}$$

while we deduce from 57 that:

$$\left|\begin{array}{l} \dfrac{\mathcal{E}_h^{n+3/2} - \mathcal{E}_h^{n+1/2}}{\Delta t} \leq \sqrt{2}\,\|\varepsilon_h^{n+1}\|_P \sqrt{\widetilde{\mathcal{E}}_h^{n+1}} \\[2ex] \dfrac{\mathcal{E}_h^{n+1/2} - \mathcal{E}_h^{n-1/2}}{\Delta t} \leq \sqrt{2}\,\|\varepsilon_h^{n}\|_P \sqrt{\widetilde{\mathcal{E}}_h^{n}} \end{array}\right. . \tag{64}$$

Finally, adding the three inequalities in (63) and (64) gives:

$$\frac{\widetilde{\mathcal{E}}_h^{n+1} - \widetilde{\mathcal{E}}_h^n}{\Delta t} \leq \sqrt{2}\Big\{\ \|\varepsilon_h^{n+1/2}\|_P \left(\sqrt{\widetilde{\mathcal{E}}_h^{n+1}} + \sqrt{\widetilde{\mathcal{E}}_h^n}\right)$$
$$+\|\varepsilon_h^{n+1}\|_P \sqrt{\widetilde{\mathcal{E}}_h^{n+1}} + \|\varepsilon_h^{n}\|_P \sqrt{\widetilde{\mathcal{E}}_h^{n}}\ \Big\}.$$

Applying a discrete Gronwall lemma, we obtain the final estimate:

$$\sqrt{\widetilde{\mathcal{E}}_h^{n+1/2}} \leq \sqrt{\widetilde{\mathcal{E}}_h^{1}} + C \sum_{n=0}^{T/\Delta t} \|\varepsilon_h^n\|_P\,\Delta t, \tag{65}$$

which does not blow up when α goes to 1 since C does not depend on α.

The case of first order problems.

The analysis of the scheme (43) by energy techniques is very similar to the one of scheme (38) explained above. We shall restrict ourselves to mentioning how one can get a discrete equivalent of the energy conservation result (30) and how one deduces from such a result the stability condition for the scheme. First note, that (43) implies (we replace the second equation of (43) at time t^n by the half sum of the same equations at times t^n and t^{n+1})

$$\begin{cases} (\dfrac{u_h^{n+1}-u_h^n}{\Delta t}, \tilde{u}_h)_P - b(\tilde{u}_h, v_h^{n+1/2}) = 0, \quad \forall\, \tilde{u}_h \in U_h, \\ (\dfrac{v_h^{n+3/2}-v_h^{n-1/2}}{\Delta t}, \tilde{v}_h)_D + b(u_h^n, \tilde{v}_h) = 0, \quad \forall\, \tilde{v}_h \in V_h. \end{cases} \tag{66}$$

We then choose $\tilde{u}_h = \dfrac{u_h^{n+1}+u_h^n}{2}$ and $\tilde{v}_h = v_h^{n+1/2}$ to obtain the conservation result:

$$\frac{E_h^{n+1}-E_h^n}{\Delta t} = 0 \quad \text{where} \quad E_h^n \stackrel{\text{def}}{=} \frac{1}{2}\{\ \|u_h^n\|_P^2 + (v_h^{n+1/2}, v_h^{n-1/2})_D\ \}. \tag{67}$$

Remark 11. By symmetry between u_h and v_h, we also have conservation of:

$$E_h^{n+1/2} \stackrel{\text{def}}{=} \frac{1}{2}\{\ (u_h^{n+1}, u_h^n)_P + \|v_h^{n+1/2}\|_D^2\ \}.$$

To get the stability condition of the scheme, one observes that, setting $v_h^n = (v_h^{n+1/2} + v_h^{n+1/2})/2$

$$E_h^n = \frac{1}{2}\left\{ \|u_h^n\|_P^2 + \|v_h^n\|_D^2 - \frac{\Delta t^2}{4}\left\|\frac{v_h^{n+1/2}-v_h^{n-1/2}}{\Delta t}\right\|_D^2 \right\},$$

that is to say, using the second equation of (44):

$$E_h^n = \frac{1}{2}\{\ \|u_h^n\|_P^2 + \|v_h^n\|_D^2 - \frac{\Delta t^2}{4}\ \|\mathbf{B}_h u_h^n\|_D^2\ \}.$$

Then, it is clear that the positivity of E_h^n is guaranteed by the stability condition:

$$\frac{\|\mathbf{B}_h\|\Delta t}{2} \leq 1. \tag{68}$$

Remark 12. Note that (68) coincides with (51) when $\mathbf{A}_h = \mathbf{B}_h^*\mathbf{B}_h$.

Links with Standard FDTD Schemes

The finite element methods can be seen also as generalisations of the finite difference method in the sense that, when applied to regular grids, they give rise to numerical schemes that can be reinterpreted in terms of finite differences. One often obtains non standard finite difference schemes that are not obvious to derive without thinking of the variational approach. However, one can also re-obtain classical schemes, especially if the mass lumping procedure is applied. In such a case, the variational approach appears as an efficient way to generalise, in a stable way, finite difference schemes to variable coefficients or to treat boundary conditions in a stable manner.

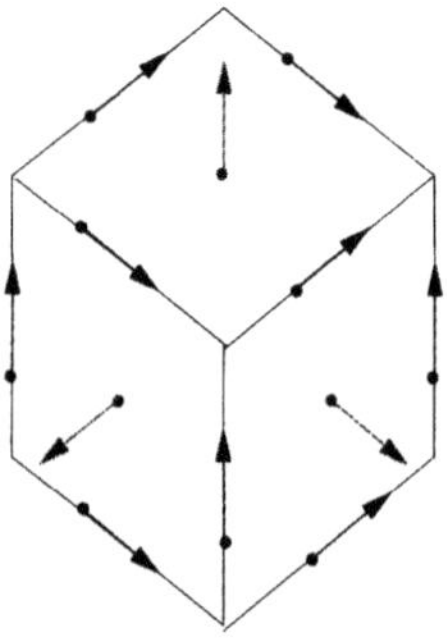

Fig. 1. The degrees of freedom for edge and face elements

Let us give the well-known example of the Yee scheme for electromagnetics in 3D. This scheme can be recovered by applying the variational scheme (43) in the case where the scalar products $(\cdot,\cdot)_P$ and $(\cdot,\cdot)_D$ are given by (8) and where:

$$U = H(\text{curl}, \Omega), \; V = L^2(\Omega)^3 \; \text{ and } \; b(e,h) = \int_\Omega \text{curl}\; e \cdot h \; \mathrm{d}x \,. \tag{69}$$

One recovers the Yee scheme with a particular choice of the approximate spaces U_h and V_h and the use of a particular quadrature formula. More precisely, we consider a uniform, infinite mesh $\mathcal{T}_h$ made of equal cubes $K \in \mathcal{T}_h$ of side h. The appropriate approximation spaces are:

$$\begin{cases} U_h = \{\; u_h \in U \;/\; \forall\, K \in \mathcal{T}_h, u_h|_K \in \mathcal{Q}_{0,1,1} \times \mathcal{Q}_{1,0,1} \times \mathcal{Q}_{1,1,0} \;\} \\ V_h = \{\; v_h \in H(\text{div}, \Omega) \;/\; \forall\, K \in \mathcal{T}_h, v_h|_K \in \mathcal{Q}_{1,0,0} \times \mathcal{Q}_{0,1,0} \times \mathcal{Q}_{0,0,1} \;\} \end{cases} \tag{70}$$

where we recall that, by definition, $\mathcal{Q}_{p_1,p_2,p_3}$ is the set of polynomials of three variables whose degree with respect to the ith variable is less or equal than p_i. Notice that, contrary to what one might expect, we do not use a space of completely discontinuous elements to approximate $V = L^2(\Omega)^3$; we use vector fields whose normal component is continuous across each face of the mesh. With this choice, we are in the situation described in (31):

$$\text{curl}\; U_h \subset V_h, \tag{71}$$

which offers a guarantee of convergence. The spaces defined by (70) are known as *edge elements* for the electric field and *face elements* for the magnetic field (see [59]). In particular a set of degrees of freedom is given by (see also Fig. 1):

- For the space U_h (degrees of freedom for the electric field): the (constant) tangential component of the vector field along each edge of the mesh.
- For the space V_h (degrees of freedom for the electric field): the (constant) tangential component of the vector field on each face of the mesh.

Remark 13. It is clear on Fig. 1, that, as far as the degrees of freedom are concerned, the roles of the electric and magnetic fields are completely symmetric (simply consider a "parallel" mesh shifted by $h/2$ in each direction). Only the interpolations of the fields are different.

Finally, in order to compute the various integrals that appear in the variational formulation, we use the following numerical quadrature formula in each cube K (which provides mass lumping):

$$\int_K f \, \mathrm{d}x = \frac{h^3}{8} \sum_{x \in S_K} f(x), \tag{72}$$

where S_K is the set of vertices of K. Using this formula leads to the Yee scheme.

3 Fictitious Domain Methods

Let us consider as a model problem the propagation of waves in an exterior domain, the complement of a bounded obstacle. The use of a standard finite difference method requires a staircase approximation of the boundary of the obstacle (see Fig. 2) and the great disadvantage of the method is that this creates spurious numerical diffractions. A possibility for avoiding this drawback is to use a finite element method (with mass lumping). The finite element mesh may follow the boundary of the object exactly (see Fig. 2).

However, other drawbacks are introduced. First of all, the numerical implementation is much more difficult and the efficiency of the computations is decreased by the unstructured nature of the data. Furthermore, meshing the whole domain of computation (with tetrahedra in 3D) is expensive. Finally, the time step has to be chosen in accordance with the space mesh size (because of the CFL condition), which sometimes leads to small time steps in the presence of small elements in the mesh.

Here we investigate an alternative method for handling the scattering problem, namely, the fictitious domain method (denoted FDM). Such methods have recently been shown to have interesting potential for solving complicated problems ([2–4, 39, 45, 55]) particularly in the stationary case. The use of the FDM for time dependent problems is relatively new [46]. However, it gives very nice properties for these kinds of problems, particularly for exterior wave propagation. In this case, the FDM, also called the domain embedding method, consists of artificially extending the solution inside the obstacle so that the new domain of computation has a very simple shape (typically a rectangle in 2D). This extension requires the introduction of a new variable defined only at the boundary of the obstacle. This auxiliary variable accounts for the boundary condition, and can be related to a singularity across the boundary of the obstacle of the extended function.

This idea will be developed in more detail in Sect. 3.1 in the case of acoustic waves. The main point is that the mesh for the solution on the enlarged domain can be chosen independently of the geometry of the obstacle. In particular, the use of regular

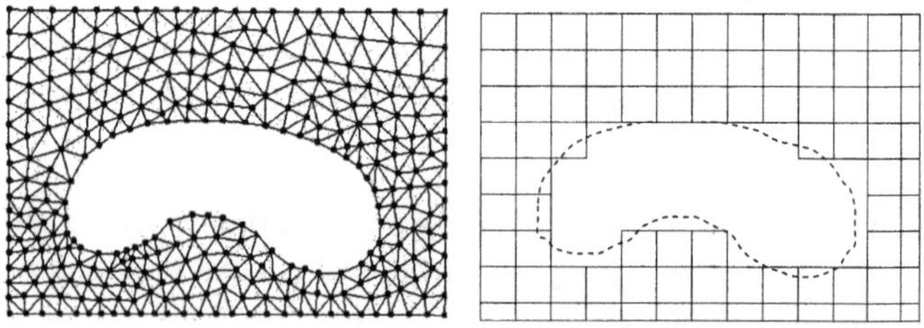

Fig. 2. Finite element mesh (left) and finite difference mesh (right)

grids or structured meshes allows for simple and efficient computations. There is an additional cost due to the determination of the new boundary unknown. However, the final numerical scheme is a slight perturbation of the scheme for the problem without an obstacle so this cost may be considered as marginal. Theoretically, the convergence of the method is linked to a uniform inf-sup condition which leads to a compatibility condition between the boundary mesh and the uniform mesh: this implies that the two mesh grids cannot be chosen completely independently, but this is not an important constraint for the applications. Another important point is that the stability condition of the resulting scheme is the same as the one of the finite difference scheme.

3.1 Presentation of the Method: The Acoustic Dirichlet Problem

Our model problem is the scattering of an acoustic wave (in a homogeneous medium with speed 1) by an obstacle $\mathcal{O}$, with a Dirichlet condition on the boundary:

$$\begin{cases} \dfrac{\partial^2 u}{\partial t^2} - \Delta u = 0, \text{ in } D = \mathbb{R}^d \setminus \mathcal{O}, \quad (d = 2 \text{ or } 3) \\ u = 0, \qquad\qquad \text{on } \gamma = \partial D. \end{cases} \tag{73}$$

We assume that the incident wave is generated by initial conditions (omitted here) at time $t = 0$. In order to have a finite computational domain, the classical technique consists of bounding the domain D and imposing absorbing conditions on the exterior boundary [38, 65]. For the sake of simplicity, a Dirichlet condition is assumed on the exterior boundary as well, and, for our purpose, we choose the geometry of the external boundary (which does not interest us here) to be rectangular (2D) or box-shaped (3D). We denote by Ω this bounded domain and by C the rectangle $\Omega \bigcup \mathcal{O}$. We want to solve the simple problem described by

$$\begin{cases} \dfrac{\partial^2 u}{\partial t^2} - \Delta u = 0, \text{ in } \Omega, \\ u = 0, \qquad\qquad \text{on } \partial\Omega = \gamma \cup \partial C. \end{cases} \tag{74}$$

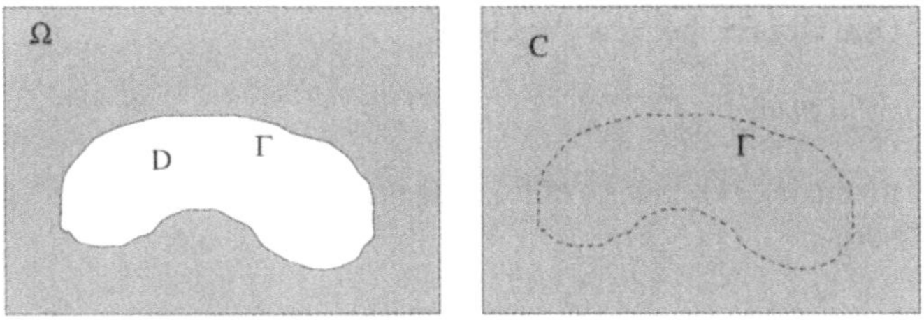

Fig. 3. Principle of the fictitious domain method

The Fictitious Domain Formulation

The main idea of the FDM is to extend u defined on domain Ω to a function (still denoted by u for simplicity) defined on the enlarged domain C, with $\mathrm{H}^1(C)$ regularity. Note that this regularity requirement implies the continuity of the trace of u across the boundary. More precisely, we look for u in the space

$$\tilde{U} = \left\{ v \in U = \mathrm{H}_0^1(C) \; ; \; v = 0 \text{ on } \gamma \right\}, \tag{75}$$

and we characterise u as the first argument of (u, λ), where $(u, \lambda) : \mathbb{R}^+ \to U \times L$ is the solution of the following variational "saddle-point like" evolution problem

$$\left\{ \begin{array}{ll} \dfrac{\mathrm{d}^2}{\mathrm{d}t^2}(u, v) + a(u, v) = b(v, \lambda), \; \forall v \in U, & \\ b(u, \mu) = 0, & \forall \mu \in L, \end{array} \right. \tag{76}$$

where $L = \mathrm{H}^{-1/2}(\gamma)$ and where we have set:

$$\left\{ \begin{array}{ll} (u, v) = \displaystyle\int_C u \, v \, \mathrm{d}x, & \forall \, (u, v) \in U \times U, \\ a(u, v) = \displaystyle\int_C \nabla u \cdot \nabla v \, \mathrm{d}x \,, & \forall \, (u, v) \in U \times U, \\ b(v, \mu) = \displaystyle\int_\gamma \mu \, v \, \mathrm{d}\gamma & \forall \, (v, \mu) \in U \times L. \end{array} \right. \tag{77}$$

Remark 14. Rigorously, $b(u, \mu) = \langle \mu, u \rangle_\gamma$ where $\langle \mu, u \rangle_\gamma$ denotes the duality product between L and L'. It becomes an integral over γ as soon $\mu \in L^2(\gamma)$, which justifies our notation.

In essence, the method consists of first extending the solution in the enlarged computational domain and then in introducing a new unknown at the boundary. One of the main differences between this approach and a standard conforming finite element approach lies in the fact that the Dirichlet condition is taken into account in a weak sense instead of being imposed in the functional space. In this formulation, there is no mention of the geometry of the problem, namely the boundary γ, in the functional space for the volume unknown u. The geometry only appears in $b(\cdot, \cdot)$ and L. We next give two different approaches to deriving (76).

How Does One Obtain the New Problem?

Via optimisation theory.

A first way to obtain (76) consists of freezing the time t (as a parameter) and considering the function

$$f = -\frac{\partial^2 u}{\partial t^2}(\cdot, t),$$

as data. Then $u = u(\cdot, t)$ is the solution of the following problem

$$\begin{cases} -\Delta u = f, \text{ in } \Omega, \\ u = 0 \qquad \text{on } \partial\Omega = \gamma \cup \partial C \end{cases} \tag{78}$$

and thus minimises the functional $J(v) = \int_C \left\{\frac{1}{2}|\nabla v|^2 - fv\right\} dx$ over the space $H_0^1(\Omega)$.

This space can be seen as the space of restrictions to Ω of functions of $\tilde{V}$ defined in (75). It is then natural to consider the enlarged minimisation problem defined by

$$\min_{\tilde{v} \in \tilde{V}} J(\tilde{v}) = \int_C \left\{\frac{1}{2}|\nabla \tilde{v}|^2 - \tilde{f}\tilde{v}\right\} dx, \tag{79}$$

where for instance $\tilde{f} = 0$ on $\mathcal{O}$ and $\tilde{f} = f$ on Ω. It is easy to verify that the restriction of the solution of problem (79) to Ω is exactly the solution of problem (76) that we are looking for. Problem (76) can be seen as a minimisation problem in U with an equality constraint on γ . Its solution is thus the first argument of the saddle point (u, λ) of the Lagrangian functional defined by $L(v, \mu) = J(v) - b(v, \mu)$. Writing that the derivatives in v and μ of this Lagrangian are equal to zero at (u, λ), we obtain:

$$\begin{cases} a(u, v) = b(v, \lambda) + (f, v) \quad \forall v \in U, \\ b(u, \mu) = 0 \qquad\qquad\qquad\quad \forall \mu \in L, \end{cases} \tag{80}$$

which gives precisely (76) with $f = -u_{tt}$. Here, the auxiliary unknown λ of problem (76) appears as the Lagrange multiplier associated with the constraint $v = 0$ on γ.

Via the theory of distributions.

Another way to understand the system of equations (76) is to say that, having extended u by continuity across γ and assuming that u still satisfies the wave equation inside $\mathcal{O}$, (this means that u solves the homogeneous Dirichlet problem inside and outside γ), we have in the sense of distributions in C

$$\frac{\partial^2 u}{\partial t^2} - \Delta u = \lambda \ \delta_\gamma , \tag{81}$$

where δ_γ is the surface measure supported on γ and λ is the jump of the normal derivative of u across γ (this is a second "physical" reinterpretation of the auxiliary unknown) :

$$\lambda = \left[\frac{\partial u}{\partial n}\right]. \tag{82}$$

We can also interpret the multiplier λ as a source term distributed on γ. If this source term were known, we simply would have to solve the wave equation in a square with a right hand side: the FDTD approach makes sense in such a geometric situation. In fact, λ is unknown and becomes a control variable in order to make u satisfy the boundary condition on γ.

Multiplying (81) by a test function $v \in U$ and integrating over C leads, at least formally, to the first equation of (76), the second one being the weak formulation of the fact that u vanishes on γ. This establishes an analogy between the FDM and integral equations for scattering problems [14]. Indeed, in this kind of method λ is typically the quantity that is chosen as the unknown. Nevertheless let us point out a very important difference between our approach and these methods. Integral equations are known to lead, after discretisation, to the solution of full linear systems in λ; as will be shown later, this is not the case for the FDM.

Finite Element Approximation and Time Discretisation

Space discretisation.

Let U_h (respectively L_H) be a finite dimensional subspace of U (respectively L). Here $h > 0$ and $H > 0$ represent two approximation parameters (a priori independent) allowed to tend to 0. In practice, they will be the step-size of a (regular) volume mesh of C (h) and of a surface mesh of γ (H). We approximate the variational problem (76) by

$$\begin{cases} \text{Find } (u_h, \lambda_H) : \mathbb{R}^+ \to U_h \times L_H \quad \text{such that} \\ \dfrac{\mathrm{d}^2}{dt^2}(u_h, v_h) + a(u_h, v_h) = b(v_h, \lambda_H), \; \forall v_h \in U_h, \\ b(u_h, \mu_H) = 0, \qquad\qquad \forall \mu_H \in L_H. \end{cases} \tag{83}$$

The spaces U_h and L_H will be assumed to satisfy the usual approximation properties. Typically, U_h can be a P1 or Q1 finite element space based on a regular mesh in C, which permits us to recover a standard finite difference scheme for the wave equation. For L_H, which is a subspace of $H^{-1/2}(\gamma)$, it makes sense to use discontinuous functions, for instance piecewise constant functions on a conforming (not necessarily regular) mesh of γ. If p denotes the dimension of U_h and q the one of L_H, one gets:

$$p = O(h^{-d}), \quad q = O(H^{-(d-1)}), \quad (h, H \to 0),$$

so that, since in practice H and h will be of the same order, we shall have $q \ll p$.

Using the same notation as in Sect. 2.3, it is easy to rewrite problem (83) in matrix form:

$$\left\{\begin{array}{l} \mathcal{M}_h \dfrac{\mathrm{d}^2 \mathbf{u}_h}{\mathrm{d}t^2} + \mathcal{A}_h \mathbf{u}_h = \mathcal{B}_H^* \boldsymbol{\lambda}_H \\ \\ \mathcal{B}_H \mathbf{u}_h = 0, \end{array}\right. \tag{84}$$

where we emphasise that :

- the mass matrix $\mathcal{M}_h$ is diagonal (typically the identity matrix), thanks to mass lumping,
- the stiffness matrix $\mathcal{A}_h$ represents a discrete Laplacian on a regular grid,
- the rectangular matrix $\mathcal{B}_H$ represents a discrete trace operator on γ.

Remark 15. The matrix $\mathcal{B}_H$ actually depends on the both approximation parameters h and H.

Problem (84) appears as a system of ordinary differential equations with an algebraic constraint. One can eliminate the time derivative of $\mathbf{u}_h$ between the two equations of (84) to obtain an equation which directly relates $\boldsymbol{\lambda}_H$ and $\mathbf{u}_h$:

$$\mathcal{Q}_H \boldsymbol{\lambda}_H = \mathcal{B}_H \mathcal{M}_h^{-1} \mathcal{A}_h \mathbf{u}_h, \quad \mathcal{Q}_H = \mathcal{B}_H \mathcal{M}_h^{-1} \mathcal{B}_H^* \mathbf{u}_h. \tag{85}$$

If Q_H, which is by construction symmetric and positive, is invertible (this issue will be discussed in Sect. 3.1), we see that $\mathbf{u}_h$ is the solution of the ordinary differential system:

$$\mathcal{M}_h \frac{\mathrm{d}^2 \mathbf{u}_h}{\mathrm{d}t^2} + \left(I - \mathcal{B}_H \mathcal{Q}_H^{-1} \mathcal{B}_H^*\right) \mathcal{A}_h \mathbf{u}_h = 0, \tag{86}$$

which proves the existence of the discrete solution. Moreover, it is easy to prove the following energy conservation result (simply take v_h as the time derivative of u_h and $\mu_h = \lambda_h$ in (83)) :

$$\frac{\mathrm{d}}{\mathrm{d}t} \left\{ \frac{1}{2} \left(\| \frac{\mathrm{d}u_h}{\mathrm{d}t}(t) \|_P^2 + a(u_h(t), u_h(t)) \right) \right\} = 0. \tag{87}$$

Time discretisation.

According to Sect. 2.4, we shall apply the standard leap frog procedure. With obvious notation, this leads to the following problem:

$$\left\{\begin{array}{l} \text{Find } (u_h^n, \lambda_H^n) \in U_h \times L_H, \quad n > 1, \quad \text{such that} \\ \left(\dfrac{u_h^{n+1} - 2u_h^n + u_h^{n-1}}{\Delta t^2}, v_h \right) + a(u_h^n, v_h) = b(v_h, \lambda_H^n), \ \forall v_h \in U_h, \\ \\ b(u_h^n, \mu_H) = 0, \quad \forall \mu_H \in L_H, \end{array}\right. \tag{88}$$

or equivalently, in a matrix form:

$$\left\{\begin{array}{l} \mathcal{M}_h \dfrac{\mathbf{u}_h^{n+1} - 2\mathbf{u}_h^n + \mathbf{u}_h^{n-1}}{\Delta t^2} + \mathcal{A}_h \mathbf{u}_h^n = \mathcal{B}_H^* \boldsymbol{\lambda}_H^n, \\ \\ \mathcal{B}_H \mathbf{u}_h^n = 0. \end{array}\right. \tag{89}$$

For practical computation, we shall replace the second equation of (89) by another, which is nothing but (85) written at time t^n and can be obtained by eliminating $\mathbf{u}_h^{n+1}$ and $\mathbf{u}_h^{n-1}$ between the two equations of (89):

$$\begin{cases} \mathcal{M}_h \dfrac{\mathbf{u}_h^{n+1} - 2\mathbf{u}_h^n + \mathbf{u}_h^{n-1}}{\Delta t^2} + \mathcal{A}_h \mathbf{u}_h^n = \mathcal{B}_H^* \boldsymbol{\lambda}_H^n, \\ \mathcal{Q}_H \boldsymbol{\lambda}_H^n = \mathcal{B}_H \mathcal{M}_h^{-1} \mathcal{A}_h \mathbf{u}_h^n. \end{cases} \tag{90}$$

Remark 16. One shows that (89) and (90) are equivalent if $\mathcal{B}_H \mathbf{u}_h^0 = \mathcal{B}_H \mathbf{u}_h^1 = 0$, which expresses in a discrete way a compatibility condition between the initial data and the boundary condition.

Assuming that the solution is known up to time t^n, the algorithm to compute the solution at t^{n+1} has two steps:

- Compute $\mathbf{u}_h^{n+1}$ via the first equation of (89) (a purely explicit step).
- Solve $\mathcal{Q}_H \boldsymbol{\lambda}_H^{n+1} = \mathcal{B}_H \mathcal{M}_h^{-1} \mathcal{A}_h \mathbf{u}_h^{n+1}$ in order to compute $\boldsymbol{\lambda}_H^{n+1}$ (the invertibility of $\mathcal{Q}_H$ is thus the only condition for the existence of the discrete solution).

We see that, in comparison with a standard FDTD procedure inside C, the only additional cost is the inversion of the matrix $\mathcal{Q}_H$. This cost is marginal, due to the properties of $\mathcal{Q}_H$ (see Sect. 3.1). In fact, from the computational point of view, the difficult step in the implementation lies in the construction of the matrix $\mathcal{B}_H$ (and then of $\mathcal{Q}_H$): in 3D for instance, it involves the determination of the intersections between a cubic mesh and a surface mesh (typically with triangular plane facets). See for instance [42] for more details.

Theoretical Issues

Properties of the matrix Q_H.

The following properties of the matrix $\mathcal{Q}_H = \mathcal{B}_H \mathcal{M}_h^{-1} \mathcal{B}_H^*$ are general, but we shall illustrate them in the case $d = 2$ when we make the Q_1–P_0 choice:

- Q_1 finite elements on a regular square mesh for the construction of U_h.
- P_0 (piecewise constant) finite element for the construction of L_H.

We introduce the respective bases of U_h and L_H:

- For U_h, $\{w_i; 1 \le i \le p\}$ where each w_j is supported by 4 squares with a common vertex.
- For L_H, $\{\varphi_j; 1 \le j \le q\}$ where each φ_j is the characteristic function of a (curvilinear) segment of the mesh of γ.

Of course the entries of the matrix $\mathcal{B}_H$ are

$$b(w_i, \varphi_j) = \int_\gamma w_i \, \varphi_j \, \mathrm{d}\sigma. \tag{91}$$

To emphasise that $\mathcal{Q}_H$ should be easy to invert, let us observe the following.

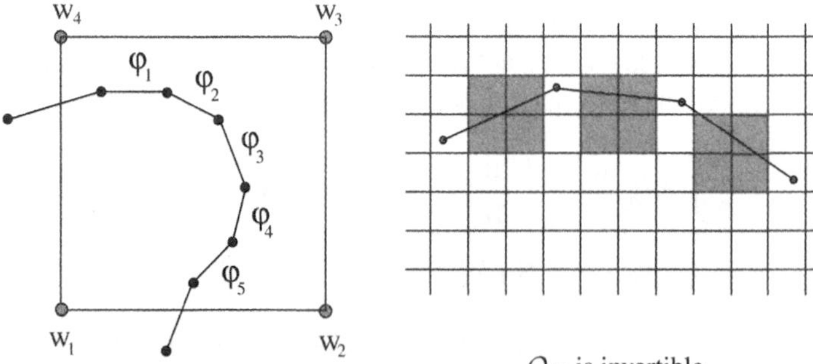

$\mathcal{Q}_H$ is not invertible

$\mathcal{Q}_H$ is invertible

Fig. 4. The two opposite cases

- $\mathcal{Q}_H$ is symmetric and positive (by construction!).
- $\mathcal{Q}_H$ is a "small" matrix: its dimension is exactly (q, q), to be compared with (p, p) for $\mathcal{A}_H$, with $p \gg q$.
- $\mathcal{Q}_H$ is a sparse matrix with narrow bandwidth. This is due to the sparsity of the matrix $\mathcal{B}_H$: indeed the coefficient $b(\varphi_i, w_j)$ vanishes as soon as the supports of the two basis functions (φ_j, w_i) do not intersect.

The (crucial) remaining question is the definiteness of $\mathcal{Q}_H$ which corresponds to the fact that the kernel of the matrix is equal to 0, or equivalently that $\mathcal{B}_H$ is surjective from U_h onto L_H. This suggests that the space L_H must not be too large, or in other words that one must not impose too many "boundary"constraints to the discrete solution. As for any mixed method, there is a compatibility condition between the two spaces U_h and L_H that can be reduced to a compatibility relation between the two meshes of C and γ: the volume mesh cannot be too large with respect to the boundary mesh or, roughly speaking, the ratio H/h must be large enough. In this sense the two meshes cannot be completely independent.

Let us illustrate this in the Q_1–P_0 choice. The invertibility of $\mathcal{Q}_H$ can be reformulated as:

$$b(v_h, \mu_H) = 0, \quad \forall\, v_h \in U_h \Longrightarrow \mu_H = 0. \tag{92}$$

Let us consider the following opposite cases (illustrated in Fig. 4):

- *A case where the volume mesh is (locally) too fine.* Assume that one square K of the 2D mesh contains 5 segments of the 1D mesh (cf. figure 4). Let $\{w_j, 1 \leq j \leq 4\}$ be the four Q_1 basis functions associated with the vertices of K and $\{\nu_l, 1 \leq l \leq 5\}$ the five characteristic functions of the above segments. Obviously one can find $\mu_H \in L_H$ of the form:

$$\mu_H = \sum_{l=1}^{5} \mu_l \nu_l\,, \quad \text{s.t.} \quad \mu_H \neq 0 \quad \text{and} \quad b(w_j, \mu_H) = 0, \quad \forall\, 1 \leq j \leq 4.$$

For any $v_h \in U_h$, $b(v_h, \mu_H)$ is a linear combination of the $b(w_j, \mu_H)$ and therefore vanishes, which contradicts (92).

- *A case where the volume mesh is large enough.* Assume that for each segment I_l (whose characteristic function is ν_l) of the 1D mesh one is able to find a Q_1 basis function w_j such that the intersection supp $w_j \cap I_j$ is included in I_j and has a non zero (surface) measure (see Fig. 4). In such a case, if $\mu_H = \sum \mu_l \, \nu_l$:

$$\left| \begin{array}{l} b(v_h, \mu_H) = 0, \quad \forall\, v_h \in U_h \Longrightarrow b(w_l, \mu_H) = 0, \quad \forall\, l, \\ \qquad\qquad\qquad\qquad\qquad\quad \Longrightarrow \mu_l \left(\int_{I_l} w_l \, \mathrm{d}\gamma \right) = 0, \quad \forall\, l, \\ \qquad\qquad\qquad\qquad\qquad\quad \Longrightarrow \mu_l = 0, \quad \forall\, l, \quad (\text{since } \int_{I_l} w_l \, \mathrm{d}\gamma > 0). \end{array} \right.$$

  which proves (92).

Stability analysis.

One very interesting property of the fictitious domain method is that the numerical scheme is stable under the same CFL condition as without the obstacle. In other words, the stability condition is independent of the geometry of the problem and coincides with the one of the pure FDTD scheme. To see that, it suffices to take:

$$v_h = \frac{u_h^{n+1} - u_h^{n-1}}{\Delta t} \quad \text{and} \quad \mu_H = \lambda_H^n$$

in the two equations of (88) to show the conservation of the discrete energy:

$$E_h^{n+1/2} = \frac{1}{2} \, \| \frac{u_h^{n+1} - u_h^n}{\Delta t} \|_P^2 + \frac{1}{2} \, a(u_h^{n+1}, u_h^n).$$

One then concludes as in Sect. 2.4.

Convergence analysis.

Let us restrict ourselves to the semi-discrete problem. The convergence of the method requires a condition which is stronger than (92), namely the uniform inf-sup condition:

$$\inf_{\mu_H \in L_H} \; \sup_{v_h \in U_h} \; \frac{b(v_h, \mu_H)}{\|\mu_H\|_L \, \|v\|_U} \geq \alpha > 0 \,. \tag{93}$$

Let us estimate the error between the approximate solution (u_h, λ_h) of the semi-discrete problem (83) and the exact solution (u, λ) of problem (76) provided that this solution is regular enough. As the proof is essentially an adaptation of the one we gave in Sect. 2.3, we simply indicate its main steps and refer to [32] for more details. Thanks to (93) and to the coercivity of $a(\cdot, \cdot)$ in U, we can introduce the elliptic projection:

$$\left| \begin{array}{l} U \times L \rightarrow U_h \times L_H \\ (u, \lambda) \;\; \rightarrow (\Pi_h u, \Pi_H \lambda) \end{array} \right. \tag{94}$$

where $(\Pi_h u, \Pi_H \lambda)$ is defined by:

$$\begin{cases} a(u - \Pi_h u, v_h) = b(v_h, \lambda - \Pi_H \lambda), \forall v_h \in U_h, \\ b(u - \Pi_h u, \mu_H) = 0, \qquad\qquad \forall \mu_H \in L_H. \end{cases} \tag{95}$$

Moreover, we have the classical result [16]:

$$\|u - \Pi_h u\|_U + \|\lambda - \Pi_H \lambda\|_L \leq C \left(\inf_{v_h \in U_h} \|u - v_h\|_U + \inf_{\mu_H \in L_H} \|\lambda - \mu_H\|_L \right). \tag{96}$$

Let us write:

$$\left| \begin{array}{l} u_h - u = \eta_h - \varepsilon_h, \quad \eta_h = u_h - \Pi_h u, \quad \varepsilon_h = u - \Pi_h u, \\ \lambda_H - \lambda = \tau_H - \theta_H, \quad \tau_H = \lambda_H - \Pi_H \lambda, \quad \theta_H = \lambda - \Pi_H \lambda. \end{array} \right. \tag{97}$$

As ε_h and θ_H tend to 0 thanks to (96), it suffices to estimate η_h and τ_H. One easily sees that:

$$\begin{cases} \left(\dfrac{\mathrm{d}^2 \eta_h}{\mathrm{d}t^2}, v_h\right) + a(\eta_h, v_h) - b(v_h, \tau_H) = \left(\dfrac{\mathrm{d}^2 \varepsilon_h}{\mathrm{d}t^2}, v_h\right), \forall v_h \in U_h, \\ b(\eta_h, \mu_H) = 0, \qquad\qquad \forall \mu_H \in L_H. \end{cases} \tag{98}$$

From the second equation, we deduce (we first differentiate in time this equation for fixed μ_H and then take $\mu_H = \tau_H$):

$$b(\frac{\mathrm{d}\eta_h}{\mathrm{d}t}, \tau_H) = 0.$$

Then taking $v_h = \dfrac{\mathrm{d}\eta_h}{\mathrm{d}t}$ in the first equation of (98) leads to:

$$\frac{\mathrm{d}}{\mathrm{d}t}\mathcal{E}_h = (\frac{\mathrm{d}^2 \varepsilon_h}{\mathrm{d}t^2}, \frac{\mathrm{d}\eta_h}{\mathrm{d}t}) \quad \text{where} \quad \mathcal{E}_h(t) \stackrel{\text{def}}{=} \frac{1}{2}\{ \, \|\frac{\mathrm{d}\eta_h}{\mathrm{d}t}(t)\|_P^2 + a(\eta_h(t), \eta_h(t)) \, \}. \tag{99}$$

One then concludes as in Sect. 2.3 to get the estimate of $\mathcal{E}_h$ and thus of η_h.

Finally, to get an estimate for τ_H, we use the uniform inf-sup condition:

$$\|\tau_H\|_L \leq \alpha^{-1} \sup_{v_h \in U_h} \frac{b(v_h, \tau_H)}{\|v_h\|_U}. \tag{100}$$

From the first equation of (98) we deduce that:

$$|b(v_h, \tau_H)| \leq C \left\{ \|\frac{\mathrm{d}^2 \eta_h}{\mathrm{d}t^2}\| + \|\eta_h\|_U + \|\varepsilon_h\| \right\} \|v_h\|_U,$$

so that finally:

$$\|\tau_H\|_L \leq C \left\{ \|\frac{\mathrm{d}^2 \eta_h}{\mathrm{d}t^2}\| + \|\eta_h\|_U + \|\varepsilon_h\| \right\}, \tag{101}$$

which means essentially that one controls τ_H in terms of quantities which have already been estimated (η_h and η_h). We refer the reader to [32] for the details.

Accuracy of the fictitious domain method.

The counterpart to the good properties of the fictitious domain method, in terms of simplicity and robustness, is its limited accuracy. As a matter of fact, the convergence proof shows that the accuracy is essentially driven by the "interpolation error" associated with the exact solution (u, λ) of the continuous problem :

$$\inf_{v_h \in U_h} \|u - v_h\|_U + \inf_{\mu_H \in L_H} \|\lambda - \mu_H\|_L$$

(and the same quantity where (u, λ) are replaced by successive time derivatives). The limitation of the accuracy is then due to the fact that the regularity of the exact solution in C – or inside an element K of the volume mesh – is limited, independently of the smoothness of the data of the problem, by the fact that the normal derivative of u presents a jump across γ. As a consequence, the maximal space regularity for $u(\cdot, t)$ is (it is the same for time derivatives):

$$u(\cdot, t) \in H^{3/2-\varepsilon}(C), \quad \forall\, \varepsilon > 0.$$

As a consequence we simply have:

$$\inf_{v_h \in U_h} \|u(\cdot, t) - v_h\|_U \leq C(\varepsilon, u)\, h^{1/2-\varepsilon}$$

while, as soon as $\lambda(\cdot, t) \in H^{1/2}(\gamma)$:

$$\inf_{\mu_H \in L_H} \|\lambda(\cdot, t) - \mu_H\|_L \leq C(\lambda)\, H.$$

Then the convergence analysis provides the following error estimates in which, for clarity, we make explicit the Sobolev norms that are used (see [44] for a more rigorous presentation):

$$\|u(\cdot,t) - u_h(\cdot,t)\|_{H^1(C)} + \|\lambda(\cdot,t) - \lambda_H(\cdot,t)\|_{H^{-1/2}(\gamma)} \leq C(\varepsilon, u, \lambda) \left(h^{1/2-\varepsilon} + H\right). \tag{102}$$

Using classical duality arguments it is possible to get an estimate of $u(\cdot, t) - u_h(\cdot, t)$ in L^2 (see again [44] in the elliptic case):

$$\|u(\cdot,t) - u_h(\cdot,t)\|_{L^2(C)} + \|\lambda(\cdot,t) - \lambda_H(\cdot,t)\|_{H^{-1/2}(\gamma)} \leq C(\varepsilon, u, \lambda)\ \left(h^{1-\varepsilon} + H\right). \tag{103}$$

This last estimate shows that the fictitious domain method is essentially of order 1.

About the uniform inf-sup condition.

It is natural to look at what kind of compatibility relation between the two meshes implies the uniform inf-sup condition. We already saw that, in the Q_1–P_0 case, the invertibility of $\mathcal{Q}_H$ would require a sufficiently fine mesh. It appears that the same kind of condition, namely:

$$H \geq C\, h, \tag{104}$$

is sufficient for the uniform inf-sup condition. One can describe a very general procedure to prove such a result (inspired in particular by the work of Babúska [5]). One wants to show that:

$$\forall \mu_H \in L_H, \exists\, v_h \in U_h \;/\; b_\Gamma(\mu_H, v_h) \geq \alpha \; \|\mu_H\|_L \; \|v_h\|_U \tag{105}$$

for some $\alpha > 0$, at least for a ratio h/H small enough.

For this, it is sufficient to have the two following properties:

1. One can construct $\mathcal{R} \in \mathcal{L}(L, U)$ such that

$$\forall \mu \in L, \quad b_\Gamma(\mu, \mathcal{R}\mu) \geq \|\mathcal{R}\mu\|_U^2, \quad \|\mathcal{R}\mu\|_U \geq \nu \; \|\mu\|_L \tag{106}$$

which obviously provides a proof for the continuous inf-sup condition.

2. One can construct an "interpolation " operator $\Pi_h \in \mathcal{L}(U, U_h)$ such that

$$\forall \mu_H \in L_H, \quad \|\mathcal{R}\mu_H - \Pi_h \mathcal{R}\mu_H\|_U \leq o(\frac{h}{H}) \; \|\mu_H\|_L^2. \tag{107}$$

Indeed, let us assume that these two criteria are fulfilled. For $\mu_H \in L_H$, to obtain (105), we choose $v_h = \Pi_h \mathcal{R}\mu_H$, then:

$$\left| \begin{aligned} b_\Gamma(\mu_H, v_h) &= b_\Gamma(\mu_H, \Pi_h \mathcal{R}\mu_H), \\ &= b_\Gamma(\mu_H, \mathcal{R}\mu_H) + b_\Gamma(\mu_H, \mu_H - \Pi_h \mathcal{R}\mu_H), \\ &\geq \|\mathcal{R}\mu_H\|_U^2 - o(\frac{h}{H}) \; \|\mu_H\|_L^2. \end{aligned} \right.$$

We then use the inequalities:

$$\left| \begin{aligned} &\|\mathcal{R}\mu_H\|_U^2 \geq \nu \; \|\mu_H\|_L \; \|\mathcal{R}\mu_H\|_U, \\ &\|\mu_H\|_L^2 \leq \nu^{-1} \; \|\mu_H\|_L \; \|\mathcal{R}\mu_H\|_U, \end{aligned} \right.$$

to conclude that:

$$b_\Gamma(\mu_H, v_h) \geq \{\nu - \nu^{-1} \; o(h/H)\ \} \; \|\mu_H\|_L \; \|v_h\|_U,$$

which proves the inf-sup condition for h/H small enough.

Let us apply this to our model problem ($U = H_0^1(C), L = H^{-1/2}(\Gamma)$ and Q_1–P_0 finite elements for $U_h \times L_H$). For criterion 1, we construct $u = \mathcal{R}\mu \in H^1(C)$ as:

$$\begin{cases} -\Delta u + u = 0, & \text{in } C \\ [u]_\Gamma = 0, \quad [\dfrac{\partial u}{\partial n}]_\Gamma = \mu \text{ on } \Gamma\,. \end{cases}$$

By Green's formula:

$$b_\Gamma(\mu, u) = \int_\Gamma \mu \, u \; \mathrm{d}\gamma = \int_C (\; |\nabla u|^2 + |u|^2) \; \mathrm{d}x = \|\mathcal{R}\mu\|_{1,C}^2\,,$$

and by the trace theorem:

$$\|\mu\|_{-1/2,\Gamma} \leq C\;(\;\|u\|_{1,C} + \|\Delta u\|_{0,C}) \leq C\;\|u\|_{1,C} = C\;\|\mathcal{R}\mu\|_{1,C},$$

which satisfies criterion 1. Moreover, we have the estimate :

$$\|u\|_{1,\Omega} = \|\mathcal{R}\mu\|_{1,\Omega} \leq C\;\|\mu\|_{-1/2,\Gamma}$$

and also, by a regularity result:

$$\mu \in H^{s-1/2}(\Gamma) \Leftrightarrow \|\mathcal{R}\mu\|_{1+s,\Omega} \leq C(s)\;\|\mu\|_{s-1/2,\Gamma}\,,$$

where $s = 1/2 - \varepsilon$ (where $\varepsilon > 0$ can be arbitrarily small). To show that criterion 2 is satisfied for the Q_1–P_0 choice, let Π_h be the orthogonal projection in $H_0^1(C)$ from U to U_h:

$$\forall u \in H^2(\Omega), \quad \|u - \Pi_h u\|_{1,\Omega} \leq C\;h\;\|u\|_{2,\Omega}\,.$$

By interpolation, for $1 \leq q \leq 2$:

$$\forall u \in H^{1+q}(\Omega), \quad \|u - \Pi_h u\|_{1,\Omega} \leq C_q\;h^q\;\|u\|_{1+q,\Omega}\,.$$

Since $L_H \subset H^{q-1/2}(\Gamma)$ for any $0 \leq q < 1$, we have, with $s = 1/2 - \varepsilon$:

$$\|\mathcal{R}\mu_H - \Pi_h\mathcal{R}\mu_H\|_{1,\Omega} \leq C_s\,h^s\;\|\mathcal{R}\mu_H\|_{1+s,\Omega}. \leq C_s\,C(s)\;h^s\;\|\mu\|_{s-1/2,\Gamma}$$

Now, we assume, that the meshes of γ form a uniformly regular family of meshes (see [5]). In such a case, we have an inverse inequality:

$$\|\mu\|_{s-1/2,\Gamma} \leq H^{-s}\;\|\mu\|_{-1/2,\Gamma}$$

so that finally:

$$\|\mathcal{R}\mu_H - \Pi_h\mathcal{R}\mu_H\|_{1,\Omega} \leq C_s\;C(s)\;\left(\frac{h}{H}\right)^s\;\|\mu\|_{s-1/2,\Gamma},$$

which is nothing but (107).

Remark 17. The main drawback of this "very general" type of proof is that it does not provide an explicit numerical value for the constant C in the inequality (104). In [44], Girault and Glowinski have obtained an explicit value for C in the case of the elliptic problem corresponding to our 2D model problem. Their proof is technically difficult and relies on the construction of a so-called Fortin's operator. This construction uses the Clement's interpolation operator [23], that is known to exist for Lagrange finite elements. That is why this proof is not directly generalisable to other equations (such as Maxwell's equations).

A Numerical Experiment

Here we show a numerical experiment to demonstrate that the fictitious domain method provides better accuracy than a staircase approximation of the boundary. We want to compute the diffraction of an incident plane wave by a "2D cone-sphere". In this case, γ is the union of one half-circle with two segments. The geometry of the problem is clear in Fig. 5 on p. 239. The incident wave is a plane wave, with Gaussian pulse, coming from the left of the picture and propagating to the right.

We represent two snapshots of the solution of the problem in Fig. 5. We clearly distinguish on these pictures the incident wave and the scattered field.

Next we want to compare the convergence of the fictitious domain method with the one of the pure FDTD approach combined with a staircase approximation of the boundary γ. To do that, we look at the solution at point a short distance to the left of the sharp tip as a function of time for $1 \leq t \leq 3$, which corresponds to the passage of the scattered field. In each case we have computed the discrete solutions for three values of the step-size h which correspond approximately to 10, 20 and 40 points per wavelength of the incident wave. The ratio h/H is kept constant.

The three curves corresponding to the fictitious domain calculations are depicted in Fig. 8. The curves are difficult to distinguish the one from the other which indicates the fact that the fictitious domain method has converged. In Fig. 9, we superpose the three curves obtained with the FDTD approach and the curve computed with the fictitious domain approach with 40 points per wavelength (considered as the reference solution). We observe that the three FDTD solutions oscillate around the exact solution. These oscillations are due to the spurious diffractions related to the staircase approximation of the boundary. Clearly, the amplitude of these oscillations decreases when h diminishes, but they are still visible with the FDTD calculation with 40 points per wavelength. In other words, the pure FDTD approach is less accurate with 40 points per wavelength than the fictitious domain method is with 10 points per wavelength.

3.2 Other Applications of the Fictitious Domain Method

In this subsection, we restrict ourselves to giving the main ideas and indicate some bibliographical pointers concerning other applications of the fictitious domain method.

The Neumann Boundary Condition

The key point of the fictitious domain method that we presented in Sect. 3.1 lies in the property that the boundary condition (in that case, the Dirichlet condition) can be considered as an equality constraint in the functional space in which the solution is searched. If one considers a Neumann boundary condition instead of the Dirichlet one, one must change suitably the formulation chosen for the interior problem (namely the problem in the extended domain C) in order to preserve such a property. The solution consists of replacing the primal variational formulation considered in

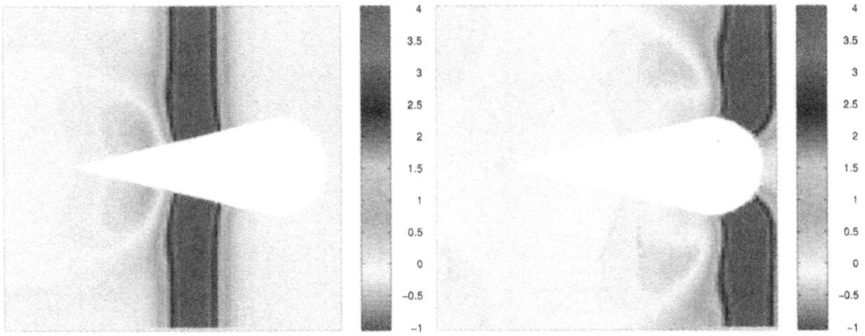

Fig. 5. The solution at two successive instants (see p. 238)

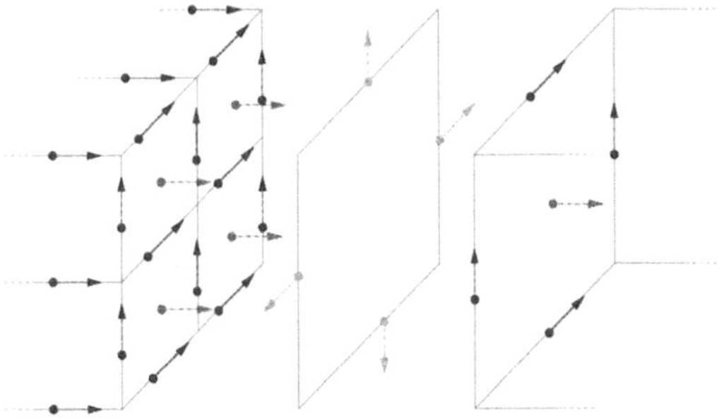

Fig. 6. The degrees of freedom matching coarse and fine grids for the Yee scheme *with* a Lagrange multiplier (see p. 253): the electric field in blue, the magnetic field in red, the electric current in green

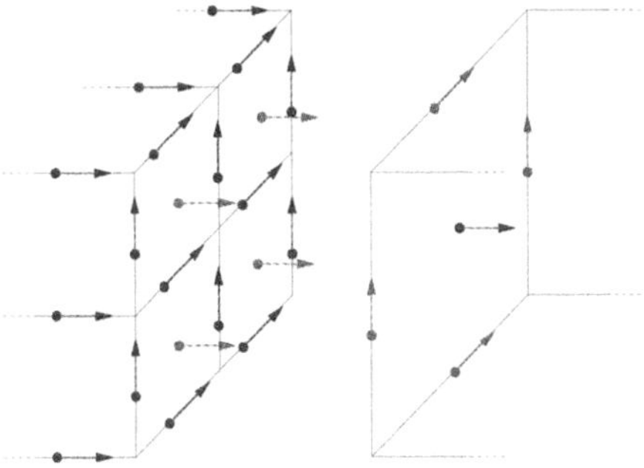

Fig. 7. As previous figure but *without* a Lagrange multiplier (see p. 256): the electric field in blue, the magnetic field in red

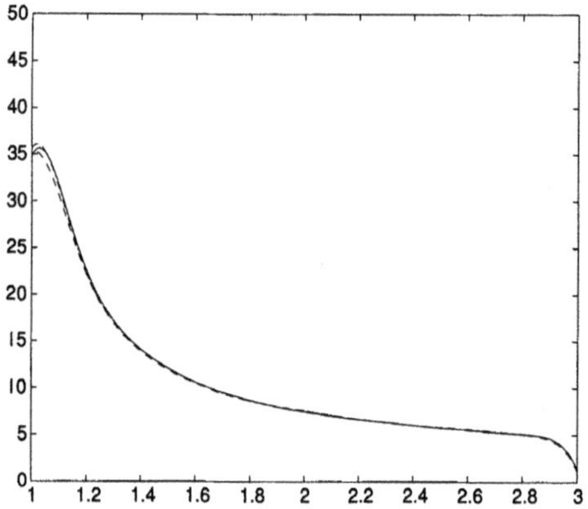

Fig. 8. Solution versus time at a point a short distance to the left of the sharp tip of the scatterer in Fig. 5 on p. 239 via the fictitious domain method for $h/\lambda = 1/10, 1/20$ and $1/40$

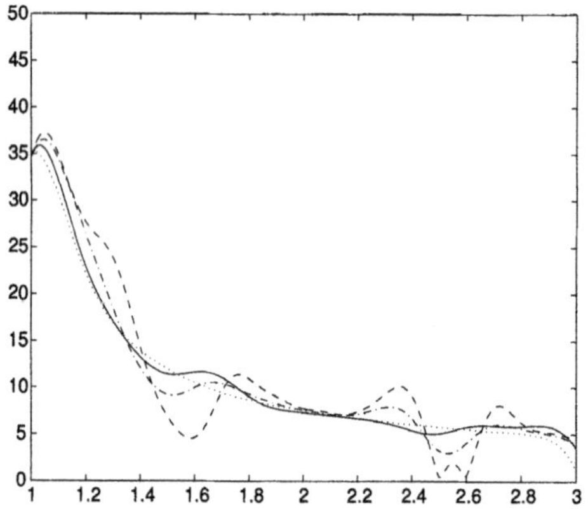

Fig. 9. Comparison with the FDTD method. The fictitious domain solution corresponds to the dotted line

Sect. 3.1 by the dual formulation or the mixed primal-dual formulation (see (16)). In both cases, the dual unknown $v = \nabla u$ has to be introduced and sought in the space $H(div; C)$ in which the Neumann boundary condition $v \cdot n = 0$ can be considered as an equality constraint. In this case, the Lagrange multiplier is nothing but the jump of the primal unknown u across Γ and is naturally sought in $H^{1/2}(\Gamma)$. We refer the reader to [49], for the inf-sup condition in this case.

Maxwell's Equations

In [31], we develop the fictitious domain method for Maxwell's equations in the case of a perfectly conducting boundary, i.e. when the boundary condition is:

$$e \times n = 0, \quad \text{on } \Gamma .$$

In this case, the primal-dual formulation (15) is used, where the electric field is sought in $H(\text{curl}; C)$ and the magnetic field in $L^2(C)^3$. The Lagrange multiplier is the jump of the magnetic field, more precisely of $h \times n$, i.e. the surface electric current.

Elastodynamics

In [10], we applied a fictitious domain approach to the scattering of elastic waves by free boundaries (which includes in particular the case of cracks). The boundary condition is :

$$\sigma\, n = 0, \quad \text{on}\, \Gamma\,.$$

We use the velocity-constraint form of the elastodynamics equations and a primal-dual mixed formulation in which the constraints are sought in a space of symmetric tensors in L^2 whose divergence is in L^2. The Lagrange multiplier is the jump $[u]$ of the displacement field across Γ. This led us to introduce a new family of mixed elements for the system of elasticity [11]. These elements take into account the symmetry constraint of the tensors in the strong sense (contrary to more traditional approaches as in [1] for instance) and are compatible with mass lumping. The extension of the method to unilateral contact conditions is considered in [8].

3.3 Scattering by Moving Obstacles

On this topic we have only very few preliminary results. A typical application is the diffraction of a wave by a rotating body. Consider for instance the same problem as in Sect. 3.1, except that the location of the obstacle now depends on t and is denoted $\mathcal{O}(t)$. Although the theory of such problem is now well established (see [61] for instance), there is little work on numerical methods for their approximation. Standard finite element methods seem difficult to apply, and would in any case require the reconstruction of the computational mesh at each time step. To avoid such complexity, the fictitious domain approach seems to be a very attractive solution. We have adapted the method of Sect. 3.1 to this case. The space discretisation leads to an ordinary differential system of the form (84) where the "boundary matrix" now depends on time:

$$\mathcal{B}_H \to \mathcal{B}_H(t)\,.$$

At this moment, we have absolutely no theory to justify the validity of the approach, but the first numerical results we have been able to obtain [66] are very promising.

3.4 Numerical Modelling in Musical Acoustics

The numerical modelling of musical instruments, namely the timpani [18] and the guitar [33], is a particular field of application where the fictitious domain method has been particularly useful. In both cases, the model appears as a coupling between various propagation models in different space dimensions: a 1D wave equation for the vibrations of the string of the guitar; a 2D propagation model for the vibrations of the soundboard of the guitar or the membrane of the timpani; and a 3D wave equation for the sound radiation inside and outside the instrument. Both models include the modelling of fluid-structure interaction that can be treated by the fictitious domain method. The boundary unknown is the jump of the pressure field across the surface of the instrument.

4 Space-time Mesh Refinement and Domain Decomposition

4.1 Introduction

In order to treat complex geometries or geometrical details in diffraction problems, another natural approach is to use local mesh refinements. Moreover, it is highly desirable to be able to treat non matching grids (this is even needed if one wants to use FDTD-like schemes in each grid). A first idea consists of using only spatial refinement (see [7] for acoustic waves, [13, 57] for Maxwell's equations). However, when a uniform time step is used, it is the stability condition on the finest mesh that will impose the time step size restriction. There are two problems with this: first the computational costs will be increased; and second the ratio $c\Delta t/h$ on the coarse grid will be much smaller than its optimal value, which will generate dispersion errors. A way to avoid these problems is to use a local time step Δt, related to h in order to keep the ratio $c\Delta t/h$ constant. This solution however raised other practical and theoretical problems that are much more intricate than in the case of a simple spatial refinement, in particular in terms of stability.

The solutions suggested in the electromagnetics literature are primarily based on interpolation techniques (in time and/or in space) especially designed to guarantee the consistency of the scheme at the coarse grid-fine grid interface [19, 51, 52, 63]. Unfortunately, the resulting schemes appear to be very difficult to analyse and may suffer from some instability phenomena.

Recently, we developed alternative solutions to these interpolation procedures. Our purpose in this section is to describe these new methods in the case of Maxwell's equations and the case of two spatial grids: one is twice as fine as the other. However, as we shall try to show by exhibiting an abstract framework, these methods are applicable to a large class of problems including acoustics, elastodynamics, fluid-structure interaction, etc. and can be generalised to stronger mesh refinement. They are based on the principle of domain decomposition and essentially consist of ensuring a priori the numerical stability of the scheme via conservation of a discrete energy.

One can distinguish two kinds of methods, depending on the choice of the variational formulation inside each grid. We shall adopt a constructive approach in our presentation: we will show that the criterion of energy conservation leads to necessary conditions for the discrete transmission equations between the two grids. We can then derive the appropriate transmission equations from consistency and accuracy considerations. We shall also describe some elements of the convergence analysis of such methods.

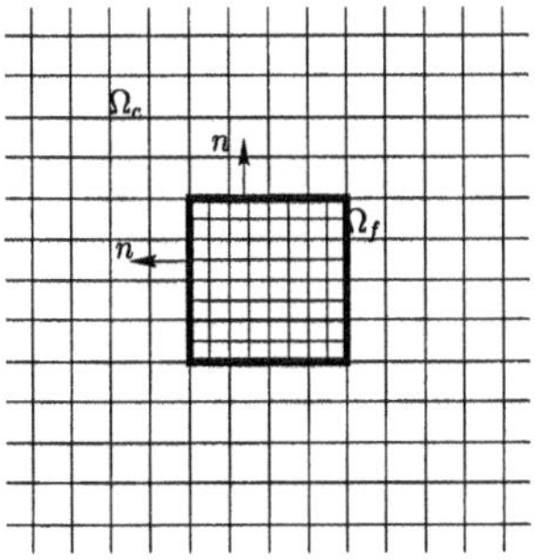

Fig. 10. 2D slice of the domains Ω_f and Ω_c

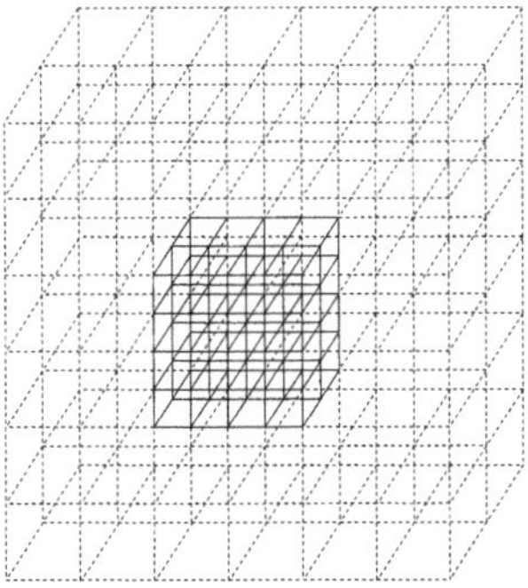

Fig. 11. 3D view of the refinement

4.2 The Domain Decomposition Approach

As a model problem, we consider Maxwell's equations in the whole space $\mathbb{R}^3$ (with $\varepsilon = \mu = 1$):

$$\begin{cases} \dfrac{\partial e}{\partial t} - \text{curl } h = 0, & x \in \mathbb{R}^3 \\ \dfrac{\partial h}{\partial t} + \text{curl } e = 0, & x \in \mathbb{R}^3 \end{cases} \tag{108}$$

Our goal is to solve this problem numerically by domain decomposition using locally a grid twice as fine as in the rest of the domain. To be more precise, we consider a box Ω_f with boundary Σ (with outward and unit normal vector n), and Ω_c denoting the exterior of Ω_f. The domain Ω_f is the one that we shall discretise with a fine grid of step-size h and Ω_c is the one that we shall discretise with a coarse grid of step $2h$ (see Fig. 4.2). In what follows, (e_c, h_c) (resp. (e_f, h_f)) will denote the restriction to Ω_c (resp. Ω_f) of (e, h). Saying that (e, h) is solution of (108) is equivalent to saying that (e_c, h_c) and $(e_f\ h_f)$ are solutions of the same equations (108) respectively in Ω_c and Ω_f and are coupled by the following transmission conditions (the continuity of the tangential traces of the two fields across Σ):

$$\begin{cases} n \times (e_c \times n) = n \times (e_f \times n) & \text{on } \Sigma\,, \\ h_c \times n = h_f \times n & \text{on } \Sigma\,. \end{cases} \tag{109}$$

4.3 Variational Formulations

A Variational Formulation with Lagrange Multiplier

This formulation is similar to the fictitious domain formulation we presented in Sect. 3.1. We introduce as an additional unknown, a Lagrange multiplier, namely the common tangential trace of the magnetic fields h_c and h_f on the interface:

$$j(x,t) = (h_f \wedge n)|_\Sigma = (h_c \wedge n)|_\Sigma. \qquad (110)$$

Note that j is nothing other than the surface electric current on Σ. We can then reformulate our problem as follows. Assuming for a moment that j is known, (e_c,h_c) and (e_f,h_f) are the respective solutions of the following decoupled problems, in which j appears as a (boundary) source term:

$$\begin{cases} \dfrac{\partial e_f}{\partial_t} - \text{curl } h_f = 0, \quad x \in \Omega_f & (i) \\ \dfrac{\partial h_f}{\partial_t} + \text{curl } e_f = 0, \quad x \in \Omega_f & (ii) \\ h_f \wedge n = j, \quad x \in \Sigma = \partial\Omega_f \end{cases} \qquad (111)$$

$$\begin{cases} \dfrac{\partial e_c}{\partial_t} - \text{curl } h_c = 0, \quad x \in \Omega_c & (i) \\ \dfrac{\partial h_c}{\partial_t} + \text{curl } e_c = 0, \quad x \in \Omega_c & (ii) \\ h_c \wedge n = j, \quad x \in \Sigma = \partial\Omega_c \end{cases} \qquad (112)$$

For a given j, by construction, the continuity of the tangential magnetic field is ensured. The idea of the method is to consider j as a control variable in order to ensure the continuity of the tangential electric field

$$n \wedge (e_f \wedge n)\,|_\Sigma = n \wedge (e_c \wedge n)\,|_\Sigma. \qquad (113)$$

To derive our variational formulation, we use the primal-dual mixed formulation of each problem (111) or (112), where the electric field is the primal variable. This means that the regularity (in this case $H(\text{curl})$ regularity) is put on the electric field and that, after multiplication by test fields and space integration, only the two equations involving the rotational of the magnetic field (cf. (111)-(ii) and (112)-(ii)) are integrated by parts: this makes j appear in the boundary term. This leads to the following abstract formulation:

Find $(e_c, h_c, e_f, h_f, j) : \mathbb{R}^+ \longrightarrow U_c \times V_c \times U_f \times V_f \times L$ such that

$$\begin{cases} \dfrac{\mathrm{d}}{\mathrm{d}t}(e_c,\, \tilde{e}_c)_c - b_c(\tilde{e}_c,\, h_c) + c_c(j,\, \tilde{e}_c) = 0, & \forall\, \tilde{e}_c \in U_c, \\ \dfrac{\mathrm{d}}{\mathrm{d}t}(h_c,\, \tilde{h}_c)_c + b_c(e_c,\, \tilde{h}_c) = 0, & \forall\, \tilde{h}_c \in V_c, \end{cases} \qquad (114)$$

$$\begin{cases} \dfrac{\mathrm{d}}{\mathrm{d}t}(e_f,\, \tilde{e}_f)_f - b_f(\tilde{e}_f,\, h_f) - c_f(j,\, \tilde{e}_f) = 0, & \forall\, \tilde{e}_f \in U_f, \\ \dfrac{\mathrm{d}}{\mathrm{d}t}(h_f,\, \tilde{h}_f)_f + b_f(e_f,\, \tilde{h}_f) = 0, & \forall\, \tilde{h}_f \in V_f, \end{cases} \qquad (115)$$

$$c_c(\tilde{j},\, e_c) = c_f(\tilde{j},\, e_f), \qquad \forall\, \tilde{j} \in L. \qquad (116)$$

The appropriate functional spaces are:

$$\begin{cases} U_f = H(\text{curl}, \Omega_f), \quad V_f = L^2(\Omega_f)^3, \\ U_c = H(\text{curl}, \Omega_c), \quad V_c = L^2(\Omega_c)^3, \\ L = \mathbf{H}_{\|}^{-1/2}(\text{div}_\Sigma, \Sigma), \quad \text{(see [17] for a precise definition).} \end{cases} \tag{117}$$

The space L is nothing but the image of the spaces V_f and V_c by the trace map

$$\gamma_\tau u = u \wedge n_{|\Sigma}.$$

The various bilinear forms appearing in (114), (115) and (116) are defined by:

$$\begin{cases} (u_c, \tilde{u}_c)_c = \int_{\Omega_c} u_c \cdot \tilde{u}_c \, \mathrm{d}x, & (u_f, \tilde{u}_f)_f = \int_{\Omega_f} u_f \cdot \tilde{u}_f \, \mathrm{d}x, \\ b_c(e_c, h_c) = \int_{\Omega_c} h_c \cdot \text{curl } e_c \, \mathrm{d}x, & b_f(e_f, h_f) = \int_{\Omega_f} h_f \cdot \text{curl } e_f \, \mathrm{d}x, \\ c_c(j, e_c) \; = < j, \pi_\tau e_c >, & c_f(j, e_f) \; = < j, \pi_\tau e_f > . \end{cases} \tag{118}$$

where π_τ denotes the trace map:

$$\pi_\tau u = n \wedge (u \wedge n)_{|\Sigma},$$

that maps the spaces V_f and V_c continuously onto the dual space of L, $L' = \mathbf{H}_{\perp}^{-1/2}(\text{curl}_\Sigma, \Sigma)$. Finally, we shall denote by $< \cdot, \cdot >$ the duality product between L and L', natural extension of the usual scalar product in $L^2(\Sigma)^3$.

A Variational Formulation Without Lagrange Multiplier

One can avoid the introduction of a Lagrange multiplier by coupling a primal-dual formulation in Ω_f (the regularity is on the electric field) with a dual-primal formulation in Ω_c (the regularity is on the magnetic field). In the domain Ω_f, we multiply equations (111) (i) and (ii) by test fields and apply an integration by parts to the first equality. This leads to:

$$\frac{\mathrm{d}}{\mathrm{d}t} \int_{\Omega_f} e_f \cdot \tilde{e}_f(x) \, \mathrm{d}x - \int_{\Omega_f} \text{curl } \tilde{e}_f \cdot h_f \, dx - < \gamma_\tau h_f, \pi_\tau \tilde{e}_f > = 0.$$

To get rid of $\gamma_\tau h_f$, which does not make sense if we look for h_f in L^2, we use the transmission condition, to replace $\gamma_\tau h_f$ by $\gamma_\tau h_c$, which makes sense since we shall look for h_c in $H(\text{curl})$:

$$\frac{\mathrm{d}}{\mathrm{d}t} \int_{\Omega_f} e_f \cdot \tilde{e}_f(x) \, \mathrm{d}x - \int_{\Omega_f} \text{curl } \tilde{e}_f \cdot h_f \, dx - < \gamma_\tau h_c, \pi_\tau \tilde{e}_f > = 0.$$

In the domain Ω_f, we proceed in the same manner except that the roles of the electric and magnetic fields are exchanged. Finally, we end up with the following variational formulation:

$$\text{Find } (e_c, h_c, e_f, h_f) : \mathbb{R}^+ \longrightarrow U_c \times V_c \times U_f \times V_f \text{ such that}$$

$$\begin{cases} \dfrac{\mathrm{d}}{\mathrm{d}t}(e_c,\, \tilde{e}_c)_c - b_c(\tilde{e}_c,\, h_c) = 0, & \forall\, \tilde{e}_c \in U_c, \\ \dfrac{\mathrm{d}}{\mathrm{d}t}(h_c,\, \tilde{h}_c)_c + b_c(e_c,\, \tilde{h}_c) + c(e_f, \tilde{h}_c) = 0, & \forall\, \tilde{h}_c \in V_c, \end{cases} \tag{119}$$

$$\begin{cases} \dfrac{\mathrm{d}}{\mathrm{d}t}(e_f,\, \tilde{e}_f)_f - b_f(\tilde{e}_f,\, h_f) - c(\tilde{e}_f, h_c) = 0, & \forall\, \tilde{e}_f \in U_f, \\ \dfrac{\mathrm{d}}{\mathrm{d}t}(h_f,\, \tilde{h}_f)_f + b_f(e_f,\, \tilde{h}_f) = 0, & \forall\, \tilde{h}_f \in V_f, \end{cases} \tag{120}$$

where we have set:

$$\begin{cases} U_f = H(\mathrm{curl}, \Omega_f), \quad V_f = L^2(\Omega_f)^3, \\ U_c = L^2(\Omega_c)^3, \quad V_c = H(\mathrm{curl}, \Omega_c), \end{cases} \tag{121}$$

$$\begin{cases} b_c(e_c, h_c) = \int_{\Omega_c} h_c \cdot \mathrm{curl}\; e_c \;\mathrm{d}x, \qquad b_f(e_f, h_f) = \int_{\Omega_f} \mathrm{curl}\; h_f \cdot e_f \;\mathrm{d}x, \\ c(e_c, h_f) \; = < \gamma_\tau h_c, \pi_\tau e_f > = < \gamma_\tau e_f, \pi_\tau h_c > . \end{cases} \tag{122}$$

4.4 Finite Element Approximation

The Case With Lagrange Multiplier

We consider here the space discretisation of ((114), (115), (116)). We introduce

$$U_{c,h} \subset U_c,\; V_{c,h} \subset V_c,\; U_{f,h} \subset U_f,\; V_{f,h} \subset V_f \text{ and } L_h \subset L, \tag{123}$$

some finite dimensional approximation subspaces. In practise these spaces will be constructed as finite element spaces associated with two different meshes of Ω_f and Ω_c, the typical situation being that the space step in the coarse grid domain Ω_c is twice as large as that in the fine grid domain Ω_f.

$$\text{Find } (e_{c,h}, h_{c,h}, e_{f,h}, h_{f,h}, j_h) : \mathbb{R}^+ \longrightarrow U_{c,h} \times V_{c,h} \times U_{f,h} \times V_{f,h} \times L \text{ such that}$$

$$\begin{cases} \dfrac{\mathrm{d}}{\mathrm{d}t}(e_{c,h},\, \tilde{e}_{c,h})_c - b_c(\tilde{e}_{c,h},\, h_{c,h}) + c_c(j_h,\, \tilde{e}_{c,h}) = 0, & \forall\, \tilde{e}_{c,h} \in U_{c,h}, \\ \dfrac{\mathrm{d}}{\mathrm{d}t}(h_{c,h},\, \tilde{h}_{c,h})_c + b_c(e_{c,h},\, \tilde{h}_{c,h}) = 0, & \forall\, \tilde{h}_{c,h} \in V_{c,h}, \end{cases} \tag{124}$$

$$\begin{cases} \dfrac{\mathrm{d}}{\mathrm{d}t}(e_{f,h},\, \tilde{e}_{f,h})_f - b_f(\tilde{e}_{f,h},\, h_{f,h}) - c_f(j_h,\, \tilde{e}_{f,h}) = 0, & \forall\, \tilde{e}_{f,h} \in U_{f,h}, \\ \dfrac{\mathrm{d}}{\mathrm{d}t}(h_{f,h},\, \tilde{h}_{f,h})_f + b_f(e_{f,h},\, \tilde{h}_{f,h}) = 0, & \forall\, \tilde{h}_{f,h} \in V_{f,h}, \end{cases} \tag{125}$$

$$c_c(\tilde{j}_h, e_{c,h}) = c_f(\tilde{j}_h, e_{f,h}), \qquad \forall\, \tilde{j}_h \in L_h, \tag{126}$$

which can be rewritten as an algebraic-differential system (we omit the subscript h):

$$\begin{cases} \mathcal{M}_f^E \dfrac{\mathrm{d}\mathbf{e}_f}{\mathrm{d}t} - \mathcal{B}_f\, \mathbf{h}_f - \mathcal{C}_f^*\, \mathbf{j} = 0, \\ \mathcal{M}_f^H \dfrac{\mathrm{d}\mathbf{h}_f}{\mathrm{d}t} + \mathcal{B}_f^*\, \mathbf{e}_f = 0, \\ \mathcal{M}_c^E \dfrac{\mathrm{d}\mathbf{e}_c}{\mathrm{d}t} - \mathcal{B}_c\, \mathbf{h}_c + \mathcal{C}_c^*\, \mathbf{j} = 0, \\ \mathcal{M}_c^H \dfrac{\mathrm{d}\mathbf{h}_c}{\mathrm{d}t} + \mathcal{B}_c^*\, \mathbf{e}_c = 0, \\ \mathcal{C}_c\, \mathbf{e}_c - \mathcal{C}_f\, \mathbf{e}_f = 0. \end{cases} \tag{127}$$

As for the fictitious domain method, this system can be (theoretically) reduced to a standard ordinary differential system, through the elimination of $\mathbf{j}$. Indeed, it is easy to derive from (127) the following stationary relation between $\mathbf{j}$ and $(\mathbf{h}_f, \mathbf{h}_c)$:

$$\left(\mathcal{C}_c \left(\mathcal{M}_c^E\right)^{-1} \mathcal{C}_c^* + \mathcal{C}_f \left(\mathcal{M}_f^E\right)^{-1} \mathcal{C}_f^*\right) \mathbf{j} = \mathcal{C}_c \left(\mathcal{M}_c^E\right)^{-1} \mathcal{B}_c \mathbf{h}_c - \mathcal{C}_f \left(\mathcal{M}_f^E\right)^{-1} \mathcal{B}_f \mathbf{h}_f. \tag{128}$$

This shows that the well-posedness of (127) is conditioned by the invertibility of the matrix:

$$\mathcal{Q} \;=\; \mathcal{C}_c \;\left(\mathcal{M}_c^E\right)^{-1} \;\mathcal{C}_c^* + \;\mathcal{C}_f \;\left(\mathcal{M}_f^E\right)^{-1} \;\mathcal{C}_f^*, \tag{129}$$

which is also equivalent to:

$$\mathrm{Ker}\; \mathcal{C}_f^* \;\cap\; \mathrm{Ker}\; \mathcal{C}_c^* = \{0\}\,. \tag{130}$$

Remark 18. The conditions (130), that can also be expressed in terms of an inf-sup condition, represent once again a compatibility between the spaces $U_{c,h}$, $U_{f,h}$ and L_h. In particular, it is clear again that the space L_h must not be too large. That is why, in practice, we shall often use as the boundary mesh for Σ the trace of the coarse grid mesh for Ω_c.

The Case Without Lagrange Multiplier

This situation is simpler. We keep here the notation (123) for the approximation spaces (except that L_h does not exist any longer). The semi-discrete problem to be solved is:

Find $(e_{c,h}, h_{c,h}, e_{f,h}, h_{f,h}) : \mathbb{R}^+ \longrightarrow U_{c,h} \times V_{c,h} \times U_{f,h} \times V_{f,h}$ such that

$$\begin{cases} \dfrac{\mathrm{d}}{\mathrm{d}t}(e_{c,h},\, \tilde{e}_{c,h})_c - b_c(\tilde{e}_{c,h},\, h_{c,h}) = 0, & \forall\, \tilde{e}_{c,h} \in U_{c,h}, \\ \dfrac{\mathrm{d}}{\mathrm{d}t}(h_{c,h},\, \tilde{h}_{c,h})_c + b_c(e_{c,h},\, \tilde{h}_{c,h}) + c(e_{f,h},\, \tilde{h}_{c,h}) = 0, & \forall\, \tilde{h}_{c,h} \in V_{c,h}, \end{cases} \tag{131}$$

$$\begin{cases} \frac{\mathrm{d}}{\mathrm{d}t}(e_{f,h},\, \tilde{e}_{f,h})_f - b_f(\tilde{e}_{f,h},\, h_{f,h}) - c(\tilde{e}_{f,h}, h_{c,h}) = 0, & \forall\, \tilde{e}_{f,h} \in U_{f,h}, \\ \frac{\mathrm{d}}{\mathrm{d}t}(h_{f,h},\, \tilde{h}_{f,h})_f + b_f(e_{f,h},\, \tilde{h}_{f,h}) = 0, & \forall\, \tilde{h}_f \in V_{f,h}, \end{cases} \tag{132}$$

which leads directly to an ordinary differential system:

$$\begin{cases} \mathcal{M}_f^E \frac{\mathrm{d}\mathbf{e}_f}{\mathrm{d}t} - \mathcal{B}_f\, \mathbf{h}_f = 0, \\ \mathcal{M}_f^H \frac{\mathrm{d}\mathbf{h}_f}{\mathrm{d}t} + \mathcal{B}_f^*\, \mathbf{e}_f - \mathcal{C}^*\, \mathbf{h}_c = 0, \\ \mathcal{M}_c^E \frac{\mathrm{d}\mathbf{e}_c}{\mathrm{d}t} - \mathcal{B}_c\, \mathbf{h}_c + \mathcal{C}\, \mathbf{e}_f = 0, \\ \mathcal{M}_c^H \frac{\mathrm{d}\mathbf{h}_c}{\mathrm{d}t} + \mathcal{B}_c^*\, \mathbf{e}_c = 0. \end{cases} \tag{133}$$

Remark 19. Note that in this case there is no particular requirement for the semi-discrete problem to be well-posed.

4.5 Time Discretisation

The Case With Lagrange Multiplier

The interior schemes.

We look for a time discretisation scheme for ((124), (125), (126)) that will use a time step Δt in Ω_f and a time step $2\Delta t$ in Ω_g. More precisely:

- In Ω_f, the discrete interior unknowns will be $e_{f,h}^n$, $h_{f,h}^{n+1/2}$.
- In Ω_c, the discrete interior unknowns will be $e_{c,h}^{2n}$, $h_{c,h}^{2n+1}$.

With such a choice, the two computational grids only meet at "even instants" $t = t^{2n}$ at which the two electric fields (in Ω_c and Ω_f) will be computed simultaneously. Concerning the interface unknown j_h, the definition (110) (this is also clear in system (127)) suggests that the electric current must be computed at the same instants than the magnetic field. This leads a priori to consider two different ways to discretise j_h:

- If j is considered as the tangential trace of h_c, in which case we denote it j_c, one would like to compute it at times t^{2n+1}: $j_{c,h}^{2n+1}$.
- If j is considered as the tangential trace of h_c, in which case we denote it j_f, one would like to compute it at times t^{2n+1}: $j_{f,h}^{n+1/2}$.

Then we use a standard staggered second order approximation in time for (114) and (115). For the time interval $[t^{2n}, t^{2n+2}]$, this amounts to considering the following set of equations:

$$\begin{cases} \left(\dfrac{e_{f,h}^{2n+1} - e_{f,h}^{2n}}{\Delta t}, \tilde{e}_{f,h} \right)_f - b_f(\tilde{e}_{f,h}, h_{f,h}^{2n+1/2}) - c_f(j_{f,h}^{2n+1/2}, \tilde{e}_{f,h}) = 0, \\ \left(\dfrac{h_{f,h}^{2n+1/2} - h_{f,h}^{2n-1/2}}{\Delta t}, \tilde{h}_{f,h} \right)_f + b_f(e_{f,h}^{2n}, \tilde{h}_{f,h}) = 0, \end{cases} \tag{134}$$

$$\begin{cases} \left(\dfrac{e_{f,h}^{2n+2} - e_{f,h}^{2n+1}}{\Delta t}, \tilde{e}_{f,h} \right)_f - b_f(\tilde{e}_{f,h}, h_{f,h}^{2n+3/2}) - c_f(j_{f,h}^{2n+3/2}, \tilde{e}_{f,h}) = 0, \\ \left(\dfrac{h_{f,h}^{2n+3/2} - h_{f,h}^{2n+1/2}}{\Delta t}, \tilde{h}_{f,h} \right)_f + b_f(e_{f,h}^{2n+1}, \tilde{h}_{f,h}) = 0, \end{cases} \tag{135}$$

$\forall\, \tilde{e}_{f,h} \in U_{f,h}, \forall \tilde{h}_{f,h} \in V_{f,h}$,

$$\begin{cases} \left(\dfrac{e_{c,h}^{2n+2} - e_{c,h}^{2n}}{\Delta t}, \tilde{e}_{c,h} \right)_c - b_c(\tilde{e}_{c,h}, h_{c,h}^{2n+1}) + c_c(j_{c,h}^{2n+1}, \tilde{e}_{c,h}) = 0, \\ \left(\dfrac{h_{c,h}^{2n+1} - h_{c,h}^{2n-1}}{\Delta t}, \tilde{h}_{c,h} \right)_c + b_c(e_{c,h}^{2n}, \tilde{h}_{c,h}) = 0, \end{cases} \tag{136}$$

$\forall\, \tilde{e}_{c,h} \in U_{c,h}, \forall\, \tilde{h}_{c,h} \in V_{c,h}$.

To complete our scheme we must explain how we couple the two grids through discrete transmission conditions that will induce:

- the continuity of the tangential trace of the magnetic field, namely the equality:

$$j_c = j_f, \tag{137}$$

- the continuity (in the weak sense) of the tangential trace of the electric field, namely the equation:

$$c_c(\tilde{j}_h, e_{c,h}) = c_f(\tilde{j}_h, e_{f,h}), \qquad \forall\, \tilde{j}_h \in L_h. \tag{138}$$

This should be achieved with a guarantee of stability for the coupled system. It is the most delicate point and, for the construction, we will be guided by an energy analysis.

Energy conservation and transmission scheme.

According to what we have seen in Sect. 2.4, one can define fine and coarse grid energies as follows:

$$\begin{cases} E_{f,h}^{n} = \dfrac{1}{2} \left(\, \|e_{f,h}^{n}\|_f^2 + (h_{f,h}^{n+1/2}, h_{f,h}^{n+1/2})_f \, \right), \\ E_{c,h}^{2n} = \dfrac{1}{2} \left(\, \|e_{c,h}^{2n}\|_c^2 + (h_{c,h}^{n+1/2}, h_{c,h}^{n+1/2})_c \, \right). \end{cases} \tag{139}$$

The total energy can only be defined at even instants:

$$E_h^{2n} = E_{f,h}^{2n} + E_{c,h}^{2n}. \tag{140}$$

We are looking for a conservative scheme, i.e. a scheme that will enforce the equality:

$$E_h^{2n+2} = E_h^{2n}. \tag{141}$$

Then, by construction, the L^2 stability of the coupled scheme will be ensured under the double strict CFL stability condition (the proof is the same as in Sect. 2.4):

$$\|B_{c,h}\|\Delta t < 1, \qquad \frac{\|B_{f,h}\|\Delta t}{2} < 1, \tag{142}$$

where we recall that:

$$\|B_{c,h}\| = \sup_{\substack{(e_{c,h}, h_{c,h}) \\ \in U_{c,h} \times V_{c,h}}} \frac{b_c(e_{c,h}, h_{c,h})}{\|e_{c,h}\|_c \; \|h_{c,h}\|_c}, \quad \|B_{f,h}\| = \sup_{\substack{(e_{f,h}, h_{f,h}) \\ \in U_{f,h} \times V_{f,h}}} \frac{b_f(e_{f,h}, h_{f,h})}{\|e_{f,h}\|_c \; \|h_{f,h}\|_c}.$$

Remark 20. Each of the two strict inequalities (142) corresponds to the stability condition of the same scheme in each grid separately. In practice, since the space step will be twice as large in the coarse grid, we shall have:

$$\|B_{f,h}\| = 2 \; \|B_{c,h}\|,$$

so that (142) reduces to one single inequality.

Following the technique presented in Sect. 2.4, one easily computes that:

$$\begin{cases} \dfrac{E_{f,h}^{2n+1} - E_{f,h}^{2n}}{\Delta t} = c_f\big(j_{f,h}^{2n+1/2}, \dfrac{e_{f,h}^{2n+1} + e_{f,h}^{2n}}{2}\big), \\[2ex] \dfrac{E_{f,h}^{2n+2} - E_{f,h}^{2n+1}}{\Delta t} = c_f\big(j_{f,h}^{2n+3/2}, \dfrac{e_{f,h}^{2n+2} + e_{f,h}^{2n+1}}{2}\big), \end{cases}$$

while, in the other hand:

$$\frac{E_{c,h}^{2n+2} - E_{c,h}^{2n}}{2\Delta t} = - \; c_c\big(j_{c,h}^{2n+1}, \frac{e_{c,h}^{2n+2} + e_{c,h}^{2n}}{2}\big).$$

Combining these three equalities, we derive the following energy identity:

$$\begin{aligned} \frac{E_h^{2n+2} - E_h^{2n}}{2\Delta t} &= \frac{1}{2} \, c_f\big(j_{f,h}^{2n+1/2}, \frac{e_{f,h}^{2n+1} + e_{f,h}^{2n}}{2}\big) \\ &+ \frac{1}{2} c_f\big(j_{f,h}^{2n+3/2}, \frac{e_{f,h}^{2n+2} + e_{f,h}^{2n+1}}{2}\big) - c_c\big(j_{c,h}^{2n+1}, \frac{e_{c,h}^{2n+2} + e_{c,h}^{2n}}{2}\big). \end{aligned}$$

To obtain our transmission scheme, we look for three additional linear equations that should be consistent with (137) and (138) and should ensure the equality:

$$c_c(j_{c,h}^{2n+1}, \frac{e_{c,h}^{2n+2}+e_{c,h}^{2n}}{2}) = \frac{1}{2}\, c_f(j_{f,h}^{2n+3/2}, \frac{e_{f,h}^{2n+2}+e_{f,h}^{2n+1}}{2}) + \frac{1}{2}\, c_f(j_{f,h}^{2n+1/2}, \frac{e_{f,h}^{2n+1}+e_{f,h}^{2n}}{2}). \tag{143}$$

There are two "natural" schemes reaching this objective:

- **(Scheme A)**. This first scheme is to compute the electric current j only with the coarse time step:

$$\begin{cases} j_{c,h}^{2n+1} = j_{f,h}^{2n+3/2} = j_{f,h}^{2n+1/2}, \qquad (\equiv j_h^{2n+1}) \\ c_c(\tilde{j}_h, \dfrac{e_{c,h}^{2n+2}+e_{c,h}^{2n}}{2}) = c_f(\tilde{j}_h \dfrac{e_{f,h}^{2n+2}+2e_{f,h}^{2n+1}+e_{f,h}^{2n}}{4}), \qquad \forall\, \tilde{j}_h \in L_h. \end{cases} \tag{144}$$

  In some sense, one writes two times the continuity of j (in an off-centred way) and one times the continuity of the tangential component of e (in a centred way).

- **(Scheme B)**. For this second scheme, we exchange the roles of e and j:

$$\begin{cases} c_f(\tilde{j}_h, \dfrac{e_{f,h}^{2n+2}+e_{f,h}^{2n+1}}{2}) = c_f(\tilde{j}_h, \dfrac{e_{f,h}^{2n+1}+e_{f,h}^{2n}}{2}) = c_c(\tilde{j}_h, \dfrac{e_{c,h}^{2n+2}+e_{c,h}^{2n}}{2}), \\ j_{c,h}^{2n+1} = \dfrac{1}{2}\left(j_{f,h}^{2n+3/2}+j_{f,h}^{2n+1/2}\right). \end{cases} \tag{145}$$

  This time one writes two times the continuity of the tangential component of e (in an off-centred way) and one times the continuity of j.

It can be shown that each scheme defines the discrete solution in a unique way (see next paragraph). At a first guess, both schemes seem to be first order accurate with respect to time. A deeper analysis shows (see Sect. 4.6) that it is not the case and the scheme A is more accurate than scheme B. That is why we shall restrict the rest of our presentation to scheme A.

Practical computations.

It is easy to see that the scheme (134),(135),(136) and (144) leads to the following algebraic problem:

$$\begin{cases} \mathcal{M}_f^E\, \dfrac{\mathbf{e}_f^{2n+1}-\mathbf{e}_f^{2n}}{\Delta t} - \mathcal{B}_f\, \mathbf{h}_f^{2n+1/2} - \mathcal{C}_f^*\, \mathbf{j}^{2n+1} = 0, \\ \mathcal{M}_f^H\, \dfrac{\mathbf{h}_f^{2n+1/2}-\mathbf{h}_f^{2n-1/2}}{\Delta t} + \mathcal{B}_f^*\, \mathbf{e}_f^{2n} = 0, \\ \mathcal{M}_f^E\, \dfrac{\mathbf{e}_f^{2n+2}-\mathbf{e}_f^{2n+1}}{\Delta t} - \mathcal{B}_f\, \mathbf{h}_f^{2n+3/2} - \mathcal{C}_f^*\, \mathbf{j}^{2n+1} = 0, \\ \mathcal{M}_f^H\, \dfrac{\mathbf{h}_f^{2n+3/2}-\mathbf{h}_f^{2n+1/2}}{\Delta t} + \mathcal{B}_f^*\, \mathbf{e}_f^{2n+1} = 0, \end{cases} \tag{146}$$

$$\left\{\begin{array}{l} \mathcal{M}_c^E \dfrac{\mathbf{e}_c^{2n+2} - \mathbf{e}_c^{2n}}{2\Delta t} - \mathcal{B}_c\, \mathbf{h}_c^{2n+1} + \mathcal{C}_c^*\, \mathbf{j}^{2n+1} = 0, \\[2ex] \mathcal{M}_c^H \dfrac{\mathbf{h}_c^{2n+1} - \mathbf{h}_c^{2n-1}}{2\Delta t} + \mathcal{B}_c^*\, \mathbf{e}_c^{2n} = 0, \end{array}\right. \tag{147}$$

$$\mathcal{C}_c \frac{\mathbf{e}_c^{2n+2} + \mathbf{e}_c^{2n}}{2} - \mathcal{C}_f \frac{\mathbf{e}_f^{2n+2} + 2\mathbf{e}_f^{2n+1} + \mathbf{e}_f^{2n}}{4} = 0. \tag{148}$$

Assuming that all the unknowns have been computed up to time t^{2n}, it is easy to see that, once $\mathbf{j}^{2n+1}$ is known, equations (146) and (147) allow us to compute explicitly

$$\mathbf{h}_f^{2n+1/2}, \mathbf{e}_f^{2n+1}, \mathbf{h}_f^{2n+3/2}, \mathbf{e}_f^{2n+2}, \mathbf{h}_c^{2n+1} \text{ and } \mathbf{e}_c^{2n+2}.$$

(Moreover $\mathbf{h}_f^{2n+1/2}$ and $\mathbf{h}_c^{2n+1}$ can be computed independently of $\mathbf{j}^{2n+1}$).

Equation (148) will be used to derive an equation allowing us to compute $\mathbf{j}^{2n+1}$. We show below that for each n, $\mathbf{j}^{2n+1}$ is computed by solving a symmetric positive linear system. Take the difference between (146)(3) and (146)(1) to eliminate $\mathbf{j}^{2n+1}$:

$$\mathcal{M}_f^E \frac{\mathbf{e}_f^{2n+2} - 2\mathbf{e}_f^{2n+1} + \mathbf{e}_f^{2n}}{\Delta t^2} - \mathcal{B}_f \frac{\mathbf{h}_f^{2n+1/2} - \mathbf{h}_f^{2n-1/2}}{\Delta t} = 0. \tag{149}$$

Thanks to (146)(2), we deduce

$$\mathcal{M}_f^E \frac{\mathbf{e}_f^{2n+2} - 2\mathbf{e}_f^{2n+1} + \mathbf{e}_f^{2n}}{\Delta t^2} + \mathcal{B}_f \left(\mathcal{M}_H^f\right)^{-1} \mathcal{B}_f^*\, \mathbf{e}_f^{2n+1} = 0, \tag{150}$$

that we can rewrite as

$$\frac{\mathbf{e}_f^{2n+2} + 2\mathbf{e}_f^{2n+1} + \mathbf{e}_f^{2n}}{4} = \left(I - \frac{\Delta t^2}{4} \left(\mathcal{M}_f^E\right)^{-1} \mathcal{B}_f \left(\mathcal{M}_f^H\right)^{-1} \mathcal{B}_f^*\right) \mathbf{e}_f^{2n+1}. \tag{151}$$

From (146)(1), we can write

$$\mathbf{e}_f^{2n+1} = T_f^{2n+1} + \left(\mathcal{M}_f^E\right)^{-1} \mathcal{C}_f^*\, \mathbf{j}^{2n+1} \tag{152}$$

where T_f^{2n+1} only depends on $\mathbf{e}_f^{2n}$, $\mathbf{h}_f^{2n+1}$ (and is thus a known quantity). Let us set:

$$\mathcal{Q}(\Delta t) = \left(\mathcal{M}_f^E\right)^{-1} - \frac{\Delta t^2}{4} \left(\mathcal{M}_f^E\right)^{-1} \mathcal{B}_f \left(\mathcal{M}_f^H\right)^{-1} \mathcal{B}_f^* \left(\mathcal{M}_f^E\right)^{-1}. \tag{153}$$

Substituting (152) into (151) leads to:

$$\frac{\mathbf{e}_f^{2n+2} + 2\mathbf{e}_f^{2n+1} + \mathbf{e}_f^{2n}}{4} = \mathcal{Q}(\Delta t)\, \mathcal{M}_f^E \left(T_f^{2n+1} + \left(\mathcal{M}_f^E\right)^{-1} \mathcal{C}_f^*\, \mathbf{j}^{2n+1}\right) \tag{154}$$

In the same way, from (147)(1) we deduce

$$\frac{\mathbf{e}_c^{2n+2} + \mathbf{e}_c^{2n}}{2} = T_c^{2n+1} - \left(\mathcal{M}_c^E\right)^{-1} \mathcal{C}_c^*\, \mathbf{j}^{2n+1} \tag{155}$$

where, once again, T_c^{2n+1} is known. Substituting (155) and (154) into (148) leads finally to:

$$\left(\mathcal{C}_c \left(\mathcal{M}_c^E\right)^{-1} \mathcal{C}_c^* + \mathcal{C}_f \mathcal{Q}(\Delta t) \mathcal{C}_f^*\right) \mathrm{j}^{2n+1} = T^{2n+1} \tag{156}$$

where

$$T^{2n+1} = \mathcal{C}_f \mathcal{M}_f^E \; \mathcal{Q}(\Delta t) \; T_f^{2n+1} - \mathcal{C}_c \; T_c^{2n+1},$$

is a known quantity. Finally solving our scheme during each interval $[t^{2n}, t^{2n+2}]$ consists of:

- a succession of explicit steps in each grid separately,
- the resolution of the linear system (156).

The existence and uniqueness of the discrete solution is guaranteed by the invertibility of the matrix of the symmetric system (156). It is easy to see that this matrix is positive definite as soon as the compatibility condition (130) is satisfied and the matrix $\mathcal{Q}(\Delta t)$ is positive definite. This last property is equivalent to the matrix inequality:

$$\frac{\Delta t^2}{4} \; \mathcal{B}_f \left(\mathcal{M}_f^H\right)^{-1} \mathcal{B}_f^* < \mathcal{M}_f^E$$

which is nothing but the (strict) CFL stability condition on the fine grid.

Application to the Yee scheme.

In [41] (see also [29] in the 1D case) , we apply the above procedure to derive a conservative space-time refinement scheme for the Yee scheme (I mean here that the Yee scheme is used as the interior scheme in each grid). According to Sect. 2.4, one obtains the Yee scheme by constructing the spaces $U_{f,h}$ and $V_{f,h}$ (respectively $U_{c,h}$ and $V_{c,h}$) as the spaces U_h and V_h in definition (70), using a regular cubic mesh of step-size h in Ω_f (respectively of step-size $2h$ in Ω_g). For the space J_h, we have chosen as the mesh $\mathcal{T}_h(\Sigma)$ of the interface Σ, the trace of the coarse grid mesh in Ω_c, i.e. a regular mesh made of squares C of side $2h$. Then, we use standard (tangential) Raviart-Thomas 2D elements, which provides a conforming subspace of $L = \mathbf{H}_{\|}^{-1/2}(\mathrm{div}_\Sigma, \Sigma)$:

$$L_h = \{ j_h \in L \;/\; \forall\, C \in \mathcal{T}_h(\Sigma), j_h|_C \in \mathcal{Q}_{1,0} \times \mathcal{Q}_{1,0} \} \tag{157}$$

The degrees of freedom in J_h are the fluxes (or equivalently the constant normal component) of the vector fields across (or on) each edge of the surface mesh. The global spatial discretisation is summarised in Fig. 6 on p. 239. It can easily be shown that:

$$\mathrm{Ker}\; \mathcal{C}_f^* = 0, \tag{158}$$

which ensures the existence and uniqueness of the discrete solution under the CFL condition (142). Moreover, one can compute that, in the case of an infinite mesh and constant coefficient, this condition is nothing but the well known stability condition for the Yee scheme namely:

$$\frac{c\Delta t}{h} < \frac{\sqrt{3}}{3}. \tag{159}$$

For more details and numerical results, I refer the reader to [41].

Remark 21. If one uses the trace of the fine grid mesh as the surface mesh, the property (158) is no longer true.

The Case Without Lagrange Multiplier

Construction of the conservative scheme.

We keep the notation of the previous section but this time we change the instants where the unknowns are computed in the fine grid (in which we exchange the roles of the two fields). The discrete unknowns are:

$$e_{f,h}^{n},\ h_{f,h}^{n+1/2} \quad \text{in } \Omega_f, \quad e_{c,h}^{2n+1},\ h_{c,h}^{2n} \quad \text{in } \Omega_c. \tag{160}$$

For the time interval $[t^{2n}, t^{2n+2}]$, we consider the following scheme:

$$\begin{cases} (\dfrac{e_{f,h}^{2n+1} - e_{f,h}^{2n}}{\Delta t},\, \tilde{e}_{f,h})_f - b_f(\tilde{e}_{f,h},\, h_{f,h}^{2n+1/2}) - c([h_{c,h}]^{2n+1/2},\, \tilde{e}_{f,h}) = 0, \\ (\dfrac{h_{f,h}^{2n+1/2} - h_{f,h}^{2n-1/2}}{\Delta t},\, \tilde{h}_{f,h})_f + b_f(e_{c,h}^{2n},\, \tilde{h}_{f,h}) = 0, \end{cases} \tag{161}$$

$$\begin{cases} (\dfrac{e_{f,h}^{2n+2} - e_{f,h}^{2n+1}}{\Delta t},\, \tilde{e}_{f,h})_f - b_f(\tilde{e}_{f,h},\, h_{f,h}^{2n+3/2}) - c([h_{c,h}]^{2n+3/2},\, \tilde{e}_{f,h}) = 0, \\ (\dfrac{h_{f,h}^{2n+3/2} - h_{f,h}^{2n+1/2}}{\Delta t},\, \tilde{h}_{f,h})_f + b_f(e_{c,h}^{2n+1},\, \tilde{h}_{f,h}) = 0, \end{cases} \tag{162}$$

$\forall\, \tilde{e}_{f,h} \in U_{f,h}, \forall\, \tilde{h}_{f,h} \in V_{f,h},$

$$\begin{cases} (\dfrac{e_{c,h}^{2n+1} - e_{c,h}^{2n-1}}{\Delta t},\, \tilde{e}_{c,h})_c - b_c(\tilde{e}_{c,h},\, h_{c,h}^{2n}) = 0, \\ (\dfrac{h_{c,h}^{2n+2} - h_{c,h}^{2n}}{\Delta t},\, \tilde{h}_{c,h})_c + b_c(e_{c,h}^{2n+1},\, \tilde{h}_{c,h}) + c([e_{f,h}]^{2n+1}, \tilde{h}_{c,h}) = 0, \end{cases} \tag{163}$$

$\forall\ \tilde{e}_{c,h} \in U_{c,h}, \forall\ \tilde{h}_{c,h} \in V_{c,h}$, where the quantities $[e_{f,h}]^{2n+1}$, $[h_{c,h}]^{2n+3/2}$ and $[h_{c,h}]^{2n+1/2}$ remain to be determined.

For this we are guided by an energy analysis. The total discrete energy is still defined by (140) but we have to change the definition of the coarse grid energy:

$$\begin{cases} E_{f,h}^{n} = \dfrac{1}{2}\left(\|e_{f,h}^{n}\|_f^2 + (h_{f,h}^{n+1/2}, h_{f,h}^{n+1/2})_f \right) \\ E_{c,h}^{2n} = \dfrac{1}{2}\left(\|h_{c,h}^{2n}\|_c^2 + (e_{c,h}^{n+1/2}, e_{c,h}^{n+1/2})_c \right). \end{cases} \tag{164}$$

Note however that the condition for the strict positivity of these two energies remains the double CFL condition (142). It is easy to obtain the following identity:

$$\frac{E_h^{2n+2}-E_h^{2n}}{2\Delta t} = -\frac{1}{2}\,c\,\Big(\frac{e_{f,h}^{2n+1}+e_{f,h}^{2n}}{2},\ [h_{c,h}]^{2n+1/2}\Big)$$
$$+\,c\,\Big(\,[e_{f,h}]^{2n+1},\ \frac{h_{c,h}^{2n+2}+\ h_{c,h}^{2n}}{2}\Big)-\frac{1}{2}\,c\,\Big(\frac{e_{f,h}^{2n+2}+e_{f,h}^{2n+1}}{2},\ [h_{c,h}]^{2n+3/2}\Big). \tag{165}$$

In particular we see that we obtain a conservative scheme if we choose:

$$\begin{cases} [e_{f,h}]^{2n+1} = \dfrac{e_{f,h}^{2n+2}+2e_{f,h}^{2n+1}+e_{f,h}^{2n}}{4}, \\[2mm] [h_{c,h}]^{2n+1/2} = [h_{c,h}]^{2n+3/2} = \dfrac{h_{c,h}^{2n+2}+h_{c,h}^{2n}}{2}. \end{cases} \tag{166}$$

The scheme ((161), (162), (163), (166)) is then L^2 stable under the strict CFL condition (142).

Practical computations.

The scheme can be written in a matrix form as:

$$\begin{cases} \mathcal{M}_f^E\ \dfrac{\mathbf{e}_f^{2n+1}-\mathbf{e}_f^{2n}}{\Delta t}+\mathcal{B}_f\ \mathbf{h}_f^{2n+1/2}-\mathcal{C}^*\ \dfrac{h_{c,h}^{2n+2}+h_{c,h}^{2n}}{2}=0\,, \\[2mm] \mathcal{M}_f^H\ \dfrac{\mathbf{h}_f^{2n+1/2}-\mathbf{h}_f^{2n-1/2}}{\Delta t}-\mathcal{B}_f^*\ \mathbf{e}_f^{2n}=0\,, \\[2mm] \mathcal{M}_f^E\ \dfrac{\mathbf{e}_f^{2n+2}-\mathbf{e}_f^{2n+1}}{\Delta t}+\mathcal{B}_f\ \mathbf{h}_f^{2n+3/2}-\mathcal{C}^*\ \dfrac{h_{c,h}^{2n+2}+h_{c,h}^{2n}}{2}=0\,, \\[2mm] \mathcal{M}_f^H\ \dfrac{\mathbf{h}_f^{2n+3/2}-\mathbf{h}_f^{2n+1/2}}{\Delta t}-\mathcal{B}_f^*\ \mathbf{e}_f^{2n+1}=0\,, \end{cases} \tag{167}$$

$$\begin{cases} \mathcal{M}_c^E\ \dfrac{\mathbf{e}_c^{2n+1}-\mathbf{e}_c^{2n-1}}{2\Delta t}+\mathcal{B}_c\ \mathbf{h}_c^{2n}=0\,, \\[2mm] \mathcal{M}_c^H\ \dfrac{\mathbf{h}_c^{2n+2}-\mathbf{h}_c^{2n}}{2\Delta t}-\mathcal{B}_c^*\ \mathbf{e}_c^{2n+1}+\mathcal{C}\ \dfrac{e_{f,h}^{2n+2}+2e_{f,h}^{2n+1}+e_{f,h}^{2n}}{4}=0\,. \end{cases} \tag{168}$$

In practice, the only non zero lines of the coupling matrix $\mathcal{C}$ will correspond to the degrees of freedom of which are associated with the boundary Σ. This means that the computation of the interior degrees of freedom remains completely explicit while the computation of the degrees of freedom associated with Σ requires the solution of a linear system. Typically, the linear system to be solved to compute $h_{c,h}^{2n+2}$ can be written (the computations are left to the reader):

$$(\mathcal{M}_c^H+\mathcal{C}^*\ \mathcal{Q}(\Delta t)\ \mathcal{C})\ h_{c,h}^{2n+2}=T, \tag{169}$$

where T is a known quantity and where $\mathcal{Q}(\Delta t)$ is the symmetric matrix defined by (153). In practice, the matrix: $\mathcal{C}^* \ \mathcal{Q}(\Delta t) \ \mathcal{C}$ has only a small non zero block associated with the degrees of freedom of the coarse grid magnetic field which are located on the interface Σ. That is why the scheme remains essentially explicit. Moreover, since the matrix $\mathcal{Q}(\Delta t)$ is positive definite under the CFL condition, the system (169) is invertible (independently of the properties of C).

Remark 22. In the case when one wants to apply the Yee scheme inside each grid one naturally exchanges the roles of the magnetic and electric fields as it is illustrated in Fig. 7 on p. 239.

4.6 About the Error Analysis

General Considerations

Up to now, there are very few complete results about the error analysis of the mesh refinement methods that we presented above. That is why we shall restrict ourselves to giving some insights about what such an analysis could be and indicate some directions of research (some of them being already under way). In fact, it is natural to distinguish two sources of errors:

- the space discretisation and the change of space step between Ω_c and Ω_f,
- the time stepping and the change of time step between Ω_c and Ω_f.

To analyse the error due to the space discretisation, it suffices to look at the semi-discrete problems of Sect. 4.4:

- The case of the "multiplier free" method is much simpler in the sense that this method can be considered as a conforming method at the interface Σ – the natural continuous spaces involved in the variational formulation accept discontinuous functions across Σ, and the two continuity conditions are natural conditions contained in the variational formulation itself. In such a case, standard results such as the ones of Sect. 2.3 can be applied.
- The case of the method with multiplier is more delicate since this time we have a "non conforming" method at the interface in the sense that one of the two continuity conditions across Σ is an essential one and appears in the continuous spaces for the variational formulation. The introduction of the Lagrange multiplier allows us to take into account in a weak way this continuity condition, but also introduces a non conformity at the interface of the numerical method. However, in this case all the methodology developed for the analysis of the mortar element method, which relies either on the second Strang's lemma (in this case the Lagrange multiplier no longer appears in the analysis, [13, 15]) or the mixed finite element technology ([12]), should be applicable. This remains however an open question. In particular, the role of the choice of the space L_H for the Lagrange multiplier (which is of obvious importance at least for the uniqueness of the discrete solution through the discrete inf-sup conditions) on the quality of the treatment of the numerical interface should be clarified.

The error due to the time discretisation appears to be a more original issue. It has already been investigated in the case of the 1D wave equation for which the question of the space approximation does not pose any difficulty, even in the case of the method with Lagrange multiplier. We summarise the results of this analysis in the next section.

The Case of the 1D Wave Equation

We consider here a very simple 1D model problem on the real line:

$$\frac{\partial u}{\partial t}+\frac{\partial v}{\partial x}=0, \qquad \frac{\partial v}{\partial t}+\frac{\partial u}{\partial x}=0, \qquad x\in\mathbb{R}, \quad t>0\,. \tag{170}$$

The two domains Ω_c and Ω_f are respectively the two half spaces $\{x<0\}$ and $\{x>0\}$ the 1D equivalent of the 3D transmission problem of Sect. 4.2 is simply:

$$\begin{cases}\dfrac{\partial u_c}{\partial t}+\dfrac{\partial v_c}{\partial x}=0,\\[2mm] \dfrac{\partial v_c}{\partial t}+\dfrac{\partial u_c}{\partial x}=0,\end{cases}\text{in } \Omega_c, \qquad \begin{cases}\dfrac{\partial u_f}{\partial t}+\dfrac{\partial v_f}{\partial x}=0,\\[2mm] \dfrac{\partial v_f}{\partial t}+\dfrac{\partial u_f}{\partial x}=0,\end{cases}\text{in } \Omega_f, \tag{171}$$

with the (artificial) interface conditions

$$u_c(0,t)=u_f(0,t), \qquad v_c(0,t)=v_f(0,t). \tag{172}$$

We apply the following procedure: in each grid, we use the primal-dual formulation with u in H^1 and v in L^2. The Lagrange multiplier λ is simply a real number. For the space discretisation, we use two regular meshes of respective sizes h in Ω_f and $2h$ in Ω_c, continuous P_1 finite elements for u and piecewise constant elements for v. For the time discretisation we use the staggered grid procedure of Sect. 2.4 with time step Δt in Ω_f and $2\Delta t$ in Ω_c. With obvious notation, the unknowns of the scheme are simply:

$$\begin{cases} u_{c,2j}^{2n}, \quad v_{c,2j-1}^{2n+1}, \quad j\le 0, \quad n\ge 0, \quad \text{in } \quad \Omega_c,\\[2mm] u_{f,j}^{n}, \quad v_{f,j+1/2}^{n+1/2}, \quad j\ge 0, \quad n\ge 0, \quad \text{in } \quad \Omega_f. \end{cases} \tag{173}$$

To understand the scheme, it is sufficient to look at Fig. 12. In each interval of time $[t^{2n},t^{2n+2}]$, the interior leap-frog scheme permits computation of the discrete solution everywhere except the three "interface values" at $x=0$ (inside the small boxes in Fig. 12), namely:

$$u_{c,0}^{2n+2}, u_{f,0}^{2n+1}, u_{f,0}^{2n+2}. \tag{174}$$

The use of the transmission scheme (A), or (144), provides, after elimination by hand of the Lagrange multiplier j^{2n+1}, the three missing equations:

$$\begin{cases} v_{c,-1}^{2n+1}-\dfrac{h}{2\Delta t}\left(u_{c,0}^{2n+2}-u_{c,0}^{2n}\right)=v_{f,1/2}^{2n+1/2}+\dfrac{h}{2\Delta t}(u_{f,0}^{2n+1}-u_{f,0}^{2n})\\[4mm] v_{c,-1}^{2n+1}-\dfrac{h}{2\Delta t}\left(u_{c,0}^{2n+2}-u_{c,0}^{2n}\right)=v_{f,1/2}^{2n+3/2}+\dfrac{h}{2\Delta t}(u_{f,0}^{2n+2}-u_{f,0}^{2n+1})\\[4mm] \dfrac{u_{f,0}^{2n+2}+2u_{f,0}^{2n+1}+u_{f,0}^{2n}}{4}=\dfrac{\left(u_{c,0}^{2n+2}+u_{c,0}^{2n}\right)}{2}. \end{cases} \tag{175}$$

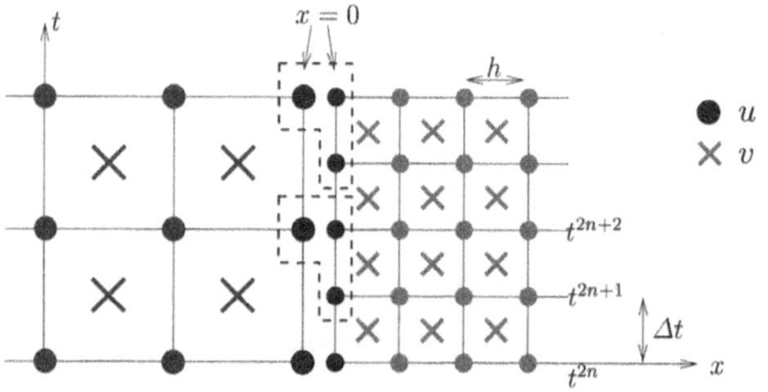

Fig. 12. The new unknowns: large (resp. small) crosses correspond to v_c (resp. v_f), large (resp. small) circles correspond to u_c (resp. u_f)

If instead one uses scheme (B) (i.e. (145)), one obtains:

$$\left\{\begin{array}{l} v_{c,-1}^{2n+1} - \dfrac{h}{2\Delta t}\left(u_{c,0}^{2n+2} - u_{c,0}^{2n}\right) = \dfrac{v_{f,1/2}^{2n+1/2} + v_{f,1/2}^{2n+3/2}}{2} + \dfrac{h}{2\Delta t}(u_{f,0}^{2n+2} - u_{f,0}^{2n}) \\ \dfrac{\left(u_{c,0}^{2n+2} + u_{c,0}^{2n}\right)}{2} = \dfrac{u_{f,0}^{2n+2} + u_{f,0}^{2n+1}}{2}, \\ \dfrac{\left(u_{c,0}^{2n+2} + u_{c,0}^{2n}\right)}{2} = \dfrac{u_{f,0}^{2n+1} + u_{f,0}^{2n}}{2}. \end{array}\right. \tag{176}$$

One can perform the L^2 error analysis of both schemes ([50]). We shall assume that the ratio:

$$\alpha = \frac{\Delta t}{h}, \tag{177}$$

is maintained constant, strictly smaller than 1 (which corresponds to the L^2 stability condition (142)). To state our result, let us introduce the discrete L^2 norm of the error committed on the variable u at time t^{2n}:

$$\left\{\begin{array}{l} \|u(\cdot,t^{2n}) - u_h^{2n}\|_h^2 = \|e_{c,h}^{2n}\|_h^2 + \|e_{f,h}^{2n}\|_h^2 \\ \|e_{c,h}^{2n}\|_h^2 = \displaystyle\sum_{j\le -1} |e_{c,h,2j}^{2n}|^2\, 2h + |e_{c,h,0}^{2n}|^2\, h, \\ \|e_{f,h}^{2n}\|_h^2 = |e_{f,h,0}^{2n}|^2\, \dfrac{h}{2} + \displaystyle\sum_{j\ge 1} |e_{f,h,j}^{2n}|^2\, h\,, \end{array}\right. \tag{178}$$

where $e_{c,h}^{2n}$ and $e_{f,h}^{2n}$ are the errors on the coarse and fine grids at time t^{2n}. We shall measure the error in a discrete $L^\infty(0,T,L^2)$ norm, namely:

$$\|u - u_h\|_{h,T} = \sup_{t^{2n}\le T} \|u(\cdot,t^{2n}) - u_h^{2n}\|_h\,. \tag{179}$$

Analogously, we can define a discrete $L^\infty(0,T,L^2)$ norm for the error $v - v_h$ on the unknown v.

Theorem 2. *Assume that the exact solution* (u, v) *has the regularity:*

$$(u, v) \in C^3(0, T; L^2), \tag{180}$$

then with scheme (B), one has the error estimate:

$$\|u - u_h\|_{h,T} + \|v - v_h\|_{h,T} \le C\ (1 - \alpha^2)^{-1/2}\ h^{1/2}\ \|(u, v)\|_{C^3(0,T;L^2)}. \tag{181}$$

Assume that the exact solution (u, v) *has the regularity:*

$$(u, v) \in C^{3+k}(0, T; L^2), \quad k \ge 0, \tag{182}$$

then with scheme (A), one has the error estimate:

$$\|u - u_h\|_{h,T} + \|v - v_h\|_{h,T} \le C_k\ (1 - \alpha^2)^{-1/2}\ h^{p_k}\ \|(u, v)\|_{C^3(0,T;L^2)} \tag{183}$$

where $p_k = 3/2 - 1/2^k$.

It is interesting to complete this result with some comments.

- The proof of this theorem is not completely standard. Let us give the main ideas. The errors $u - u_h$ and $v - v_h$ satisfy the same interior and transmission schemes as u_h and v_h except that truncation errors appear at the right hand sides:
  - an interior truncation error which is $O(h^2)$
  - an interface truncation error, which is located at the numerical interface and is $O(h)$, due to the off-centred nature of the discrete transmission conditions (175) or (176).

  Thus a priori we expect an $O(h)$ error estimate. However, in order to apply the same type of energy analysis as in Sect. 2.4, and more particularly to handle the interface truncation error, we have to use some kind of discrete trace estimate, typically:

  $$|u^{2n}_{c,0}| \le h^{-1/2}\ \|u^{2n}_{c,h}\|_h\,. \tag{184}$$

  This drops an additional half power of h in the estimate obtained by the energy analysis which thus provides (for both schemes):

  $$\|u - u_h\|_{h,T} + \|v - v_h\|_{h,T} = O(h^{1/2}). \tag{185}$$

  The proof stops here with scheme (176) for which it can be proven that the $O(h^{1/2})$ estimate is optimal (see the next item). With scheme (175), one can improve the estimate thanks to a boot strap argument (this is the most delicate point) which exploits the precise nature of the interface truncation error associated with scheme (175). One can show (more details will be given in [50]) that, if $(u, v) \in C^{3+k}(0, T; L^2)$:

  $$\|u - u_h\|_{h,T} + \|v - v_h\|_{h,T} = O(h^{p_{k-1}}) \Longrightarrow \|u - u_h\|_{h,T} + \|v - v_h\|_{h,T} = O(h^{p_k}), \tag{186}$$

  where $p_k = 1/2\ p_{k-1} + 3/4$ which yields $p_k = 3/2 - 1/2^k$.

- Passing from a discrete $L^\infty(0,T;L^2)$ norm to an $L^\infty(0,T;L^\infty)$, one looses one half power of h in the error estimates. This means that the method is first order accurate in L^∞ with scheme (175), but that maybe not convergent in L^∞ with scheme (176). This is in fact clearly the case since, from (176), one deduces in particular that:
$$u_{c,0}^{2n+2} = u_{c,0}^{2n},$$
In other words the trace on the interface of the coarse grid solution is constant at even instants (which is, of course, not the case of the exact solution). This also shows the optimality of the $O(h^{1/2})$ L^2-estimate for scheme (176)
- The result of the theorem essentially says that the method is of order $1/2$ with scheme (176) and of order $3/2$ (take k arbitrarily large in (183)) with scheme (175). In both cases, one loses some accuracy with respect to the interior scheme which is second order accurate. A challenging and still open question is to find more accurate conservative transmission schemes.
- The constants appearing in the error estimates (181) and (183) blow up when α tends to 1, similarly to what we obtain with the estimate (58) (see Sect. 2.4 and remark 10). It is not possible here to improve the result by applying a technique similar to the one that we presented in the remark 10: indeed, due to the change of time step between the two grids, the scheme is not invariant by translation of Δt. Moreover, the numerical experiments indicate that the method is not strongly convergent when $\alpha = 1$ (however, we conjecture that, in this case, the discrete solution converges weakly in L^2 to the continuous one).
- In [30], we made a different analysis which completes (and some how confirms) the results of Theorem 2. This analysis consists of looking at what happens to an incident harmonic wave of amplitude 1 and frequency ω in the coarse grid. This incident wave gives rise to:
  - a reflected wave of frequency ω and amplitude R in the fine grid.
  - a transmitted wave of frequency ω and amplitude T in the fine grid.
  - a parasitic transmitted wave of frequency $\omega + \pi/\Delta t$ in the fine grid. This wave is evanescent, i.e. exponentially decaying with the distance to the interface, and has an amplitude T' on the interface. The penetration depth is proportional to h.

Remark 23. The existence of the high frequency parasitic wave is due to the so-called aliasing phenomenon: the two frequencies ω are $\omega + \pi/\Delta t$ are "equal" for the coarse grid but distinct for the fine grid.

An asymptotic analysis for small h shows that:
- With scheme (175), $R = 0(h^2), T = 1 - 0(h^2), T' = 0(h)$.
- With scheme (176), $R = 0(h^2), T = 1 - 0(h^2), T' = 0(1)$.

This illustrates once again the superiority of scheme (175) (or scheme (A)) with respect to scheme (176) (or scheme (B)). It also confirms in some sense the optimality of the estimates (181) and (183). Roughly speaking, the amplitude of the parasitic wave is a function of the form (γ is a given positive constant):

$$T' \exp(-\gamma \frac{x}{h}), \quad x > 0,$$

and the L^2 norm of such a function is equal to $|T'| \sqrt{h/\gamma}$ which gives $O(h^{1/2})$ with scheme (176) and $O(h^{3/2})$ with scheme (175).

Acknowledgements. I would like to thank my collaborators Eliane Bécache, Gary Cohen, Francis Collino, Julien Diaz, Grégoire Derveaux, Thierry Fouquet, Sylvain Garcès, Florence Millot, Leila Rhaouti, Jeronimo Ródriguez, Chrysoula Tsogka for their collaboration and their essential contributions to some of the works presented in this article. I also thank particularly Jeronimo Ródriguez and Houssem Haddar for their careful reading of the manuscript and the improvements they have proposed.

References

1. D.N. Arnold, F. Brezzi and J. Douglas. PEERS: A new mixed finite element for plane elasticity. *Japan J. Appl. Math.*, 1:347–367, 1984.
2. G.P. Astrakmantev. Methods of fictitious domains for a second order elliptic equation with natural boundary conditions. *U.S.S.R Comp. Math. and Math. Phys.*, 18:114–221, 1978.
3. C. Atamian, R. Glowinski, J. Périaux, H. Stève and G. Terrason. Control approach to fictitious domain in electro-magnetism. *Conférence sur l'approximation et les methodes numeriques pour la resolution des equations de Maxwell, Hotel Pullmann, Paris*, 1989.
4. C. Atamian and P. Joly. An analysis of the method of fictitious domains for the exterior Helmholtz problem. *RAIRO Modèl. Math. Anal. Numér*, 27(3):251–288, 1993.
5. I. Babúska. The Finite Element Method with Lagrangian Multipliers. *Numer. Math.*, 20:179–192, 1973.
6. G.A. Baker and V.A. Dougalis. The Effect of Quadrature Errors on Finite Element Approximations for the Second-Order Hyperbolic Equations. *SIAM. Journ. Numer. Anal.*, 13(4):577–598, 1976.
7. A. Bamberger, R. Glowinski and Q. H. Tran. A domain decomposition method for the acoustic wave equation with discontinuous coefficients and grid change. *SIAM J. on Num. Anal.*, 34(2):603–639, April 1997.
8. E. Bécache, P. Joly and G. Scarella. A fictitious domain method for unilateral contact problems in non destructive testing. In *Computational and Fluid ans Solid Mechanics*, pp. 65–67, 2001.
9. E. Bécache, P. Joly and C. Tsogka. An analysis of new mixed finite elements for the approximation of wave propagation problems. *SIAM J. Numer.Anal.*, 37(4):1053–1084, 2000.
10. E. Bécache, P. Joly and C. Tsogka. Fictitious domains, mixed FE and PML for 2-D elastodynamics. *Journal of Computational Acoustics*, 9(3):1175–1201, 2001.
11. E. Bécache, P. Joly and C. Tsogka. A new family of mixed finite elements for the linear elastodynamic problem. *SIAM J. Numer.Anal.*, 39(6):2109–2132, 2002.
12. F. Ben Belgacem. The mortar finite element method with Lagrange multipliers. *Numer. Math.*, 84(2):173–197, 1999.

13. F. Ben Belgacem, A. Buffa and Y. Maday. The mortar finite element method for 3d maxwell equations: First results. *SIAM J. Numer. Anal.*, 39(3):880–901, 2001.
14. A. Bendali. *Approximation par éléments finis de surface de problèmes de diffraction des ondes électromagnétiques*. PhD thesis, *Université Paris VI*, 1984.
15. C. Bernardi and Y. Maday. Raffinement de maillage en éléments finis par la méthode des joints. *C. R. Acad. Sci. Paris Sér. I Math.*, 320(3):373–377, 1995.
16. F. Brezzi and M. Fortin. *Mixed and Hybrid Finite Element Methods*. Springer Series in Computational Mathematics (15). Springer Verlag, 1991.
17. A. Buffa and P. Ciarlet Jr. On traces for functional spaces related to Maxwell's equations. I. an integration by parts formula in Lipschitz polyhedra. *Mathematical Methods in the Applied Sciences*, 24(1):9–30, 2001.
18. A. Chaigne, P. Joly and L. Rhaouti. Time-domain modeling and numerical simulation of a kettledrum. *Journal of the Acoustical Society of America*, 6:3545–3562, 1999.
19. M.W. Chevalier and R.J. Luebbers. FDTD local grid with material traverse. *IEEE Transactions on Antennas and Propagation*, 45(3):411–421, March 1997.
20. M.J.S. Chin-Joe-Kong, W.A. Mulder and M. Van Veldhuizen. Higher-order triangular and tetrahedral finite elements with mass lumping for solving the wave equation. *J. Engrg. Math.*, 35(4):405–426, 1999.
21. P.G. Ciarlet. *The finite element method for elliptic problems*. Reprinted by SIAM in 2002, original by North-Holland, 1982.
22. M. Clemens, P. Thoma, T. Weiland and U. van Rienen. Computational electromagnetic-field calculation with the finite-integration method. *Surveys Math. Indust.*, 8(3-4):213–232, 1999.
23. P. Clément. Approximation by finite element functions using local regularization. *Rev. Fran. Automat. Informat. Rech. Opér. Sér. Rouge Anal. Numér.*, 9:77–84, 1975.
24. G. Cohen. *Higher-order numerical methods for transient wave equations*. Scientific Computation, 2002.
25. G. Cohen, P. Joly and N. Torjman. Higher-order finite elements with mass-lumping for the 1D wave equation. *Finite Elements in Analysis and Design*, 16:329–336, 1994.
26. G. Cohen, P. Joly, N. Torjman and J. Roberts. Higher order triangular finite elements with mass lumping for the wave equation. *SIAM J. Numer. Anal.*, 38(6):2047–2078, 2001.
27. G. Cohen and P. Monk. Gauss point mass lumping schemes for Maxwell's equations. *Numer. Methods Partial Differential Equations*, 14(1):63–88, 1998.
28. G. Cohen and P. Monk. Mur-Nédélec finite element schemes for Maxwell's equations. *Comput. Methods Appl. Mech. Engrg.*, 169(3-4):197–217, 1999.
29. F. Collino, P. Joly and T. Fouquet. A Conservative Space-time Mesh Refinement Method for the 1-D Wave Equation. Part I : Construction. *Num. Math.*, To appear.
30. F. Collino, P. Joly and T. Fouquet. A Conservative Space-time Mesh Refinement Method for the 1-D Wave Equation. Part II : Analysis. *Num. Math.*, To appear.
31. F. Collino, P. Joly and S. Garcès. A fictitious domain method for conformal modeling of the perfect electric conductors in the FDTD method. *IEEE Trans. on Antennas and Propagation*, 46:1519–1527, 1997.
32. F. Collino, P. Joly and F. Millot. Fictitious domain method for unsteady problems: Application to electromagnetic scattering. *Journal of Comp. Physics*, 138:907–938, 1997.
33. G. Derveaux. *Modélisation Numérique de la Guitare Acoustique*. PhD thesis, *Ecole Polytechnique* , 2002.
34. T. Dupont. L^2 estimates or Galerkin methods for second-order hyperbolic equations. *SIAM J. Numer.Anal.*, 10:880–889, 1973.
35. A. Elmkiès. *Eléments finis et condensation de masse pour les équations de Maxwell*. PhD thesis, Université Paris XI, Orsay, June 1998.

36. A. Elmkiès and P. Joly. Eléments finis et condensation de masse pour les équations de Maxwell : le cas 2D. *C.R. Acad. Sci. Paris*, t. 324(Série I):1287–1292, 1997.
37. A. Elmkiès and P. Joly. Eléments finis et condensation de masse pour les équations de Maxwell : le cas 3D. *C.R. Acad. Sci. Paris*, t. 325(Série I):1217–1222, 1997.
38. B. Engquist and A. Majda. Absorbing boundary conditions for the numerical simulation of waves. *Math. of Comp.*, 31:629–651, 1977.
39. S.A Finogenov and Y.A. Kuznetsov. Two stage fictitious components methods for solving the Dirichlet boundary value problem. *Sov. J. Num. Anal. Math. Modelling*, 3:301–323, 1988.
40. G.J. Fix. Effects of quadrature errors in finite element approximation of steady state, eigenvalue and parabolic problems. In *The Mathematical Fondations of the Finit Element Method with Applications to Partial Differential Equations*, pp. 525–556, 1972.
41. T. Fouquet. *Raffinement de maillage spatio-temporel pour les équations de Maxwell.* PhD thesis, Université Paris IX Dauphine, June 2000.
42. S. Garcés. *Application des méthodes de domaines fictifs à la modélisation des structures rayonnantes tridimensionnelles Etude mathématique et numérique d'un modèle.* PhD thesis, *ENSAE (Toulouse)*, 1997.
43. T. Geveci. On the application of mixed finite element methods to the wave equations. *RAIRO Modél. Math. Anal. Numér.*, 22(2):243–250, 1988.
44. V. Girault and R. Glowinski. Error analysis of a fict. domain method applied to a Dirichlet problem. *Japan J. Ind. Appl. Math.*, 12(3):487–514, 1995.
45. R. Glowinski, T.W. Pan, and J. Periaux. A fictitious domain method for Dirichlet problem and applications. *Comp. Meth. in Appl. Mech. and Eng.*, 111:283–303, 1994.
46. R. Glowinski, T.W. Pan and J. Périaux. A fictitious domain method for external incompressible viscous flow modeled by Navier-Stokes equations. *Comp. Meth. Appl. Mech. Eng.*, 112:133–148, 1994.
47. P. Havé, M. Kern and C. Lemuet. Résolution numérique de l'équation des ondes par une méthode d'éléments finis d'ordre élevé sur calculateur parallèle. Technical Report 4381, I.N.R.I.A., Février 2002.
48. P. Joly and C. Poirier. A new second order edge element on tetrahedra for time dependent maxwell's equations. In *Mathematical and Numerical Aspects of Wave Propagation*, pp. 842–847, 2000.
49. P. Joly and L. Rhaouti. Domaines fictifs, éléments finis $h(div)$ et condition de neumann: le problème de la condition inf-sup. *C. R. Acad. Sci. Paris, Série I*, 328:1225–1230, 1999.
50. P. Joly and J. Ródriguez. l^2 error analysis of space-time mesh refinement schemes for the 1d wave equation. *in preparation.*
51. I. S. Kim and W. J. R. Hoefer. A local mesh refinement algorithm for the time-domain finite-difference method to solve maxwell's equations. *IEEE Trans. Microwave Theory Tech.*, 38(6):812–815, June 1990.
52. K.S. Kunz and L. Simpson. A technique for increasing the resolution of finite-difference solutions to the maxwell equations. *IEEE Trans. Electromagn. Compat.*, EMC-23:419–422, Nov. 1981.
53. J.L. Lions and E. Magenès. *Problèmes aux Limites non Homogènes et Applications.* Dunod, 1968.
54. Ch.G. Makridakis. On mixed finite element methods for linear elastodynamics. *Numer. Math.*, 61(2):235–260, 1992.
55. G.I. Marchuk, Y.A. Kuznetsov and A.M. Matsokin. Fictitious domain and domain decomposition methods. *Sov. J. Num. Anal. Math. Modelling*, 1:3–35, 1986.
56. P. Monk. Analysis of a finite element method for Maxwell's equations. *SIAM J. Numer.Anal.*, 29(3):714–729, 1992.

57. P. Monk. Sub-gridding FDTD schemes. *ACES Journal*, 11:37–46, 1996.
58. P. Monk and O. Vacus. Error estimates for a numerical scheme for ferromagnetic problems. *SIAM J. Numer.Anal.*, 36(3):696–718, 1999.
59. J.C. Nédélec. Mixed finite elements in $\mathbb{R}^3$. *Num. Math.*, 35:315–341, 1980.
60. J.C. Nédélec. A new family of mixed finite elements in $\mathbb{R}^3$. *Num. Math.*, 50:57–81, 1986.
61. V. Petkov. *Scattering theory for hyperbolic operators.* Studies in Mathematics and its Applications, 1989.
62. S. Piperno, M. Remaki and L. Fezoui. A nondiffusive finite volume scheme for the three-dimensional Maxwell's equations on unstructured meshes. *SIAM J. Numer. Anal.*, 9(6):2089–2108, 2002.
63. D.T. Prescott and N.V. Shuley. A method for incorporating different sized cells into the finite-difference time-domain analysis technique. *IEEE Microwave Guided Wave Lett.*, 2:434–436, Nov. 1992.
64. P.A. Raviart. The use of numerical integration in finite element methods for parabolic equations. In *Topics in Numerical Analysis Proc.*, pp. 233–264, 1973.
65. A. Taflove. *Computational Electrodynamics, The Finite-Difference Time Domain method.* Artech House, London, 1995.
66. A. Ben Haj Yedder, E. Bécache, P. Joly and A. Komech. A fictitious domain approach to the scattering of waves by mobile obstacles. *in preparation.*
67. K.S. Yee. Numerical solution of initial boundary value problems involving Maxwell's equations in isotropic media. *IEEE Trans. on Antennas and propagation*, pp. 302–307, 1966.

Some Numerical Techniques for Maxwell's Equations in Different Types of Geometries

Bengt Fornberg*

University of Colorado, Department of Applied Mathematics, 526 UCB, Boulder, CO 80309, USA
fornberg@colorado.edu

Summary. Almost all the difficulties that arise in finite difference time domain solutions of Maxwell's equations are due to material interfaces (to which we include objects such as antennas, wires, etc.) Different types of difficulties arise if the geometrical features are much larger than or much smaller than a typical wave length. In the former case, the main difficulty has to do with the spatial discretisation, which needs to combine good geometrical flexibility with a relatively high order of accuracy. After discussing some options for this situation, we focus on the latter case. The main problem here is to find a time stepping method which combines a very low cost per time step with unconditional stability. The first such method was introduced in 1999 and is based on the ADI principle. We will here discuss that method and some subsequent developments in this area.

Key words: Maxwell's equations, FDTD, ADI, split step, pseudospectral methods, finite differences, spectral elements.

1 Introduction

The main difficulties that arise when solving Maxwell's equations with finite differences usually come from the (often irregular) shapes of material interfaces. There are two different length scales present in CEM (computational electromagnetics) problems:

- the size of geometrical features, and
- a typical wave length.

In many problems the two length scales are of comparable size. The (partly conflicting) goals that then need to be met by an effective numerical method include:

- good geometric flexibility (to allow for interfaces with corners or with high curvatures);
- high order of spatial accuracy (to keep the number of points per wavelength low);

* The work was supported by NSF grants DMS-0073048, DMS-9810751 (VIGRE), and also by a Faculty Fellowship from the University of Colorado at Boulder.

- guaranteed (conditional) time stepping stability; and
- low computational cost.

There are also many important applications in which the first length scale (the size of geometrical features) is far smaller than a typical wave length – maybe by five orders of magnitude, or more. Examples of such situations include the interactions between components on an integrated circuit, the effect of cellular phone signals on brain cells, and those of a lightning strike on an aircraft. In order to capture the geometry, we then need to use grids which (at least in some areas) feature an extremely high number of points per wavelength (PPW). High formal order of accuracy in the spatial discretisation is then less critical. On the other hand, the method to advance in time should feature:

- explicit (or effectively explicit) time stepping (since grids tend to be extremely large); and
- the complete absence of any CFL-type stability condition (since such conditions would force time step sizes many orders of magnitude smaller than what is needed in order to accurately resolve the wave).

The first time stepping method that met both these criteria was introduced in 1999 [46]. Since then, a second (different, but related) method has been proposed [29]. Furthermore, both of these methods have been enhanced to feature higher than second order of accuracy in time [30]. As their only nontrivial step, they both require the solution of tridiagonal linear systems. One of the approaches is based on the alternating direction implicit method (ADI), and the other one on a split step (SS) concept.

In this article, we will first state the 3D Maxwell's equations (formulated in 1873 by James Clark Maxwell, [32]), and then summarise the classical Yee scheme [44]. Following that, we will discuss the issues that have been mentioned above. The main focus of this review will be on fast and unconditionally stable time stepping procedures.

2 Maxwell's Equations and the Yee Scheme

Assuming no free charges or currents, the 3D Maxwell's equations can be written

$$\left\{\begin{array}{ll} \dfrac{\partial E_x}{\partial t} = \dfrac{1}{\varepsilon}\left(\dfrac{\partial H_z}{\partial y} - \dfrac{\partial H_y}{\partial z}\right) & \dfrac{\partial H_x}{\partial t} = \dfrac{-1}{\mu}\left(\dfrac{\partial E_z}{\partial y} - \dfrac{\partial E_y}{\partial z}\right) \\ \dfrac{\partial E_y}{\partial t} = \dfrac{1}{\varepsilon}\left(\dfrac{\partial H_x}{\partial z} - \dfrac{\partial H_z}{\partial x}\right) & \dfrac{\partial H_y}{\partial t} = \dfrac{-1}{\mu}\left(\dfrac{\partial E_x}{\partial z} - \dfrac{\partial E_z}{\partial x}\right) \\ \dfrac{\partial E_z}{\partial t} = \dfrac{1}{\varepsilon}\left(\dfrac{\partial H_y}{\partial x} - \dfrac{\partial H_x}{\partial y}\right) & \dfrac{\partial H_z}{\partial t} = \dfrac{-1}{\mu}\left(\dfrac{\partial E_y}{\partial x} - \dfrac{\partial E_x}{\partial y}\right) \end{array}\right. \tag{1}$$

where E_x, E_y, E_z and H_x, H_y, H_z denote the components of the electric and magnetic fields respectively. The permittivity ε and permeability μ will in general depend on the spatial location within the medium. If these electric and magnetic fields (multiplied by ε and μ respectively) start out divergence free, they will remain so when advanced forward in time by (1):

$$\begin{aligned}\frac{\partial}{\partial t}(\operatorname{div}(\varepsilon E)) &= \frac{\partial}{\partial t}\left(\frac{\partial(\varepsilon E_x)}{\partial x}+\frac{\partial(\varepsilon E_y)}{\partial y}+\frac{\partial(\varepsilon E_z)}{\partial z}\right)\\ &= \frac{\partial}{\partial x}\left(\frac{\partial H_z}{\partial y}-\frac{\partial H_y}{\partial z}\right)+\frac{\partial}{\partial y}\left(\frac{\partial H_x}{\partial z}-\frac{\partial H_z}{\partial x}\right)+\frac{\partial}{\partial z}\left(\frac{\partial H_y}{\partial x}-\frac{\partial H_x}{\partial y}\right)\\ &= 0\end{aligned} \tag{2}$$

and similarly for div (μH). This implies that the relations div $(\varepsilon E) = \rho$ (where ρ is the local charge density) and div $(\mu H) = 0$ need not to be imposed as additional constraints. Neither of these quantities will change during wave propagation according to (1).

Arguably, the simplest possible finite difference approximation to (1) is obtained by approximating each derivative (whether in space or time) by centred second order accurate finite differences, i.e.

$$\begin{aligned}
\frac{E_x|_{i,j,k}^{n+1}-E_x|_{i,j,k}^{n-1}}{2\Delta t} &= \frac{1}{\varepsilon}\left(\frac{H_z|_{i,j+1,k}^{n}-H_z|_{i,j-1,k}^{n}}{2\Delta y}-\frac{H_y|_{i,j,k+1}^{n}-H_y|_{i,j,k-1}^{n}}{2\Delta z}\right)\\
\frac{E_y|_{i,j,k}^{n+1}-E_y|_{i,j,k}^{n-1}}{2\Delta t} &= \frac{1}{\varepsilon}\left(\frac{H_x|_{i,j,k+1}^{n}-H_x|_{i,j,k-1}^{n}}{2\Delta z}-\frac{H_z|_{i+1,j,k}^{n}-H_z|_{i-1,j,k}^{n}}{2\Delta x}\right)\\
\frac{E_z|_{i,j,k}^{n+1}-E_z|_{i,j,k}^{n-1}}{2\Delta t} &= \frac{1}{\varepsilon}\left(\frac{H_y|_{i+1,j,k}^{n}-H_y|_{i-1,j,k}^{n}}{2\Delta x}-\frac{H_x|_{i,j+1,k}^{n}-H_x|_{i,j-1,k}^{n}}{2\Delta y}\right)\\
\frac{H_x|_{i,j,k}^{n+1}-H_x|_{i,j,k}^{n-1}}{2\Delta t} &= \frac{-1}{\mu}\left(\frac{E_z|_{i,j+1,k}^{n}-E_z|_{i,j-1,k}^{n}}{2\Delta y}-\frac{E_y|_{i,j,k+1}^{n}-E_y|_{i,j,k-1}^{n}}{2\Delta z}\right)\\
\frac{H_y|_{i,j,k}^{n+1}-H_y|_{i,j,k}^{n-1}}{2\Delta t} &= \frac{-1}{\mu}\left(\frac{E_x|_{i,j,k+1}^{n}-E_x|_{i,j,k-1}^{n}}{2\Delta z}-\frac{E_z|_{i+1,j,k}^{n}-E_z|_{i-1,j,k}^{n}}{2\Delta x}\right)\\
\frac{H_z|_{i,j,k}^{n+1}-H_z|_{i,j,k}^{n-1}}{2\Delta t} &= \frac{-1}{\mu}\left(\frac{E_y|_{i+1,j,k}^{n}-E_y|_{i-1,j,k}^{n}}{2\Delta x}-\frac{E_x|_{i,j+1,k}^{n}-E_x|_{i,j-1,k}^{n}}{2\Delta y}\right)
\end{aligned} \tag{3}$$

In the style of (2), we can see that (3) at each grid point exactly preserves the value of discrete analogs of div (εE) and div (μH).

2.1 Space Staggering

A key to the long-standing popularity of the Yee scheme (3) [44] is the concept of grid staggering. We illustrate this first in a simpler case, viz. for the scalar one-way wave equation $u_t + u_x = 0$. Centred approximations in space and time, on a Cartesian grid, result in two entirely separate interlaced computations over the grid points marked "x" and over those marked "o" in Fig. 1. By computing over only one of the sets, say, the point set marked "o", we save a factor of two in computational effort. This also avoids trouble with high-frequency oscillations, which would otherwise be

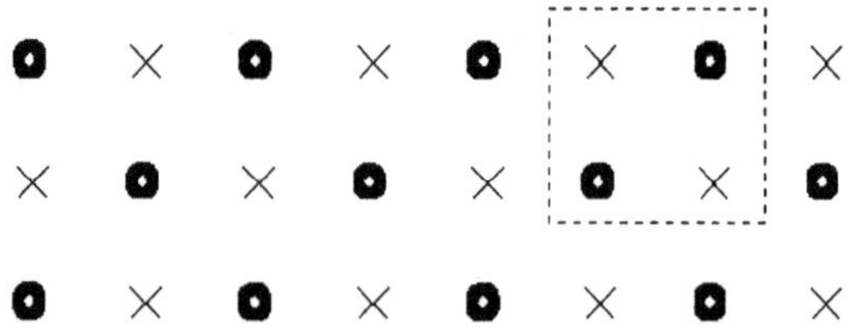

Fig. 1. Illustration of grid staggering in the $x, t-$plane for the one-way wave equation $u_t + u_x = 0$

the apparent manifestation of the two independent solutions over time most likely having drifted somewhat apart.

The same concept of staggering applies also to Maxwell's equations in 3D, but gives then far larger savings – a factor of 16 rather than of two. Just like the grid in Fig. 1 is made up of lots of translates of a 'basic grid unit' (as displayed within the dotted frame), the 3D spatial lattice for the Yee scheme is made up of translates of the block shown in Fig. 2, stacked in 3D as indicated by Fig. 3. These figures show the spatial layout only (due to the difficulty of simultaneously displaying graphically time and three spatial dimensions). On alternate time levels, only the three E-components or only the three H-components are present, respectively, in the positions as shown in Fig. 2. Considering how data is coupled by (3), this very sparse data layout suffices for a complete calculation. Each variable appears only at one of eight corner nodes, and furthermore only at every second time level. Having all variables present at all nodes at all times would amount to carrying out 16 separate independent Yee calculations.

If there were no concerns about irregular geometries, it would be an easy matter to greatly improve the computational efficiency of the Yee scheme by just increasing the order of accuracy in both time and space (while maintaining the staggering of the variables). With a fixed grid spacing, standard finite difference approximations for the first derivative (in general for odd derivatives) are much more accurate halfway between grid points than at grid points [14] (with the advantage increasing with the order of the approximation). In space, we can therefore employ highly accurate finite difference (FD) approximations of increasing orders/stencil widths. Extensions to implicit staggered approximations are derived and analysed in Fornberg and Ghrist [16]. The limits of increasing accuracy (stencil width) will in all cases correspond to the pseudospectral (PS) method (see also [11]).

2.2 Time Staggering

The stencils for a few standard classes of linear multistep schemes are shown in the left part of Fig. 4. Applied to an ODE (or system of ODEs) $y' = f(t, y)$, these require values of $f(t, y)$ to be available at the same time levels as are the y-values. Ghrist et al. [22] introduced recently a new class of time staggered ODE solvers. The staggered generalisations of Adams–Bashforth (AB), Adams–Moulton (AM) and back-

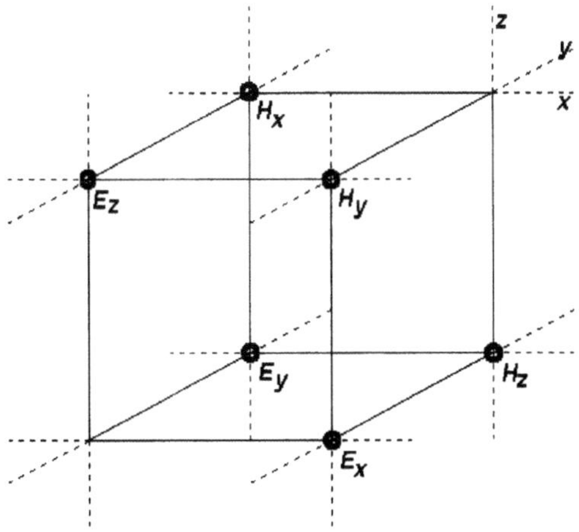

Fig. 2. Basic computational cell in the Yee scheme for Maxwell's equations. The E's and H's appear only at alternate time levels

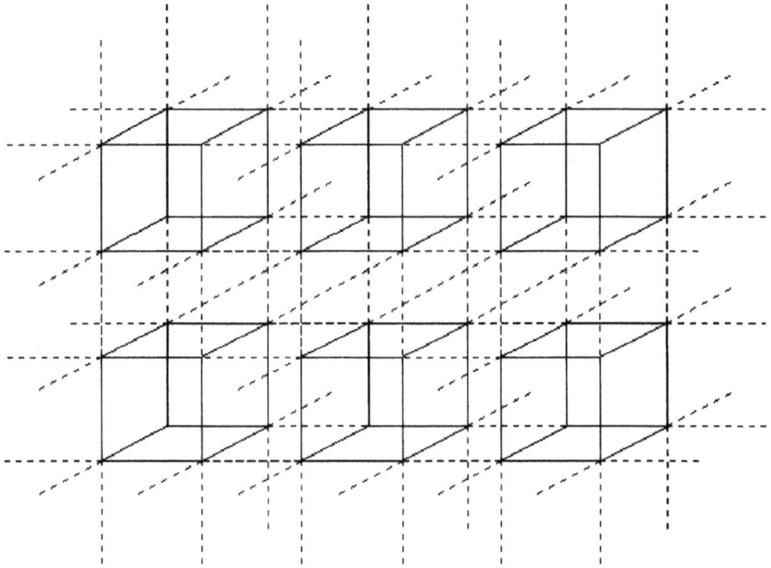

Fig. 3. Stacking of Yee cells of the form shown in the previous figure, in order to form a complete 3D Cartesian grid

ward differentiation (BD) become the explicit ABS and BDS schemes shown to the right. In the case of second order, these agree with the standard leap-frog scheme, but generalise this for higher orders of accuracy. To be applicable for wave equations (after method-of-lines discretisation), the stability domains of the ODE solvers need

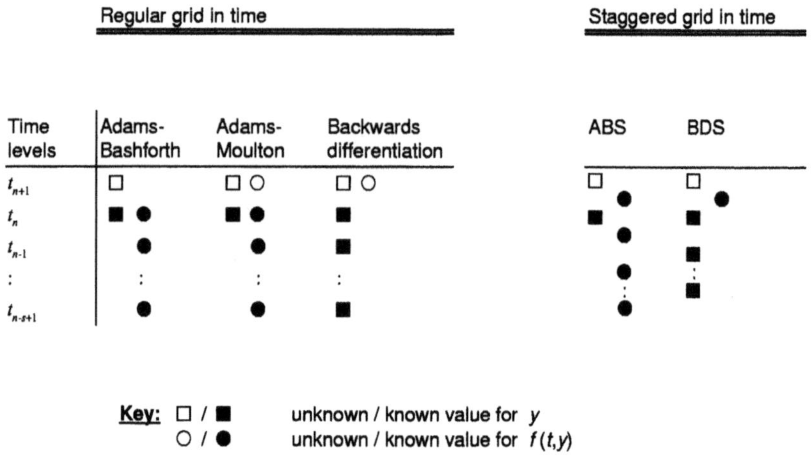

Fig. 4. Structure of some linear multistep methods for $y' = f(t, y)$

to cover a section of the imaginary axis. Both for the regular AB and staggered ABS methods this occurs with orders 3,4, 7,8, 11,12, etc. Bounds on the imaginary axis coverage (which translates directly into CFL stability restrictions) and the leading truncation error constants are close to a factor of ten more favourable for the staggered than for the non-staggered methods. Since the ABS methods require no more operations or storage than the AB methods, they are generally much preferable. A possible exception can occur in cases of wave equations with damping. Like for the leap-frog scheme, all staggered ODE solvers lack negative real axis coverage in the stability domain. As a rule, if leap-frog discretisation can be applied, so can also these (more effective) higher order generalisations.

The weights in the stencils for both the space approximations and the time stepping methods (implicit or explicit, staggered or not) are most conveniently obtained from the two-line symbolic algebra algorithm described in [12].

The lack of geometrical constraints in the time direction makes it particularly easy to use high order (big stencil) methods in that direction. The main reason this is not routinely utilised has to do with stability. For the Yee scheme, it can be shown (for example by von Neumann analysis) that computations will be unstable unless $\Delta t < 1 \,/\, (c\,\sqrt{1/(\Delta x)^2 + 1/(\Delta y)^2 + 1/(\Delta z)^2}\,)$, where $c = 1/\sqrt{\varepsilon\mu}$ is the wave speed. In this case, the actual stability constraint agrees exactly with the (often not sharp) upper bound on the time step imposed by the CFL (Courant–Friedrichs–Levy) condition. This condition usually makes it pointless to try to use higher order accuracy in time than what is used in space. Although doing so would increase the temporal accuracy, stability constraints would prevent this from being utilised to gain computational efficiency through the use of significantly larger time steps.

3 Situations Where Geometrical Features are Similar in Size to, or Larger Than a Typical Wave Length

As mentioned above, the choice of spatial discretisation method is usually dictated by the complexity of material interfaces. In this first scenario – with geometrical features similar to or larger than a typical wave length – the main problem is one of approximating Maxwell's equations at curved boundaries accurately and economically. Boundary integral-type methods (see for example [1] and [39]) offer a potentially very powerful approach. Focusing here on discretisation of Maxwell's equations in the form (1), we will next make some very brief comments on four different implementation ideas.

3.1 Yee Scheme

We have already briefly described this scheme; much more detail can be found e.g. in Taflove and Hagness [42] or Kunz and Luebbers [28]. It can probably be said, without exaggeration, that this has been the main tool for FDTD (finite difference time domain) calculations over the last 30 years. Only recently has it started to give way to higher order methods and, in particular, to methods that adapt more flexibly to irregular geometries (rather than just relying on 'staircasing' – the approximation of all domains simply as subsets of the regularly stacked Yee cells). However, thanks to its ease-of-use in cases when accuracy and computational efficiency are not critical, it will probably remain of importance for the foreseeable future.

3.2 Finite Elements

Like for the Yee scheme, many books, as well as commercial program packages, have been entirely devoted to finite element (FE) methods for CEM. We will here make no attempt to survey this discipline, but refer the readers to [26] and [27]. In general, FE methods for CEM are particularly well suited for use in software packages, where high computational efficiency has been traded for user-friendliness and convenience in applying different geometries, boundary conditions, etc. FEMLAB and ANSYS are two examples of widely available FE packages which specialise in structural mechanics, but for which very capable toolboxes for CEM are available. FE methods which can dynamically alter both gridding and element orders (*hp*-adaptive methods) can be very effective, but are very complicated to implement [5].

3.3 Finite Element–Finite Difference Hybrid Methods

Hybrid methods use different numerical methods in different regions of the computational domain. In one notable such development, Edelvik and Ledfelt [8] combine a geometrically flexible and unconditionally stable finite element discretisation near boundaries with the much simpler and more cost-effective Yee scheme throughout the bulk of the domain. In a layer near the boundaries, the two grids share edges, as is indicated in the 2D illustration in Fig. 5. The FE part gives rise to a positive definite system that is effectively solved, at each time step, by pre-conditioned conjugate gradients (typically converging in around 10–20 iterations). For the region of overlap, a procedure is used that combines the results from the two domains in a way that ensures both accuracy and overall stability [38]. Fig. 6 illustrates a very small section of the 3D grid used in a simulation of electromagnetic fields that would arise inside the cockpit area of a SAAB 2000 aircraft, if the plane was struck by lightning. In the cockpit area, a tetrahedral-type FE grid is used for both the outside skin and for interior structures. Shortly outside the skin, we see a transition to a box-like Yee grid which continues (not shown) to fill a volume surrounding the plane. Since, in this particular application, the interest is confined to the cockpit area, objects further back (like wings, engines, etc. outside the section displayed in Fig. 6) were all staircased. Computations with around 10^6 cells and 7000 time steps take around 2 days on a typical single processor workstation.

Many variations are possible. For example El Hachemi [9] uses interpolation between the FD and FE areas (as an alternative option to letting the grids share edges).

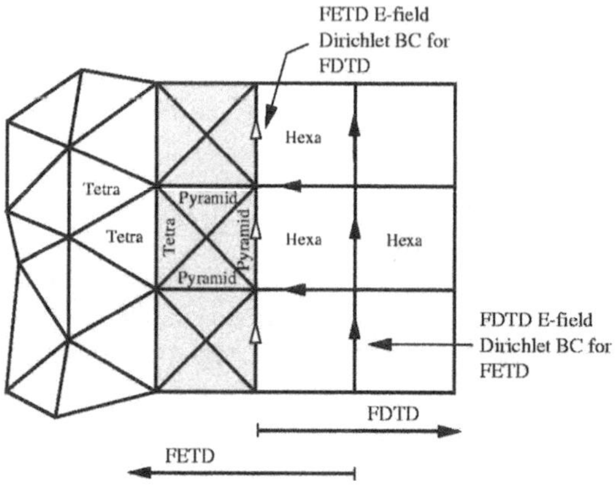

Fig. 5. Schematic illustration of transition between FE and FD areas of the FE-FD hybrid method

3.4 Spectral Elements

Several variations of spectral element-type methods have been proposed, in particular for computational fluid dynamics, but also for Maxwell's equations. The method by Hesthaven and Warburton [25] is notable in several ways: arbitrary order of accuracy, stability and convergence are strictly proven, and an element coupling is used

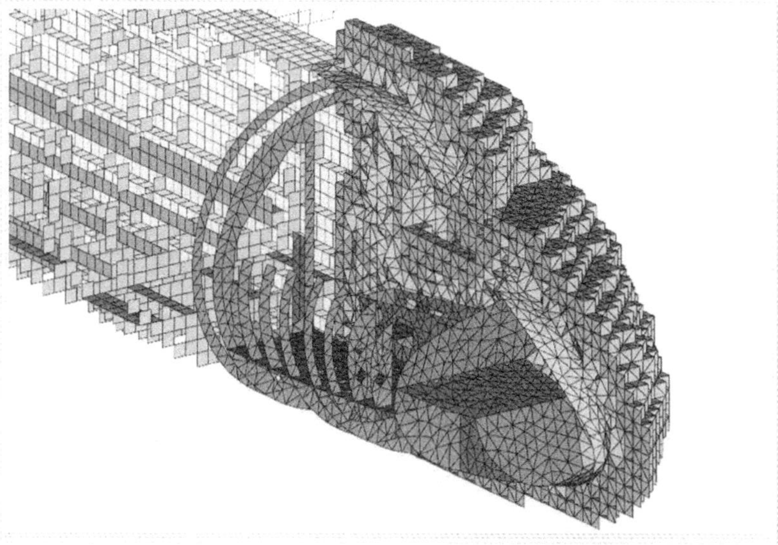

Fig. 6. Example of FE-FD gridding of the cockpit area of a Saab 2000 aircraft

which permits very effective distributed memory parallel implementation. Furthermore, very large-scale computations have been successfully demonstrated in 3D (in particular for radar scattering from aircraft). In 3D, the volume of interest is divided up into tetrahedra (some of which can be curvilinear for best fit with objects). Within each such element, the unknowns are represented by a single polynomial in the three space variables. If the degree is chosen as n, this polynomial will have $\frac{1}{6}(n+1)(n+2)(n+3)$ coefficients. That is then also the number of node points for such a tetrahedron. Their positions have been optimised to provide a particularly good interpolation capability, resulting in a certain distribution with $\frac{1}{2}(n+1)(n+2)$ node points on each face (including one at each tetrahedral corner and some along each edge), and the rest inside. While some spectral element methods couple different elements in ways to achieve a high degree of overall smoothness, the approach taken here is one of discontinuous Galerkin form, and with boundary conditions enforced only weakly through a penalty term. Data exchanges between the elements are based on characteristic information at their outer surfaces (a key to effective parallelism based on domain decomposition). Although a hybrid approach (involving a simpler gridding and numerical scheme away from interfaces) has not yet been implemented with this method, the developers are considering such a change (for which conditional stability will remain assured). For more specific information on the method, see [25].

A large part of the modelling effort in FDTD is often associated with grid generation. One of the strengths of this spectral element approach (like for *hp*-adaptive FE methods) is that it can utilise even a coarse and skewed grid, and then reach a required accuracy by means of increasing the order within each element (rather than requiring a new and better gridding; typically a more expensive proposition). Fig. 7 is a 2D illustration of this. The starting point is a deliberately skewed and coarse grid, as shown in the top left subplot (with the internal mesh in each triangle, corresponding to order $n = 10$, also displayed). The additional three subplots show how bringing up the order within the spectral elements produces convergence to the vertically symmetric scattered field that results when plane waves arrive from the left.

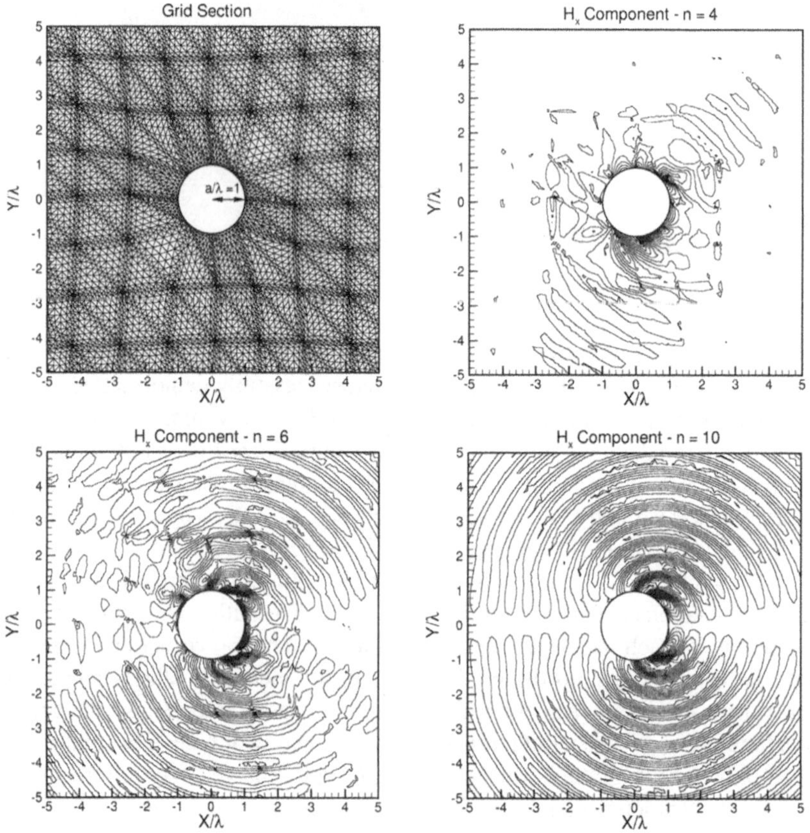

Fig. 7. Intentionally skewed gridding around a cylinder (with internal nodes for $n = 10$ shown; top left), and the scattered fields for $n = 4,\ 6$, and 10

3.5 Block-Pseudospectral Method

Of the approaches we are commenting on here, this one is the least well tested. It was introduced by Driscoll and Fornberg [7], and was shown to be highly effective in quite simple 2D geometries, such as the one shown in Fig. 8. The basic idea is somewhat similar to the FE–FD hybrid approach described above, but pushed a lot further towards high orders of accuracy in exchange for reduced geometric flexibility and a less clear stability situation. In the main part of the domain – away from objects – all spatial derivatives are approximated on an equi-spaced Cartesian grid by the implicit and staggered formula

$$\frac{9}{80}u'(x-h)+\frac{31}{40}u'(x)+\frac{9}{80}u'(x+h)=\frac{1}{h}\left[-\frac{17}{240}u(x-\frac{3}{2}h)\right.$$
$$\left.-\frac{63}{80}u(x-\frac{1}{2}h)+\frac{63}{80}u(x+\frac{1}{2}h)+\frac{17}{240}u(x+\frac{3}{2}h)\right]+O(h^6)\,.$$

This approximation (one example of a wide class of similar formulas derived and analysed in [16]) features a particularly small constant within the $O(h^6)$ error term, is quite compact, and leads only to tridiagonal systems to solve. In the circular strips that fit the outer shape of the conductors, Chebyshev-like pseudospectral discretisation was used across the strips (implemented via a differentiation matrix and not via FFTs, since there are only 6 grid points in total in that direction), and the Fourier pseudospectral method was used around the strips. (For an overview of pseudospectral methods, see [11].) Data were interpolated between the grids in the regions of overlap, and the standard fourth order Runge–Kutta scheme was used for the time integration. The simulation shown in Fig. 9 was achieved by putting together these quite standard numerical ingredients. The grid densities in the different regions were precisely as shown by dots and lines in Fig. 8. The resolution in this example was approximately 3–4 PPW, which can be compared to around the 50–60 PPW that would have been required to reach a similar accuracy with a Yee scheme. With so many fewer points needed for the higher order method *in each space dimension* (together with a low computational cost per grid point), the savings in both memory and computer time become very large.

In another test, the objects were inclined flat plates (rather than cylinders). The boundary fitting grids then used Chebyshev-type node distributions in both directions within rectangular patches. By including large numbers of such patches – which themselves can overlap – it is anticipated (but as yet not tested) that generalisations to more complex regions and to 3D will become possible. Although the 2D test cases were found to be stable, entirely without the inclusion of any artificial damping, stability has not been strictly proven. This could possibly become a significant issue in more complex settings.

Concluding this brief discussion of some numerical approaches, we would like to re-emphasise that efficient computation in free space and in the immediate vicinity of interfaces pose very different numerical challenges. Quite different numerical methods are usually preferable, suggesting that hybrid methods are a natural approach to pursue.

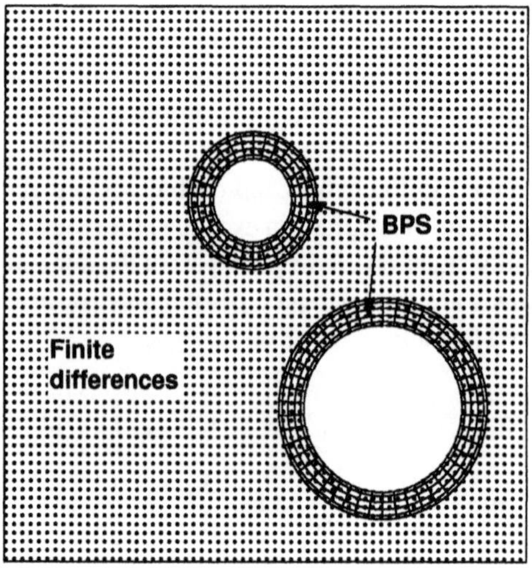

Fig. 8. Grids, with actual discretisation sizes shown (by dots and by lines) for the test case of a wave front impinging on two perfect conductors

4 Situations Where Geometrical Features are Much Smaller Than a Typical Wave Length

In the cases mentioned in the Introduction (integrated circuits, cellular phones, and lightning strikes), the spatial scales are about 10^{-5} wave lengths in size. The spatial gridding then needs to be correspondingly fine in order to capture the geometry properly, i.e. we are forced to an extreme over-sampling in space compared to what would have been needed if the goal was only to resolve a wave. Such very detailed grids, especially in 3D, will contain vast numbers of grid points (even if used only in small areas of the overall computational domain). To keep computational costs manageable, the time stepping procedure needs to be explicit (or nearly so). The CFL stability condition then tells that the time step also will need to be extremely small. The dilemma in this situation is that, while the accuracy in time can be met by using a time step that is some moderate fraction of the wave length, stability would seem to impose upper time steps bounds of maybe only 10^{-5} of this size. In order not to incur the vast expense of using such minute time steps, some approach needs to be found which essentially bypasses the restriction that is usually imposed by the CFL condition. In cases when the grids are refined only in very small areas, it can be very convenient to still be able to use the same (long) time steps everywhere.

Two discretisation methods have very recently been proposed which feature unconditional stability. Although this property has been strictly proven mainly for periodic problems in the case of a constant medium, the practical experience is very favourable also for variable medium initial-boundary value problems, as well as in combination with PML (perfectly matched layer) far-field boundary conditions. Compared to the Yee scheme, each of the new methods costs about four times as

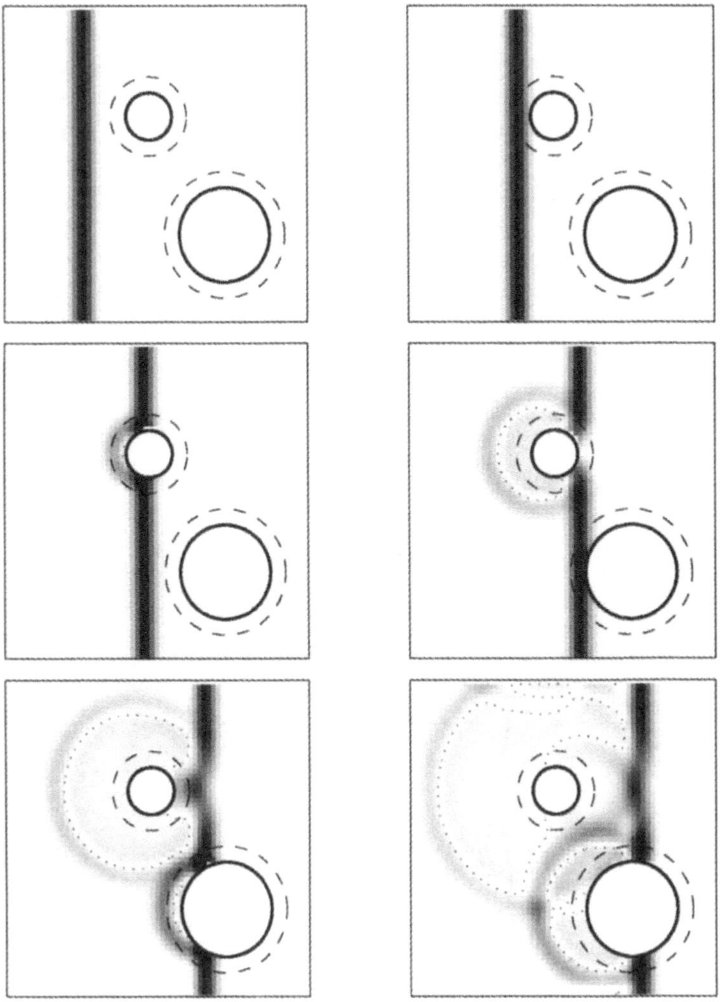

Fig. 9. Simulation of an electromagnetic wave front being scattered from two perfectly conducting cylinders

much as per time step – a small price to pay for being able to use a time step many orders of magnitude larger than would otherwise be possible. We will describe these two methods in the remainder of this section.

4.1 Alternating Direction Implicit (ADI) Method

The ADI approach has been very successful for parabolic and elliptic PDEs for the last 50 years. Seminal papers in the area include e.g. [6] and [35]. Similar 3-stage dimensional splittings for the 3D Maxwell's equations have been repeatedly tried in various forms since then, but have invariably fallen short of the goal of unconditional time stability. However, a 2-stage splitting introduced in 1999 by Zheng et al. [46, 47] does achieve this goal. The original way to state this scheme includes introducing a half-way time level $n+1/2$ between the adjacent time levels n and $n+1$. We advance our six variables as follows. **Stage 1:**

$$
\begin{aligned}
\frac{E_x|_{i,j,k}^{n+1/2} - E_x|_{i,j,k}^{n}}{\Delta t/2} &= \frac{1}{\varepsilon}\left(\frac{H_z|_{i,j+1,k}^{n+1/2} - H_z|_{i,j-1,k}^{n+1/2}}{2\Delta y} - \frac{H_y|_{i,j,k+1}^{n} - H_y|_{i,j,k-1}^{n}}{2\Delta z}\right)\\
\frac{E_y|_{i,j,k}^{n+1/2} - E_y|_{i,j,k}^{n}}{\Delta t/2} &= \frac{1}{\varepsilon}\left(\frac{H_x|_{i,j,k+1}^{n+1/2} - H_x|_{i,j,k-1}^{n+1/2}}{2\Delta z} - \frac{H_z|_{i+1,j,k}^{n} - H_z|_{i-1,j,k}^{n}}{2\Delta x}\right)\\
\frac{E_z|_{i,j,k}^{n+1/2} - E_z|_{i,j,k}^{n}}{\Delta t/2} &= \frac{1}{\varepsilon}\left(\frac{H_y|_{i+1,j,k}^{n+1/2} - H_y|_{i-1,j,k}^{n+1/2}}{2\Delta x} - \frac{H_x|_{i,j+1,k}^{n} - H_x|_{i,j-1,k}^{n}}{2\Delta y}\right)\\
\frac{H_x|_{i,j,k}^{n+1/2} - H_x|_{i,j,k}^{n}}{\Delta t/2} &= \frac{1}{\mu}\left(\frac{E_y|_{i,j,k+1}^{n+1/2} - E_y|_{i,j,k-1}^{n+1/2}}{2\Delta z} - \frac{E_z|_{i,j+1,k}^{n} - E_z|_{i,j-1,k}^{n}}{2\Delta y}\right)\\
\frac{H_y|_{i,j,k}^{n+1/2} - H_y|_{i,j,k}^{n}}{\Delta t/2} &= \frac{1}{\mu}\left(\frac{E_z|_{i+1,j,k}^{n+1/2} - E_z|_{i-1,j,k}^{n+1/2}}{2\Delta x} - \frac{E_x|_{i,j,k+1}^{n} - E_x|_{i,j,k-1}^{n}}{2\Delta z}\right)\\
\frac{H_z|_{i,j,k}^{n+1/2} - H_z|_{i,j,k}^{n}}{\Delta t/2} &= \frac{1}{\mu}\left(\frac{E_x|_{i,j+1,k}^{n+1/2} - E_x|_{i,j-1,k}^{n+1/2}}{2\Delta y} - \frac{E_y|_{i+1,j,k}^{n} - E_y|_{i-1,j,k}^{n}}{2\Delta x}\right)
\end{aligned}
\tag{4}
$$

Stage 2:

$$
\begin{aligned}
\frac{E_x|_{i,j,k}^{n+1} - E_x|_{i,j,k}^{n+1/2}}{\Delta t/2} &= \frac{1}{\varepsilon}\left(\frac{H_z|_{i,j+1,k}^{n+1/2} - H_z|_{i,j-1,k}^{n+1/2}}{2\Delta y} - \frac{H_y|_{i,j,k+1}^{n+1} - H_y|_{i,j,k-1}^{n+1}}{2\Delta z}\right)\\
\frac{E_y|_{i,j,k}^{n+1} - E_y|_{i,j,k}^{n+1/2}}{\Delta t/2} &= \frac{1}{\varepsilon}\left(\frac{H_x|_{i,j,k+1}^{n+1/2} - H_x|_{i,j,k-1}^{n+1/2}}{2\Delta z} - \frac{H_z|_{i+1,j,k}^{n+1} - H_z|_{i-1,j,k}^{n+1}}{2\Delta x}\right)\\
\frac{E_z|_{i,j,k}^{n+1} - E_z|_{i,j,k}^{n+1/2}}{\Delta t/2} &= \frac{1}{\varepsilon}\left(\frac{H_y|_{i+1,j,k}^{n+1/2} - H_y|_{i-1,j,k}^{n+1/2}}{2\Delta x} - \frac{H_x|_{i,j+1,k}^{n+1} - H_x|_{i,j-1,k}^{n+1}}{2\Delta y}\right)\\
\frac{H_x|_{i,j,k}^{n+1} - H_x|_{i,j,k}^{n+1/2}}{\Delta t/2} &= \frac{1}{\mu}\left(\frac{E_y|_{i,j,k+1}^{n+1/2} - E_y|_{i,j,k-1}^{n+1/2}}{2\Delta z} - \frac{E_z|_{i,j+1,k}^{n+1} - E_z|_{i,j-1,k}^{n+1}}{2\Delta y}\right)\\
\frac{H_y|_{i,j,k}^{n+1} - H_y|_{i,j,k}^{n+1/2}}{\Delta t/2} &= \frac{1}{\mu}\left(\frac{E_z|_{i+1,j,k}^{n+1/2} - E_z|_{i-1,j,k}^{n+1/2}}{2\Delta x} - \frac{E_x|_{i,j,k+1}^{n+1} - E_x|_{i,j,k-1}^{n+1}}{2\Delta z}\right)\\
\frac{H_z|_{i,j,k}^{n+1} - H_z|_{i,j,k}^{n+1/2}}{\Delta t/2} &= \frac{1}{\mu}\left(\frac{E_x|_{i,j+1,k}^{n+1/2} - E_x|_{i,j-1,k}^{n+1/2}}{2\Delta y} - \frac{E_y|_{i+1,j,k}^{n+1} - E_y|_{i-1,j,k}^{n+1}}{2\Delta x}\right)
\end{aligned}
\tag{5}
$$

Several things can be noted.

- The stages differ in that we swap which of the two terms in each right hand side (RHS) that is discretised on the new and on the old time level.

- On each new time level, we can obtain tridiagonal linear systems for E_x, E_y, E_z. For example in Stage 1, on the new time level, we can eliminate H_z between the first and the last equation, giving a tridiagonal system for E_x. Once we similarly get (and solve) the tridiagonal systems also for E_y and E_z, the remaining variables H_x, H_y, H_z follow explicitly.
- Yee-type staggering can again be applied, but only in space, giving savings of a factor of 8 (rather than 16) compared to the case when all variables are represented at all grid points.
- The solution at the intermediate time level $n + 1/2$ is only first order accurate. However, the accuracy is second order after each completed pair of stages (i.e. at all integer-numbered time levels).

Shortly after this ADI scheme was first proposed, Namiki [33] demonstrated its practical advantages for two test problems (a monopole antenna near a thin dielectric wall, and a stripline with a narrow gap). This scheme has also, by Liu and Gedney [31], been found to work well together with PML far field boundary conditions.

Proof of Unconditional Stability

The original stability proof (in the case of a constant medium in a periodic or infinite domain) was first given in [47] and reproduced in [42]. It uses von Neumann analysis. This leads to the demanding task of analytically determining the eigenvalues of a certain 6×6 matrix, whose entries are functions of the grid steps and wave numbers. This does prove feasible, but only through some quite heavy use of computational symbolic algebra. It transpires that all the eigenvalues have magnitude one, which establishes the unconditional stability. The following simpler energy-based stability proof was given by Fornberg [13]. A third proof, based on the alternate ADI description in Sect. 4.1, is given in [4].

As we noted above, the ADI scheme is best laid out on a staggered Yee-type grid in space, but with no staggering in time. However, for the sake of simplifying the notation, we apply it here on a regular grid in space (i.e. a grid with every one of the six quantities E_x, E_y, E_z, H_x, H_y, H_z represented at each grid location, rather than at only one out of eight such locations). If we can prove unconditional stability in this regular grid case, we have of course also proven it for the 8 uncoupled sub-problems that it contains. Our energy method for showing that no Fourier mode can diverge as time increases starts by considering the ADI scheme over an arbitrary-sized periodic box. We then demonstrate that the sum of the squares of all the unknowns remain bounded for all times.

We again assume that ε and μ are constants, let $\alpha_x = \frac{\Delta t}{2\varepsilon \Delta x}$, $\beta_x = \frac{\Delta t}{2\mu \Delta x}$, and introduce similarly α_y, α_z, β_y and β_z. Separating the terms in (4) according to their time level gives

$$\begin{aligned}
E_x|_{i,j,k}^{n+1/2} - \alpha_y\left(H_z|_{i,j+1,k}^{n+1/2} - H_z|_{i,j-1,k}^{n+1/2}\right) &= E_x|_{i,j,k}^{n} - \alpha_z\left(H_y|_{i,j,k+1}^{n} - H_y|_{i,j,k-1}^{n}\right)\\
E_y|_{i,j,k}^{n+1/2} - \alpha_z\left(H_x|_{i,j,k+1}^{n+1/2} - H_x|_{i,j,k-1}^{n+1/2}\right) &= E_y|_{i,j,k}^{n} - \alpha_x\left(H_z|_{i+1,j,k}^{n} - H_z|_{i-1,j,k}^{n}\right)\\
E_z|_{i,j,k}^{n+1/2} - \alpha_x\left(H_y|_{i+1,j,k}^{n+1/2} - H_y|_{i-1,j,k}^{n+1/2}\right) &= E_z|_{i,j,k}^{n} - \alpha_y\left(H_x|_{i,j+1,k}^{n} - H_x|_{i,j-1,k}^{n}\right)\\
H_x|_{i,j,k}^{n+1/2} - \beta_z\left(E_y|_{i,j,k+1}^{n+1/2} - E_y|_{i,j,k-1}^{n+1/2}\right) &= H_x|_{i,j,k}^{n} - \beta_y\left(E_z|_{i,j+1,k}^{n} - E_z|_{i,j-1,k}^{n}\right)\\
H_y|_{i,j,k}^{n+1/2} - \beta_x\left(E_z|_{i+1,j,k}^{n+1/2} - E_z|_{i-1,j,k}^{n+1/2}\right) &= H_y|_{i,j,k}^{n} - \beta_z\left(E_x|_{i,j,k+1}^{n} - E_x|_{i,j,k-1}^{n}\right)\\
H_z|_{i,j,k}^{n+1/2} - \beta_y\left(E_x|_{i,j+1,k}^{n+1/2} - E_x|_{i,j-1,k}^{n+1/2}\right) &= H_z|_{i,j,k}^{n} - \beta_x\left(E_y|_{i+1,j,k}^{n} - E_y|_{i-1,j,k}^{n}\right)
\end{aligned} \tag{6}$$

We next take the square of both sides of each equation above; then multiply the first three equations by ε and the next three by μ. For example, in the case of the first and the fifth equations of (6), this gives

$$\begin{aligned}
&\varepsilon\alpha_y^2\left(H_z|_{i,j+1,k}^{n+1/2} - H_z|_{i,j-1,k}^{n+1/2}\right)^2 - \frac{\Delta t}{\Delta y} E_x|_{i,j,k}^{n+1/2}\left(H_z|_{i,j+1,k}^{n+1/2} - H_z|_{i,j-1,k}^{n+1/2}\right)\\
&\quad + \varepsilon\left(E_x|_{i,j,k}^{n+1/2}\right)^2 = \varepsilon\left(E_x|_{i,j,k}^{n}\right)^2 + \varepsilon\alpha_z^2\left(H_y|_{i,j,k+1}^{n} - H_y|_{i,j,k-1}^{n}\right)^2\\
&\quad - \frac{\Delta t}{\Delta z} E_x|_{i,j,k}^{n}\left(H_y|_{i,j,k+1}^{n} - H_y|_{i,j,k-1}^{n}\right)
\end{aligned}$$

$$\begin{aligned}
&\mu\beta_x^2\left(E_z|_{i+1,j,k}^{n+1/2} - E_z|_{i-1,j,k}^{n+1/2}\right)^2 - \frac{\Delta t}{\Delta x} H_y|_{i,j,k}^{n+1/2}\left(E_z|_{i+1,j,k}^{n+1/2} - E_z|_{i-1,j,k}^{n+1/2}\right)\\
&\quad + \mu\left(H_y|_{i,j,k}^{n+1/2}\right)^2 = \mu\left(H_y|_{i,j,k}^{n}\right)^2 + \mu\beta_z^2\left(E_x|_{i,j,k+1}^{n} - E_x|_{i,j,k-1}^{n}\right)^2\\
&\quad - \frac{\Delta t}{\Delta z} H_y|_{i,j,k}^{n}\left(E_x|_{i,j,k+1}^{n} - E_x|_{i,j,k-1}^{n}\right)
\end{aligned}$$

If we add these two equations together, one of the expressions on the right hand side will become

$$-\frac{\Delta t}{\Delta z}\left\{E_x|_{i,j,k}^{n}\left(H_y|_{i,j,k+1}^{n} - H_y|_{i,j,k-1}^{n}\right) + H_y|_{i,j,k}^{n}\left(E_x|_{i,j,k+1}^{n} - E_x|_{i,j,k-1}^{n}\right)\right\}.$$

When summing this expression over the full 3D periodic volume, it cancels out to become zero (as can be seen directly, or by summation by parts; already summing in the z-direction makes it zero, and further summation in the x- and y-directions of zeros remain zero). In the same way, when we add the squares of *all* the relations in (6) over the full volume, all the products that mix E and H-terms will cancel out on both of the time levels $n+1/2$ and n. Hence, we get $\sum_{(1)}^{n+1/2} = \sum_{(2)}^{n}$, where

$$\begin{aligned}\sum\nolimits_{(1)}^{n+1/2} = \sum_{i,j,k} \Bigg\{ \varepsilon & \left[\left(E_x|_{i,j,k}^{n+1/2} \right)^2 + \left(E_y|_{i,j,k}^{n+1/2} \right)^2 + \left(E_z|_{i,j,k}^{n+1/2} \right)^2 \right] \\ & + \mu \left[\left(H_x|_{i,j,k}^{n+1/2} \right)^2 + \left(H_y|_{i,j,k}^{n+1/2} \right)^2 + \left(H_z|_{i,j,k}^{n+1/2} \right)^2 \right] \\ + \varepsilon \Bigg[\alpha_y^2 & \left(H_x|_{i,j,k+1}^{n+1/2} - H_x|_{i,j,k-1}^{n+1/2} \right)^2 + \alpha_z^2 \left(H_y|_{i+1,j,k}^{n+1/2} - H_y|_{i-1,j,k}^{n+1/2} \right)^2 \\ + \alpha_x^2 & \left(H_z|_{i,j+1,k}^{n+1/2} - H_z|_{i,j-1,k}^{n+1/2} \right)^2 \Bigg] + \mu \Bigg[\beta_z^2 \left(E_x|_{i,j+1,k}^{n+1/2} - E_x|_{i,j-1,k}^{n+1/2} \right)^2 \\ & + \beta_x^2 \left(E_y|_{i,j,k+1}^{n+1/2} - E_y|_{i,j,k-1}^{n+1/2} \right)^2 + \beta_y^2 \left(E_z|_{i+1,j,k}^{n+1/2} - E_z|_{i-1,j,k}^{n+1/2} \right)^2 \Bigg] \Bigg\}\end{aligned}$$

and

$$\begin{aligned}\sum\nolimits_{(2)}^{n} = \sum_{i,j,k} \Bigg\{ \varepsilon & \left[\left(E_x|_{i,j,k}^{n} \right)^2 + \left(E_y|_{i,j,k}^{n} \right)^2 + \left(E_z|_{i,j,k}^{n} \right)^2 \right] \\ & + \mu \left[\left(H_x|_{i,j,k}^{n} \right)^2 + \left(H_y|_{i,j,k}^{n} \right)^2 + \left(H_z|_{i,j,k}^{n} \right)^2 \right] \\ + \varepsilon \Bigg[\alpha_z^2 & \left(H_x|_{i,j+1,k}^{n} - H_x|_{i,j-1,k}^{n} \right)^2 + \alpha_x^2 \left(H_y|_{i,j,k+1}^{n} - H_y|_{i,j,k-1}^{n} \right)^2 \\ + \alpha_y^2 & \left(H_z|_{i+1,j,k}^{n} - H_z|_{i-1,j,k}^{n} \right)^2 \Bigg] + \mu \Bigg[\beta_y^2 \left(E_x|_{i,j,k+1}^{n} - E_x|_{i,j,k-1}^{n} \right)^2 \\ & + \beta_z^2 \left(E_y|_{i+1,j,k}^{n} - E_y|_{i-1,j,k}^{n} \right)^2 + \beta_x^2 \left(E_z|_{i,j+1,k}^{n} - E_z|_{i,j-1,k}^{n} \right)^2 \Bigg] \Bigg\} .\end{aligned}$$

The superscript on each of the each of the sums $\sum_{(1)}^{n+1/2}$ and $\sum_{(2)}^{n}$ denotes the time level, and the subscript indicates that the two sums furthermore differ a bit subtly in the indices and coefficients for the terms that contain differences.

Exactly in the same way as we have arrived at $\sum_{(1)}^{n+1/2} = \sum_{(2)}^{n}$ starting from the Stage 1 equations (4), we could instead have started from the Stage 2 equations (5) to obtain $\sum_{(2)}^{n+1} = \sum_{(1)}^{n+1/2}$. This tells that $\sum_{(2)}^{n+1} = \sum_{(2)}^{n}$, i.e. after each completed full time step, $\sum_{(2)}$ remains unchanged. The expression

$$\begin{aligned}\sum\nolimits^{n} = \sum_{i,j,k} \Bigg\{ & \varepsilon \left[\left(E_x|_{i,j,k}^{n} \right)^2 + \left(E_y|_{i,j,k}^{n} \right)^2 + \left(E_z|_{i,j,k}^{n} \right)^2 \right] \\ & + \mu \left[\left(H_x|_{i,j,k}^{n} \right)^2 + \left(H_y|_{i,j,k}^{n} \right)^2 + \left(H_z|_{i,j,k}^{n} \right)^2 \right] \Bigg\}\end{aligned}$$

appears in $\sum_{(2)}^{n}$ together with additional terms that are all squares of differences, and therefore are guaranteed to be positive. Since $\sum_{(2)}^{n}$ is preserved for all times (all integer values of n), the sum $\sum^{n}$ will be uniformly bounded for all times by the initial value of $\sum_{(2)}^{n}$. This implies that the ADI method is unconditionally stable.

A major advantage with 'energy-type' proofs, like the present, is that they can often be extended to hold also in the case of variable coefficients (i.e. variable media) and for different types of boundary conditions. This has already been carried out for a box-shaped cavity with perfectly conducting walls by Gao et al. [20].

Alternative Description of the ADI Method

The 3D Maxwell's equations (1) can be written

$$\frac{\partial}{\partial t}\begin{bmatrix} E_x \\ E_y \\ E_z \\ H_x \\ H_y \\ H_z \end{bmatrix} = \begin{bmatrix} \frac{1}{\varepsilon}\frac{\partial H_z}{\partial y} \\ \frac{1}{\varepsilon}\frac{\partial H_x}{\partial z} \\ \frac{1}{\varepsilon}\frac{\partial H_y}{\partial x} \\ \frac{1}{\mu}\frac{\partial E_y}{\partial z} \\ \frac{1}{\mu}\frac{\partial E_z}{\partial x} \\ \frac{1}{\mu}\frac{\partial E_x}{\partial y} \end{bmatrix} + \begin{bmatrix} -\frac{1}{\varepsilon}\frac{\partial H_y}{\partial z} \\ -\frac{1}{\varepsilon}\frac{\partial H_z}{\partial x} \\ -\frac{1}{\varepsilon}\frac{\partial H_x}{\partial y} \\ -\frac{1}{\mu}\frac{\partial E_z}{\partial y} \\ -\frac{1}{\mu}\frac{\partial E_x}{\partial z} \\ -\frac{1}{\mu}\frac{\partial E_y}{\partial x} \end{bmatrix} \tag{7}$$

or, more briefly,

$$\frac{\partial \underline{u}}{\partial t} = A\,\underline{u} + B\,\underline{u}\,.$$

The standard Crank–Nicolson discretisation in time is

$$\frac{\underline{u}^{n+1}-\underline{u}^{n}}{\Delta t} = \frac{1}{2}(A\underline{u}^{n+1}+A\underline{u}^{n}) + \frac{1}{2}(B\underline{u}^{n+1}+B\underline{u}^{n}) + O(\Delta t)^2$$

$$\Rightarrow \left(1-\frac{\Delta t}{2}A-\frac{\Delta t}{2}B\right)\underline{u}^{n+1} = \left(1+\frac{\Delta t}{2}A+\frac{\Delta t}{2}B\right)\underline{u}^{n} + O(\Delta t)^3$$

$$\Rightarrow \underbrace{\left(1-\frac{\Delta t}{2}A\right)\left(1-\frac{\Delta t}{2}B\right)\underline{u}^{n+1} = \left(1+\frac{\Delta t}{2}A\right)\left(1+\frac{\Delta t}{2}B\right)\underline{u}^{n}}_{\text{One-step approximation}} \tag{8}$$

$$+\underbrace{\frac{(\Delta t)^2}{4}A\,B(\underline{u}^{n+1}-\underline{u}^{n}) + O(\Delta t)^3}_{\text{Error }=\ O(\Delta t)^3}\,.$$

The 'one–step approximation'

$$\left(1-\frac{\Delta t}{2}A\right)\left(1-\frac{\Delta t}{2}B\right)\underline{u}^{n+1} = \left(1+\frac{\Delta t}{2}A\right)\left(1+\frac{\Delta t}{2}B\right)\underline{u}^{n} \tag{9}$$

can be written in two stages

$$\begin{cases} \left(1-\frac{\Delta t}{2}A\right)\underline{u}^{n+1/2} = \left(1+\frac{\Delta t}{2}B\right)\underline{u}^{n} \\ \left(1-\frac{\Delta t}{2}B\right)\underline{u}^{n+1} \;\;= \left(1+\frac{\Delta t}{2}A\right)\underline{u}^{n+1/2}\,. \end{cases} \tag{10}$$

To verify this, we multiply the first equation of (10) by $\left(1+\frac{\Delta t}{2}A\right)$ and the second one by $\left(1-\frac{\Delta t}{2}A\right)$. The LHS of the first equation then equals the RHS of the second equation, and (9) follows. It is straightforward to verify that the two equations in (10) – after space derivatives in A and B have been replaced by centred finite differences – become equivalent to the two stages (4) and (5) of the ADI scheme. This way of deriving the ADI method offers us several advantages.

- We recognise that $\underline{u}^{n+1/2}$ more naturally can be seen just as an intermediate computational quantity rather than as some specific intermediate time level (at which we have reduced accuracy).
- The second order accuracy in time for the overall procedure has become obvious.
- It becomes easy to determine the precise form of the local temporal error (which will be of importance to us later for implementing deferred correction in order to reach higher orders of accuracy in time).

4.2 Crank–Nicolson Based Split-Step Method (CNS)

We start this subsection by briefly reviewing the general concept and history of split step methods, and we then note how they can be applied to the Maxwell's equations.

Concept

In the simplest form of split step, featuring only first order accuracy in time, one would advance an ODE (or a system of ODEs)

$$u_t = A(u) + B(u) \tag{11}$$

from time t to time $t + \Delta t$ by successively solving

$$\begin{array}{ll} u_t = 2A(u) & \text{from } t \text{ to } t + \frac{1}{2}\Delta t, \text{ followed by} \\ u_t = 2B(u) & \text{from } t + \frac{1}{2}\Delta t \text{ to } t + \Delta t \end{array}$$

Here, $A(u)$ and $B(u)$ can be very general nonlinear operators (in particular, there is no requirement that A and B commute). The two time increments are each of length $\frac{1}{2}\Delta t$. We therefore denote this splitting by $\{\frac{1}{2}, \frac{1}{2}\}$. One obtains second order accuracy in time by instead alternating the two equations in the pattern A, B, A while using the time increments $\{\frac{1}{4}, \frac{1}{2}, \frac{1}{4}\}$ – known as 'Strang splitting' [40]. In 1990 Yoshida [45] devised a systematic way to obtain similar split step methods of still higher orders. From an implementation standpoint, one simply chooses certain longer time increment sequences, while again alternating A, B, A, $B, \ldots$. Table 1 shows the coefficients of methods of orders 1, 2, 4, 6, and 8. The coefficients for the methods of order 6 and above are not unique.

The split step approach is especially interesting for PDEs. If such an equation is of the form $u_t + A(u, u_x) + B(u, u_y) = 0$, immediate dimensional splitting leads to two 1D problems. Splitting is also of significant interest for certain nonlinear wave equations (see e.g. [15] for comparisons between split step and two other approaches). For example, the Korteweg–de Vries (KdV) equation $u_t + uu_x + u_{xxx} = 0$ can be split into $u_t + 2uu_x = 0$ (which features few numerical difficulties over brief times) and $u_t + 2u_{xxx} = 0$ (which is linear, and can be solved analytically, thereby bypassing otherwise severe stability restrictions).

It can be shown (Suzuki, [41]) that methods of orders above two will need to feature at least some negative time increments. Although this is of little concern in

Method	Time increment sequence		
SS1	0.50000 00000 00000 00000	0.50000 00000 00000 00000	
SS2	0.25000 00000 00000 00000	0.50000 00000 00000 00000	0.25000 00000 00000 00000
SS4	0.33780 17979 89914 40851	0.67560 35959 79828 81702	-0.08780 17979 89914 40851
	-0.85120 71919 59657 63405	-0.08780 17979 89914 40851	0.67560 35959 79828 81702
	0.33780 17979 89914 40851		
SS6	0.19612 84026 19389 31595	0.39225 68052 38778 63191	0.25502 17059 59228 84938
	0.11778 66066 79679 06684	-0.23552 66927 04878 21832	-0.58883 99920 89435 50347
	0.03437 65841 26260 05298	0.65759 31603 41955 60944	0.03437 65841 26260 05298
	-0.58883 99920 89435 50347	-0.23552 66927 04878 21832	0.11778 66066 79679 06684
	0.25502 17059 59228 84938	0.39225 68052 38778 63191	0.19612 84026 19389 31595
SS8	0.22871 10615 57447 89169	0.45742 21231 14895 78337	0.29213 43956 99000 73022
	0.12684 66682 83105 67707	-0.29778 97250 73598 45089	-0.72242 61184 30302 57885
	-0.40077 32180 57163 83322	-0.07912 03176 84025 08760	0.44497 46255 63618 95284
	0.96906 95688 11262 99329	-0.00561 77738 38196 20526	-0.98030 51164 87655 40380
	-0.46445 25958 95878 59173	0.05139 99246 95898 22035	0.45281 32300 44769 50634
	0.85422 65353 93640 79233	0.45281 32300 44769 50634	0.05139 99246 95898 22035
	-0.46445 25958 95878 59173	-0.98030 51164 87655 40380	-0.00561 77738 38196 20526
	0.96906 95688 11262 99329	0.44497 46255 63618 95284	-0.07912 03176 84025 08760
	-0.40077 32180 57163 83322	-0.72242 61184 30302 57885	-0.29778 97250 73598 45089
	0.12684 66682 83105 67707	0.29213 43956 99000 73022	0.45742 21231 14895 78337
	0.22871 10615 57447 89169		

Table 1. Coefficients of some Split-Step methods

our context of Maxwell's equations, it does make the splitting idea problematic in cases when an equation is partly dissipative, such as for example the Navier-Stokes equations.

To heuristically explain why splitting works, we can for simplicity replace (11) by $u_t = A\,u + B\,u$ where A and B are constant matrices (the procedure allows for more general nonlinear operators, but in our present context of Maxwell's equations, A and B will become matrices after we have discretised in space). In this simplified case we can write the analytic solution of the system (11) of ODEs as

$$u(t) = e^{(A+B)t}\,u(0)$$

where $e^{(A+B)t}$ is to be understood as the matrix

$$\begin{aligned} e^{(A+B)t} &= I + t(A+B) + \tfrac{t^2}{2!}(A+B)^2 + \ldots \\ &= I + t(A+B) + \tfrac{t^2}{2!}(A^2 + AB + BA + B^2) + \ldots \end{aligned} \tag{12}$$

(taking note of the fact that A and B in general do not commute). The solution operator using SS1 amounts to replacing the exact operator $e^{(A+B)t}$ by

$$\begin{aligned} e^{2B\frac{t}{2}}e^{2A\frac{t}{2}} &= (I + tB + \tfrac{t^2}{2!}B^2 + \ldots)(I + tA + \tfrac{t^2}{2!}A^2 + \ldots) \\ &= I + t(A+B) + \tfrac{t^2}{2!}(A^2 + 2BA + B^2) + \ldots . \end{aligned} \tag{13}$$

The expansion in (13) differs from the one in (12) in the t^2−term. This tells that this particular splitting is only first order accurate. Carrying out the same expansion for the SS2 scheme gives

$$\begin{aligned}&e^{2A\frac{t}{4}}e^{2B\frac{t}{2}}e^{2A\frac{t}{4}}\\ &=(I+\tfrac{t}{2}A+\tfrac{t^2}{8}A^2+\ldots)(I+tB+\tfrac{t^2}{2}B^2+\ldots)(I+\tfrac{t}{2}A+\tfrac{t^2}{8}A^2+\ldots)\\ &=I+t(A+B)+\tfrac{t^2}{2!}(A^2+AB+BA+B^2)+\ldots.\end{aligned}$$

This agrees with (12) throughout the t^2−term, thereby ensuring that SS2 (Strang splitting) indeed is accurate to second order.

A similar verification in the case of SS4 will produce an expansion that reproduces (12) also through the next two powers of t (not displayed in (12)). This SS4 scheme was originally found (via direct algebraic expansions similar to the ones above) by Neri in 1987 [34]. Closed form expressions are in this case available for the fractional step lengths c_i and d_i in the expansion

$$e^{(A+B)t}=e^{c_1At}e^{d_1Bt}e^{c_2At}e^{d_2Bt}e^{c_3At}e^{d_3Bt}e^{c_4At}+O(t^5)\,,$$

namely

$$c_1=c_4=\frac{1}{2(2-2^{1/3})}\,,\ c_2=c_3=(1-2^{1/3})c_1\,,\ d_1=d_3=2c_1\,,\ d_2=-2^{4/3}c_1\,.$$

The key contributions by Yoshida in 1990 [45] were to

- demonstrate that it is possible to find sequences of time increments that give time stepping accuracies of any order;
- devise a practical algorithm for computing these sequences of increments; and
- note that, in order to get a high order in time, special time step sequences can be applied as an addition to any scheme that is second order accurate (i.e. if we have any second order scheme for (7) – irrespective of if it is itself based on splitting or not – we can use certain time stepping sequences to bring it up to any order in time). This will be discussed later.

A couple of comments regarding the last point above.

- We can use either ADI or CNS (Strang split 3D Maxwell's equations, using Crank-Nicolson for the sub-problems – as will be described next) as our basic scheme, and then use certain time step sequences to enhance it to higher orders. This will become one of the three enhancement techniques we will consider later for increasing the time accuracy of the unconditionally stable ADI and CNS schemes.
- The methods SS4, SS6, SS8 in Table 1 are just special cases of this more general observation. These schemes arise if we choose, as our basic second order scheme in this process the CNS scheme (here also denoted SS2).

Application of Split Step to Maxwell's Equations

Immediate dimensional splitting of the 3D Maxwell's equations would lead us to consider a PDE of the form $\frac{\partial \underline{u}}{\partial t} = A\underline{u} + B\underline{u} + C\underline{u}$ where $\underline{u}$ denotes the vector $(E_x, E_y, E_z, H_x, H_y, H_z)^T$. Although 3-way splitting is feasible, the particular structure of the 3D Maxwell's equations permits a much more effective alternative. We start this by writing the Maxwell's equations as we did earlier in (7) and, like then, we abbreviate them as

$$\frac{\partial \underline{u}}{\partial t} = A\underline{u} + B\underline{u} \tag{14}$$

The split-step approach leads us to repeatedly advance $\frac{\partial \underline{u}}{\partial t} = 2A\underline{u}$ and $\frac{\partial \underline{u}}{\partial t} = 2B\underline{u}$ by certain time increments. These two subproblems can be written out explicitly as shown below. As first noted by Lee and Fornberg [29], each of the two subproblems obtained in this manner amount to three pairs of mutually entirely uncoupled 1D equations:

$$\left\{\begin{array}{l} \left\{\begin{array}{l} \frac{\partial E_x}{\partial t} = \frac{2}{\varepsilon}\frac{\partial H_z}{\partial y} \\ \frac{\partial H_z}{\partial t} = \frac{2}{\mu}\frac{\partial E_x}{\partial y} \end{array}\right\} \\ \\ \left\{\begin{array}{l} \frac{\partial E_y}{\partial t} = \frac{2}{\varepsilon}\frac{\partial H_x}{\partial z} \\ \frac{\partial H_x}{\partial t} = \frac{2}{\mu}\frac{\partial E_y}{\partial z} \end{array}\right\} \\ \\ \left\{\begin{array}{l} \frac{\partial E_z}{\partial t} = \frac{2}{\varepsilon}\frac{\partial H_y}{\partial x} \\ \frac{\partial H_y}{\partial t} = \frac{2}{\mu}\frac{\partial E_z}{\partial x} \end{array}\right\} \end{array}\right\}, \quad \left\{\begin{array}{l} \left\{\begin{array}{l} \frac{\partial E_x}{\partial t} = -\frac{2}{\varepsilon}\frac{\partial H_y}{\partial z} \\ \frac{\partial H_y}{\partial t} = -\frac{2}{\mu}\frac{\partial E_x}{\partial z} \end{array}\right\} \\ \\ \left\{\begin{array}{l} \frac{\partial E_y}{\partial t} = -\frac{2}{\varepsilon}\frac{\partial H_z}{\partial x} \\ \frac{\partial H_z}{\partial t} = -\frac{2}{\mu}\frac{\partial E_y}{\partial x} \end{array}\right\} \\ \\ \left\{\begin{array}{l} \frac{\partial E_z}{\partial t} = -\frac{2}{\varepsilon}\frac{\partial H_x}{\partial y} \\ \frac{\partial H_x}{\partial t} = -\frac{2}{\mu}\frac{\partial E_z}{\partial y} \end{array}\right\} \end{array}\right\}. \tag{15}$$

Each of the 1D subsystems in (15) can very easily be solved numerically. If we choose a method which preserves the L^2−norm for each 1D sub-problem, the sum of the squares of all the unknowns will be preserved through each sub-step, and therefore also throughout the complete computation. Unconditional numerical stability is then assured. In particular, this will be the case if we approximate each of the 1D subsystems in (15) with a Crank-Nicolson type approximation. For example, to advance the first of the six sub-problems a distance $\Delta t/4$ in time, we would use

$$\frac{E_x|_{i,j,k}^{n+1/4} - E_x|_{i,j,k}^{n}}{\Delta t/4} = \frac{2}{\varepsilon}\frac{1}{2}\left\{\frac{H_z|_{i,j+1,k}^{n+1/4} - H_z|_{i,j-1,k}^{n+1/4}}{2\Delta y} + \frac{H_z|_{i,j+1,k}^{n} - H_z|_{i,j-1,k}^{n}}{2\Delta y}\right\}$$

$$\frac{H_z|_{i,j,k}^{n+1/4} - H_z|_{i,j,k}^{n}}{\Delta t/4} = \frac{2}{\mu}\frac{1}{2}\left\{\frac{E_x|_{i,j+1,k}^{n+1/4} - E_x|_{i,j-1,k}^{n+1/4}}{2\Delta y} + \frac{E_x|_{i,j+1,k}^{n} - E_x|_{i,j-1,k}^{n}}{2\Delta y}\right\}.$$

If we here use the second equation to eliminate H_z on the new time level from the first equation, we get a tridiagonal system to solve for E_x. Advancing (14) using Strang splitting together with these Crank-Nicolson approximations gives the scheme that we denote by CNS, which is second order accurate in time and space.

Comparison Between Different Split Step Sequences

The possibility of using split step together with (15) for numerical time integration of Maxwell's equations was first explored in [29]. In that study, a periodic domain was used, and all the 1D subproblems were solved analytically in (discrete) Fourier space. All errors that arose were therefore due to the time stepping, and it became possible to clearly compare the effectiveness of SS schemes of different orders. One of the observations that was made there was that different split step methods of the same order can have very different leading error coefficients. Seven different SS8 methods have been found. Fig. 10 shows, for a typical test problem, how the accuracy improves with increased temporal resolution. In this log-log plot, the slopes of the curves confirm the 8th order of accuracy in all cases, but the errors nevertheless differ by a full $3\frac{1}{2}$ orders of magnitude (for details about the test, see the original paper). The scheme that performs the best here – SS8d – is the one given in Table 1. Simply looking at the coefficients for the different schemes gives no clear indication of their difference in accuracy.

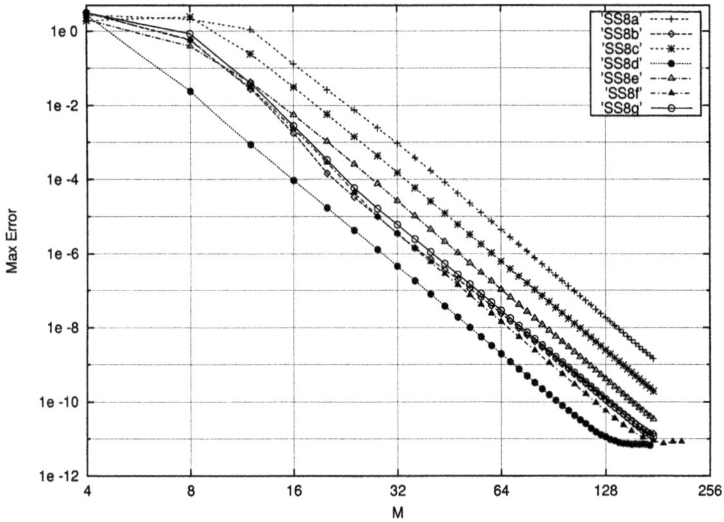

Fig. 10. Log-log plot comparison of error vs. number of subintervals in time for seven different SS8 methods

4.3 Enhancements to Reach Higher Orders of Accuracy in Time

We have just described two possible ways to obtain second order accuracy in time combined with unconditional stability (at least for pure initial value problems): ADI and CNS. High order methods are usually more effective than low order ones. In the present case of space being over-resolved (with respect to the wave length – in order to capture the fine geometrical features), we are in the (unusual) situation that it is not important to bring up the spatial order of accuracy. The situation in time is fundamentally different. Only one scale is present – a large one imposed by the wave length. A high order time stepping method can give cost savings by means of allowing significantly longer time steps than can be used with a second order method – as long as the unconditional stability is preserved.

Some approaches to reach higher order in time are certain *not* to work. For example, explicit Runge–Kutta and linear multistep methods can never feature unbounded stability domains. Because of the second Dahlquist stability barrier [2, 3] there are no prospects either in trying to add more back time levels to the ADI scheme; no implicit linear multistep method can combine A-stability with higher than second order accuracy in time. Following the treatment in Lee and Fornberg [30], we will next discuss three enhancements to ADI and CNS that will reach higher orders in time without running into the difficulties just mentioned. As it will transpire, two of them (later to be denoted EX and DC) will preserve unconditional stability up to all orders, at least for pure initial value problems in the case of constant media (the only situation for which stability has been rigorously established for the ADI and CNS methods). It appears that the third approach (TS) will preserve stability up to fourth order, but strong evidence for this is still lacking. For orders higher than four, cases of instability have been found. We will next introduce the three approaches (in the order TS, EX and DC).

Special Time–Step Sequences

The pioneering paper – as well as the main reference – on this procedure is Yoshida [45]. We start by assuming that we have some numerical procedure which advances our solution with second order accuracy in time. Expressed in the form of an operator, we write the advancement of the solution from time t to the time $t+\tau$ as $\underline{u}(t+\tau) = S_2(\tau)\ \underline{u}(t)$. We furthermore assume that the local error in the $S_2(\tau)-$operator is expandable in terms of the time step τ in the form $c_1\tau^3 + c_2\tau^5 + c_3\tau^7 + \ldots$. It then transpires that the composite operator

$$S_4(\tau) = S_2(w_1\tau)\ S_2(w_0\tau)\ S_2(w_1\tau) \tag{16}$$

becomes (globally) fourth order accurate in time if the constants w_0 and w_1 are chosen as $w_0 = -\frac{2^{1/3}}{2-2^{1/3}}$ and $w_1 = \frac{1}{2-2^{1/3}}$. This idea can be continued indefinitely, with the general result stating that

$$S_{2p}(\tau) = S_2(w_k\tau)\ \ldots\ S_2(w_1\tau)\ S_2(w_0\tau)\ S_2(w_1\tau)\ \ldots\ S_2(w_k\tau)$$

Order $(2p)$	Coefficients $\mathbf{w}_0, \mathbf{w}_1, \ldots \mathbf{w}_k$ where $k = 2^{p-1} - 1$.		
4	- 1.70241 43839 19315 26810	1.35120 71919 59657 63405	
6	1.31518 63206 83911 21888	- 1.17767 99841 78871 00695	0.23557 32133 59358 13368
	0.78451 36104 77557 26382		
8	1.70845 30707 87281 58467	0.10279 98493 91796 44070	- 1.96061 02329 75310 80761
	1.93813 91376 22525 98658	-0.15824 06353 68050 17520	-1.44485 22368 60605 15769
	0.25369 33365 66211 35415	0.91484 42462 29791 56675	

Table 2. Coefficients of some time increment sequencies

is accurate of order $2p$ if the constants $w_k, k = 0, 1, \ldots, 2^{p-1} - 1$ satisfy certain nonlinear systems of algebraic equations. For the $p = 2$ case the system becomes

$$\begin{cases} w_0 + 2w_1 = 1 \\ w_0^3 + 2w_1^3 = 0 \end{cases} \tag{17}$$

and in case of $p = 3$

$$\begin{cases} w_0+2(w_1+w_2+w_3) = 1 \\ w_0^3+2(w_1^3+w_2^3+w_3^3) = 0 \\ w_0^5+2(w_1^5+w_2^5+w_3^5) = 0 \\ -w_0^4w_1-w_0^3w_1^2+w_0^2w_1^3+w_0w_1^4-w_0^4w_2-2w_0^3w_1w_2-2w_0w_1^3w_2-4w_1^4w_2-w_0^3w_2^2-2w_1^3w_2^2+ \\ +w_0^2w_2^3+4w_0w_1w_2^3+4w_1^2w_2^3+w_0w_2^4+2w_1w_2^4-w_0^4w_3-2w_0^3w_1w_3-2w_0w_1^3w_3-4w_1^4w_3- \\ -2w_0^3w_2w_3-4w_1^3w_2w_3-2w_0w_2^3w_3-4w_1w_2^3w_3-4w_2^4w_3-w_0^3w_3^2-2w_1^3w_3^2-2w_2^3w_3^2+w_0^2w_3^3+ \\ +4w_0w_1w_3^3+4w_1^2w_3^3+4w_0w_2w_3^3+8w_1w_2w_3^3+4w_2^2w_3^3+w_0w_3^4+2w_1w_3^4+2w_2w_3^4 = 0 \end{cases} \tag{18}$$

Convenient recursive expressions to create the algebraic system for the general order of accuracy $2p$ are given in [45]. These are very well suited for numerical computation of the coefficients. Since the number of steps in these time-step sequences increase exponentially with the order, fairly low orders are probably of most interest. Table 2 will therefore suffice for most needs.

If we apply these sequences to the SS2 method, we get the methods we earlier described as SS4, SS6, etc. But the sequences can just as well be applied to other second order time stepping methods, such as the ADI method. In the following, we will sometimes denote the original ADI method as ADI2, and the time sequence enhancements of it as ADI-TS4, ADI-TS6, etc.

Table 2 gives one example of coefficients for each order. The eighth order scheme that is listed here corresponds to the split step scheme that was most effective in the comparison shown in Fig. 10. Coefficients for all presently known schemes up to and including 8th order can be found in [29].

We conclude this section by briefly explaining how the systems (17), (18) etc. can be obtained. If X and Y are operators (or matrices) that commute, then $e^X \cdot e^Y = e^Z$ with $Z = X + Y$. If X and Y do not commute, the expression for Z becomes far more complicated. By the Baker-Campbell-Hausdorff (BCH) formula [43]

$$
\begin{aligned}
Z = (X+Y) &+ [X,Y] \\
&+ \tfrac{1}{12}\left([X,X,Y]+[Y,Y,X]\right) \\
&+ \tfrac{1}{24}\,[X,Y,Y,X] \\
&+ \left\{\begin{array}{l}\text{commutators of successi-}\\ \text{vely increasing orders}\end{array}\right\}
\end{aligned}
\tag{19}
$$

where $[X,Y] = X\,Y - Y\,X$, $[X,Y,V] = [X,[Y,V]]$, etc. Repeated application of (19) gives $e^X \cdot e^Y \cdot e^X = e^W$ where

$$
\begin{aligned}
W = (2X+Y) &+ \tfrac{1}{6}\left([Y,Y,X]-[X,X,Y]\right) \\
&+ \left\{\begin{array}{l}\text{commutators of successi-}\\ \text{vely increasing orders}\end{array}\right\}
\end{aligned}
\tag{20}
$$

With our assumption that the operator $S_2(\tau)$ advances the equation $u' = Au$ the distance τ in time, with a local error of $c_1\tau^3 + c_2\tau^5 + \ldots$, it can be written as $S_2(\tau) = e^{\tau A + \tau^3 C + O(\tau^5)}$. The RHS of (16) then becomes

$$
e^{\overbrace{(w_1\tau)A+(w_1\tau)^3C+O(\tau^5)}^{X}}\; e^{\overbrace{(w_0\tau)A+(w_0\tau)^3C+O(\tau^5)}^{Y}}\; e^{\overbrace{(w_1\tau)A+(w_1\tau)^3C+O(\tau^5)}^{X}} \;.
$$

By (20) this becomes

$$
e^{\overbrace{(2w_1+w_0)\tau A+(2w_1^3+w_0^3)\tau^3C+O(\tau^5)}^{2X+Y} \;\; + \;\; \overbrace{O(\tau^5)}^{\text{higher terms}}} \;.
$$

This represents a fourth order method if it is of the form $e^{\tau A + O(\tau^5)}$, i.e. if (17) holds. To reach higher orders, we need to extend (20) to products of still more exponentials. When the commutators also come into play for some of the relations that need to be satisfied, the complexity of the resulting algebraic equations rapidly increases, as is seen in (18). However, as we noted just above, the original reference [45] shows that the higher order systems can nevertheless be obtained recursively very conveniently.

For some further developments on 'composition methods' for time stepping, see Hairer et al. [24].

Richardson Extrapolation

The idea, going back to Lewis Fry Richardson in 1927 [37], has been used since then in many applications, e.g. in extrapolation methods for ODEs and as Romberg's method for quadrature. If we have a numerical procedure for which the error of a computed variable u depends on a step size h in the following way

$$
u_h = \text{Exact} + c_1 h^2 + c_2 h^4 + \ldots \quad ,
$$

then repeating this calculation using a step size $h/2$ will give

$$u_{h/2} = \text{Exact} + c_1 \tfrac{1}{4} h^2 + c_2 \tfrac{1}{16} h^4 + \ldots$$

These two results can be linearly combined in order to eliminate $c_1 h^2$, giving the more accurate result

$$v_{h/2} = \frac{4\, u_{h/2} - u_h}{3} = \text{Exact} - c_2 \tfrac{1}{4} h^4 + \ldots$$

This idea can be continued repeatedly, and the results are conveniently laid out in triangular form

Directly computed	Extrapolated		
order 2	order 4	order 6	order 8
u_h			
$u_{h/2}$	$v_{h/2} = \frac{4u_{h/2}-u_h}{3}$		
$u_{h/4}$	$v_{h/4} = \frac{4u_{h/4}-u_{h/2}}{3}$	$w_{h/4} = \frac{16v_{h/4}-v_{h/2}}{15}$	
$u_{h/8}$	$v_{h/8} = \frac{4u_{h/8}-u_{h/4}}{3}$	$w_{h/8} = \frac{16v_{h/8}-v_{h/4}}{15}$	$x_{h/8} = \frac{64w_{h/8}-w_{h/4}}{63}$
...	...	...	...

In the context of quadrature, it is common practice to halve the step between each calculation (in order to be able to re-use as many old function values as possible). The drawback with that strategy is that the number of function evaluations then grows exponentially with the order. In the context of ODEs – which is our situation when time stepping Maxwell's equations – it is cheaper to make smaller changes in the time step between the different computations. The extrapolation procedure will still increase the order by two for each new original computation, but without the need for each new computation to be twice as expensive as the previous one.

Deferred Correction

This concept of deferred correction was introduced by Pereyra [36], first in the context of solving 2-point boundary value problems for ODEs, and subsequently for solving PDEs. Frank and collaborators [17–19] use the method for increasing the order of accuracy in the time stepping of ODEs. This was considered again recently by Gustafsson and Kress [23] who also illustrate the effectiveness of this for a methods-of-lines solution of a 1D heat equation. We will describe it here in the context of the ADI scheme. The procedure to increase the temporal order of accuracy from 2 to 4 consists of the following steps.

- Run the ADI2 scheme over some time interval $[0, T]$.
- From the numerical values of this solution, evaluate an approximation of the local truncation error $\underline{E}^{n+1/2}$ at each time level.
- Re-run the ADI2 scheme over the time $[0, T]$, but this time include $\underline{E}^{n+1/2}$ as a RHS (a forcing function) to the equation.

The last two steps can be repeated in order to reach still higher orders of accuracy (6, 8, etc.). Two orders of accuracy will be gained each time the basic second order scheme is re-run.

To apply this idea to the ADI scheme, we start by noting that the local error in (9) becomes

$$\begin{aligned} \underline{E}^{n+1/2} &= \left(1 - \tfrac{\Delta t}{2} A\right)\left(1 - \tfrac{\Delta t}{2} B\right)\underline{u}^{n+1} - \left(1 + \tfrac{\Delta t}{2} A\right)\left(1 + \tfrac{\Delta t}{2} B\right)\underline{u}^{n} \\ &= \left(1 + \tfrac{(\Delta t)^2}{4} AB\right)\left(\underline{u}^{n+1} - \underline{u}^{n}\right) - \tfrac{\Delta t}{2}\left(\underline{u}^{n+1} + \underline{u}^{n}\right) \end{aligned}$$

From the expansions

$$\begin{aligned} \underline{u}^{n+1} - \underline{u}^{n} &= \Delta t\, \underline{u}_t^{n+1/2} + \tfrac{(\Delta t)^3}{24}\underline{u}_{ttt}^{n+1/2} + \tfrac{(\Delta t)^5}{1920}\underline{u}_{ttttt}^{n+1/2} + \ldots \\ \underline{u}^{n+1} + \underline{u}^{n} &= 2\underline{u}^{n+1/2} + \tfrac{(\Delta t)^2}{4}\underline{u}_{tt}^{n+1/2} + \tfrac{(\Delta t)^4}{192}\underline{u}_{tttt}^{n+1/2} + \ldots \end{aligned}$$

follow

$$\underline{E}^{n+1/2} = (\Delta t)\left(\underline{u}_t^{n+1/2} - (A+B)\underline{u}^{n+1/2}\right) + \qquad \text{Vanishes because of the PDE}$$

$$+(\Delta t)^3\left(\tfrac{AB}{4}\underline{u}_t^{n+1/2} - \tfrac{A+B}{8}\underline{u}_{tt}^{n+1/2} + \tfrac{1}{24}\underline{u}_{ttt}^{n+1/2}\right) + \qquad \text{Use to get DC of order 4}$$

$$+(\Delta t)^5\left(\tfrac{AB}{96}\underline{u}_{ttt}^{n+1/2} - \tfrac{A+B}{384}\underline{u}_{tttt}^{n+1/2} + \tfrac{1}{1920}\underline{u}_{ttttt}^{n+1/2}\right) + \qquad \text{Use to get DC of order 6}$$

$$+\ldots \qquad \ldots$$

To proceed from ADI2 to ADI-DC4 (deferred correction to 4th order), we approximate $\underline{E}^{n+1/2}$ by

$$\begin{aligned} \underline{E}^{n+1/2} \approx{}& \tfrac{(\Delta t)^2}{4} AB(\underline{u}^{n+1} - \underline{u}^{n}) - \tfrac{\Delta t}{16}(A+B)(\underline{u}^{n+2} - \underline{u}^{n+1} - \underline{u}^{n} + \underline{u}^{n-1}) \\ &+ \tfrac{1}{4}(\underline{u}^{n+2} - 3\underline{u}^{n+1} + 3\underline{u}^{n} - \underline{u}^{n-1})\,. \end{aligned}$$

In the same manner as we showed that (9) was equivalent to (10), we can show that

$$\left(1 - \tfrac{\Delta t}{2} A\right)\left(1 - \tfrac{\Delta t}{2} B\right)\underline{u}^{n+1} = \left(1 + \tfrac{\Delta t}{2} A\right)\left(1 + \tfrac{\Delta t}{2} B\right)\underline{u}^{n} + \underline{E}^{n+1/2} \tag{21}$$

is equivalent to

$$\left\{ \begin{aligned} \left(1 - \frac{\Delta t}{2} A\right)\underline{u}^{n+1/2} &= \left(1 + \frac{\Delta t}{2} B\right)\underline{u}^{n} + \tfrac{1}{2}\underline{E}^{n+1/2} \\ \left(1 - \frac{\Delta t}{2} B\right)\underline{u}^{n+1} &= \left(1 + \frac{\Delta t}{2} A\right)\underline{u}^{n+1/2} + \tfrac{1}{2}\underline{E}^{n+1/2}\,. \end{aligned} \right. \tag{22}$$

Multiplying the second equation in (22) by $\left(1 - \frac{\Delta t}{2} A\right)$ and then using the first equation in (22) leads in a few steps to (21)

$$\begin{aligned}
&\left(1 - \tfrac{\Delta t}{2} A\right)\left(1 - \tfrac{\Delta t}{2} B\right) \underline{u}^{n+1} \\
&\quad = \left(1 - \tfrac{\Delta t}{2} A\right)\left(1 + \tfrac{\Delta t}{2} A\right) \underline{u}^{n+1/2} + \tfrac{1}{2}\left(1 - \tfrac{\Delta t}{2} A\right) \underline{E}^{n+1/2} \\
&\quad = \left(1 + \tfrac{\Delta t}{2} A\right)\left(1 - \tfrac{\Delta t}{2} A\right) \underline{u}^{n+1/2} + \tfrac{1}{2}\left(1 - \tfrac{\Delta t}{2} A\right) \underline{E}^{n+1/2} \\
&\quad = \left(1 + \tfrac{\Delta t}{2} A\right)\left\{\left(1 + \tfrac{\Delta t}{2} B\right) \underline{u}^{n} + \tfrac{1}{2}\underline{E}^{n+1/2}\right\} + \tfrac{1}{2}\left(1 - \tfrac{\Delta t}{2} A\right) \underline{E}^{n+1/2} \\
&\quad = \left(1 + \tfrac{\Delta t}{2} A\right)\left(1 + \tfrac{\Delta t}{2} B\right) \underline{u}^{n} + \underline{E}^{n+1/2} .
\end{aligned}$$

The corrections in this deferred correction procedure can therefore be very conveniently implemented by applying half the amount of the approximative local error to each of the two ADI stages.

Enhancements by Re-starts

During computations over long times, *re-starts* can improve the accuracy of both Richardson extrapolation and deferred correction calculations. Instead of running the calculations over $[0, T]$, we run at first over $[0,\ T/2]$ to obtain an accurate value at $T/2$. Then, re-starting from that point, we compute up to time T. This amounts to a '2 subinterval' calculation. A '4 subinterval' calculation would similarly split $[0, T]$ into $[0, T/4]$,$[T/4, T/2]$,$[T/2, 3T/4]$,$[3T/4, T]$, etc. The idea is to avoid ever running the underlying low-order method across a long time interval, as it would then accumulate very large errors. Subsequent extrapolation or deferred correction would then have little chance of working well. With the re-start approach, the low-order computations will never be given the time to drift too far off course, and the subsequent corrections will therefore be correspondingly more effective. For additional discussion and test results, see [30].

Abbreviations for the Time Stepping Methods

The time stepping methods we have just introduced and which we next will compare are

ADI	Alternating Direction Implicit FDTD method
CNS	Split step method of order 2 using Crank-Nicolson
ADI-TS	Time sequence enhanced ADI
CNS-TS	Time sequence enhanced CNS
ADI-EX	Richardson extrapolation enhanced ADI
ADI-DC	Deferred correction enhanced ADI
ADI-REX	The ADI-EX method further enhanced by restarts
ADI-RDC	The ADI-DC method further enhanced by restarts

Since it was found in [30] that the ADI-based methods might be marginally more effective than the CNS-based ones, we will here not include CNS-EX, CNS-DC, CNS-REX and CNS-RDC. At the end of each of the abbreviations, we also add a number specifying the order in time.

Test Problem

We consider the following exact periodic solution to (1) over the unit cube with $\varepsilon = \mu = 1$:

$$\begin{array}{ll} E_x = \cos(2\pi(x+y+z) - 2\sqrt{3}\pi t) & H_x = \sqrt{3}E_x \\ E_y = -2E_x & H_y = 0 \\ E_z = E_x & H_z = -\sqrt{3}E_x \,. \end{array} \tag{23}$$

We discretise all spatial derivatives by centred second order finite differences. Instead of quoting the size of the time and space steps explicitly, we instead give *points per time interval* (PPT, number of time steps / total time T for test problem) and *points per wave length* (PPW, with the wave length here equal to $\sqrt{3}$). By converting the spatial variables over to the Fourier domain, the system to be solved can be written as $\widehat{\underline{u}}_t = A\widehat{\underline{u}}$ where A is a 6×6 matrix (independently of how fine we choose our spatial discretisation). On this new system of ODEs, we then carry out all our different time stepping procedures. This conversion over to the Fourier domain allows us to observe the influence of the PPW quantity also for very high values without this leading to any increased computational cost per time step.

Computational Cost Comparisons

Fig. 11 illustrates how the L_2 error at a final time $t = 1$ varies with PPW and PPT. The value of PPW determines the spatial discretisation error level. If PPW is held fixed and PPT is increased without bound, time stepping errors will decrease to zero, and the total error will come down to the level that is set by the spatial errors for the particular value of PPW. In the limit of PPW= ∞ we will only see time stepping errors. The errors for the different methods will then decrease indefinitely, as is indicated by the dotted extrapolations in Fig. 12 (near to where we have labelled the different methods).

In Fig. 12 we have fixed PPW to 10^5, but have also indicated in the right margin what the asymptotic error levels becomes for other PPW choices. Again, extrapolations corresponding to PPW= ∞ are also shown. When comparing the relative computational cost between different time stepping methods, we need to consider not only the number of time steps but also the number of operations per time step. The three subplots show the errors for our time stepping methods of orders 2, 4, and 6 respectively, when this cost has been factored in – thus with relative cost replacing PPT along the horizontal axis. We can make a number of observations from Fig. 12.

- Unless PPW is quite large, there is little reason to increase the order in time beyond what ADI2 offers.
- For very high PPW (the situation in the class of problems we are considering; with geometrical features much smaller than a typical wave length), the benefits of higher order time stepping can be substantial.
- Of the three different enhancement approaches, DC (deferred correction) appears the least effective and EX (Richardson extrapolation) the most effective.

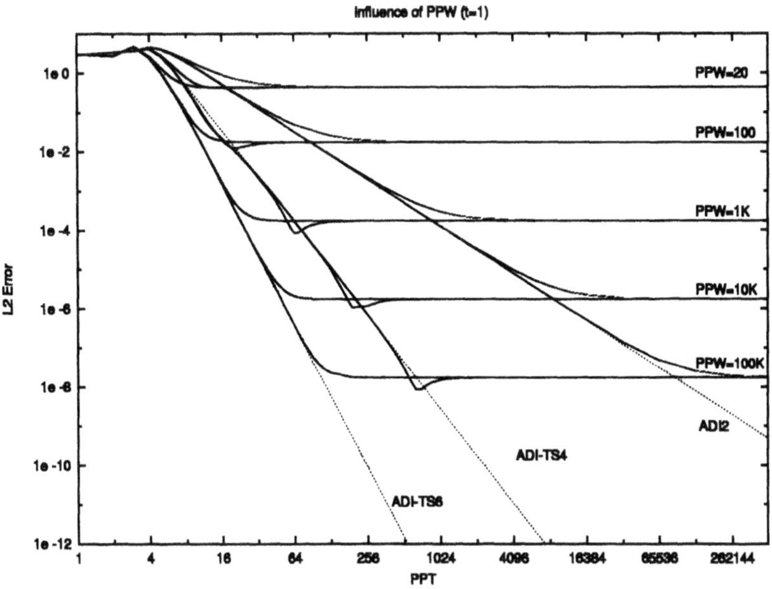

Fig. 11. L_2 error at time $t = 1$ as function of PPW and PPT in the case of the ADI method and time-sequence enhanced versions of it

The situation with regard to unconditional stability is somewhat unclear in the case of the TS methods. It appears to hold for TS4 but, in certain cases, does not hold for TS6. For the DC and EX methods, it will hold for all orders as long as the number of restarts are held finite (rather than being increased with PPT). Re-started, high order extrapolations of the ADI approach (ADI-REX) appears to be particularly attractive for high PPW calculations.

For more details on these higher order enhancements (such as how the number of re-starts influence the resulting accuracy), see Lee and Fornberg [30]. In conclusion, regarding this class of time stepping methods, it needs to be added that they so far have been applied only to very simple periodic problems. Tests with variable media and irregular interfaces should be carried out before any firm recommendations can be made.

4.4 Conclusions

Problems in CEM feature two length scales: the size of geometric features, and a typical wave length. The first part of this article focused on the case when the length scales are similar. Thanks to its simplicity, the Yee scheme has been popular in many applications. Its drawbacks are its low (second order) accuracy and lack of geometric flexibility. We noted that it can be enhanced to high orders in both space and time by using wider FD stencils and by incorporating more back levels respectively. The novel, high-order time-staggered linear multistep methods that were briefly described

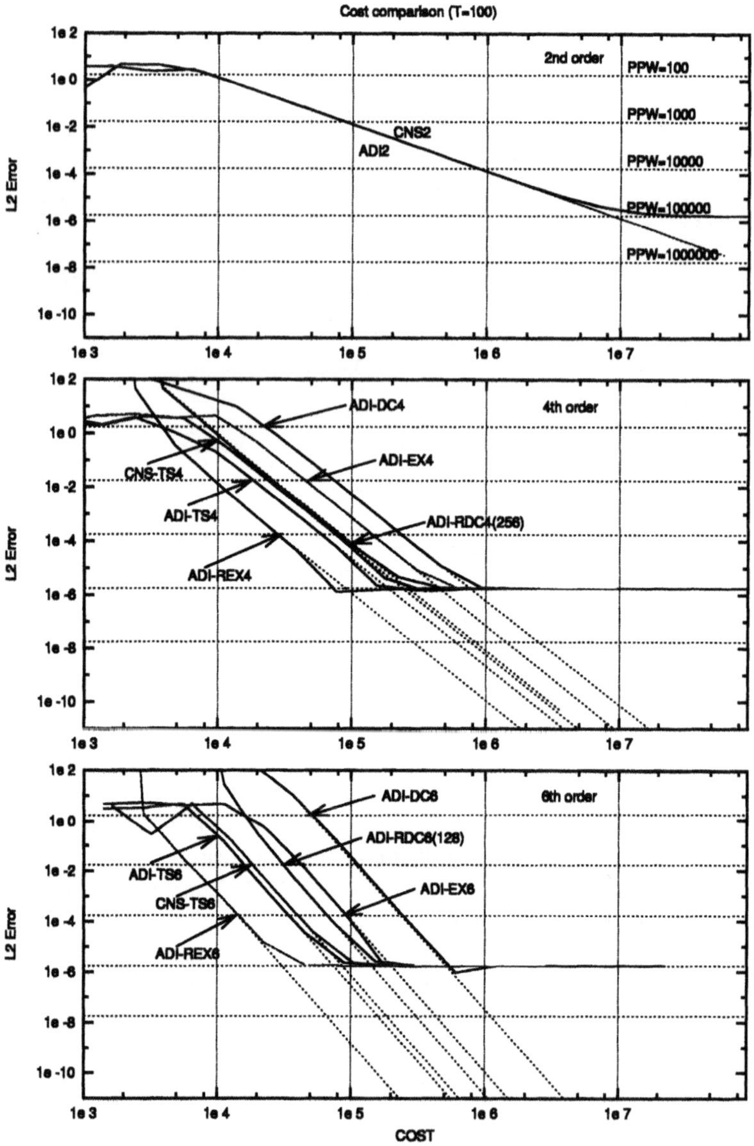

Fig. 12. L_2 error at time $T = 100$ as function of PPW and of relative computational cost for time stepping methods of different orders

have better stability and accuracy properties than their classical Adams-type non-staggered counterparts. The difficulty with geometric flexibility can be met by taking a hybrid approach, such as switching to a FE scheme near interfaces or to a FD scheme on 'patches' that are mapped to follow curved interfaces. Boundary integral methods can in some cases give very high efficiency, not only for time harmonic cases but also for fully time dependent ones.

After mentioning a few implementations, we turned our attention to cases where the geometrical features are many orders of magnitude smaller than a typical wave length. The primary issue then becomes how to effectively bypass the CFL stability condition. Two second order accurate methods with unconditional stability have recently been found, ADI and CNS. The remainder of the paper has been devoted to a description of these and a discussion about how they can be enhanced to feature higher order accuracies in time (without losing their unconditional stability). Mixed with reports of successful implementations and insights are also some tentative concerns, e.g. the

- significance of the lack of exact pointwise conservation of div (εE) and div (μH) is unclear;
- possibility of large errors arising from the $\frac{(\Delta t)^2}{4} A\, B(\underline{u}^{n+1} - \underline{u}^n)$ term in (8) in cases of variable coefficients (and which constant media dispersion analysis does not seem to reveal [21]);
- stability situation with regard to ADI-type methods in combination with certain boundary conditions;
- accuracy of some of the high order time stepping approaches in cases when the equations have explicit time dependence due to the boundary conditions or to forcing.

All these issues are at present under study by different research groups. These open questions notwithstanding, ADI-type methods form an exciting new direction in CEM, now on the verge of moving from test problems to production applications.

4.5 Acknowledgements

Several of the results described here have been obtained in collaboration with Toby Driscoll and Jongwoo Lee. In addition, very helpful discussions with Fredrik Edelvik, Jan Hesthaven, and Gunnar Ledfelt are gratefully acknowledged.

References

1. O. Bruno, New high–order, high–frequency methods in computational electromagnetism, Topics in Computational Wave Propagation 2002, Springer (2003).
2. G. Dahlquist, Convergence and stability in the numerical integration of ordinary differential equations, Math. Scand. 4 (1956), 33–53.
3. G. Dahlquist, 33 years of numerical instability, part I, BIT, 25 (1985), 188–204.
4. M. Darms, R. Schuhmann, H. Spachmann and T. Weiland, Asymmetry effects in the ADI–FDTD algorithm, to appear in IEEE Microwave Guided Wave Lett.
5. L. Demkowicz, Fully automatic hp–adaptive finite elements for the time–harmonic Maxwell's equations, Topics in Computational Wave Propagation 2002, Springer (2003).
6. J. Douglas, Jr, On the numerical integration of $\frac{\partial^2 u}{\partial x^2} + \frac{\partial^2 u}{\partial y^2} = \frac{\partial u}{\partial t}$ by implicit methods, J. Soc. Indust. Appl. Math., 3 (1955), 42–65.

7. T.A. Driscoll and B. Fornberg, Block pseudospectral methods for Maxwell's equations: II. Two–dimensional, discontinuous–coefficient case, SIAM Sci. Comput. 21 (1999), 1146–1167.
8. F. Edelvik and G. Ledfelt, A comparison of time–domain hybrid solvers for complex scattering problems, Int. J. Numer. Model 15 (5–6) (2002), 475–487.
9. M. El Hachemi, Hybrid methods for solving electromagnetic scattering problems on overlapping grids, in preparation.
10. E. Forest and R.D. Ruth, Fourth order symplectic integration, Physica D 43 (1990), 105–117.
11. B. Fornberg, A Practical Guide to Pseudospectral Methods, Cambridge University Press (1996).
12. B. Fornberg, Calculation of weights in finite difference formulas, SIAM Review, 40 (1998), 685–691.
13. B. Fornberg, A short proof of the unconditional stability of the ADI–FDTD scheme, University of Colorado, Department of Applied Mathematics Technical Report 472 (2001).
14. B. Fornberg, High order finite differences and the pseudospectral method on staggered grids, SIAM J. Numer. Anal. 27(1990), 904–918.
15. B. Fornberg and T.A. Driscoll, A fast spectral algorithm for nonlinear wave equations with linear dispersion, JCP, 155 (1999), 456–467.
16. B. Fornberg and M. Ghrist, Spatial finite difference approximations for wave–type equations, SIAM J. Numer. Anal. 37 (1999), 105–130.
17. R. Frank and C.W. Ueberhuber, Iterated defect correction for diferential–equations 1. Theoretical results, Computing, 20 Nr 3 (1978), 207–228.
18. R. Frank, F. Macsek and C.W. Ueberhuber, Iterated defect correction for diferential–equations 2. Numerical experiments, Computing, 33 Nr 2 (1984), 107–129.
19. R. Frank, J. Hertling and H. Lehner,Defect correction algorithms for stiff ordinary differential equations, Computing, Supplement 5 (1984), 33–41.
20. L. Gao, B. Zhang and D. Liang, Stability and convergence analysis of the ADI–FDTD algorithm for 3D Maxwell equation. In preparation.
21. S.G. García, T.-W. Lee and S.C. Hagness, On the accuracy of the ADI–FDTD method, IEEE Antennas and Wireless Propagation Letters 1 No 1 (2002), 31–34.
22. M. Ghrist, T.A. Driscoll and B. Fornberg, Staggered time integrators for wave equations, SIAM J. Num. Anal. 38 (2000), 718–741.
23. B. Gustafsson and W. Kress, Deferred correction methods for initial value problems, BIT 41 (2001), 986–995.
24. E. Hairer, C. Lubich and G. Wanner, Geometric Numerical Integration, Springer Verlag (2002).
25. J. Hesthaven and T. Warburton, Nodal high–order methods on unstructured grids I. Time–domain solution of Maxwell's equations, JCP, 181 (2002), 186–221.
26. R. Hiptmair, Finite elements in computational electromagnetism, Acta Numerica 2002 (2002), 237–339.
27. J. Jin, The Finite Element Method in Electromagnetics, Wiley, New York (1993).
28. K.S. Kunz and J. Luebbers, The Finite Difference Time Domain Method for Electromagnetics, CRC Press, Inc. (1993).
29. J. Lee and B. Fornberg, A split step approach for the 3D Maxwell's equations, University of Colorado, Department of Applied Mathematics Technical Report 471 (2001), submitted to Journal of Computational and Applied Mathematics (2002).
30. J. Lee and B. Fornberg, Some unconditionally stable time stepping methods for the 3D Maxwell's equations, Submitted to IMA journal of Applied Mathematics (2002).

31. G. Liu and S.D. Gedney, Perfectly matched layer media for an unconditionally stable three–dimensional ADI–FDTD method, IEEE Microwave Guided Wave Lett. 10 (2000), 261–263.
32. J.C. Maxwell, A Treatise on Electricity and Magnetism, Clarendon Press, Oxford (1873).
33. T. Namiki, 3D ADI–FDTD method – Unconditionally stable time–domain algorithm for solving full vector Maxwell's equations, IEEE Transactions on Microwave Theory and Techniques, 48, No 10 (2000), 1743–1748.
34. F. Neri, Lie algebras and canonical integration, Dept. of Physics, University of Maryland, preprint (1987).
35. D. Peaceman and J.H.H. Rachford, The numerical solution of parabolic and elliptic differential equations, J. Soc. Indust. Appl. Math. 3 (1955), 28–41.
36. V. Pereyra, Accelerating the convergence of discretization algorithms, SIAM J. Numer. Anal. 4 (1967), 508–532.
37. L.F. Richardson, The deferred approach to the limit, Phil. Trans. A, 226 (1927), 299–349.
38. T. Rylander and A. Bondeson, Stable FEM–FDTD hybrid method for Maxwell's equations, Computer Physics Communications, 125 (2000), 75–82.
39. B. Shanker, A.A. Ergin, K. Aygun, E Michielssen, Analysis of transient electromagnetic scattering phenomena using a two–level plane wave time–domain algorithm, IEEE Trans. Antennas Propagation 48 (2000), 510–523.
40. G. Strang, On construction and comparison of difference schemes, SIAM J. Numer. Anal. 5 (1968), 506–516.
41. M. Suzuki, General theory of fractal path integrals with applications to many–body theories and statistical physics, J. Math. Phys. 32 (1991), 400–407.
42. A. Taflove and S.C. Hagness, Computational Electrodynamics: The Finite–Difference Time–Domain Method, 2nd ed., Artech House, Norwood (2000).
43. V.S. Varadarajan, Lie groups, Lie algebras and their representation, Prentice Hall, Englewood Cliffs (1974).
44. K.S. Yee, Numerical solution of initial boundary value problems involving Maxwell's equations in isotropic media, IEEE Trans. Antennas Propagation, 14 (1966), 302–307.
45. H. Yoshida, Construction of higher order symplectic integrators, Physics Letters A, 150 (1990), 262–268.
46. F. Zheng, Z. Chen, J. Zhang, A finite–difference time–domain method without the Courant stability conditions, IEEE Microwave Guided Wave Lett. 9 (1999), 441–443.
47. F. Zheng, Z. Chen, J. Zhang, Toward the development of a three–dimensional unconditionally stable finite–difference time–domain method, IEEE Transactions on Microwave Theory and Techniques, 48, No 9 (2000), 1550–1558.

On Retarded Potential Boundary Integral Equations and their Discretisation

Tuong Ha-Duong

Université de Technologie de Compiègne, BP 20529, 60205 Compiègne Cedex, France
tuong.ha-duong@utc.fr

Summary. The paper deals with the retarded potential boundary integral equations (RPBIE) used in the numerical resolution of transient scattering problems (the so-called time domain boundary element methods). We propose here a review and update of the mathematical analysis of the involved RPBIE. Our approach, via Laplace transform, is described in some details for the classical acoustic scattering problems. The main results are: (i) existence and uniqueness theorems on a functional framework closely linked to the energy of the scattered waves; (ii) space-time variational formulations for the so-called "first kind" RPBIE, with coerciveness obtained by energy estimates. That leads us to advocate choosing these first kind RPBIE and their Galerkin approximations, instead of the second kind RPBIE and the collocation approximations. The actual space-time boundary elements are described in some detail. Examples of numerical experiments that confirm the unconditional stability of our schemes are reported, as well as references for similar results in other related work.

Key words: Boundary integral equations; Time domain; Galerkin approximation; Stability; Acoustic scattering

1 Introduction

This paper is concerned with *retarded potential boundary integral equations* (RPBIEs) and their discretisation, applied in transient scattering problems. Initiated in the early 1960s by Friedman and Shaw [37], time domain boundary element methods as applied to wave propagation problems now have a forty year history of development and applications. A large majority of papers dealing with these methods use collocation discretisation schemes (see [16, 20, 28, 49, 56] etc. and other references cited in these papers) and one can say with Birgisson et al. [16], that "*there is growing evidence of numerical instabilities*" in those schemes. Actually, various methods are proposed to improve the stability of these schemes: techniques of time-averaging (see [25, 26, 28, 56]), of shifted time-steps (cf. [16, 52]) or by having recourse to some implicitness (cf. [20, 30]). The main idea is to manage to kill the high frequencies of the algebraic systems obtained in the discretisation process. Nevertheless, even if the "interior resonances" are mentioned to explain instabilities (see [54–56]), a general proof of stability of these improved schemes is not really available (see however [29]), especially when the surface is not simple. On the other hand, one notes the absence of mathematical analysis of the "integral equations" involved in these papers. One is then led to think that some essential feature may be missing and that the proposed techniques are, in some sense, insufficiently well-grounded.

Actually, the difficulties appear as soon as one looks at the generally preferred "second kind" RPBIEs, such as (14) below or the well-known electric and magnetic field integral equations (EFIE and MFIE) in electromagnetics applications (see e.g. [49]): all the "integral operators" involved in fact contain surface integrals of some terms where both the function concerned and its time derivative are present, and are evaluated with arguments composed of mixtures of space and time variables. Thus, there is really no hope for these equations to satisfy the appellation of "second kind integral equation" in the sense of Fredholm. It is then not enough to evoke the Fredholm framework to say whether these equations admit a unique solution continuous with respect to the data or not. Yet, those questions are essential to any attempt to analyse the numerical schemes that can be set to solve the equations.

The main part of this paper responds to this necessity for mathematical analysis of the RPBIE. Because of the limited space, only acoustic scattering problems will be considered. Hopefully, that restriction permits us to present a sufficiently detailed account of the method and main results (the ones that are directly useful in the construction of discretisation schemes!). For other type of waves, we refer to the literature cited in the bibliography and commented in Sect. 6.

Following the methodology indicated in [9] (see also [39, 40]) to deal with the so called "first kind" RPBIEs used in Dirichlet or Neumann scattering problems, we propose an approach based on the Fourier-Laplace transform of the RPBIE, and the analysis of the transformed equations with respect to the frequency. We also update the results of [9] by the use of better indexed norms in the spatial Sobolev spaces and by giving some new results in finite time intervals. This is done in Sects. 3 and 4, both for "first kind" and "second kind" RPBIEs in the three basic acoustic problems, namely the Dirichlet, Neumann and Robin problems. The different RPBIEs that can be used for solving these problems are recalled in Sect. 2. This analysis leads us to advocate the use of first kind RPBIEs and their Galerkin approximations, instead of second kind RPBIEs and collocation approximations. This preference is justified by the fact that a space-time variational formula can be set up for these first kind RPBIEs in a natural manner suggested by energy identities, giving a coerciveness property that ensures unconditional stability properties for conforming Galerkin approximation schemes. Sect. 5 is devoted to the presentation of the space-time Galerkin methods and some numerical experiments confirming the expected stability properties.

This methodology has been developed with success for other acoustic, electromagnetic and elastodynamic scattering problems. However, most of the work is reported in mathematical theses only available in French, and is generally unknown to the community of workers in engineering (see for example the review papers of

Beskos [15] and Bonnet et al. [17,49]). So, in Sect. 6, we give a rapid survey of these papers. Finally, conclusions are drawn in Sect. 7.

2 From the Governing Equations to the Associated RPBIE

We consider the scattering problem of transient acoustic waves in a fluid medium with a submerged object (the scatterer) which could be *soft, hard* or *absorbing*.

Let Ω^{i} be the bounded domain occupied by the scatterer in the usual space ($\mathbb{R}^3$) and $\Omega^{\mathrm{e}} = \mathbb{R}^3 \setminus \bar{\Omega}^{\mathrm{i}}$ the exterior domain occupied by the fluid medium. The domain Ω^{e} is supposed connected but Ω^{i} is not. Their common boundary $\Gamma = \partial\Omega^{\mathrm{i}} = \partial\Omega^{\mathrm{e}}$ is supposed regular enough. We denote by u^{e} the scattered acoustic pressure created in the fluid medium by an incident field u^{inc} (the wave propagating without the obstacle). We suppose that the perturbation u^{inc} comes from the exterior medium and that at the instant $t = 0$ it has not reached the target yet. Therefore, we have the following initial boundary value problem:

$$\frac{1}{c^2}\frac{\partial^2 u^{\mathrm{e}}(x,t)}{\partial t^2} - \Delta u^{\mathrm{e}}(x,t) = 0 \quad \text{in } \Omega^{\mathrm{e}} \times \mathbb{R}^+ \tag{1}$$

$$u^{\mathrm{e}}(x,0) = 0 \quad \text{in } \Omega^{\mathrm{e}} \tag{2}$$

$$\frac{\partial u^{\mathrm{e}}}{\partial t}(x,0) = 0 \quad \text{in } \Omega^{\mathrm{e}} \tag{3}$$

$$\mathcal{B}u^{\mathrm{e}}(x,t) = f(x,t) \quad \text{on } \Gamma \times \mathbb{R}^+ \tag{4}$$

where n denotes the unit normal vector to Γ, oriented from Ω^{i} to Ω^{e}, c is the speed of sound in the medium, which can (and, from now on, will) be set to be 1 by choosing appropriate units of measurement, and $\mathcal{B}$ the boundary operator:

$$\mathcal{B}u^{\mathrm{e}}(x,t) = \begin{cases} u^{\mathrm{e}}(x,t) & \text{for a soft scatterer} \\ \dfrac{\partial u^{\mathrm{e}}}{\partial n} - \dfrac{\alpha}{c}\dfrac{\partial u^{\mathrm{e}}}{\partial t} & \text{for a hard or absorbing scatterer} \end{cases} \tag{5}$$

where f is related to the incident waves by

$$f(x,t) = -\mathcal{B}u^{\mathrm{inc}}(x,t). \tag{6}$$

and α is the impedance function of the surface Γ, with

$$\alpha(x) \geq 0, \quad \forall x \in \Gamma, \tag{7}$$

($\alpha(x) = 0, \quad \forall x$ for the hard scatterer).

The problem u^{e} is well-posed. A method for proving that is to look at the energy of the total pressure wave $u^T = u^{\mathrm{e}} + u^{\mathrm{inc}}$ in Ω^{e} (assuming as usual that the incident wave is of finite energy). Actually, if

$$E(u^T,t) = \frac{1}{2}\int_{\Omega^{\mathrm{e}}} \left(|\nabla u^T(x,t)|^2 + |u^T(x,t)|^2\right) \mathrm{d}x\,,$$

then, using Green's formula and the boundary conditions (4), (6), one obtains:

$$\frac{\mathrm{d}E(u^T,t)}{\mathrm{d}t} = \begin{cases} 0 & \text{for soft scatterers} \\ -\int_\Gamma \alpha(x)(\frac{\partial u^T}{\partial t})^2(x,t)\,\mathrm{d}\sigma_x & \text{for hard or absorbing cases.} \end{cases}$$

That proves that this energy is non-increasing (decreasing if $\alpha(x) > 0, \ \forall x \in \Gamma$).

Now, there are different ways to solve the problem $(P^\mathrm{e}) = (1-4)$ using boundary integral equations. Each of them consists in looking for the solution as a prescribed combination of *surface retarded potentials* and expressing the boundary condition as a function of the densities of the potentials. Equivalently, this corresponds to associating with the *exterior* problem (P^e) an appropriate *interior* problem (P^i). Each choice of (P^i) induces a different integral equation. Below, in Sect. 2.2, we will present two different RPBIEs for each of the problems considered. Before that, let us recall some known facts and also fix the notation used throughout.

2.1 Surface Retarded Potentials

Let us denote by u a solution of the wave equation in $\mathbb{R}^3\backslash\Gamma$, equal to u^e in $\Omega^\mathrm{e} \times \mathbb{R}^+$ and u^i in $\Omega^\mathrm{i} \times \mathbb{R}^+$. Then u satisfies the following representation formula (see e.g. the book of Stratton [58])

$$\begin{aligned} u(x,t) = &-\frac{1}{4\pi}\int_\Gamma \frac{n_y.(x-y)}{|x-y|}\left(\frac{\varphi(y,\tau)}{|x-y|^2} + \frac{\dot\varphi(y,\tau)}{|x-y|}\right)\mathrm{d}\sigma_y \\ &+\frac{1}{4\pi}\int_\Gamma \frac{p(y,\tau)}{|x-y|}\,\mathrm{d}\sigma_y \qquad \forall (x,t) \in (\mathbb{R}^3\backslash\Gamma)\times\mathbb{R}^+, \end{aligned} \tag{8}$$

where

$$\varphi = u^\mathrm{i} - u^\mathrm{e}, \quad p = \frac{\partial u^\mathrm{i}}{\partial n} - \frac{\partial u^\mathrm{e}}{\partial n} \quad \text{on } \Gamma\times\mathbb{R}^+,$$

are the jumps of u and its normal derivative across Γ, and

$$\tau = t - |x-y|$$

is the *retarded time* ($\tau = t - |x-y|/c$ in the general case). The dot on φ designates, as usual, the time derivative.

On the other hand, each of the integrals on the RHS of (8) also represents a solution of the wave equation in both domains $\Omega^\mathrm{e}\times\mathbb{R}^+$ and $\Omega^\mathrm{i}\times\mathbb{R}^+$: we have the so-called *single layer retarded potential*

$$Lp(x,t) = \frac{1}{4\pi}\int_\Gamma \frac{p(y,\tau)}{|x-y|}\,\mathrm{d}\sigma_y, \quad (x,t)\in(\mathbb{R}^3\backslash\Gamma)\times\mathbb{R}^+; \tag{9}$$

and the *double layer retarded potential*

$$M\varphi(x,t) = \frac{1}{4\pi}\int_\Gamma \frac{n_y.(x-y)}{|x-y|}\left(\frac{\varphi(y,\tau)}{|x-y|^2} + \frac{\dot\varphi(y,\tau)}{|x-y|}\right)\mathrm{d}\sigma_y\,, \tag{10}$$

for $(x,t) \in (\mathbb{R}^3 \backslash \Gamma) \times \mathbb{R}^+$.

It is also known that these potentials, defined for regular enough density functions p and φ, satisfy the following limits when the point $x \in \Omega^{\rm i}$ (resp. $\Omega^{\rm e}$) tends to a point on the surface Γ:

$$\begin{cases} (Lp)^i(x,t) & = (Lp)^{\rm e}(x,t) & = Sp(x,t) \\ \dfrac{\partial (Lp)^i}{\partial n}(x,t) & = (I/2+K)p(x,t) & \\ \dfrac{\partial (Lp)^e}{\partial n}(x,t) & = (-I/2+K)p(x,t) & \\ (M\varphi)^{\rm i}(x,t) & = (-I/2+K')\varphi(x,t) & \\ (M\varphi)^{\rm e}(x,t) & = (I/2+K')\varphi(x,t) & \\ \dfrac{\partial (M\varphi)^i}{\partial n}(x,t) = & \dfrac{\partial (M\varphi)^e}{\partial n}(x,t) & = D\varphi(x,t) \end{cases} \tag{11}$$

where the *retarded potential operators* K, K', S, D are defined for regular functions on $\Gamma \times \mathbb{R}^+$ by:

$$\begin{cases} Sp(x,t) = \dfrac{1}{4\pi} \displaystyle\int_\Gamma \dfrac{p(y,\tau)}{|x-y|} \,\mathrm{d}\sigma_y \\ Kp(x,t) = \dfrac{1}{4\pi} \displaystyle\int_\Gamma n_x.\nabla_x (\dfrac{p(y,\tau)}{|x-y|}) \,\mathrm{d}\sigma_y \\ \qquad = \dfrac{1}{4\pi} \displaystyle\int_\Gamma \dfrac{n_x.(x-y)}{|x-y|} \left(\dfrac{p(y,\tau)}{|x-y|^2} + \dfrac{\dot{p}(y,\tau)}{|x-y|} \right) \mathrm{d}\sigma_y \\ K'\varphi(x,t) = \dfrac{1}{4\pi} \displaystyle\int_\Gamma - n_y.\nabla_x (\dfrac{\varphi(y,\tau)}{|x-y|}) \,\mathrm{d}\sigma_y \\ \qquad = \dfrac{1}{4\pi} \displaystyle\int_\Gamma \dfrac{n_y.(x-y)}{|x-y|} \left(\dfrac{\varphi(y,\tau)}{|x-y|^2} + \dfrac{\dot{\varphi}(y,\tau)}{|x-y|} \right) \mathrm{d}\sigma_y \\ D\varphi(x,t) = \lim\limits_{x' \in \Omega^+ \to x} n_x.\nabla_{x'} (-\dfrac{1}{4\pi} \displaystyle\int_\Gamma n_y.\nabla_{x'} (\dfrac{\varphi(y,t-|x'-y|)}{|x'-y|}) \,\mathrm{d}\sigma_y) \,. \end{cases} \tag{12}$$

Notice that all the integrals in (12) are taken in a classical sense (weakly integrable kernels). However, if one simply puts the limit inside the integral on the RHS of $D\varphi$, a singular kernel of order 3 (i.e. of the same order as $|x-y|^{-3}$) would appear. Thus, the operator D contains a *hypersingular kernel* and must be properly defined. Since we will present in this paper the Galerkin method for the RPBIEs, a natural method to deal with this hypersingularity is to use distribution (or duality) theory. That is done in [9] and gives rise to the following formula:

$$\int_0^\infty \int_\Gamma D\varphi(x,t)\eta(x,t) \,\mathrm{d}\sigma_x \,\mathrm{d}t = \frac{1}{4\pi} \int_0^\infty \iint_{\Gamma \times \Gamma} \mathcal{I} \,\mathrm{d}\sigma_y \,\mathrm{d}\sigma_x \,\mathrm{d}t \tag{13}$$

where

$$\mathcal{I} = \frac{n_x.n_y}{|x-y|} \dot{\varphi}(y,\tau)\dot{\eta}(x,t) + \frac{\mathrm{curl}_\Gamma\,(\varphi(y,\tau)).\,\mathrm{curl}_\Gamma\,\eta(x,t)}{|x-y|} \,,$$

η is a test function, the tangential curl operator is defined by

$$\mathrm{curl}_\Gamma\,\varphi = n \wedge \nabla \tilde{\varphi}$$

and $\tilde{\varphi}$ is defined in a tubular neighbourhood of Γ, constant along each normal line to Γ and equal to φ at the intersection point.

Now, only a weakly singular kernel $|x-y|^{-1}$ is present in (13) and one sees that the integrals in its RHS make sense when φ and η have their first derivatives square integrable (on $\mathbb{R}^+ \times \Gamma$) and η is compactly supported with respect to the time variable. Our choice of the discretisation functions in the following will take into account this fact.

2.2 Examples of RPBIEs

The Dirichlet Problem

In the frequency domain, the most popular integral equation used for solving the Dirichlet problem is a "second kind" one. That corresponds to using a double layer potential to represent the solution, or equivalently, to associate with the exterior problem a Neumann interior problem so that there is no discontinuity of the normal derivative across the surface Γ.

In the transient scattering problem, the same idea leads to the following RPBIE

$$(I/2 + K')\varphi(x,t) = f(x,t), \quad (x,t) \in \Gamma \times \mathbb{R}\,. \tag{14}$$

Instead of that, if one considers a Dirichlet interior problem with the same boundary values of u as in the exterior, then one gets a "first kind" RPBIE:

$$Sp(x,t) = f(x,t), \quad (x,t) \in \Gamma \times \mathbb{R} \tag{15}$$

Although these choices are common, they are not the only ones that can be made. We will discuss them in the Sect. 4.

The Robin and Neumann Problems

In the case of an absorbing boundary condition, if the solution is represented as a single layer retarded potential, the following RPBIE is obtained:

$$(-I/2 + K)p - \alpha S\dot{p} = f(x,t) \tag{16}$$

A discretisation scheme and numerical experiments for this RPBIE are reported in [35].

Another idea is to consider an interior problem with the following absorbing boundary condition:

$$\frac{\partial u^{\mathrm{i}}}{\partial n} + \alpha \frac{\partial u^{\mathrm{i}}}{\partial t} = g(x,t) \quad \text{on } \Gamma \times R^+ \tag{17}$$

where

$$g(x,t) = -\left(\frac{\partial u^{\mathrm{inc}}}{\partial n} + \alpha \frac{\partial u^{\mathrm{inc}}}{\partial t}\right) \quad \text{on } \Gamma \times R^+, \quad \text{with } g(x,0) = 0\,.$$

One can remark on the change of the sign before α in the boundary condition (17): this is necessary to have a formula for energy decay like the one given above for the exterior problem. This choice can be explained by a desire to keep some symmetry between the two problems.

Now, using the definition of f and g, one obtains the following system of equations (see [41] and also [53] for a 2-dimensional problem) where the right hand sides explicitly involve the incident wave u^{inc}:

$$\begin{cases} 2(Kp - D\varphi) + \alpha\dot{\varphi} = -2\partial_n u^{\mathrm{inc}} , \\ p + 2\alpha(S\dot{p} - K'\dot{\varphi}) = -2\alpha\partial_t u^{\mathrm{inc}} . \end{cases} \tag{18}$$

When $\alpha = 0$, the system (18) is reduced to $p = 0$ with the equation

$$D\varphi = \partial_n u^{\mathrm{inc}}. \tag{19}$$

This is the "first kind" RPBIE obtained when one uses a double layer retarded potential to solve the Neumann problem. Also when $\alpha = 0$, the equation (16) becomes

$$(-I/2 + K)p = f, \tag{20}$$

the more frequently used "second kind" RPBIE for this hard scatterer problem.

2.3 Some Preliminary Remarks

We are concerned with the numerical resolution of RPBIEs like (14), (15), (16) and the system (18), particularly the method of discretisation, and the properties of the resulting schemes. For a mathematical analysis of these schemes it is useful to do some functional analysis on the operators involved. That is the subject of the following two sections. Here we bring out some preliminary facts.

We first notice that the operators S, K, K', D are not merely time-dependent integral operators on Γ – even for the simplest of them, namely S. This means that one *cannot* write $Sp(x,t)$ as an integral of the form

$$\int_\Gamma K(x,y;t)p(y,t)\,\mathrm{d}\sigma_y .$$

In fact, one has formally:

$$Sp(x,t) = \frac{1}{4\pi}\int_{\mathbb{R}}\int_\Gamma \frac{\delta(t-s-|x-y|)}{|x-y|}p(y,s)\,\mathrm{d}\sigma_y\,\mathrm{d}s$$

where δ is the Dirac function. Thus, time and space variables are intertwined in the kernel of the operators. Further, a look at (12) shows that for the operators K, K' and D, the time derivative of the density functions are present under the surface integrals, again preventing any analysis of these operators as time-dependent integral operators on Γ.

Actually, as pointed out in [9, 39, 40], the best point of view is to consider the involved operators as time convolution of integral operators on Γ. One can see that with the following (formal) calculations

$$\begin{aligned} Sp(x,t) &= \frac{1}{4\pi}\int_\Gamma \frac{p(y,t-|x-y|)}{|x-y|}\,\mathrm{d}\sigma_y \\ &= \Delta t \sum_j \frac{1}{4\pi\Delta t}\int_{\{y\in\Gamma;\ t_{j-1}<|x-y|<t_j\}} (\cdots)\qquad \mathrm{d}\sigma_y \end{aligned}$$

where $t_j = j\Delta t$. Then, letting $\Delta t \to 0$ and $j\Delta t \to s$ in the last integral, one gets

$$Sp(x,t) = \int_0^t S(s)p(\cdot,t-s)(x)\,\mathrm{d}s$$

where

$$[S(s)f](x) = \frac{1}{4\pi s}\int_{\{y\in\Gamma;\ |x-y|=s\}} f(y)\,\mathrm{d}l(y)$$

for every function f defined on Γ where $\mathrm{d}l(y)$ is the length measure on the curve $\{y\in\Gamma;\ |x-y|=s\}$ (see Fig. 1). This can also be seen by resorting to the Fourier-Laplace transform with respect to the time variable. Since, if one writes

$$\begin{aligned} (\mathcal{L}f)(\omega) = \hat{f}(\omega) &= \int_{-\infty}^{\infty} \mathrm{e}^{\mathrm{i}\omega t} f(t)\,\mathrm{d}t \\ &= \mathcal{F}(\mathrm{e}^{-\sigma t} f)(\eta) \qquad \text{where } \omega = \eta + \mathrm{i}\sigma \end{aligned}$$

for a Laplace transformable function f (we say that f is *Laplace transformable* if it is *causal*, i.e. $f(t) = 0$ for $t < 0$ and if there is some real σ such that $\mathrm{e}^{-\sigma t} f$ is temperate, so that its Fourier transform $\mathcal{F}$ is defined), then one gets:

$$\mathcal{L}(Sp)(x,\omega) = \frac{1}{4\pi}\int_\Gamma \frac{\mathrm{e}^{\mathrm{i}\omega|x-y|}}{|x-y|}\hat{p}(y,\omega)\,\mathrm{d}\sigma_y = \hat{S}_\omega \hat{p}(x,\omega),$$

where $\hat{S}_\omega$ is, for each frequency ω, an integral operator on Γ. Similar formulae can be easily written down for operators K, K', D.

In the transform, the causal assumption permits us to place ω in the complex half-plane $\mathbb{C}_{\omega_0} = \{\omega; \operatorname{Im}\omega \geq \omega_0 > 0\}$. Then, assuming that all functions involved are Laplace transformable, and applying the transform to the RPBIE, one gets a BIE on Γ depending on the frequency ω. Corresponding to this BIE is both an exterior and an interior problem for the Helmholtz equation, which can be analysed with respect to $\omega \in \mathbb{C}_{\omega_0}$, using classical functional tools. The results obtained can then be transferred to the RPBIE by the inverse transform, using the Paley-Wiener theorem. The main results are *existence* and *uniqueness* of the solutions (in some *ad hoc* Sobolev spaces on $\Gamma \times \mathbb{R}^+$), as well as their *estimates* with respect to the data, and finally, *the property of finite velocity of propagation* which implies that the solutions at time t depend only on the data at times $s \leq t$.

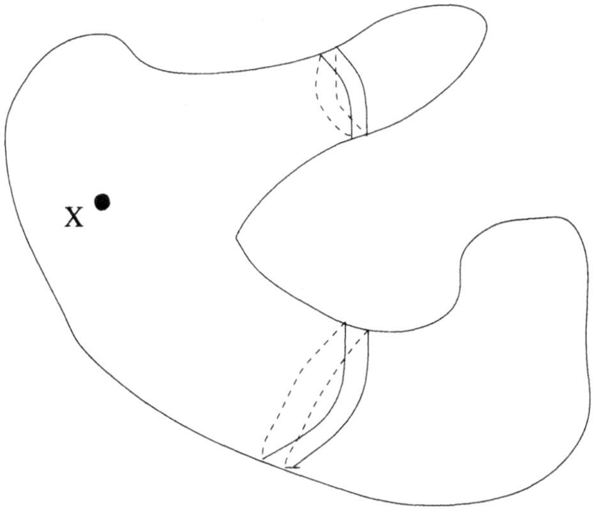

Fig. 1. Geometry of RPO

3 The Helmholtz Equation With Complex Frequency

3.1 Indexed Norms in Sobolev Spaces

Let u be a wave function in a domain Ω, where

$$E(t;u,\Omega) = \frac{1}{2}\int_{\Omega} \left(|\nabla u(x,t)|^2 + \dot{u}^2(x,t)\right)\,\mathrm{d}x$$

is the energy of u in Ω at time t. Assume that for some real σ, the function $\mathrm{e}^{-\sigma t}E(t)$ is in $L^2(\mathbb{R})$. Applying the Parseval equality to the Laplace transform, one gets:

$$\int_{-\infty}^{\infty} \mathrm{e}^{-2\sigma t}E(t;u,\Omega)\,\mathrm{d}t = \frac{1}{4\pi}\int_{-\infty+\mathrm{i}\sigma}^{\infty+\mathrm{i}\sigma}\int_{\Omega}\left(|\nabla\hat{u}(x,\omega)|^2 + |\mathrm{i}\omega\hat{u}(x,\omega)|^2\right)\,\mathrm{d}x$$

This suggests using the following indexed norm in $H^1(\Omega)$

$$\|f\|_{1,\omega,\Omega} = \int_{\Omega}(|\nabla f(x)|^2 + |\omega f(x)|^2)\,\mathrm{d}x\,,$$

which is equivalent (resp. equal) to the usual norm of $H^1(\Omega)$ for $\omega \neq 0$ (resp. $\omega = 1$). This is why we always restrict ourselves to ω with $\operatorname{Im}\omega = \sigma > 0$.

As for the traces of elements of $H^1(\Omega)$, one needs to introduce an atlas of Γ as usual. Consider a finite covering of Ω by open sets $(\mathcal{O}_i)_{0\leq i\leq I}$ (with $(\mathcal{O}_{1\leq i\leq I})$ covering Γ and $\Gamma\cap\mathcal{O}_0 = \emptyset$), a smooth partition of unity (α_i) subordinate to this cover and diffeomorphisms (φ_i) mapping each $\mathcal{O}_i$, $(1 \leq i \leq I)$ into the cube $Q = \{x = (x_1,x_2,x_3);\ -1 < x_j < 1\}$ and $\mathcal{O}_i\cap\Omega$ into $Q^+ = \{x\in Q\ x_3 > 0\}$, thus $\mathcal{O}_i\cap\Gamma$ into $\Sigma = \{x\in Q\ x_3 = 0\}$. Now, for f defined on Γ, one sets

$$(\theta_i f)(x') = (\alpha_i f) \circ \varphi_i^{-1}(x', 0); \quad x' \in \Sigma$$

and

$$|f|_{1/2,\omega,\Gamma} = \left(\sum_{i=1}^{p} \int_{R^2} (|\omega|^2 + |\xi|^2)^{1/2} |\widehat{\theta_i f}(\xi)|^2 \mathrm{d}\xi \right)^{1/2} \tag{21}$$

where $\widehat{\theta_i f}(\xi)$ denotes the Fourier transform of this function. Then, the well-known trace theorem in $H^1(\Omega)$ can be precisely stated with these norms in the following.

Lemma 1. *(See [14].)*

1. *There exists a positive constant C depending only on Ω and σ_0 such that*

$$|\gamma u|_{1/2,\omega,\Gamma} \leq C \|u\|_{1,\omega,\Omega} \tag{22}$$

for all $u \in H^1(\Omega)$ and $\omega \in \{\omega \in \mathbb{C};\ \operatorname{Im} \omega = \sigma \geq \sigma_0 > 0\}$, where γ denotes the trace operator in $H^1(\Omega)$.

2. *Conversely, one can construct an extension operator $\mathcal{R}$ from $H^{1/2}(\Gamma)$ to $H^1(\Omega)$ such that*

$$\|\mathcal{R}\varphi\|_{1,\omega,\Omega} \leq C |\varphi|_{1/2,\omega,\Gamma} \tag{23}$$

for all $\varphi \in H^{1/2}(\Gamma)$ and $u \in H^1(\Omega)$ and $\omega \in \{\omega \in \mathbb{C};\ \operatorname{Im} \omega = \sigma \geq \sigma_0 > 0\}$.

From (23) and Green's formula, one easily gets the following estimate for the normal derivative $\gamma_1 u \stackrel{\text{def}}{=} \dfrac{\partial u}{\partial n}$ of a solution of the Helmholtz equation:

Lemma 2. *There exists a positive constant C depending only on Ω and σ_0 such that, for all $\omega \in \{\omega \in \mathbb{C};\ \operatorname{Im} \omega = \sigma \geq \sigma_0 > 0\}$, and $u \in H^1(\Omega)$ satisfying $\Delta u + \omega^2 u = 0$ in Ω, one has*

$$|\gamma_1 u|_{-1/2,\omega,\Gamma} \leq C \|u\|_{1,\omega,\Omega} \tag{24}$$

where $|.|_{-1/2,\omega,\Gamma}$ is the dual norm of $|.|_{1/2,\omega,\Gamma}$.

Remark 1. 1. In [59], considering the energy of electromagnetic waves, Terrasse introduced the norms

$$|||f|||_{1,\omega,\Omega} = \int_\Omega \left(\frac{|\nabla f(x)|^2}{|\omega|^2} + |f(x)|^2 \right) \mathrm{d}x$$

and

$$\|f\|_{1/2,\omega,\Gamma} = \left(\sum_{i=1}^{p} \int_{R^2} \left(1 + \frac{|\xi|^2}{|\omega|^2} \right)^{1/2} |\widehat{\theta_i f}(\xi)|^2 \,\mathrm{d}\xi \right)^{1/2}$$

proportional to those used here, with coefficients depending on $|\omega|$. She also considered similar norms in the spaces $H^{-1/2}(\operatorname{div}, \Gamma)$, $H^{-1/2}(\operatorname{curl}, \Gamma)$ as well as in higher degree Sobolev spaces (necessary for regularity results, but not treated below). On the other hand, using characterisations of Sobolev norms on Γ by the coefficients of the expansion of the functions involved in a basis of eigenfunctions of the Laplace-Beltrami operator on Γ, she obtained the same estimates as in Lemmas 1 and 2 above.

2. These estimates are optimal in the sense that the constants C cannot be replaced by $C|\omega|^{-s}$ for any positive real s. In [9], we showed that the best constant is of the form $C|\omega|^{1/2}$ in (23) when the classical norm of $H^{1/2}(\Gamma)$ is used on the RHS instead of the norm $|.|_{1/2,\omega,\Gamma}$. By the correspondence $\frac{\partial}{\partial t} \leftrightarrow -\mathrm{i}\omega$, this question is relevant to the regularity with respect to the time variable in the transient problems.

Now, we can use these norms to make estimates *as a function of* ω for the integral operators $S_\omega, K_\omega, K'_\omega, D_\omega$, the Fourier-Laplace transforms of S, K, K', D (we drop the hat sign when there is no possible confusion). For example:

$$S_\omega p(x) = \int_\Gamma \frac{\mathrm{e}^{\mathrm{i}\omega|x-y|}}{4\pi|x-y|} p(y)\,\mathrm{d}\sigma_y \quad x \in \Gamma \tag{25}$$

where, if $x \notin \Gamma$, the integral on the RHS of (25) is the single layer potential for the Helmholtz equation. It is the transform of the single layer retarded potential Lp, and consequently, will be denoted as $L_\omega p$.

We fix here some $\omega \in \mathbb{C}$ with imaginary part greater than a fixed positive number: $\omega \in \mathbb{C}_{\sigma_0} = \operatorname{Im}\omega = \sigma \geq \sigma_0 > 0$.

Proposition 1. *Let* $p \in H^{-1/2}(\Gamma)$. *Then*

$$\|L_\omega p\|_{1,\omega,\Omega} \leq C|\omega p|_{-1/2,\omega,\Gamma} \tag{26}$$

$$|S_\omega p|_{1/2,\omega,\Gamma} \leq C|\omega p|_{-1/2,\omega,\Gamma} \tag{27}$$

$$|K_\omega p|_{-1/2,\omega,\Gamma} \leq C|\omega p|_{-1/2,\omega,\Gamma} \tag{28}$$

where $\Omega = \Omega^i \cup \Omega^e$ *and positive constant* C *depends only on* Γ *and* σ_0.

Proof. We consider a smooth p, the passage to the limits being justified by the estimates. Then, it is easy to verify that $u(x) = L_\omega p(x)$ is solution of the problem:

$$\begin{cases} \Delta u + \omega^2 u = 0 \text{ in } \Omega \overset{\text{def}}{=} \Omega^{\mathrm{i}} \cup \Omega^{\mathrm{e}} \\ [u] \qquad\quad = 0 \text{ on } \Gamma \\ [\frac{\partial u}{\partial n}] \qquad = p \text{ on } \Gamma \end{cases} \tag{29}$$

where $[u]$ stands for the jump $u^{\mathrm{i}} - u^{\mathrm{e}}$ of u across the boundary Γ, and $S_\omega p$ is the common value of the interior and exterior traces of u on Γ. By Green's formula:

$$\int_\Omega (|\nabla u|^2 - \omega^2|u|^2)\,\mathrm{d}x = \int_\Gamma \overline{u}\,p\,\mathrm{d}\sigma_x = < p, u >$$

Multiplying this formula by $i\bar{\omega}$ and taking the real parts, one gets:

$$\sigma \int_\Omega (|\nabla u|^2 + |\omega u|^2)\,\mathrm{d}x = Re < p, -\mathrm{i}\omega u > \tag{30}$$

That proves the well-posedness of problem (29) in the following closed subspace of $H^1(\Omega)$:

$$\Xi = \{u \in H^1(\Omega); [u] = 0 \text{ on } \Gamma\},$$

and the estimate

$$\|u\|^2_{1,\omega,\Omega} \leq \frac{1}{\sigma} |p|_{-1/2,\omega,\Gamma} |\mathrm{i}\omega S_\omega p|_{1/2,\omega,\Gamma}$$

Using inequalities (22) and (24), and the trace formula for the potential, one gets the desired estimates. □

Similarly, for the double layer potential $M_\omega \varphi$ and its traces, one has:

Proposition 2. *Let $\varphi \in H^{1/2}(\Gamma)$. Then*

$$\|M_\omega \varphi\|_{1,\omega,\Omega} \leq C|\omega\varphi|_{1/2,\omega,\Gamma} \tag{31}$$

$$|D_\omega \varphi|_{-1/2,\omega,\Gamma} \leq C|\omega\varphi|_{1/2,\omega,\Gamma} \tag{32}$$

$$|K'_\omega \varphi|_{1/2,\omega,\Gamma} \leq C|\omega\varphi|_{1/2,\omega,\Gamma} \tag{33}$$

where $\Omega = \Omega^i \cup \Omega^e$ and positive constant C depends only on Γ and σ_0.

Here, we have to replace the space Ξ by

$$\Xi_1 = \{u \in H^1(\Omega); \Delta u + \omega^2 u = 0, [\frac{\partial u}{\partial n}] = 0 \text{ on } \Gamma\}$$

which is itself a closed subspace of

$$\Xi_0 = \{u \in H^1(\Omega); \Delta u + \omega^2 u = 0 \text{ in } \Omega\}.$$

It is worth noting that the kernels of K_ω, K'_ω and D_ω contain a term with ω explicitly as a multiplying factor. For example

$$K_\omega(x,y) = \mathrm{i}\omega \left(\mathrm{e}^{\mathrm{i}\omega|x-y|} \frac{n_x.(x-y)}{4\pi|x-y|^2} \right) + \mathrm{e}^{\mathrm{i}\omega|x-y|} \frac{1}{4\pi|x-y|^3}$$

Thus, the presence of factor $|\omega|$ in the estimates for K_ω, K'_ω and D_ω cannot be avoided. As for the operator S_ω, it is not so clear! Actually, if one considers S_ω as an operator from $L^2(\Gamma)$ to $L^2(\Gamma)$, one sees that this factor disappears. However, the estimate stated in (27) concerns the norm of S_ω as an operator from $H^{-1/2}(\Gamma)$ to $H^{1/2}(\Gamma)$: in terms of the operator S, one can say that, while gaining a degree of regularity in space, it loses a degree of regularity in time. This loss of regularity prevents us from using the Fredholm theory to directly deal with the RPBIE, since we will not have the required compactness properties.

3.2 Solving the Helmholtz BIE With Index Norms

The Second Kind BIE for the Dirichlet Problem

We begin with the equation

$$(I/2 + K'_\omega)\varphi = f, \tag{34}$$

which is the transform of (14). It is well-known that (34) is a second kind integral equation in the sense of Fredholm (in $L^2(\Gamma)$ as well as $H^{1/2}(\Gamma)$), and since the "spurious frequencies" ω are real, thus do not belong to the complex half-plane $\{\omega \in \mathbb{C},\ \mathrm{Im}\,\omega > 0\}$ considered here, the kernel of $I/2 + K'_\omega$ is $\{0\}$. Equation (34) is then well-posed. Moreover, the theory of its discretisation is also well developed: see e.g. the book of Atkinson ([3]).

However, estimation of the solution of (34) as a function of the frequency variable ω, which is required to give estimates of the solution of the RPBIE (14) with respect to the time variable, is not so easy. To do this, we can proceed as follows:

1. First, we solve the exterior problem

$$\begin{cases} \Delta u^{\mathrm{e}} + \omega^2 u^{\mathrm{e}} &= 0 \quad \text{in } \Omega^{\mathrm{e}} \\ u^{\mathrm{e}} &= f \quad \text{on } \Gamma \\ u^{\mathrm{e}} \in H^1(\Omega^{\mathrm{e}}) \end{cases} \tag{35}$$

Classically, we pass by the homogeneous problem

$$\begin{cases} \Delta v^{\mathrm{e}} + \omega^2 v^{\mathrm{e}} &= -(\Delta + \omega^2) w^{\mathrm{e}} \quad \text{in } \Omega^{\mathrm{e}} \\ v^{\mathrm{e}} &= 0 \quad \text{on } \Gamma \\ v^{\mathrm{e}} \in H^1(\Omega^{\mathrm{e}}) \end{cases} \tag{36}$$

where $w^{\mathrm{e}} = \mathcal{R}f$ as defined in Lemma 1, and $v^{\mathrm{e}} = u^{\mathrm{e}} - w^{\mathrm{e}}$. Standard variational formula for the last problem yields v^{e} and then u^{e}:

$$\|u^{\mathrm{e}}\|_{1,\omega,\Omega^{\mathrm{e}}} \leq C|\omega||f|_{1/2,\omega,\Gamma}$$

with a constant C depending only on σ_0 for all ω such that $\mathrm{Im}\,\omega \geq \sigma_0 > 0$. From Lemma 2, one also obtains the following estimate for $g \overset{\text{def}}{=} \partial_n u^{\mathrm{e}}$:

$$|g|_{-1/2,\omega,G} \leq C|\omega||f|_{1/2,\omega,\Gamma}.$$

2. Then we solve the interior Neumann problem

$$\begin{cases} \Delta u^{\mathrm{i}} + \omega^2 u^{\mathrm{i}} &= 0 \quad \text{in } \Omega^{\mathrm{i}} \\ \partial_n u^{\mathrm{i}} &= g \quad \text{on } \Gamma \\ u^{\mathrm{i}} \in H^1(\Omega^{\mathrm{i}}) \end{cases} \tag{37}$$

One gets

$$\|u^{\mathrm{i}}\|_{1,\omega,\Omega^{\mathrm{i}}} \leq C|\omega||g|_{-1/2,\omega,\Gamma}$$

3. Finally, since $\varphi = (u^{\mathrm{e}} - u^{\mathrm{i}})_{|\Gamma}$, the trace Lemma 1 yields the estimate

$$|\varphi|_{1/2,\omega,\Gamma} \leq C|\omega^2||f|_{1/2,\omega,\Gamma}$$

for the solution of (34).

The First Kind BIE for the Dirichlet Problem

Now, we want to solve

$$S_\omega p = f \tag{38}$$

the transform of the RPBIE (14). The calculations in the proof of Proposition 1, via the single layer potential $u = L_\omega p$, show that p is a solution of the variational problem

$$\begin{cases} p \in H^{-1/2}(\Gamma) \text{ s.t.} \\ a_\omega(p,q) \stackrel{\text{def}}{=} < -\mathrm{i}\omega S_\omega p, q > = < -\mathrm{i}\omega f, q > \quad \forall q \in H^{-1/2}(\Gamma) \end{cases} \tag{39}$$

where the sesquilinear form a is coercive, since by (30) and (24)

$$\mathrm{Re} a_\omega(p,p) = \sigma \|u\|_{1,\omega,\Omega^{\mathrm{e}} \cup u^{\mathrm{i}}} \geq C(\sigma_0)|p|^2_{-1/2,\omega,\Gamma} \tag{40}$$

$\forall p \in H^{-1/2}(\Gamma)$ and $\omega \in \mathbb{C}_{\sigma_0}$. Thus, equation (38) is well-posed in $H^{-1/2}(\Gamma)$ and its solution satisfies the estimate

$$|p|_{-1/2,\omega,\Gamma} \leq C|\omega||f|_{1/2,\omega,\Gamma} \,. \tag{41}$$

A System of BIEs for the Robin Problem

The treatment of the second kind BIE for Neumann or Robin problems is similar to that in the Dirichlet case, so we conclude this subsection with the following system of BIEs, the transform of (18), in the case of a Robin condition:

$$\begin{cases} 2(K_\omega p - D_\omega \varphi) - \mathrm{i}\omega\alpha\varphi = f \stackrel{\text{def}}{=} -2\partial_n u^{\text{inc}} \,, \\ p - 2\mathrm{i}\omega\alpha(S_\omega p - K'_\omega \varphi) = g \stackrel{\text{def}}{=} 2\mathrm{i}\omega\alpha u^{\text{inc}} \,. \end{cases} \tag{42}$$

We consider the combined Helmholtz potential $u = L_\omega p - M_\omega \varphi$. Trace formulae for potentials show that the LHS of the equations in (42) are respectively

$$A_1 \stackrel{\text{def}}{=} 2(K_\omega p - D_\omega \varphi) - \mathrm{i}\omega\alpha\varphi = \partial_n u^{\mathrm{i}} + \partial_n u^{\mathrm{e}} - \mathrm{i}\omega\alpha(u^{\mathrm{i}} - u^{\mathrm{e}}) \,,$$
$$A_2 \stackrel{\text{def}}{=} p - 2\mathrm{i}\omega\alpha(S_\omega p - K'_\omega \varphi) = \partial_n u^{\mathrm{i}} - \partial_n u^{\mathrm{e}} - \mathrm{i}\omega\alpha(u^{\mathrm{i}} + u^{\mathrm{e}}) \,.$$

If $\alpha > 0$, Green's formula yields:

$$\mathrm{Re} \int_\Gamma \left(A_1(\overline{-\mathrm{i}\omega\varphi}) + \frac{A_2}{\alpha}\overline{p} \right) \mathrm{d}\sigma_x = 2\sigma \int_{\Omega^{\mathrm{i}} \cup \Omega^{\mathrm{e}}} \left(|\nabla u|^2 + |\omega u|^2 \right) \mathrm{d}x + \int_\Gamma \left(\alpha|\omega\varphi|^2 + \frac{1}{\alpha}|p|^2 \right) \mathrm{d}\sigma_x \,. \tag{43}$$

We are led to the following variational formulation of (42):

$$\text{Find } (p,\varphi) \in V \text{ s.t. } A_\omega((p,\varphi),(q,\psi)) = L(q,\psi) \;\; \forall (q,\psi) \in V \tag{44}$$

where $V = L^2(\Gamma) \times H^{1/2}(\Gamma)$,

$$A_\omega((p,\varphi),(q,\psi)) \stackrel{\text{def}}{=} \int_\Gamma \left(A_1(\overline{-\mathrm{i}\omega\psi}) + \frac{A_2}{\alpha}\bar{q} \right) \mathrm{d}\sigma_x$$

and

$$L(q,\psi) \stackrel{\text{def}}{=} 2\int_\Gamma \left(\partial_n u^{\text{inc}}(\overline{-\mathrm{i}\omega\psi}) + \mathrm{i}\omega u^{\text{inc}}.\bar{q}\right) \mathrm{d}\sigma_x$$

Clearly, if

$$\exists \alpha_0 > 0 \text{ and } \alpha_1 > 0 \ \text{ s.t. } \ \alpha_0 \leq \alpha(x) \leq \alpha_1 \ \forall x \in G \tag{45}$$

all the integrals involved are well defined and, by (43), the variational problem (44) is well-posed in V. Moreover, one has:

$$\frac{1}{\alpha_1}|p|^2_{0,\Gamma} + \alpha_0|\omega\varphi|^2_{0,\Gamma} + 2\sigma\int_{\Omega^{\mathrm{i}}\cup\Omega^{\mathrm{e}}} \left(|\nabla u|^2 + |\omega u|^2\right) \mathrm{d}\Omega$$
$$\leq |\omega f|_{-1/2,\omega,\Gamma}|\varphi|_{1/2,\omega,\Gamma} + |g|_{0,\Gamma}|p|_{0,\Gamma}\,,$$

yielding the following estimates for the solution of this problem:

$$\begin{cases} |\omega\varphi|_{1/2,\omega,\Gamma} \leq C(|\omega f|_{-1/2,\omega,\Gamma} + |g|_{0,\Gamma}) \\ |p|_{0,\Gamma} \qquad\quad \leq C(|\omega f|_{-1/2,\omega,\Gamma} + |g|_{0,\Gamma})\,, \end{cases} \tag{46}$$

where the constants depend only on σ_0 for all $\omega \in \mathbb{C}_{\sigma_0}$ (and naturally, on the geometry).

4 Analysis of the RPBIE

4.1 Sobolev Spaces Associated With the Wave Energy

Let E be a Hilbert space and define

$$LT(\sigma, E) = \{f \in \mathcal{D}'_+(E)\ ;\ \mathrm{e}^{-\sigma t} f \in \mathcal{S}'_+(E)\}$$

where $\mathcal{D}'_+(E)$ and $\mathcal{S}'_+(E)$ denote, as usual, the sets of distributions and temperate distributions on $\mathbb{R}$, with values in E and support in $[0,\infty[$. It is clear that $LT(\sigma, E) \subset LT(\sigma', E)$ if $\sigma < \sigma'$. We denote by $\sigma(f)$ the infimum of all σ such that $f \in LT(\sigma, E)$. For $f \in LT(E) \stackrel{\text{def}}{=} \bigcup_{\sigma\in\mathbb{R}} LT(\sigma, E)$, the set of Laplace transformable distributions with values in E, we define its Fourier-Laplace transform (as in the scalar case) by

$$\hat{f}(\omega) = \mathcal{F}(\mathrm{e}^{-\sigma t} f)(\eta)$$

for $\sigma > \sigma(f)$ and $\omega = \eta + \mathrm{i}\sigma$.

We recall here for convenience the main results on the transform:

Theorem 1. *(cf. [60] or [24, Chap. 16]))*

1. (Paley-Wiener theorem). *An E-valued function $\hat{f}(\omega)$ is the Fourier-Laplace transform of $f \in LT(E)$ if and only if it is holomorphic in some half plane $C_{\sigma_0} = \{\omega \in \mathbb{C};\ \mathrm{Im}\,\omega > \sigma_0\}$ and of temperate growth in some closed half plane of C_{σ_0}. This last condition means that there exist $\sigma_1 > \sigma_0$, $C > 0$ and $k \in \mathbb{N}^*$ such that*

$$||\hat{f}(\omega)||_E \leq C(1 + |\omega|)^k \text{ for all } \omega \text{ s.t. } \mathrm{Im}\,\omega \geq \sigma_1 \tag{47}$$

2. *Moreover, the support of $f \in LT(E)$ is in $[T, \infty[$ if and only if the inequality (47) is replaced by*

$$||\hat{f}(\omega)||_E \leq C(1 + |\omega|)^k \mathrm{e}^{-(\mathrm{Im}\,\omega)T} \text{ for all } \omega \text{ s.t. } \mathrm{Im}\,\omega \geq \sigma_1 \tag{48}$$

3. (Parseval theorem). *On the other hand, if $f, g \in L^1_{loc}(\mathbb{R}, E) \cap LT(E)$, one has the following formula*

$$\frac{1}{2\pi} \int_{\mathbb{R}+i\sigma} (\hat{f}(\omega), \hat{g}(\omega))_E \,\mathrm{d}\omega = \int_{-\infty}^{\infty} \mathrm{e}^{-2\sigma t} (f(t), g(t))_E \,\mathrm{d}t \tag{49}$$

where $(\cdot, .)_E$ is the hermitian product of E and $\sigma > \max(\sigma(f), \sigma(g))$.

Remark 2. 1. For E-valued functions of a complex variable, strong and weak holomorphy coincide (see [60]), so that, for our purposes in the following, the verification of this point does not pose any particular problem (and will be omitted!): the integral operators $S_\omega, K_\omega, K'_\omega, D_\omega$ are analytical operator-valued functions of $\omega \in \mathbb{C}$, and then their inverses are analytical on their domains of definition.
2. Part 2, an immediate corollary of part 1, associated with the estimates in the last section, permits us to find again an important property of propagation: the solutions of RPBIEs at time t will only depend on the data at earlier times.

Parts 1 and 3 of Theorem 1 lead us to define the following Hilbert spaces (see [9]):

$$\mathcal{H}^s_\sigma(\mathbb{R}^+, E) = \{f \in LT(\sigma, E);\ \mathrm{e}^{-\sigma t} \Lambda^s f \in L^2(\mathbb{R}, E)\}$$

where $s \in \mathbb{R}$, $\sigma > 0$, and Λ^s denotes the fractional derivative with respect to the variable t, defined by $\widehat{\Lambda^s f}(\omega) = (-\mathrm{i}\omega)^s \hat{f}(\omega)$.

However, in view of the results in last section, we will specially consider the following space:

$$H^{1,1}_{\sigma,\Omega} = \left\{ u \in LT(\sigma, H^1(\Omega));\ \int_{\mathbb{R}+i\sigma} \|\hat{u}\|^2_{1,\omega,\Omega} \,\mathrm{d}\omega < +\infty \right\}$$

which is equipped with a hilbertian structure for the norm

$$||u||^2_{1,1;\sigma} = \int_{\mathbb{R}+i\sigma} \|\hat{u}\|^2_{1,\omega,\Omega} \,\mathrm{d}\omega$$

By the Parseval equality, we have $u \in H^{1,1}_{\sigma,\Omega}$ if and only if

$$u \in LT(\sigma, H^1(\Omega)) \text{ and } \int_{-\infty}^{\infty} \mathrm{e}^{-2\sigma t} \int_{\Omega} (|\nabla u(x,t)|^2 + |\dot{u}(x,t)|^2)\,\mathrm{d}x\,\mathrm{d}t < \infty .$$

That is, $u \in H^{1,1}_{\sigma,\Omega}$ if and only if $\mathrm{e}^{-\sigma t}\nabla u \in L^2(\mathbb{R}, (L^2(\Omega))^3)$ and $\mathrm{e}^{-\sigma t}\dot{u} \in L^2(\mathbb{R}, L^2(\Omega))$. This last condition implies that $u \in \mathcal{C}(\mathbb{R}, L^2(\Omega))$, and then, by the assumption that u is causal, $u(\cdot, 0) = 0$ *a.e.*. Then

$$|u(x,t)|^2 \le t \int_0^t |\dot{u}(x,s)|^2 \mathrm{d}s \tag{50}$$

for all $x \in \Omega$ except possibly a negligible set independent of t, and consequently, for all $\sigma' > \sigma$, $\mathrm{e}^{-\sigma' t} u \in L^2(\mathbb{R}, H^1(\Omega)) \cap H^1(\mathbb{R}, L^2(\Omega))$. Thus, for all $T > 0$,

$$u \in {}_0H^{1,1}(\Omega \times (0,T)) = \{u \in L^2((0,T), H^1(\Omega)) \cap H^1((0,T), L^2(\Omega));\ u(\cdot,0) = 0\}.$$

See Lions-Magenes [43, Chap.4] for a thorough study of this space. We only note here that the equality $u(\cdot, 0) = 0$ can be taken in $H^{1/2}(\Omega)$ (instead of $L^2(\Omega)$ as one can deduce from $u \in H^1((0,T), L^2(\Omega))$. On the other hand, it is worth noticing, from (50), that the "energy norm" $\int_0^T E(u)(t)\,\mathrm{d}t$ is equivalent to the usual norm $(||u||^2_{L^2(0,T;H^1(\Omega))} + ||u||^2_{H^1(0,T;L^2(\Omega))})^{1/2}$ in this space.

As for the functions defined on the boundary Γ, we define:

$$H^{1/2,1/2}_{\sigma,\Gamma} = \left\{ u \in LT(\sigma, H^{1/2}(\Gamma));\ \int_{\mathbb{R}+i\sigma} |\hat{u}|^2_{1/2,\omega,\Gamma}\,\mathrm{d}\omega < +\infty \right\}$$

We note that, for all ω such that $\operatorname{Im}\omega \ge \sigma_0 > 0$, and $a \ge 0$:

$$(|\omega|^2 + a)^{1/2} \le (1 + |\omega|^2)^{1/2} + (1+a)^{1/2} \le 2\left(\frac{1+\sigma_0^2}{\sigma_0^2}\right)^{1/2} (|\omega|^2 + a)^{1/2} \tag{51}$$

Plugging (51) with $a = |\xi|^2$ into the definition (21) of the norm $|.|_{1/2,\omega,\Gamma}$, one sees that $u \in H^{1/2,1/2}_{\sigma,\Gamma}$ if and only if $\mathrm{e}^{-\sigma t} u \in L^2(\mathbb{R}, H^{1/2}(\Gamma)) \cap H^{1/2}(\mathbb{R}, L^2(\Gamma))$. We must emphasise that, as was the case for the functions in $H^{1,1}_{\sigma}(\Omega)$, the causality condition implies a behaviour of the functions of $H^{1/2,1/2}_{\sigma,\Gamma}$ at time $t = 0$, which coincides with $u(\cdot, 0) = 0$ if u is more regular in time. And, for the restriction of u to the finite time interval $(0,T)$, one has

$$u \in {}_0H^{1/2,1/2}(\Gamma \times (0,T)) = L^2(0,T; H^{1/2}(\Gamma)) \cap {}_0H^{1/2}(0,T; L^2(\Gamma)),$$

which is the trace space of ${}_0H^{1,1}(\Omega \times (0,T))$. The left subscript 0 emphasises the fact that u is a restriction of causal function (again, if it is regular enough in time, it must satisfy the condition $u(\cdot, 0) = 0$).

Similar considerations can be made for

$$H^{-1/2,-1/2}_{\sigma,\Gamma} = \left\{ u \in LT(\sigma, H^{-1/2}(\Gamma));\ \int_{\mathbb{R}+i\sigma} |\hat{u}|^2_{-1/2,\omega,\Gamma}\,\mathrm{d}\omega < +\infty \right\}$$

but it is simpler to use the characterisation of the dual of intersections:

$$u \in H^{-1/2,-1/2}_{\sigma,\Gamma} \Leftrightarrow \mathrm{e}^{-\sigma t} u \in L^2(\mathbb{R}, H^{-1/2}(\Gamma)) + H^{-1/2}(\mathbb{R}, L^2(\Gamma))$$

and

$$u \in H^{-1/2,-1/2}(\Gamma \times (0,T)) = L^2(0,T; H^{-1/2}(\Gamma)) + H^{-1/2}(0,T; L^2(\Gamma)) .$$

No specific behaviour of u at $t = 0$ is necessary here.

4.2 The Second Kind RPBIE

Consider the RPBIE (14) for the Dirichlet problem:

$$(I/2 + K')\varphi = f .$$

We suppose that $f \in LT(\sigma_0, H^{1/2}(\Gamma))$ for some $\sigma_0 > 0$. We already noted that, by Proposition 2, $K'\varphi$ loses a degree of regularity with respect to φ. Thus, compactness (in the space-time functional space) is missing, and one cannot consider the operator $I/2 + K'$ as a second kind integral operator in the sense of Fredholm. However, one can use the inverse Fourier-Laplace transform to obtain some useful results from Sect. 3.2. We sum them up here.

Theorem 2. *1. K' is a bounded operator from $H^{1,1/2,1/2}_{\sigma,\Gamma} = \{\varphi; \dot{\varphi} \in H^{1/2,1/2}_{\sigma,\Gamma}\}$ into $H^{1/2,1/2}_{\sigma,\Gamma}$.*

2. If $f \in H^{2,1/2,1/2}_{\sigma,\Gamma} = \{f; \ddot{f} \in H^{1/2,1/2}_{\sigma,\Gamma}\}$, then the retarded potential equation (14) has a unique solution $\varphi \in H^{1/2,1/2}_{\sigma,\Gamma}$.

We note that $H^{k,1/2,1/2}_{\sigma,\Gamma}$ is a Hilbert space with the norm

$$||u||_{H^{k,1/2,1/2}_{\sigma,\Gamma}} = \int_{\mathbb{R}+i\sigma} |\omega|^k \, |\hat{u}|^2_{1/2,\omega,\Gamma} \, \mathrm{d}\omega$$

and, by (51), coincides with

$$\{\varphi; \mathrm{e}^{-\sigma t}\varphi \in H^k(\mathbb{R}, H^{1/2}(\Gamma)) \cap H^{k+1/2}(\mathbb{R}, L^2(\Gamma))\}.$$

The solution of (14) then satisfies the estimate

$$||\varphi||_{H^{1/2,1/2}_{\sigma,\Gamma}} \leq C||f||_{H^{2,1/2,1/2}_{\sigma,\Gamma}} . \tag{52}$$

Remark 3. The theorem underlines a special character of RPBIEs, which will be met again for other equations: a loss of regularity (with respect to the time variable) for both the operator involved (here $I/2 + K'$) and its "inverse". That may be due to the chosen functional framework where time and space variables are treated separately. However, if one thinks about the discretisation processes, it is not clear how to avoid this separation. On the other hand, the fact that our operators do not operate on Sobolev spaces of $\Gamma \times \mathbb{R}$, as indicated by Bachelot (cf. [8]), renders the search for a better framework to analyse the equations and their discretisation much more difficult.

The estimate (33) and the second point of Theorem 1 show also that the value of $K'\varphi(\cdot,t)$ $(0 \le t \le T)$ does not depend on the values of $\varphi(\cdot,t')$ for $t' > T$, and that if φ is solution of (14), then $\varphi(\cdot,t)$ $(0 \le t \le T)$ does not depend on the values of $f(t')$, $t' > T$. Thus, if $\ddot{f}$ is the restriction to $(0,T)$ of some function $\tilde{f} \in H^{2,1/2,1/2}_{\sigma,\Gamma}$ (for some $\sigma > 0$), (this implies that $f \in H^2(0,T;H^{1/2}(\Gamma)) \cap H^{5/2}(0,T;L^2(\Gamma))$), then the solution of

$$(I/2 + K')\tilde{\varphi} = \tilde{f} \text{ on } \Gamma \times \mathbb{R}$$

yields the solution of

$$(I/2 + K')\varphi = f \text{ on } \Gamma \times (0,T)$$

by simply restricting $\tilde{\varphi}$ to $(0,T)$. One also has the estimate

$$||\varphi||_{H^{1/2,1/2}(\Gamma\times(0,T))} \le C||\ddot{f}||_{H^{1/2,1/2}(\Gamma\times(0,T))}$$

with some positive constant (depending on T).

The analysis of the second kind RPBIE for Neumann or Robin problem is similar and we do not elaborate more.

4.3 The First Kind RPBIE and its Space-Time Variational Formulation

Now consider the "first kind" RPBIE (15): $Sp = f$. From the results in Sect. 3, one deduces:

Theorem 3. *1. S is a bounded operator from $H^{1,-1/2,-1/2}_{\sigma,\Gamma} = \{p; \dot{p} \in H^{-1/2,-1/2}_{\sigma,\Gamma}\}$ into $H^{1/2,1/2}_{\sigma,\Gamma}$.*
2. If $f \in H^{1,1/2,1/2}_{\sigma,\Gamma} = \{f; \dot{f} \in H^{1/2,1/2}_{\sigma,\Gamma}\}$, then the retarded potential equation (15) has a unique solution p satisfying

$$||p||_{H^{-1/2,-1/2}_{\sigma,\Gamma}} \le C||\dot{f}||_{H^{1/2,1/2}_{\sigma,\Gamma}}.$$

However, the main difference with the "second kind" RPBIE (14) consists of a direct link of the function Sp with the energy of the associated single layer retarded potential Lp. We will see that this link can be exploited to develop numerical schemes for (15) that can be proved stable – and which really are!

So, let $p \in H^{-1/2,-1/2}_{\sigma,\Gamma}$ for some $\sigma > 0$ and u be the single layer retarded potential Lp. Denote by $E_{e,i}(t)$ its energy in the domains $\Omega^{e,i}$ and $E = E_e + E_i$ the total energy. Green's formula yields

$$\frac{\mathrm{d}E}{\mathrm{d}t} = \int_\Gamma p(x,t)\, S\dot{p}(x,t)\, \mathrm{d}\sigma_x \tag{53}$$

the integral being taken in the sense of duality between $H^{-1/2}(\Gamma)$ and $H^{1/2}(\Gamma)$. Consequently, for all $\sigma > 0$:

$$\int_0^\infty \mathrm{e}^{-2\sigma t} E(t)\,\mathrm{d}t = \frac{1}{2\sigma}\int_0^\infty \mathrm{e}^{-2\sigma t}\int_\Gamma p(x,t)\, S\dot{p}(x,t)\,\mathrm{d}\sigma_x\,\mathrm{d}t$$

On the other hand, by Parseval's formula and (24), one has

$$\begin{aligned}\int_0^\infty \mathrm{e}^{-2\sigma t} E(t)\,\mathrm{d}t &= \frac{1}{4\pi}\int_{\mathbb{R}+\mathrm{i}\sigma} \|\hat{u}\|^2_{1,\omega,\Omega^{\mathrm{i}}\cup\Omega^{\mathrm{e}}}\,\mathrm{d}\omega \\ &\geq C\int_{\mathbb{R}+\mathrm{i}\sigma} \|\hat{p}\|^2_{-1/2,\omega,\Gamma}\,\mathrm{d}\omega \\ &= C\|p\|^2_{H^{-1/2,-1/2}_{\sigma,\Gamma}}\end{aligned}$$

Thus

$$\int_0^\infty \mathrm{e}^{-2\sigma t}\int_\Gamma p(x,t)\, S\dot{p}(x,t)\,\mathrm{d}\sigma_x\,\mathrm{d}t \geq C\|p\|^2_{H^{-1/2,-1/2}_{\sigma,\Gamma}} \tag{54}$$

We are led to replace the RPBIE (15) by the space-time variational problem

$$a_\sigma(p,q) \stackrel{\text{def}}{=} \int_0^\infty \mathrm{e}^{-2\sigma t}\int_\Gamma S\dot{p}(x,t)\, q(x,t)\,\mathrm{d}\sigma_x\,\mathrm{d}t = \int_0^\infty \mathrm{e}^{-2\sigma t}\int_\Gamma \dot{f}(x,t)\, q(x,t)\,\mathrm{d}\sigma_x\,\mathrm{d}t$$
$$\text{for all test functions } q, \tag{55}$$

which is exactly the F-L transform of the variational problem (39). This correspondence with the frequency problem permits us to prove that the solution of (55) is independent of σ (such that $f \in H^{1,1/2,1/2}_{\sigma,\Gamma}$) and satisfies the equation (15), results that one cannot obtain directly from the Lax Milgram theorem since the space on which the bilinear form a_σ is continuous is not the one on which it satisfies the coerciveness property (54). However, the coerciveness (54) is sufficient for proving unconditional stability of conforming Galerkin approximations by boundary finite element methods. As recalled in the introduction, this technique was first introduced in [9], and was the starting point of several works on the RPBIE for the scattering problems by acoustic as well as electromagnetic, elastic waves.

However, a somewhat perturbing fact remains unresolved: in all of this work, for simplicity, the calculations were done with the constant σ set to zero, and while the numerical experiments remained stable, this stability was not really proved because coerciveness estimates like (54) do not pass to the limit when $\sigma \to 0$: the constant C actually tends to 0 in this limit. On the other hand, if one can take $\sigma = 0$ in (55) for test functions q *with bounded support* with respect to the time variable, we cannot even guarantee that the integrals in $a_\sigma(p,q)$ are finite if we replace both q by p and σ by 0. We answer this question below.

First, since calculations are made for finite times, we shall extend the estimates in Theorem 3 to finite time intervals. For this purpose, we assume that the data f of (15) are given with $\dot{f} \in H^{1/2,1/2}(\Gamma \times (0,T))$ and null conditions at $t = 0$ so that there exists a causal extension $\tilde{f}$ of f in $\mathbb{R}$ such that

$$\|\tilde{f}\|_{H^{1,1/2,1/2}_{\sigma,\Gamma}} \leq C\|\dot{f}\|_{H^{1/2,1/2}(\Gamma\times(0,T))} \tag{56}$$

for all $\sigma > 0$ and some constant C independent of σ. Now, if p is solution of (55) (with $\tilde{f}$ replacing f), then $p(\cdot,t)$ does not depend on $\tilde{f}(\cdot,t')$, $(t' > t)$, so that p satisfies

$$a(p,q;t) \stackrel{\text{def}}{=} \int_0^t \int_\Gamma S\dot{p}(x,s)\, q(x,s)\, \mathrm{d}\sigma_x\, \mathrm{d}s = \int_0^t \int_\Gamma q(x,s)\, \dot{\tilde{f}}(x,s)\, \mathrm{d}\sigma_x\, \mathrm{d}s$$
for all test functions q and for all t. (57)

Conversely, if p is a solution of (57) for all $t > 0$, multiplying the two members by $\mathrm{e}^{-2\sigma t}$ and integrating on $\mathbb{R}^+$, one gets (55). Thus, the variational problem (57) for $0 \le t \le T$ admits a unique solution equal to the restriction to $(0,T)$ of the solution of (55) with $\tilde{f}$ replacing f. This is also the solution of RPBIE

$$Sp(x,t) = f(x,t) \quad \text{on} \quad \Gamma \times (0,T).$$

In particular, for all $t \in (0,T)$:

$$\int_0^t \int_\Gamma p(x,s)\, S\dot{p}(x,s)\, \mathrm{d}\sigma_x\, \mathrm{d}s = \int_0^t \int_\Gamma p(x,s)\, \dot{f}(x,s)\, \mathrm{d}\sigma_x\, \mathrm{d}s \tag{58}$$

The LHS of (58) is $E(t)$, so that, if we integrate the two sides of this equality, we get

$$\begin{aligned}\int_0^T E(t)\, \mathrm{d}t &= \int_0^T \int_0^t \int_\Gamma p(x,s)\, \dot{f}(x,s)\, \mathrm{d}\sigma_x\, \mathrm{d}s\, \mathrm{d}t \\ &= \int_0^T (T-s) \int_\Gamma p(x,s)\, \dot{f}(x,s)\, \mathrm{d}\sigma_x\, \mathrm{d}s \\ &\le C(T) \|p\|_{H^{-1/2,-1/2}(\Gamma\times(0,T))} \|\dot{f}\|_{H^{1/2,1/2}(\Gamma\times(0,T))}\end{aligned}$$

Now, by standard duality methods, one can estimate the normal trace of any solution of the wave equation with initial conditions (see for example [38]) with respect to the energy norm:

$$\|p\|^2_{H^{-1/2,-1/2}(\Gamma\times(0,T))} \le C \int_0^T E(t)\, \mathrm{d}t \tag{59}$$

and one finally gets the stability estimate for the solution of (57):

$$\|p\|_{H^{-1/2,-1/2}(\Gamma\times(0,T))} \le C \|\dot{f}\|_{H^{1/2,1/2}(\Gamma\times(0,T))} \tag{60}$$

The stability of conforming Galerkin discretisation follows: let V_h be a (finite-dimensional) subspace of $H^{-1/2,-1/2}(\Gamma \times [0,T])$, and the discretised unknown p_h satisfy:

$$\begin{cases} (\dot{p}_h) \in V_h \quad \text{s.t.} \\ a(p_h, q_h, T) = \displaystyle\int_0^T \int_\Gamma q_h(x,s)\, \dot{f}(x,s)\, \mathrm{d}\sigma_x\, \mathrm{d}s \;\; \forall (q_h) \in V_h. \end{cases}$$

then

$$\|p_h\|_{H^{-1/2,-1/2}(\Gamma\times(0,T))} \le C \|\dot{f}\|_{H^{1/2,1/2}(\Gamma\times(0,T))}\,.$$

One sees above that the energy formula (53) is directly responsible for our space-time variational formulation of the RPBIE (15). For the "second kind" RPBIE (14), this is not so simple. Indeed, since $(I/2+K')\varphi = u^{\mathrm{e}}$ for the double retarded potential $u = M\varphi$, one can write the following energy formula

$$\frac{\mathrm{d}E_{\mathrm{e}}(t)}{\mathrm{d}t} = \int_\Gamma \frac{\partial u^{\mathrm{e}}(x,t)}{\partial n}\dot{u}^{\mathrm{e}}(x,t)\,\mathrm{d}\sigma_x = \int_\Gamma D\varphi(x,t)(I/2+K')\varphi(x,t)\,\mathrm{d}\sigma_x\,.$$

Thus, to obtain a coercive variational formulation for (14), one should multiply the two sides by $D\dot{\varphi}$ before integrating. This complication explains our preference for the "first kind" RPBIE.

4.4 The Robin Problem

As already explained at the end of Sect. 4.1, we will not consider the "second kind" RPBIE (16) here, but will concentrate on the system (18). The basic and simplest way to obtain a variational formula for it is to multiply each equation by a test function and to integrate on the domain $\Gamma \times \mathbb{R}^+$ where the unknown functions are defined. However, as in the case of the RPBIE (15) for the Dirichlet problem, we prefer to start from an energy formula. Using Green's formula, one obtains after some computations the following identity relating E to the density functions (φ, p) in the retarded potential representation (8) of u:

$$\frac{\mathrm{d}E}{\mathrm{d}t} = \int_\Gamma (p(S\dot{p} - K'\dot{\varphi}) + (Kp - D\varphi)\dot{\varphi})\,\mathrm{d}\sigma_x$$

And since our functions are null for $t = 0$, one gets:

$$E(t) = \int_0^t \int_\Gamma (p(S\dot{p} - K'\dot{\varphi}) + (Kp - D\varphi)\dot{\varphi})\,\mathrm{d}\sigma_x\,\mathrm{d}s \tag{61}$$

for any $T > 0$.

From that, one easily gets a space-time variational formulation of (16) in $\Gamma \times \mathbb{R}$, that corresponds to the F-L transform of the variational formulation (44) of the frequency BIE (42). Following the analysis in Sect. 4.2, we can also go directly to a formulation on finite time intervals:

$$\begin{cases} \displaystyle\int_0^t \int_\Gamma (2(Kp - D\varphi) + \alpha\dot{\varphi})\,\dot{\psi}\,\mathrm{d}\sigma_x\,\mathrm{d}s = -2\int_0^t \int_\Gamma \partial_n u^{\mathrm{inc}}\dot{\psi}\,\mathrm{d}\sigma_x\,\mathrm{d}s, \\ \displaystyle\int_0^t \int_\Gamma \left(\frac{p}{\alpha} + 2(S\dot{p} - K'\dot{\varphi})\right) q\,\mathrm{d}\sigma_x\,\mathrm{d}s = -2\int_0^t \int_\Gamma \partial_t u^{\mathrm{inc}} q\,\mathrm{d}\sigma_x\,\mathrm{d}s, \end{cases} \tag{62}$$

where (ψ, q) are (sufficiently regular) test functions – adjoint variables of the unknowns (φ, p). We denote by b_t the bilinear form of the system (62):

$$\begin{aligned} b_t((\varphi,p),(\psi,q)) = &\int_0^t \int_\Gamma (2(Kp - D\varphi) + \alpha\dot{\varphi})\,\dot{\psi}\,\mathrm{d}\sigma_x\,\mathrm{d}s \\ &+ \int_0^t \int_\Gamma \left(\frac{p}{\alpha} + 2(S\dot{p} - K'\dot{\varphi})\right) q\,\mathrm{d}\sigma_x\,\mathrm{d}s. \end{aligned} \tag{63}$$

This bilinear form now matches the energy identity (61) in the following sense:

$$b_t((\varphi,p),(\varphi,p)) = 2E(t) + \int_0^t \int_\Gamma \left(\alpha(x)\dot{\varphi}^2(x,t) + \frac{1}{\alpha} p^2(x,t) \right) \mathrm{d}\sigma_x \, \mathrm{d}s \,. \quad (64)$$

To make sense for that, with the unknown function φ having its first derivatives square integrable, and p being itself square integrable on $\Gamma \times [0,T]$, we will assume (this hypothesis can be weakened, as shown below) that

$$\exists\, \alpha_0 > 0 \text{ and } \alpha_1 > 0 \text{ such that } \alpha_0 \leq \alpha(x) \leq \alpha_1 . \quad (65)$$

Formula (64) constitutes the main advantage of our variational formulation. It clearly proves the uniqueness of the solutions of problem (62) (we recall that $\varphi(0) = 0$), and above all, it permits us to prove the unconditional stability for Galerkin conforming approximations of (62) when the space and time steps tends to zero. For that, we note first the following estimate, obviously resulting from (64):

$$b_T((\varphi,p),(\varphi,p)) \geq C(||\dot{\varphi}||^2 + ||p||^2) \quad (66)$$

(with norms in $L^2(\Gamma \times [0,T])$, where T is the final time of computations). Now, let V_h and W_h be some (finite-dimensional) subspaces of $L^2(\Gamma \times [0,T])$, and suppose that the discretised unknowns (φ_h, p_h) satisfy:

$$\begin{cases} (\dot{\varphi_h}, p_h) \in V_h \times W_h \quad \text{s.t.} \\ b_T((\varphi_h,p_h),(\psi_h,q_h)) = -2 \displaystyle\int_0^T \int_\Gamma (\partial_n u^{\mathrm{inc}} \dot{\psi_h} + \partial_t u^{\mathrm{inc}} q_h) \, \mathrm{d}\sigma_x \, \mathrm{d}t \\ \qquad\qquad \forall (\psi_h, q_h) \ \text{ s.t. } \ (\dot{\psi_h}, q_h) \in V_h \times W_h . \end{cases} \quad (67)$$

Then, one has from (66) and Cauchy-Schwarz inequality:

$$\begin{aligned} C(||\dot{\varphi_h}||^2 + ||p_h||^2) &\leq b_T((\varphi_h,p_h),(\varphi_h,p_h)) \\ &\leq \left(||\partial_n u^{\mathrm{inc}}||^2 + ||\partial_t u^{\mathrm{inc}}||^2\right)^{1/2} \left(||\dot{\varphi_h}||^2 + ||p_h||^2)\right)^{1/2} . \end{aligned}$$

That proves the boundedness of $||\dot{\varphi_h}||$ and $||p_h||$ regardless of how fine the subspaces V_h and W_h are made.

Remark 4. 1. To derive error bounds for the approximated solution, more thorough functional analysis shall be necessary. It was done for similar problems in [9].

2. Let us show why one can weaken the hypothesis (65). Let Γ_0 be the set $\{x \in \Gamma,\ \alpha(x) = 0\}$ and suppose that $\Gamma = \Gamma_0 \cup \Gamma_1$ and that

$$\alpha > 0 \ \text{ on } \Gamma_1, \quad \alpha \text{ and } (1/\alpha) \in L^\infty(\Gamma_1) . \quad (68)$$

Then, since $p = \frac{\partial u^{\mathrm{i}}}{\partial n} - \frac{\partial u^{\mathrm{e}}}{\partial n} = 0$ on Γ_0 (see the boundary conditions (4) and (17)), we can write the variational problem (62) without change. The formula (64) remains valid, with only a minor change: Γ_1 instead of Γ in the integral on its RHS.

3. To find the equivalent of the coerciveness property (66) and its application to the stability estimate, we must also use the term $E(T)$ in $b_T((\varphi,p),(\varphi,p))$ and relations between the energy of u and its potential densities (φ,p). The same problem occurs for the Neumann condition ($\alpha = 0$). These relations can be written down as in the case of the single layer potential intervening in the RPBIE (15).

5 Galerkin Schemes for the Space-Time Variational Problems

We follow the presentation of Galerkin boundary element methods introduced by Nédélec [51] for BIEs in stationary problems, and Bamberger–Ha-Duong [9] for RPBIEs in acoustic scattering with perfectly reflecting obstacles.

We will take the variational problem (62) as a model to present the different steps of our discretisation, since we have there all the four retarded potential operators. Let us carry out a convenient change of variables before doing that. From the boundary conditions (4) and (17) and the definition of the function p, one sees that

$$p(x,t) = -\alpha(x)\partial_t\left(u^{\mathrm{i}}(t) + u^{\mathrm{e}}(t) - 2u^{\mathrm{inc}}(t)\right).$$

Thus, an integration of this equality yields

$$\lambda(x,t) = \int_0^t p(x,s)ds = \alpha(x)\left(u^{\mathrm{i}}(x,t) + u^{\mathrm{e}}(x,t) - 2u^{\mathrm{inc}}(x,t)\right)$$

and suggests the use of λ in place of p as the unknown. The advantage is that λ has the same degree of regularity (in time) as $\varphi = u^{\mathrm{i}} - u^{\mathrm{e}}$, and then can be approximated with the same temporal elements. On the other hand, integrating by parts the time integrals of the second equation in (62), recalling that all test functions have bounded time support, one obtains a variational problem equivalent to (62) for the unknown functions (φ,λ) (we write the time integrals on $(0,+\infty)$ to leave open the interval of computations):

$$\begin{cases} \displaystyle\int_0^{+\infty}\left(\int_\Gamma \alpha\dot{\varphi}\dot{\psi} + 2\int_\Gamma K\dot{\lambda}\dot{\psi} - 2\int_\Gamma D\varphi\dot{\psi}\right) = -2\int_0^{+\infty}\int_\Gamma \partial_n u^{\mathrm{inc}}\dot{\psi}, \\ \displaystyle\int_0^{+\infty}\left(\int_\Gamma \frac{1}{\alpha}\lambda\dot{q} + 2\int_\Gamma S\dot{\lambda}\dot{q} - 2\int_\Gamma K'\varphi\dot{q}\right) = -2\int_0^{+\infty}\int_\Gamma u^{\mathrm{inc}}\dot{q}, \end{cases} \tag{69}$$

Clearly, one can write all that with new notation $(\eta,r) = (\dot{\psi},\dot{q})$:

$$\begin{cases} \displaystyle\int_0^{+\infty}\left(\int_\Gamma \alpha\dot{\varphi}\eta - 2\int_\Gamma D\varphi\eta + 2\int_\Gamma K\dot{\lambda}\eta\right) = -2\int_0^{+\infty}\int_\Gamma \partial_n u^{\mathrm{inc}}\eta, \\ \displaystyle\int_0^{+\infty}\left(\int_\Gamma -2\int_\Gamma K'\varphi r + \frac{1}{\alpha}\lambda r + 2\int_\Gamma S\dot{\lambda}r\right) = -2\int_0^{+\infty}\int_\Gamma u^{\mathrm{inc}}r. \end{cases} \tag{70}$$

Therefore, our schemes will be presented in terms of the unknowns (φ,λ) (instead of (φ,p)) and test functions (η,r) as in (70).

5.1 Spatial Discretisation

First, we approximate surface Γ by a piecewise polynomial surface Γ_h. For simplicity, we will use here a surface Γ_h composed of N_h triangular facets constructed from a grid M_h of N_s points on Γ:

$$\Gamma_h = \bigcup_{i=1}^{N_h} T_{h_i}$$

where each T_{h_i} is a planar triangle with vertices belonging to M_h, with maximal length smaller than h. Clearly, one can have $\Gamma = \Gamma_h$ for a polyhedral surface. In the general case, we suppose that the meshing is fine enough so that the orthogonal projection $\mathcal{P}$ from Γ to Γ_h defines a bijection from Γ to Γ_h. Every function ψ on Γ_h can then be associated with a function $\tilde{\psi} = \psi \circ \mathcal{P}^{-1}$ on Γ. In the following construction of the schemes, we will not recall this fact and will directly deal with the functions on Γ_h. Thus, one can say that a function ψ defined on Γ is approximated by a computed function ψ_h on Γ_h (instead of $\psi_h \circ \mathcal{P}^{-1}$).

Now, to approximate functions on Γ, we will use the classical P^0 and P^1 finite element spaces on Γ_h (denoted below by $\tilde{W_h}$ and $\tilde{V_h}$ respectively). The later is necessary for the function φ since, as we have noted, we suppose that its first derivatives are square integrable.

$$\begin{aligned}\tilde{W_h} &= \{f : \Gamma_h \longrightarrow \mathbf{R} \text{ s.t. } f_{T_h} \in P^0 \quad \forall T_h \in M_h\} \\ &= \{f : \Gamma_h \longrightarrow \mathbb{R} \text{ s.t. } f(x) = \sum_{j=1}^{N_h} b_j \lambda_j(x), (b_j)_{j=1,N_h} \in \mathbb{R}^{N_h}\}\end{aligned}$$

and

$$\begin{aligned}\tilde{V_h} &= \{f : \Gamma_h \longrightarrow \mathbf{R} \text{ continuous and s.t. } \ f_{T_h} \in P^1 \ \forall T_h \in M_h\} \\ &= \{f : \Gamma_h \longrightarrow \mathbb{R} \text{ s.t. } f(x) = \sum_{j=1}^{N_s} a_j \varphi_j(x), \ (a_j)_{j=1,N_s} \in \mathbb{R}^{N_s}\}.\end{aligned}$$

In those formulae, $(\lambda_j)_{1 \le j \le N_h}$ are the basis of piecewise constant functions associated with the triangles (T_{h_j})

$$\lambda_j(x) = \delta_j^k \text{ if } x \in T_{h_k}$$

and $(\varphi_j)_{1 \le j \le N_s}$ the basis of piecewise P^1 and continuous functions associated with the vertices S_j (i.e. $\varphi_j(S_k) = \delta_j^k$). One can give an explicit expression for (φ_j):

$$\varphi_j(x) = \begin{cases} E^j_{T_h}(x) = -\frac{(x - S_{j+}).\nu}{2|T_h|} |S_{j+} - S_{j-}| & \text{if } x \in T_h \text{ and } T_h \in \{T_h\}_{S_j}, \\ 0 \text{ otherwise,} \end{cases}$$

where $\{T_h\}_{S_j}$ is the set of triangles admitting S_j as one vertex, and for each triangle $T_h \in \{T_h\}_{S_j}$, the 3 vertices counted in the direct sense are (S_{j+}, S_j, S_{j-}), with ν

the outward normal unit vector to the segment $[S_{j+}, S_{j-}]$. Finally, $|T_h|$ is the area of T_h.

Thus, the unknowns φ and λ can be approximated spatially as:

$$\varphi(x,t) \approx \sum_{j=1}^{N_s} a_j(t)\varphi_j(x)\,, \quad \lambda(x,t) \approx \sum_{j=1}^{N_h} b_j(t)\lambda_j(x)$$

where a_j and b_j are functions with square integrable first derivatives and null for $t \leq 0$.

5.2 Time Discretisation

Now, let $\{t_n = n\Delta t; n \in \mathbb{N}\}$ be a regular subdivision of the time axis $\mathbb{R}^+$ where $\Delta t \geq 0$. We will approximate a_j and b_j by functions in

$$\begin{aligned} H^1(\Delta t, \mathbb{R}) &= \{f : \mathbb{R}^+ \longrightarrow \mathbb{R} \text{ continuous s.t. } f_{/[n\Delta t,(n+1)\Delta t]} \in P^1, n \in \mathbb{N}\} \\ &= \{f : R^+ \longrightarrow R \text{ s.t. } f(t) = \sum_{m\geq 1} a_m \beta^m(t)\} \end{aligned}$$

where β^m is the hat function:

$$\beta^m(t) = \begin{cases} (t - t_{m-1})/\Delta t & \text{if } t \in]t_{m-1}, t_m], \\ (t_{m+1} - t)/\Delta t & \text{if } t \in]t_m, t_{m+1}], \\ 0 & \text{otherwise.} \end{cases}$$

The unknown functions φ and λ are then approximated in the space and time variables by:

$$\varphi_h = \sum_{m\geq 1} \sum_{j=1}^{N_s} a_j^m \varphi_j \beta^m\,, \quad \lambda_h = \sum_{m\geq 1} \sum_{j=1}^{N_h} b_j^m \lambda_j \beta^m\,. \tag{71}$$

On the other hand, we recall that the test functions η and r in the variational problem (70) correspond to the time derivative of ψ and q, dual functions of φ and $p = \dot{\lambda}$ respectively. Thus, to be consistent with the above choice of the approximations of unknowns, we will take in (70) the following test functions r_i^n and η_i^n for each time step n and each degree of freedom i (corresponding to a triangle in the case of r_i^n, and a vertex in the case of η_i^n) :

$$\begin{cases} \eta_i^n = \varphi_i \gamma^n & n \in \mathbb{N}, i \in [1, N_s]\,, \\ r_i^n = \lambda_i \partial_t \gamma^n & n \in \mathbb{N}, i \in [1, N_h]\,, \end{cases} \tag{72}$$

where γ^n is proportional to the derivative of β^n in the interval $]t_{n-1}, t_n]$:

$$\gamma^n(t) = \begin{cases} 1 & \text{if } t \in]t_{n-1}, t_n], \\ 0 & \text{otherwise.} \end{cases}$$

If one fixes the run time T for our computations, and $\Delta t = T/N_T$, then the spaces V_h and W_h of (67) are generated respectively by

$$\{(\varphi_j \gamma^n),\ 1 \le j \le N_s,\ 1 \le n \le N_T\}$$

and

$$\{(\lambda_j \gamma^n),\ 1 \le j \le N_h,\ 1 \le n \le N_T\}.$$

5.3 Some Integration Problems

First, the test function r_i^n contains Dirac distributions at times t_{n-1} and t_n, so that we must ensure that when one puts the discretised functions φ_h, λ_h, η_i^n and r_i^n into the integrals in (70) in place of φ, λ, η and r, the integrals remain well defined. It is clear that this is the case for $\int_0^\infty \int_{\Gamma_h} \frac{1}{\alpha} \lambda_h r_i^n$ because the function λ_h is continuous with respect to the time variable. Other integrals, like the following

$$\begin{aligned} I &= \int_0^\infty \int_{\Gamma_h} S\dot{\lambda}_h r_i^n \, \mathrm{d}\sigma \, \mathrm{d}t \\ &= \frac{1}{4\pi} \int_0^\infty \iint_{\Gamma_h \times \Gamma_h} \frac{\dot{\lambda}_h(y, t - |x-y|)}{|x-y|} r_i^n(x,t) \, \mathrm{d}\sigma_y \, \mathrm{d}\sigma_x \, \mathrm{d}t, \end{aligned}$$

have to be defined through some approximation processes. This is described in some details in [41]. One gets finally the following formula where all integrals are in the usual sense (with no more Dirac):

$$I = \sum_{m=1}^{n} \sum_{j=1}^{N_h} \frac{b_j^m}{4\pi \Delta t} \left(\iint_{E_{n-m-2}} -2 \iint_{E_{n-m-1}} + \iint_{E_{n-m}} J_{ij}^3(x,y) \, \mathrm{d}\sigma_x \, \mathrm{d}\sigma_y \right) \tag{73}$$

where

$$J_{ij}^3(x,y) = \frac{\lambda_j(y)\lambda_i(x)}{|x-y|},$$

and

$$E_l = \{(x,y) \in \Gamma_h \times \Gamma_h \ \ s.t.\ \ t_l \le |x-y| \le t_{l+1}\}.$$

One can also observe in (73) that

1. the sum over m is limited by n ;
2. each b_j^m is multiplied by a coefficient which contains only spatial integrals and which only depends on the time indices m, n through their difference $n - m$ (see the integrals on the RHS).

The first point confirms that the test function r_i^n has no influence on the unknown function at later times t_m $(m > n)$. The second point constitutes the discrete equivalence of the fact that the retarded potential operator S is a *time convolution of an integral operator on* Γ as we pointed out in Sect. 2.3.

Finally, we see in (73) several surface integrals of the form

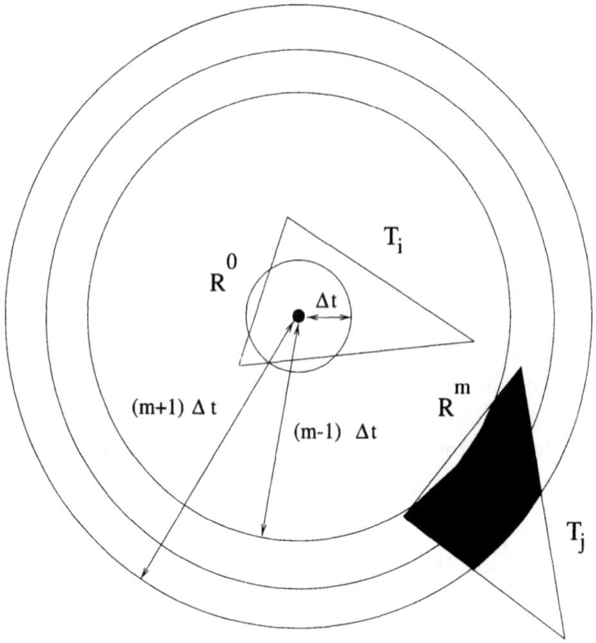

Fig. 2. The domain of spatial integration

$$\iint_{t_l \leq |x-y| \leq t_{l+1}} (\cdots)\, \mathrm{d}\sigma_x\, \mathrm{d}\sigma_y,$$

where the function to be integrated contains a singular term $0(|x-y|^{-1})$. We decompose one such integral as

$$\sum_{i,j=1}^{N_h} \int_{x \in T_i} \left(\int_{y \in T_j ; t_l \leq |x-y| \leq t_{l+1}} (\cdots)\, \mathrm{d}\sigma_y \right) \mathrm{d}\sigma_x .$$

The domain of integration in the last integral is sketched out in Fig.2. When the triangles T_i and T_j are close together (including the case $i = j$), we use an analytical formulation for the first integral (with respect to y), giving a regular function of x. A numerical integration formula is then used for the integration on T_i. Otherwise, numerical integration formulae are used for both integrals. The details of these computations can be seen in [34]. In [32] one can find a multigrid method to compute these integrals. One can also mention the so-called "plane wave time domain algorithm" recently introduced by Ergin, Shanker and Michielssen and co-workers, that can significantly reduce the computational costs of the resulting schemes (see for example [33]).

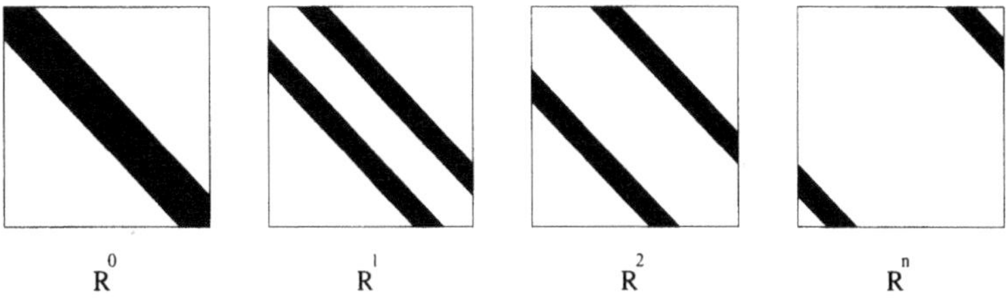

Fig. 3. The profiles of matrices M

5.4 Properties of the Schemes

After computation of all the integrals, one gets the final discretised system for (70), which can be written in the matrix form as:

$$\sum_{m=1}^{n} R^{n-m} \begin{pmatrix} A^m \\ B^m \end{pmatrix} = \begin{pmatrix} C^n \\ D^n \end{pmatrix}$$

where A^m and B^m are the unknown vectors $(a_i^m)_{i=1,Ns}$ and $(b_j^m)_{j=1,Nh})$, C^m and D^m are the discretised right hand sides and R^p is the block matrix

$$R^p = \begin{pmatrix} M^p & N^p \\ \tilde{M}^p & \tilde{N}^p \end{pmatrix} .$$

Thus, one obtains a *time-marching scheme* that respects the character of time convolution of the RPBIE (18). One can also remark that the matrices R^p are sparse and symmetric. Moreover, they vanish for values of p such that ($p \geq 2 + Lt/(c\Delta t)$ where Lt is the diameter of the object Ω_i - here we reintroduce the speed c of sound to show the dependence of the discrete RPBIE on it). The profiles of the matrices are shown in Fig. 3.

This scheme can be described as *semi-explicit* since one has to do only one inversion of matrix, namely R^0. To do that, we compute once and for all the LU-decomposition of R^0. The expressions of the coefficients of $M_{i,j}^{n-m}$, $\tilde{M}_{i,j}^{n-m}$ ... as well as the proof of invertibility of R^0 are given in [41].

5.5 Numerical Experiments

We have given in [41] a large number of computational results showing different aspects of our code. Below, we only present some principal facts pulled out of this work:

1. First, the validation of the code is verified through the comparison of the so-called acoustic backscattering cross section of an absorbing sphere (discretised into 752 triangles, obtained on the one hand by a frequency code (resolution of the BIE for Helmholtz equation) for different values of the frequency, and on

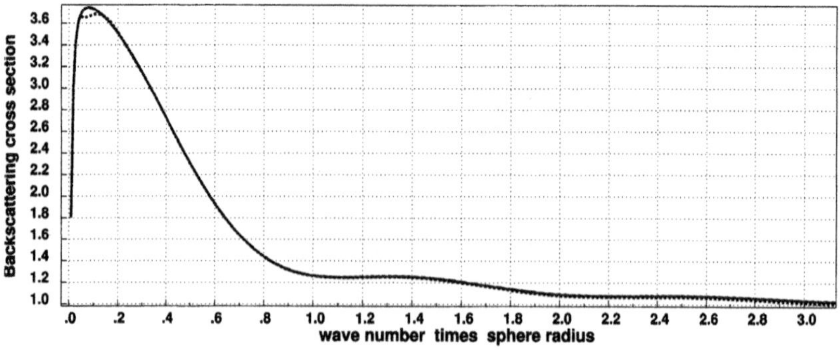

Fig. 4. Validation test (impedance 0.01). Backscatter cross section versus the wave number × sphere radius. The frequency domain result is solid, time domain result is dashed

the other hand by our code and inverse Fourier transform calculations. We show here only one figure taken from those experiments. We would like to emphasise the fact that we have extracted frequency results from time computations but not conversely. Actually, in many industrial problems, the incident signals are so perturbed that it is not possible to model them with a reasonable number of values of the frequency. This is also the case when the incident signals are impulsional (e.g. a very narrow Gaussian pulse).

2. The theoretical stability is verified numerically through a double checking process:
   - For a fixed spatial discretisation of the scatterer, we did computations with different values of the CFL coefficient (CFL $= c\Delta t/\Delta x$ where c is the wave speed and Δt and Δx are respectively the time and space steps). Computations remain stable when we decrease the CFL from 1.2 to 0.3 (the value 1.2 is tried just to show the stability of the calculations, even if, physically, it violates the causality principle). See Figs. 7–10 in [41].
   - We also run our code for a very long time (more than 7000 time steps, corresponding to a characteristic ratio cT/a, where a is the diameter of the scatterer, larger than 500). In these tests, we used both a Gaussian incidence plane wave and a sinus profiled plane wave. In the later case, incident signals continue to be poured over the scatterer as far as we run the computations. Our experiments show that in both cases numerical solutions are not spoiled by instabilities even after very large time steps. See Figs. 11–16 in [41].
3. The convergence of the code was checked using a fixed CFL coefficient and a scatterer (actually, a cube) discretised into smaller and smaller triangle pieces (from 48 to 768 elements).
4. Finally, simulations of the sound waves scattered by a cavity representing a nacelle of an airplane, discretised into 3488 elements, were given at time steps t_n with different values of n up to 800.

The last experiments having confirmed the good behaviour of our code, we have applied it as the direct part of an *inverse solver*, using an optimisation method to detect sources of sound or to find an impedance profile– which can depend on the frequency, thus, is modelled in the time problem by a boundary condition containing a time convolution term. The results are presented in the thesis of Ludwig [46]. See also [47, 48].

6 Related Work

As stated in the Introduction, our method was applied successfully in other acoustic, electromagnetic and elastodynamic scattering problems. In this section, we give a rapid survey of that work.

6.1 Acoustics

In this paper, since we concentrated on the analysis of the question of stability, the mathematical apparatus necessary to analyse the approximation properties of the schemes is neglected. The basis of that is interpolation theorems in the Sobolev type spaces introduced in the text (and their higher degree fellow creatures). Some of these theorems can be seen in the paper of Bamberger and Ha-Duong [9], which can certainly be improved in the framework of the functional spaces used in this paper. The first numerical paper using Galerkin methods for RPBIE is, as far as we know it, the one of Ding et al. in 1989 [31]. It concerned a second kind equation for the Neumann problem. For the Robin problem, some numerical results of the Galerkin schemes for the second kind RPBIE are presented in [35]. Filipe has also implemented in her thesis [34] Galerkin schemes for the first kind RPBIE (15), and proved their stability with respect to the CFL coefficient.

One of the main difficulties in the implementation of our method is the computation of many integrals on the product of two triangles (their influence coefficients). The singularity of the integrals requires very careful computations when the two triangles are identical or close together. On the other hand, when they are far enough one from the other, it is not necessary to have recourse to high precision computations. To overcome these contradictory demands, El Gharib has developed a very efficient multigrid method in his thesis [32], to compute these influence coefficients. One can also see in this thesis a method associating the retarded potential integrals with a high frequency approximation of the scattered waves. These improvements are put into application by the company IMACS in its (unpublished) packages SONATE (acoustics) and ZEUS (electromagnetics). The main ideas and several examples of numerical experiments from these packages can however be seen in a conference presentation by Abboud [1], and also in [2].

6.2 Elastodynamics

The first work on space-time variational method for elastic retarded potential boundary integral equations is the thesis of Bécache [11]. The main difficulty, compared to the case of scalar waves, is due to the complexity of the Green's tensors. In the frequency domain, as in the scalar case, a hypersingularity appears in the first kind BIE for the Neumann problem and needs some regularisation process to be properly defined (and computed). While several such processes were known before, an important result in Bécache's work is that she obtained a new one which respects the causality character of the corresponding retarded potential operator (see [13]).

She could then apply the Galerkin method to some crack problems, and obtained stable schemes as well. Numerical experiments for 2D straight cracks are presented in [12, 14]. Bécache's work is extended to 3D problems in the thesis of Barbier [10], where one can find numerous formulae for very complicated integrals and comparisons of numerical results with those existing in the literature (mostly in the frequency domain). At the same time, Chudinovich [21] carried out the analysis of elastic RPBIEs in [21], but no numerical application was given.

One important problem in elastodynamics is the coupling problem (an elastic object immersed in a fluid medium). The thesis of Filipi [34] is devoted to the time domain fluid-structure problems. Theoretical results for well-posedness and stability are obtained, associated with a particular energy of the coupled system. However, the whole coupling Finite Element- Boundary Element schemes is not implemented yet (see [36]).

6.3 Electromagnetics

Clearly, it is in electromagnetic scattering problems that one meets the most extensive use of boundary element methods. The references cited in the Introduction gives an idea, hopefully reliable enough, of the literature in this field. Again, collocation schemes out play all the others, both for the electric field and the magnetic field integral equations (EFIE and MFIE), with the same drawbacks as in the acoustic problems: lack of mathematical analysis of the equations themselves and lack of stability for their discretised schemes.

Pujols [53] was the first person to investigate the EFIE for a perfect conductor along the lines of our method. Her computations, even if restricted to 2D problems, proved that the recovery of the radar cross section from time domain calculations is possible. This result is confirmed by Terrasse [59] in the 3D case, for such real scatterers as a missile discretised into 700 elements.

However, the main and most important contribution of Terrasse is her thorough study of the functional spaces necessary to analyse the electromagnetic RPBIE: the indexed norms of $H^{-1/2}(\mathrm{div}, \Gamma)$, $H^{-1/2}(\mathrm{curl}\,, \Gamma)$, their temporal counterparts and their approximation by Raviart-Thomas elements etc.. Her functional framework is intensively used in later studies of other electromagnetic problems, by different authors: Lange [5, 6, 42] for scattering by an absorbing obstacle, Lubet [7, 44] for a conductor coated by a heterogeneous material, treated by a coupling of boundary element and finite element methods. Bounhoure [4, 18] revisited this last problem more thoroughly, with an analysis of four possible coupling procedures, two of which are stable and the others not. It is worth noticing that the paper [4] contains an extended summary of the functional framework of Terrasse (with all useful results clearly stated, without demonstration). Finally, Sayah [57] analyses the coupling problem in a general situation where the conductor and the heterogeneous material exist side by side, the first one not necessarily coated by the second. He also study a general impedance condition obtained as a limit when the scatterer is thinly coated. The implementation of his schemes is incorporated into the packages ZEUS of the company IMACS.

7 Conclusion

We have presented an overview of the literature on numerical methods for RPBIEs. A rather detailed exposition of the space-time variational method for basic acoustic scattering problems is given, with functional results that hopefully can be used for analysing different discretisation procedures. Standard conforming Galerkin approximation methods have our preference, since via an energy identity, we are able to prove in a simple manner their unconditional stability. The method is described and the results of its implementation is reported briefly (with references to a recent paper). These results confirm the stability in a wide range of situations and for large run times never reached before.

We would like to emphasise that, as far as we know, direct mathematical analysis of the retarded potential operators (i.e. based on the property of the kernel of the integrals) does not exist yet. Such mathematical analysis could possibly help to justify some discretisation procedures of the RPBIE that are set up without recourse to the wave problems they issue from. In any case, in all basic scattering problems reviewed in Sect. 6, some energy identity does exist, permitting construction of Galerkin schemes that can be proved stable.

Dedication

This article is dedicated to Jean-Claude Nédélec on the occasion of his sixtieth birthday.

References

1. T. Abboud, Recent Developments in Retarded Potential Methods, slides from a presentation at the PSCI Workshop on Computational Electromagnetics, KTH Stockholm, December 1998. `http://imacs.polytechnique.fr/presentations/retarded/toc_b.html`
2. T. Abboud, J. El Gharib and B. Zhou, Retarded Potentials for Acoustic Impedance Problems, Proc. of the 5th Intern. Conf. on Math. Numer. Aspects of Wave Propagations, Santiago de Compostela, Spain 2001, p. 703-708.
3. K.E. Atkinson, The Numerical Solutions of Integral Equations of the Second Kind, Cambridge University Press, 1997.
4. A. Bachelot, L. Bounhoure and A. Pujols, Couplage éléments finis- potentiels retardés pour la diffraction électromagnétique par un obstacle hétérogène, Numer. Math., **89**, (2001), 257-306.

5. A. Bachelot, V. Lange, Time Dependent Integral Method for Maxwell's System, Proc. of the 3rd Intern. Conf. on Math. and Numer. Aspects of Wave Propagation Phenomena, Mandela (France), (1995), 151-159.
6. A. Bachelot, V. Lange, Time Dependent Integral Method for Maxwell's System with Impedance Boundary Condition, Proc. of the 10th Intern. Conf. on Boundary Element Technology, (1995), 137-144.
7. A. Bachelot, V. Lubet, On the Coupling of Boundary Element and Finite Element Methods for a Time Problem, Proc. of the 3rd Intern. Conf. on Math. and Numer. Aspects of Wave Propagation Phenomena, Mandela (France), (1995), 130-139.
8. A. Bachelot, A. Pujols, Equations intégrales Espace-Temps pour le système de Maxwell, CRAS, série I, **t.314**, (1992), 639-644.
9. A. Bamberger, T. Ha-Duong, Formulation variationnelle espace-temps pour le calcul par potentiel retardé d'une onde acoustique, Math. Meth. Appl. Sci., **8**, (1986), 405-435 and 598-608.
10. D. Barbier, Méthodes des potentiels retardés pour la simulation de la diffraction d'onde élastodynamique par une fissure tridimensionnelle, Thèse de l'école Polytechnique, 1999.
11. E. Bécache, Résolution par une méthode d'équations intégrales d'un problème de diffraction d'ondes élastiques transitoires par une fissure, Thèse de l'Université Paris 6, 1991.
12. E. Bécache, A Variational Boundary Integral Equation Method for an Elastodynamic Antiplane Crack, Int. J. for Numer. Meth. in Eng. **36**, (1993), 969-984.
13. E. Bécache, J-C. Nédélec and N. Nishimura, Regularization in 3D for Anistropic Elastodynamic Crack and Obstacle Problems, J. of Elasticity **31**, (1993), 25-46.
14. E. Bécache, T. Ha-Duong, A Space-Time Variational Formula for the Boundary Integral Equation in a 2D Elastic Crack Problem, Math. Modelling and Numer. Anal. **28(2)**, (1994), 141-176.
15. D. Beskos, Boundary Element Methods in Dynamic Analysis: Part II (1986-1996), Appl. Mech. Rev. **50 (3)**, 149-197.
16. B. Birgisson, E. Siebrits and A.P. Pierce, Elastodynamic Direct Boundary Element Methods with Enhanced Numerical Stability Properties, Int. J. Numer. Meth. Eng. **46**, (1999), 871-888.
17. M. Bonnet, G. Maier and C. Polizzotto, Symmetric Galerkin Boundary Element Methods, Appl. Mech. Rev. **51 (11)**, 669- 704.
18. L. Bounhoure, Couplage éléments finis- potentiels retardés pour la diffraction électromagnétique par un obstacle hétérogène, Thèse de l'Université de Bordeaux I, 1998.
19. J.J. Bowman, T.B.A. Senior and P.L.E. Uslenghi, Electromagnetic and Acoustic Scattering by Simple Shapes, A SUMMA Book, 1987.
20. M. J. Bluck and S. P. Walker, Analysis of Three-Dimensional Transient Acoustic Wave Propagation using the Boundary Integral Equation Method, Int. J. Numer. Meth. Eng. **39**, (1996), 1419-1431.
21. I. Yu. Chudinovich, The Boundary Integral Equation Method in the Third Boundary Value Problem of the Theory of Elasticity, Math. Meth. Appl. Sci., **16**, (1994), 203-215 and 217-227.
22. D. Colton and R. Kress, Inverse Acoustic and Electromagnetic Scattering Theory, Springer Verlag, 1992.
23. M. Costabel, Developments in Boundary Element Methods for Time-Dependent Problems, in L. Jentsch, F. Trlzsch (eds), Problems and Methods in Mathematical Physics, B.G. Teubner, Leipzig 1994, pp. 17-32.
24. R. Dautray, J.L. Lions, Analyse mathématiques et calcul numérique pour les sciences et techniques, Masson, Paris 1985.

25. P.J. Davies, Numerical stability and convergence of approximations of retarded potential integral equations, SIAM J. Numer. Anal. 31: 856-875 (1994).
26. P.J. Davies, D.B. Duncan, Averaging techniques for time marching schemes for retarded potential integral equations, Applied Numerical Mathematics **23** (1997), 291-310.
27. P.J. Davies, A stability analysis of a time marching scheme for the general surface electric field integral equation, Applied Numerical Mathematics **27 (1)** (1998), 33-57 .
28. P.J. Davies, D.B. Duncan, On the behaviour of time discretisations of the electric field integral equation, Applied Mathematics and Computation **107** (2000), 1-26.
29. P.J. Davies, D.B. Duncan, Stability and convergence of collocation schemes for retarded potential integral equations, Strathclyde Mathematics Report 2001/23.
30. S.J. Dodson, S.P. Walker and M.J. Bluck, Implicitness and stability of time domain integral equation scattering analysis, ACES J., **13** (1998), 291-301.
31. Y. Ding, A. Forestier, T. Ha-Duong, A Galerkin Scheme for the Time Domain Integral Equation of Acoustic Scattering from a Hard Surface, J. Acoust. Soc. Am., **86 (4)** (1989), 1566-1572.
32. J. El Gharib, Problèmes de potentiels retardés pour l'acoustique, Thèse de l'école Polytechnique, 1999.
33. A.A. Ergin, B. Shanker and E. Michielssen, Fast analysis of transient acoustic wave scattering from rigid bodies using the multilevel plane wave time domain algorithm, J. Acoust. Soc. Am, **117 (3)**, (2000), 1168-1178.
34. M. Filipe, Etude mathématique et numérique d'un problème d'intéraction fluide-structure dépendant du temps par la méthode de couplage Eléments Finis-Equations Intégrales, Thèse de l'école Polytechnique, 1994.
35. M. Filipe, A. Forestier, T. Ha-Duong, A Time Dependent Acoustic Scattering Problem, Proc. of the 3rd Intern. Conf. on Math. and Numer. Aspects of Wave Propagation Phenomena, Mandelieu (France), (1995), 140-150.
36. M. Filipe, T. Ha-Duong, A Coupling of FEM-BEM for Elastic Structure in a Transient Acoustic Field, Proc. of the 15th Biennial Conference on Mechanical Vibration and Noise, ASME Design Engineering Technical Conf., Boston (USA), (1995), **3**, part B, 33-38.
37. M.B. Friedman and R.P. Shaw, Diffraction of Pulses by Cylindrical Obstacles of Arbitrary Cross Section, J. Appl. Mech., **29**, (1962), 40-46.
38. T. Ha-Duong, Equations integrales pour la résolution numérique de problèmes de diffraction d'ondes acoustiques dans R^3, Thèse de l'Université de Paris VI, 1987.
39. T. Ha-Duong, A Mathematical Analysis of Boundary Integral Equations in the Scattering Problems of Transient Waves, in *Boundary Element Methods*, vol. IX, C.A. Brebia, W.L. Wendland and G. Kuhn (eds), (1987), 101-114.
40. T. Ha-Duong, On the Transient Acoustic Scattering by a Flat Object, Japan J. Appl. Math., **7** (1990), 489-513.
41. T. Ha-Duong, B. Ludwig and I. Terrasse, A Galerkin BEM for Transient Acoustic Scattering by an Absorbing Obstacle, to appear in Intern. J. Numer. Meth. Eng. (2002)
42. V. Lange, Equations intégrales espace-temps pour les équations de Maxwell. Calcul du champ diffracté par un obstacle dissipatif, Thèse de l'Université de Bordeaux I, 1995.
43. J.L. Lions, E. Magenes, Problèmes aux limites non homogènes, Dunod, Paris 1968.
44. V. Lubet, Couplage potentiels retardés-éléments finis pour la résolution d'un problème de diffraction d'ondes par un obstacle inhomogène, Thèse de l'Université de Bordeaux I, 1994.
45. C. Lubich, On the multistep time discretization of linear initial-boundary value problems and their boundary integral equations, Numer. Math., **67**, (1994), 365-390

46. B. Ludwig, Quelques problèmes inverses en acoustique transitoire, Thèse de l'Université de Compiègne, 2000.
47. B. Ludwig, I. Terrasse, S. Alestra and T. Ha-Duong, Résolution numérique du problème de reconstruction de sources acoustiques transitoires par les méthodes d'éléments de frontière, Proc. of the 4th Intern. Conf. on Acoustic and Vibratory Surv. Meth. and Diagnostic Tech., Compigne, France 2001, p. 565-574
48. B. Ludwig, I. Terrasse, S. Alestra and T. Ha-Duong, Inverse Acoustic Impedance Problems transitoire, Proc. of the 5th Intern. Conf. on Math. Numer. Aspects of Wave Propagations, Santiago de Compostela, Spain 2001, p. 667-671.
49. E.K. Miller, An overview of time-domain integral equations models in electromagnetics, J. of Electromagnetic Waves and Appl. **1** (1987), 269-293.
50. C.S. Morawetz, Decay for solutions of the exterior problem for the wave equation, Comm. on Pure and Appl. Math. **28** (1975), 229-264.
51. J.C. Nédélec, Curved finite element methods for the solution of the singular integral equations on surface in R^3, Comp. Meth. Appl. Mech. Eng. **8** (1976), 61-80.
52. A.P. Pierce and E. Siebrits, Stability Analysis and Design of Time-Stepping Schemes for General Elastodynamic Boundary Element Models, Int. J. Numer. Meth. Eng. **40**, (1997), 319-342.
53. A. Pujols, Equations intégrales espaces-temps pour le système de Maxwell. Application au calcul de la surface équivalente radar, Thèse de l'Université de Bordeaux I, 1991.
54. B.P. Rynne, Stability and convergence of time marching methods in scattering problems, IMA J. of Appl. Math., **35**, (1985), p. 297-310
55. B.P. Rynne, Instabilities in time marching methods for scattering problems, Electromagnetics,**6**, (1986), p. 129-144.
56. B.P. Rynne, P.D. Smith, Stability of Time Marching Algorithms for the Electric Field Integral Equation, J. Electromagnetic Waves and Appl. **4**, (1990), 1181-1205.
57. T. Sayah, Méthodes des Potentiels Retardés pour les milieux hétérogènes et l'approximation des couches minces par conditions d'impédance généralisées en électromagnétisme, Thèse de l'Université Paris 6, 1998.
58. J.A. Stratton, Electromagnetic Theory, McGraw-Hill, New York 1941.
59. I. Terrasse, Résolution mathématique et numérique des équations de Maxwell instationnaires par une méthode de potentiels retardés, Thèse de l'école Polytechnique, 1993.
60. F. Trèves, Basic linear partial differential equations, Academic Press, New York 1975.

Inverse Scattering Theory for Time–Harmonic Waves

Andreas Kirsch

Mathematisches Institüt II, Englerstr. 2, Universität Karlsruhe, Engelstr. 2, D-76128 Karlsruhe, Germany
Andreas.Kirsch@math.uni-karlsruhe.de

1 An Introduction to Scattering Theory for Time-Harmonic Waves

First, we will briefly recall the physical models for the (linearised) acoustic and electromagnetic wave propagation and how they reduce to boundary value problems for the Helmholtz equation. For more details we refer to, e.g. [17, 18, 25, 50].

1.1 Acoustic Wave Propagation

Let $v(x,t)$ be the velocity of a particle at $x \in \mathbb{R}^3$ and time t, $p(x,t)$ be the pressure (the difference with respect to the static case), $\rho(x,t)$ be the density, and $c(x)$ be the speed of sound. We assume that $\rho(x,t) \equiv \rho_0(x)$ and $c(x) \equiv c_0$ in the background medium where c_0 is constant. The basic equations of hydrodynamics lead, after linearisation, to the existence of a potential U with

$$v = \frac{1}{\rho_0} \nabla U \,, \qquad p = -\frac{\partial U}{\partial t} - \alpha U \,,$$

(where $\alpha \geq 0$ denotes a damping constant) and

$$c^2 \Delta U = \frac{\partial^2 U}{\partial t^2} + \alpha \frac{\partial U}{\partial t} \tag{1}$$

in the medium. Here, the gradient ∇ and Laplacian Δ are given by

$$\nabla U = \left(\frac{\partial U}{\partial x_1}, \frac{\partial U}{\partial x_2}, \frac{\partial U}{\partial x_3} \right)^\top \in \mathbb{R}^3 \quad \text{and} \quad \Delta U = \sum_{j=1}^{3} \frac{\partial^2 U}{\partial x_j^2} \,, \quad \text{respectively.}$$

We assume that the fields are periodic with respect to time t (time-harmonic waves), i.e. of the form

$$U(x,t) = u_1(x)\,\cos(\omega t) \;+\; u_2(x)\,\sin(\omega t)$$
$$= \mathrm{Re}\,\left[u(x)\,e^{-i\omega t}\right] \quad \text{where } u := u_1 + iu_2\,.$$

Here, ω denotes the frequency and $\lambda = 2\pi/\omega$ the wave length. The wave equation (1) leads to the Helmholtz equation

$$\Delta u(x) \;+\; \frac{\omega^2}{c^2}\left(1 + i\frac{\alpha}{\omega}\right) u(x) \;=\; 0\,, \quad \text{i.e.} \quad \Delta u(x) \;+\; k^2 n(x)\, u(x) \;=\; 0 \quad (2)$$

with wave number $k = \omega/c_0$ and index of refraction

$$n(x) \;=\; \frac{c_0^2}{c(x)^2}\left(1 + i\frac{\alpha}{\omega}\right).$$

This differential equation has to be completed with boundary conditions or transmission conditions on the boundary $\partial\Omega$ of the scattering obstacle Ω which we assume to be imbedded in the background medium. In the sound-soft case we require that the pressure p vanishes on $\partial\Omega$. This leads to the Dirichlet condition $u = 0$ on $\partial\Omega$. In the sound-hard case the normal component of the velocity v vanishes which leads to the Neumann boundary condition $\partial u/\partial\nu = 0$ on $\partial\Omega$. Here, $\nu = \nu(x)$ denotes the unit normal vector at x on $\partial\Omega$, and $\partial u/\partial\nu$ the normal derivative. In the penetrable case we require that the normal component of the velocity and the pressure have to be continuous which leads to the transmission conditions $\left(\partial u/\partial\nu\right)_+/\rho_0 = \left(\partial u/\partial\nu\right)_-/\rho$ and $u_+ = u_-$ where the subscripts $+$ and $-$ indicate the limits from the outside or inside, respectively.

1.2 Electromagnetic Wave Propagation

In electromagnetism, the following fields are relevant:

$H = H(x,t)$	magnetic field	(in A/m)
$B = B(x,t)$	magnetic induction	(in $Vs/m^2 = T$ Tesla)
$E = E(x,t)$	electric field	(in V/m)
$D = D(x,t)$	electric induction	(in As/m^2)
$J = J(x,t)$	current	(in A/m^2).

They are coupled by Maxwell's Equations (in differential form):

$$\mathrm{curl}\, H = \frac{\partial D}{\partial t} + J \quad \text{(Ampère's Law)}$$
$$\mathrm{curl}\, E = -\frac{\partial B}{\partial t} \quad \text{(Faraday's Law of Induction)}$$

and by the constitutive relations

$$\begin{aligned} D &= \varepsilon E \ \text{ with permittivity } \varepsilon(x), \\ B &= \mu H \ \text{ with permeability } \mu \equiv \mu_0 \text{ constant}, \\ J &= \sigma E \ \text{ with conductivity } \sigma(x) \text{ (Ohm's law)}. \end{aligned}$$

Here, the curl of a vector field F is defined by

$$\operatorname{curl} F \;=\; \nabla \times F \;=\; \left(\frac{\partial F_3}{\partial x_2} - \frac{\partial F_2}{\partial x_3}, \frac{\partial F_1}{\partial x_3} - \frac{\partial F_3}{\partial x_1}, \frac{\partial F_2}{\partial x_1} - \frac{\partial F_1}{\partial x_2} \right)^\top \in \mathbb{R}^3 .$$

Again, we consider time-harmonic waves:

$$E(x,t) \;=\; \operatorname{Re}\left[E(x)e^{-i\omega t}\right], \quad B(x,t) \;=\; \operatorname{Re}\left[B(x)e^{-\omega t}\right], \quad \text{etc.}$$

Replacing H and D by $\mu_0^{-1}B$ and εE, respectively, yields

$$\operatorname{curl} B \;=\; \mu_0(\sigma - i\omega\varepsilon)E , \quad \operatorname{curl} E \;=\; i\omega B .$$

Note that, in general, ε and σ depend on x! From the vector identity div curl $= 0$ the second equation yields $\operatorname{div} B = 0$, where

$$\operatorname{div} B = \sum_{j=1}^{3} \partial B_j / \partial x_j$$

denotes the divergence of the vector field B. Eliminating B from the system yields

$$\operatorname{curl}^2 E \;=\; i\omega \operatorname{curl} B \;=\; \mu_0(\omega^2\varepsilon + i\omega\sigma)E , \quad \text{i.e.} \quad \operatorname{curl}^2 E \;=\; k^2 n(x)\, E$$

with wave number $k = \omega / c_0$, speed of light in vacuum $c_0 = 1/\sqrt{\varepsilon_0 \mu_0}$, and index of refraction

$$n(x) \;=\; \frac{\varepsilon(x)}{\varepsilon_0} \left[1 + i \frac{\sigma(x)}{\omega \varepsilon(x)} \right] .$$

Here, ε_0 denotes the permittivity of vacuum.

On the other hand, solving for B leads to

$$\operatorname{curl} \left[\frac{1}{n(x)} \operatorname{curl} B \right] \;=\; k^2\, B .$$

In the special case that ε and σ are constant, thus n is constant, the equations reduce to vector-Helmholtz equations

$$\Delta E \,+\, k^2 n\, E \;= 0 \quad \text{and} \quad \Delta B \,+\, k^2 n\, B , \quad \text{respectively.}$$

These differential equations must be completed with boundary or with transmission conditions. On interfaces the tangential components of E and H are continuous. Formulated in terms of E this yields that $\nu \times E$ and $\nu \times \operatorname{curl} E$ are continuous. The following cases lead to two-dimensional problems:

(a) TM-mode (Transversal-Magnetic) or E-mode:

Let the index of refraction n not depend on one variable (say x_3), i.e. $n = n(x_1, x_2)$, and the electric field be of the form $E = (0, 0, u)^\top$ with some scalar component $u = u(x)$. From $\operatorname{div}(nE) = 0$ we conclude that also u depends on x_1 and x_2 only. Then $\Delta u + k^2 n u = 0$ in $\mathbb{R}^2$. The magnetic field B has the form

$$B = \frac{1}{i\omega} \operatorname{curl} E = \frac{1}{i\omega} \left(\frac{\partial u}{\partial x_2}, -\frac{\partial u}{\partial x_1}, 0 \right)^\top .$$

We observe that in this case B and E are orthogonal to each other!

(b) TE-mode or M-mode:

Let the index of refraction n be constant, and the magnetic field be of the form $B = (0, 0, v)^\top$. Then $\Delta v + k^2 n v = 0$ and

$$E = \frac{1}{\mu_0(\sigma - i\omega\varepsilon)} \operatorname{curl} B = \frac{1}{\mu_0(\sigma - i\omega\varepsilon)} \left(\frac{\partial v}{\partial x_2}, -\frac{\partial v}{\partial x_1}, 0 \right)^\top .$$

On an interface with constant cross section along the x_3−axis the transmission conditions reduce to the continuity of u and $\partial u / \partial \nu$ in the TM-mode and of v and $\partial v / \partial \nu$ in the TE-mode.

1.3 Plane and Spherical Waves

The simplest solutions of the Helmholtz equation

$$\Delta u + k^2 u = 0 \quad \text{in } \mathbb{R}^d , \quad d = 2 \text{ or } 3 , \tag{3}$$

are plane and spherical waves.

(a) Plane waves are of the form

$$u(x) = a \, e^{ikx \cdot \hat{\theta}} , \quad x \in \mathbb{R}^d ,$$

with some unit vector $\hat{\theta} \in \mathbb{R}^d$, describing the direction of the plane wave, and some polarisation constant $a \in \mathbb{C}$. Recall that, for k being real valued and positive, the physical wave is given by

$$\begin{aligned} \operatorname{Re} \left[u(x) e^{-i\omega t} \right] &= \operatorname{Re} \left[a \, e^{i(kx \cdot \hat{\theta} - \omega t)} \right] \\ &= a_1 \cos(k\, x \cdot \hat{\theta} - \omega t) - a_2 \sin(k\, x \cdot \hat{\theta} - \omega t) \end{aligned}$$

where $a = a_1 + i a_2$.

(b) Spherical waves in $\mathbb{R}^3$ are of the form

$$u(x) = a \frac{\exp(ik|x-y|)}{|x-y|}, \quad x \in \mathbb{R}^3 \setminus \{y\},$$

with $a \in \mathbb{C}$ and source point $y \in \mathbb{R}^3$. Thus for real valued and positive wave number

$$\text{Re}\left[u(x)e^{-i\omega t}\right] = \text{Re}\left[a \frac{\exp(i(k|x-y| - \omega t))}{|x-y|}\right]$$

$$= a_1 \frac{\cos(k|x-y| - \omega t)}{|x-y|} - a_2 \frac{\sin(k|x-y| - \omega t)}{|x-y|}.$$

"Spherical" waves in $\mathbb{R}^2$ are cylindrical waves. They make use of the Hankel function $H_0^{(1)}$ of first kind and order zero and have the form

$$u(x) = a\, H_0^{(1)}(k|x-y|), \quad x \in \mathbb{R}^2 \setminus \{y\}.$$

1.4 Formulation of the Scattering Problems

For the (direct) scattering problem, we are given the wave number $k \in \mathbb{C} \setminus \{0\}$ with $\text{Re}\, k \geq 0$, $\text{Im}\, k \geq 0$, a bounded region $\Omega \subset \mathbb{R}^d$ ($d = 2$ or 3) with boundary $\partial\Omega \in C^2$, and a solution u^{inc} of the Helmholtz equation $\Delta u + k^2 u = 0$ in a region containing $\overline{\Omega}$ in its interior (i.e. an incident wave). We assume that $\mathbb{R}^d \setminus \overline{\Omega}$ is connected.

The scattering of u^{inc} by Ω produces a scattered wave u^s with corresponding total wave $u = u^{inc} + u^s$. The direct scattering problem is to determine u^s. This leads to the following boundary value problems where we distinguish between impenetrable and penetrable scatterers.

(A) Impenetrable scatterer: (Fig. 1)

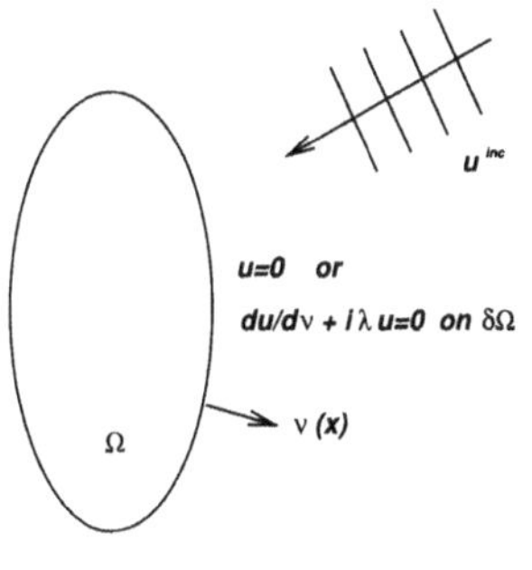

Fig. 1. Impenetrable scatterer

Determine $u \in C^2(\mathbb{R}^d \setminus \overline{\Omega}) \cap C^1(\mathbb{R}^d \setminus \Omega)$ with
(A-a) $\Delta u + k^2\, u = 0$ in $\mathbb{R}^d \setminus \overline{\Omega}$,

(A-b) $u^s = u - u^{inc}$ satisfies the **Sommerfeld radiation condition**

$$\lim_{|x|\to\infty} |x|^{(d-1)/2} \left(\frac{\partial u^s(x)}{\partial r} - ik\, u^s(x) \right) = 0 \tag{4}$$

uniformly with respect to $\hat{x} = x/|x|$.

(A-c) u satisfies a *boundary condition on* $\partial\Omega$, e.g. $u = 0$ on $\partial\Omega$ (Dirichlet boundary condition) or $\partial u/\partial\nu + i\lambda u = 0$ on $\partial\Omega$ (impedance boundary condition).

We call this the **obstacle scattering case.**

(B) Penetrable scatterer: (Fig. 2) Determine $u \in C^2(\mathbb{R}^d \setminus \overline{\Omega}) \cap C^2(\Omega) \cap C^1(\mathbb{R}^d)$

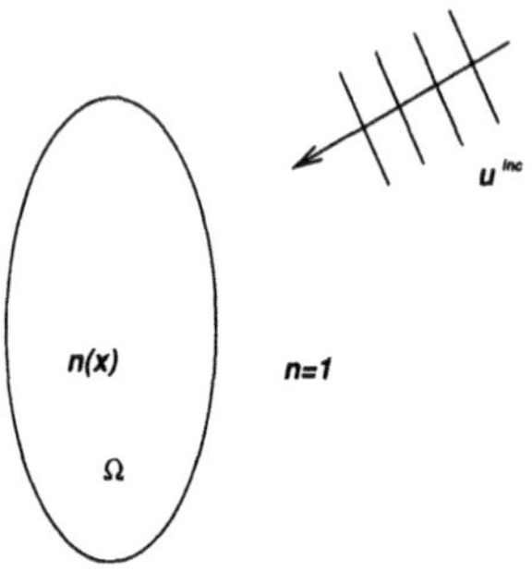

Fig. 2. Penetrable scatterer

with

(B-a) $\Delta u + k^2 n(x)\, u = 0$ in $\mathbb{R}^d$ with $n = n(x) \in C^2(\mathbb{R}^d)$ such that the support of $n - 1$ is contained in $\overline{\Omega}$.

(B-b) $u^s = u - u^{inc}$ satisfies *Sommerfeld's radiation condition* (4).

We call this scattering by an **inhomogeneous medium.**

A solution of the Helmholtz equation in the exterior of a bounded set is called a radiating solution if it satisfies the radiation condition (4).

1.5 Uniqueness and Existence

The boundary value problems in scattering theory are always set up in an unbounded region. Sommerfeld's radiation condition (4) ensures uniqueness of the direct scattering problems. The basic ingredient is the following result which is due to Rellich [72].

Lemma 1. *(Rellich's Lemma) Let v be a radiating solution of the Helmholtz equation $\Delta v + k^2 v = 0$ for $|x| > R_0$ with real and positive wave number k. If, furthermore,*

$$\lim_{R\to\infty} \int_{|x|=R} |v|^2 ds = 0$$

then $v \equiv 0$ *for* $|x| > R_0$.

As announced, application of this result and Green's formula yield:

Theorem 1. *(Uniqueness)*
Let $k \in \mathbb{C} \setminus \{0\}$ *with* $\operatorname{Re} k \geq 0$ *and* $\operatorname{Im} k \geq 0$.

(i) The boundary value problems (A-a), (A-b), (A-c) for the Dirichlet or impedance boundary condition (with impedance λ *such that* $\operatorname{Re}\left[\overline{k}\lambda(x)\right] \geq 0$*) have at most one solution.*
(ii) The scattering problem (B-a), (B-b) has at most one solution.

Proof. (Sketch proof for the Dirichlet problem). The difference $v = u_1 - u_2 = u_1^s - u_2^s$ satisfies the radiation condition (4) and $\Delta v + k^2 v = 0$ in $\mathbb{R}^d \setminus \overline{\Omega}$. Furthermore, $v = 0$ on $\partial\Omega$. The first Green's formula in the region $D_R = \left\{x \in \mathbb{R}^d \setminus \overline{\Omega} : |x| < R\right\}$ yields:

$$\iint\limits_{D_R} \left[|\nabla v|^2 - k^2 |v|^2\right] dx = \underbrace{-\int\limits_{\partial\Omega} \overline{v}\,\frac{\partial v}{\partial\nu}\,ds}_{=0} + \int\limits_{|x|=R} \overline{v}\,\frac{\partial v}{\partial r}\,ds\,.$$

Furthermore,

$$\begin{aligned}
\int\limits_{|x|=R} \left|\frac{\partial v}{\partial r} - ikv\right|^2 ds &= \int\limits_{|x|=R} \left[\left|\frac{\partial v}{\partial r}\right|^2 + |kv|^2\right] ds \;-\; 2\operatorname{Im}\left[\overline{k} \int\limits_{|x|=R} \overline{v}\,\frac{\partial v}{\partial r}\,ds\right] \\
&= \int\limits_{|x|=R} \left[\left|\frac{\partial v}{\partial r}\right|^2 + |kv|^2\right] ds + 2\operatorname{Im} k \iint\limits_{D_R} |\nabla v|^2\,dx \;+ \\
&\quad + 2|k|^2 \operatorname{Im} k \iint\limits_{D_R} |v|^2\,dx\,.
\end{aligned}$$

All terms on the right hand side are non-negative and the sum converges to zeros as $R \to \infty$ by the radiation condition. Therefore, all terms on the right hand side converge to zero separately.

Case 1: $\operatorname{Im} k > 0$. Then $\iint_{D_R} |v|^2\,dx \longrightarrow 0$, $R \to \infty$, thus $v \equiv 0$.
Case 2: k real and positive. Lemma 1 implies $v \equiv 0$ for $|x| \geq R_0$ and then unique continuation yields $v \equiv 0$ in $\mathbb{R}^d \setminus \Omega$.

Existence can be shown by different methods. We briefly recall the **integral equation methods** since we will later come back to the boundary integral operators. The integral equation methods rely on the **fundamental solution** Φ of the Helmholtz equation which is defined by

$$\Phi(x,y) = \begin{cases} \dfrac{\exp(ik|x-y|)}{4\pi\,|x-y|} & \text{in } \mathbb{R}^3\,, \\[2ex] (i/4)\, H_0^{(1)}(k|x-y|) \text{ in } \mathbb{R}^2\,. \end{cases}$$

Again, $H_0^{(1)}$ denotes the Hankel function of first kind and order zero.

The "direct" approach to an integral equation is made with the following representation formula, see [17].

Theorem 2. *(Representation Theorem) Let $u^s \in C^2(\mathbb{R}^d \setminus \overline{\Omega}) \cap C^1(\mathbb{R}^d \setminus \Omega)$ be a radiating solution of the Helmholtz equation in $\mathbb{R}^d \setminus \overline{\Omega}$. Then*

$$u^s(x) = \int\limits_{\partial\Omega} \left[u^s(y)\, \frac{\partial\Phi(x,y)}{\partial\nu(y)} - \Phi(x,y)\, \frac{\partial u^s(y)}{\partial\nu} \right] ds(y)\,, \quad x \notin \overline{\Omega}\,.$$

Application of Green's second theorem to u^{inc} and $\Phi(x,\cdot)$ in the region Ω yields

$$0 = \int\limits_{\partial\Omega} \left[u^{inc}(y)\, \frac{\partial\Phi(x,y)}{\partial\nu(y)} - \Phi(x,y)\, \frac{\partial u^{inc}(y)}{\partial\nu} \right] ds(y)\,, \quad x \notin \overline{\Omega}\,.$$

Summation of both formulas leads to

$$u^s(x) = \int\limits_{\partial\Omega} \left[u(y)\, \frac{\partial\Phi(x,y)}{\partial\nu(y)} - \Phi(x,y)\, \frac{\partial u(y)}{\partial\nu} \right] ds(y)\,, \quad x \notin \overline{\Omega}\,. \tag{5}$$

So far, we have not made use of the boundary conditions.

1. In the case of an impedance boundary condition we just let x tend to a boundary point. The classical continuity results of the single and double layer potentials with densities $\partial u/\partial\nu$ and u, respectively, (see again [17]) and the replacement of $\partial u/\partial\nu$ by $-i\lambda u$ yields

$$u^s(x) = \frac{1}{2}\, u(x) + \int\limits_{\partial\Omega} u(y) \left[\frac{\partial\Phi(x,y)}{\partial\nu(y)} + i\lambda\,\Phi(x,y) \right] ds(y)\,, \quad x \in \partial\Omega\,, \tag{6}$$

   i.e. by adding u^{inc} on both sides,

$$\frac{1}{2}\, u(x) - \int\limits_{\partial\Omega} u(y) \left[\frac{\partial\Phi(x,y)}{\partial\nu(y)} + i\lambda\,\Phi(x,y) \right] ds(y) = u^{inc}(x)\,, \quad x \in \partial\Omega\,. \tag{7}$$

2. In the case of a Dirichlet boundary condition we take the normal derivative of (5) at the boundary. The boundary condition $u = 0$ on ∂D and the jump conditions of the normal derivatives of the single and double layer potentials yield

$$\frac{\partial u^s(x)}{\partial\nu} = \frac{1}{2}\, \frac{\partial u(x)}{\partial\nu} - \int\limits_{\partial\Omega} \frac{\partial u(y)}{\partial\nu}\, \frac{\partial\Phi(x,y)}{\partial\nu(x)}\, ds(y)\,, \quad x \in \partial\Omega\,, \tag{8}$$

i.e. by adding $\partial u^{inc}/\partial\nu$,

$$\frac{1}{2}\frac{\partial u(x)}{\partial\nu} + \int_{\partial\Omega} \frac{\partial u(y)}{\partial\nu}\frac{\partial\Phi(x,y)}{\partial\nu(x)}\,ds(y) = \frac{\partial u^{inc}(x)}{\partial\nu}, \quad x\in\partial\Omega. \tag{9}$$

Equations (7) and (9) are Fredholm integral equations of the second kind for $u|_{\partial\Omega}$ and $\partial u/\partial\nu$, respectively, since the integral operators

$$S\varphi(x) = \int_{\partial\Omega} \varphi(y)\,\Phi(x,y)\,ds(y), \quad x\in\partial\Omega, \tag{10}$$

$$D\varphi(x) = \int_{\partial\Omega} \varphi(y)\,\frac{\partial\Phi(x,y)}{\partial\nu(y)}\,ds(y), \quad x\in\partial\Omega, \tag{11}$$

$$D^*\varphi(x) = \int_{\partial\Omega} \varphi(y)\,\frac{\partial\Phi(x,y)}{\partial\nu(x)}\,ds(y), \quad x\in\partial\Omega, \tag{12}$$

are all compact in the space $C(\partial\Omega)$ of continuous functions on $\partial\Omega$. Equations (7) and (9) can be written in short form as

$$\frac{1}{2}u - Du - i\,S(\lambda u) = u^{inc} \quad\text{and}\quad \frac{1}{2}\frac{\partial u}{\partial\nu} + D^*\frac{\partial u}{\partial\nu} = \frac{\partial u^{inc}}{\partial\nu} \quad\text{on } \partial\Omega.$$

Furthermore, the operators $\frac{1}{2}I - D - iS\lambda$ and $\frac{1}{2}I + D^*$ are isomorphisms from X onto itself provided $\mathrm{Re}\,[\overline{k}\lambda] \geq 0$ and k^2 is not a Dirichlet or Neumann eigenvalue, respectively, of $-\Delta$ in Ω. Here X is either $C(\partial\Omega)$ or $L^2(\partial\Omega)$.

There exists an alternative approach which is sometimes called the indirect integral equation method. We formulate it for the Dirichlet problem only. One searches for a solution in the form of a combined single and double layer, i.e. in the form

$$u^s(x) = \int_{\partial\Omega} \varphi(y)\left[\alpha\frac{\partial}{\partial\nu(y)}\Phi(x,y) + i\beta\,\Phi(x,y)\right]ds(y), \quad x\notin\overline{\Omega},$$

for some $\alpha,\beta\in\mathbb{R}$ and $\varphi\in C(\partial\Omega)$. Then $u^s + u^{inc}$ solves the scattering problem (A-a)–(A-c) under Dirichlet boundary conditions if, and only if,

$$\frac{\alpha}{2}\varphi(x) + \int_{\partial\Omega} \varphi(y)\left[\alpha\frac{\partial}{\partial\nu(y)}\Phi(x,y) + i\beta\,\Phi(x,y)\right]ds(y) = -u^{inc}(x), \quad x\in\partial\Omega,$$

i.e.

$$\frac{\alpha}{2}\varphi + \alpha\,D\varphi + i\beta\,S\varphi = -u^{inc} \text{ on } \partial\Omega.$$

For $\alpha = 0$ this is a Fredholm equation of the first kind, while for $\alpha \neq 0$ this is a Fredholm equation of the second kind. The solvability conditions are given as follows:

$\alpha = 1, \beta = 0$: Solvable iff k^2 is not a Neumann eigenvalue of $-\Delta$ in Ω.
$\alpha = 0, \beta = 1$: Solvable iff k^2 is not a Dirichlet eigenvalue of $-\Delta$ in Ω.
$\alpha = 1, \beta < 0$: Always solvable (see [6, 62, 66]).

Different modifications of the double layer ansatz (for the Dirichlet problem) have been suggested by Jones [46], Ursell [75, 76], Kleinman and Roach [60], and others.

Finally, we consider the scattering problem (B-a), (B-b). From the jump conditions of the volume potential one derives the following **Lippmann-Schwinger** integral equation for the total field u (see [18])

$$u(x) \;=\; u^{inc}(x) \;+\; k^2 \iint\limits_{\Omega} (1 - n(y))\, u(y)\, \Phi(x,y)\, dy\,, \quad x \in \overline{\Omega}\,. \tag{13}$$

Again, this is a Fredholm integral equation of the second kind. Since uniqueness holds by Theorem 1 existence follows by the Fredholm alternative.

2 The Scattering Amplitude and the Inverse Problem

2.1 The Scattering Amplitude

For every radiating solution of the Helmholtz equation in some region $\{x \in \mathbb{R}^d : |x| > R\}$ exterior of a ball the following Atkinson-Wilcox expansion holds (see [17]):

$$u^s(x) \;=\; \frac{e^{ik|x|}}{|x|^{(d-1)/2}} \sum_{n=0}^{\infty} \frac{F_n(\hat{x})}{|x|^n}\,, \quad \hat{x} = \frac{x}{|x|}\,, \quad |x| > R\,.$$

The series converges absolutely and uniformly in every set of the form $\{x \in \mathbb{R}^d : |x| \geq R'\}$ for any $R' > R$.

Definition 1. *Let S^{d-1} be the unit sphere in $\mathbb{R}^d$. The function $F_0 : S^{d-1} \to \mathbb{C}$ is called the* **far field pattern** *or* **scattering amplitude** *or* **radiation pattern** *of u^s. We denote F_0 by u^∞, i.e. u^∞ is defined by*

$$u^s(x) \;=\; \frac{e^{ik|x|}}{|x|^{(d-1)/2}}\, u^\infty(\hat{x}) \left[1 \,+\, \mathcal{O}\left(\frac{1}{|x|}\right)\right], \quad \hat{x} = \frac{x}{|x|}\,,$$

as $|x| \to \infty$, uniformly with respect to $\hat{x} \in S^{d-1}$.

Remark: The far field pattern u^∞ determines u^s uniquely since, by Rellich's Lemma and unique continuation, $u^\infty(\hat{x}) \equiv 0$ implies that u^s vanishes in $\mathbb{R}^d \setminus \Omega$.

We recall the Representation Theorem 2, i.e.

$$u^s(x) \;=\; \int\limits_{\partial\Omega} \left[u(y)\, \frac{\partial \Phi(x,y)}{\partial \nu(y)} - \Phi(x,y)\, \frac{\partial u(y)}{\partial \nu}\right] ds(y)\,, \quad x \notin \overline{\Omega}\,.$$

The asymptotic behaviour of the fundamental solution

$$\Phi(x,y) = \gamma_d \frac{\exp(ik|x|)}{|x|^{(d-1)/2}} e^{-ik\hat{x}\cdot y} \left[1 + \mathcal{O}\left(\frac{1}{|x|}\right)\right], \quad |x| \to \infty, \quad \text{and} \tag{14}$$

$$\frac{\partial \Phi(x,y)}{\partial \nu(y)} = \gamma_d \frac{\exp(ik|x|)}{|x|^{(d-1)/2}} \frac{\partial}{\partial \nu(y)} e^{-ik\hat{x}\cdot y} \left[1 + \mathcal{O}\left(\frac{1}{|x|}\right)\right], \tag{15}$$

with $\gamma_3 = 1/(4\pi)$ and $\gamma_2 = e^{i\pi/4}/\sqrt{8k\pi} = (1+i)/(4\sqrt{k\pi})$ yields the following representations of the far field patterns:

- In the case of Dirichlet boundary conditions we have

$$u^\infty(\hat{x}) = -\gamma_d \int\limits_{\partial\Omega} \frac{\partial u(y)}{\partial \nu} e^{-ik\hat{x}\cdot y} \, ds(y), \quad \hat{x} \in S^{d-1}. \tag{16}$$

- In the case of Neumann boundary conditions we have

$$u^\infty(\hat{x}) = \gamma_d \int\limits_{\partial\Omega} u(y) \frac{\partial}{\partial \nu} e^{-ik\hat{x}\cdot y} \, ds(y), \quad \hat{x} \in S^{d-1}. \tag{17}$$

- In the case of impedance boundary conditions we have

$$u^\infty(\hat{x}) = \gamma_d \int\limits_{\partial\Omega} u(y) \left[\frac{\partial}{\partial \nu} e^{-ik\hat{x}\cdot y} + i\lambda(y)\, e^{-ik\hat{x}\cdot y}\right] ds(y), \quad \hat{x} \in S^{d-1}. \tag{18}$$

The functions u, u^s, u^∞ depend on the wave number k, the direction of incidence $\hat{\theta}$, the region Ω and/or the index of refraction $q(x)$. We indicate this dependence by writing $u(x) = u(x;\hat{\theta})$, etc.

The following **conclusions** can be drawn from the representations (16)–(18):

- The far field u^∞ is analytic with respect to $\hat{x}$ and $\hat{\theta}$ and $k > 0$.
- The following reciprocity principle holds: $u^\infty(\hat{\theta};\hat{x}) = u^\infty(-\hat{x};-\hat{\theta})$ for all $\hat{x},\hat{\theta} \in S^{d-1}$.

2.2 The Inverse Scattering Problems and Uniqueness

In this section we will formulate the inverse scattering problems and present the ideas of two proofs of uniqueness for the inverse problem. From now on we restrict ourselves to the Dirichlet or Neumann boundary value problem.

The **inverse problem** that we consider here is to determine the shape of Ω from the knowledge of $u^\infty(\hat{x};\hat{\theta})$ for all $\hat{x},\hat{\theta} \in S^{d-1}$ and for one fixed wave number k.

Theorem 3. *(Uniqueness) Let u_1 and u_2 be the total fields corresponding to Ω_1 and Ω_2, respectively. If $u_1^\infty(\hat{x},\hat{\theta}) = u_2^\infty(\hat{x},\hat{\theta})$ for all $\hat{x},\hat{\theta} \in S^{d-1}$ then $\Omega_1 = \Omega_2$.*

The classical **proofs** of this result begin with an application of Rellich's Lemma and analytic continuation. This yields that $u_1(\cdot,\hat\theta)$ and $u_2(\cdot,\hat\theta)$ coincide on the unbounded component G' of $\mathbb{R}^d \setminus (\Omega_1 \cup \Omega_2)$ for every $\hat\theta \in S^{d-1}$. The proofs are indirect, i.e. one assumes that $\Omega_1 \neq \Omega_2$. Let, e.g. $\Omega_2 \not\subset \Omega_1$. We present the ideas of two different approaches.

(A) Schiffer's idea (see [61]) was to consider the function $u_1(\cdot,\hat\theta)$ in the region $D := \left(\mathbb{R}^d \setminus \overline{G'}\right) \setminus \overline{\Omega}_1$ (the shaded region in Fig. 3). Since $u_1(\cdot,\hat\theta) = 0$ on ∂D we observe that $u_1(\cdot,\hat\theta)$ is a Dirichlet eigenfunction of $-\Delta$ in D corresponding to the eigenvalue k^2. This holds for every $\hat\theta \in S^{d-1}$. Since $u_1(\cdot,\hat\theta)$, $\hat\theta \in S^{d-1}$, are linearly independent, this contradicts the fact that the eigenspace of $-\Delta$ has finite dimension – which holds independently on the smoothness of ∂D. (Note that D does not necessarily satisfy a cone condition.) These arguments are questionable in the case of Neumann boundary conditions because the finiteness of the dimension of the eigenspace relies on the compactness of the imbedding of the Sobolev space $H^1(D)$ into $L^2(D)$ which holds only under certain regularity assumptions on ∂D.

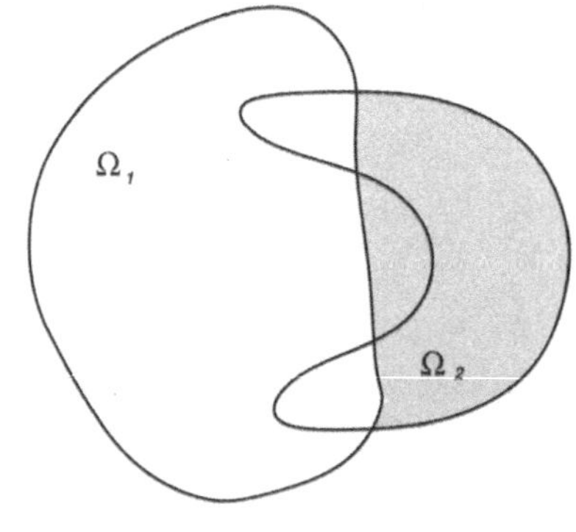

Fig. 3. Configuration used in Schiffer's proof of uniqueness

(B) A second proof of uniqueness was given by Kress and Kirsch in [56] and uses the mixed reciprocity principle. Again we begin by noting that $u_1(\cdot,\hat\theta)$ and $u_2(\cdot,\hat\theta)$ coincide in G'. To formulate the mixed reciprocity principle we introduce the total field $v(\cdot;z)$ of the scattering problem (A-a)–(A-c) corresponding to incident wave $v^{inc}(x;z) = \Phi(x,z)/\gamma_d$ with source point $z \notin \overline{\Omega}$. The corresponding scattered fields and far field pattern are denoted by $v^s(\cdot;z)$ and $v^\infty(\cdot;z)$, respectively. Then the following relation holds where again $u^s(\cdot;\hat\theta)$ denotes the scattered field corresponding to plane wave incidence of direction $\hat\theta \in S^{d-1}$ (see [70]):

$$v^\infty(\hat x;z) = u^s(z;-\hat x) \quad \text{for all } \hat x \in S^{d-1} \text{ and } z \in \mathbb{R}^d \setminus \overline{\Omega}. \tag{19}$$

Since $u_1^s(z,\hat\theta) = u_2^s(z,\hat\theta)$ for all $z \in G'$ and $\hat\theta \in S^{d-1}$ this yields $v_1^\infty(\hat\theta,z) = v_2^\infty(\hat\theta,z)$ for all $z \in G'$ and $\hat\theta \in S^{d-1}$. A second application of Rellich's Lemma and analytic continuation yields that $v_1(x,z) = v_2(x,z)$ for all $x,z \in G'$.

We now fix some $\hat{z} \in \partial\Omega_2 \setminus \overline{\Omega}_1$ and choose a sequence $z_j \in G'$ with $z_j \to \hat{z}$ (see Fig. 4). With respect to Ω_1 and incident fields $v^{inc}(\cdot; z_j)$ the boundary value problem (A-a) – (A-c) is a regular perturbation problem and thus $v_1^s(\hat{z}; z_j) \to v_1^s(\hat{z}; \hat{z})$ as j tends to infinity.

However, with respect to Ω_2 we have, by the Dirichlet boundary condition, $v_2^s(\hat{z}; z_j) = -\Phi(\hat{z}, z_j)/\gamma_d$ which is unbounded as j tends to infinity. This is a contradiction. This type of proof carries over also to a wide class of boundary conditions,

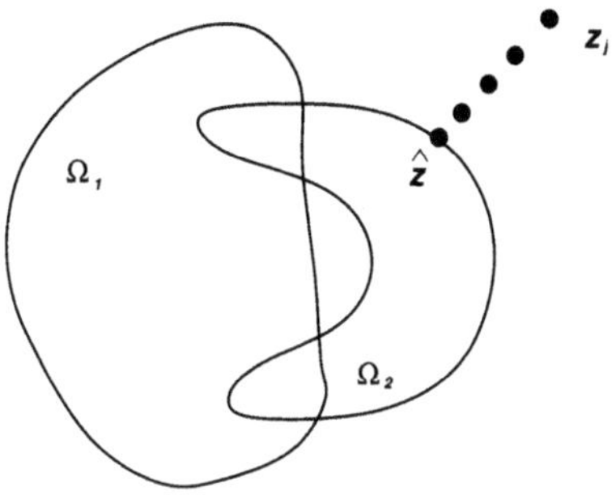

Fig. 4. Configuration used in second proof of uniqueness

see [29, 37, 44, 56, 58].

2.3 The Far Field Operator

From now on for the sake of notation we restrict ourselves to the case $d = 3$. We define the **far field operator** $F : L^2(S^2) \to L^2(S^2)$ by

$$F\psi(\hat{x}) = \int_{S^2} u^\infty(\hat{x}; \hat{\theta})\, \psi(\hat{\theta})\, ds(\hat{\theta}), \quad \hat{x} \in S^2 .$$

The operator

$$\mathcal{S} := I + \frac{ik}{2\pi} F$$

is called the **scattering matrix** or the **scattering operator**. The following result has been proven in, e.g. [18], for the case of an inhomogeneous medium:

Lemma 2. *For $g \in L^2(S^2)$ define the* **Herglotz wave function** u_g^{inc} *by*

$$u_g^{inc}(x) = \int_{S^2} e^{ikx\cdot\hat{\theta}}\, g(\hat{\theta})\, ds(\hat{\theta}), \quad x \in \mathbb{R}^3 .$$

Let u_g be the corresponding solution of the scattering problem (A-a) – (A-c) with Dirichlet or impedance boundary conditions. Then

$$F - F^* - \frac{ik}{2\pi} F^* F = 2iR \quad \textit{with}$$

$$R := \begin{cases} 0 & \textit{in the Dirichlet case,} \\ \frac{1}{4\pi} L^* (\operatorname{Re} \lambda) L & \textit{in the impedance case} \end{cases}$$

where $L : L^2(S^2) \to L^2(\partial\Omega)$ *maps* g *onto* $u_g\big|_{\partial\Omega}$. *Here,* F^* *denotes the* L^2*-adjoint of* F.

From this the following properties of F can easily be shown:

Theorem 4. *(i)* $\mathcal{S}$ *is subunitary, i.e.* $\mathcal{S}^*\mathcal{S} = I - \frac{k}{\pi} R$, *and* R *is non-negative.*
(ii) For Dirichlet or Neumann boundary condition (or if λ *is purely imaginary) the scattering matrix* $\mathcal{S}$ *is unitary, i.e.* $\mathcal{S}^*\mathcal{S} = \mathcal{S}\mathcal{S}^* = I$, *and* F *is normal, i.e.* $F^*F = FF^*$.
(iii) F *is of trace class and* $\operatorname{Im} F := \frac{1}{2i}(F - F^*)$ *is non-negative.*
(iv) F *has a complete set of eigenfunctions, and the eigenvalues are contained in the disc with centre* $2\pi i/k$ *and radius* $2\pi/k$.

2.4 The Factorisation Method

In this section we present an alternative approach for proving uniqueness of the inverse problem. In contrast to the proofs indicated in the previous section this one is direct and constructive. We will see in Section 3 that it can also be used to visualise the obstacle, i.e. to solve the inverse problem numerically.

One of the basic tools is a factorisation of the far field operator. Before we do this we define the **Herglotz operator** $H : L^2(S^2) \to H^{1/2}(\partial\Omega)$ by[1]

$$Hg(x) = \int_{S^2} e^{ikx\cdot\hat{\theta}} g(\hat{\theta})\, ds(\hat{\theta}), \quad x \in \partial\Omega. \tag{20}$$

Its adjoint $H^* : H^{-1/2}(\partial\Omega) \to L^2(S^2)$ is given by

$$H^*\varphi(\hat{x}) = \int_{\partial\Omega} e^{-iky\cdot\hat{x}} \varphi(y)\, ds(y), \quad \hat{x} \in S^2.$$

We characterise the region Ω by the range $\mathcal{R}(H^*)$ of H^* in the following theorem:

Theorem 5. *Assume that* k^2 *is not a Dirichlet eigenvalue in* Ω. *For* $z \in \mathbb{R}^3$ *we define* $\phi_z \in L^2(S^2)$ *by*

$$\phi_z(\hat{x}) = \exp(-ikz\cdot\hat{x}), \quad \hat{x} \in S^2.$$

Then:

$$z \in \Omega \iff \phi_z \in \mathcal{R}(H^*).$$

[1] We denote by $H^{\pm 1/2}(\partial\Omega)$ the Sobolev space of order $\pm 1/2$.

Proof. First we note that $\frac{1}{4\pi} H^*\varphi$ is the far field pattern v^∞ of the single layer potential

$$v(x) = \int_{\partial\Omega} \Phi(x,y)\,\varphi(y)\,ds(y)\,, \quad x \notin \overline{\Omega}\,,$$

and ϕ_z is the far field pattern of $4\pi\Phi(\cdot,z)$, see (14).

Let first $z \in \Omega$. Since the boundary operator S from (10) is an isomorphism from $H^{-1/2}(\partial\Omega)$ onto $H^{1/2}(\partial\Omega)$ since k^2 is not a Dirichlet eigenvalue there exists $\varphi \in H^{-1/2}(\partial\Omega)$ with $S\varphi = \Phi(\cdot,z)$ on $\partial\Omega$. The uniqueness of the exterior Dirichlet problem yields $v \equiv \Phi(\cdot,z)$ in $\mathbb{R}^3 \setminus \overline{\Omega}$ and thus $\frac{1}{4\pi} H^*\varphi = v^\infty = \Phi^\infty(\cdot,z) = \frac{1}{4\pi}\phi_z$ on S^2 which proves $\phi_z \in \mathcal{R}(H^*)$.

Let now $z \notin \Omega$ and assume, on the contrary, that $\phi_z = H^*\varphi$ for some $\varphi \in H^{-1/2}(\partial\Omega)$. Rellich's Lemma yields $v \equiv \Phi(\cdot,z)$ outside of every ball containing $\Omega \cup \{z\}$ in its interior. Since $\Phi(x,z)$ is singular at $x = z$ of order $1/|x-z|$ but v is in $H^1_{loc}(\mathbb{R}^3 \setminus \overline{\Omega})$, an analytic continuation argument yields a contradiction.

The second step is to link the far field operator F to the Herglotz operator H. This is done in the following theorem which justifies the name of the method.

Theorem 6. *Assume that k^2 is not an eigenvalue of $-\Delta$ with respect to the Dirichlet boundary condition and, in addition, with respect to the boundary condition of the scattering problem (A-a) – (A-c). Then there exists an isomorphism T from $H^{1/2}(\partial\Omega)$ onto $H^{-1/2}(\partial\Omega)$ with*

$$-4\pi\,F = H^*\,T\,H\,.$$

In particular, in the case of Dirichlet boundary conditions, T is given by $T = S^{-1}$ with single layer operator $S : H^{-1/2}(\partial\Omega) \to H^{1/2}(\partial\Omega)$.

In the case of Neumann boundary conditions, T is given by $T = \Lambda N^{-1}\Lambda$ where $N : H^{1/2}(\partial\Omega) \to H^{-1/2}(\partial\Omega)$ is the normal derivative of the double layer potential and $\Lambda : H^{1/2}(\partial\Omega) \to H^{-1/2}(\partial\Omega)$ is the interior Dirichlet-to-Neumann operator which exists since we assumed that k^2 is not a Dirichlet eigenvalue in Ω.

In the case of an impedance boundary condition, the form of T is more complicated and we refer to [32].

Proof. (For the Dirichlet boundary condition.) We introduce the auxiliary operator $G : C(\partial\Omega) \to L^2(S^2)$ which maps functions $f \in C(\partial\Omega)$ into the far field pattern w^∞ of the solution w of the exterior boundary value problem

$$\Delta w + k^2 w = 0 \text{ in } \mathbb{R}^3 \setminus \overline{\Omega}\,, \quad w = f \text{ on } \partial\Omega\,, \quad w \text{ satisfies (4)}\,.$$

We observe that by superposition $Fg = v^\infty$ where v^∞ is the far field pattern of the solution of the scattering problem (A-a) – (A-c) for incident field

$$v^{inc}(x) = \int_{S^2} g(\hat\theta)\, e^{ikx\cdot\hat\theta}\, d\hat\theta\,, \quad x \in \mathbb{R}^3\,.$$

Therefore, from the definitions of H and G we conclude that $F = -GH$. Furthermore, we have seen already that $\frac{1}{4\pi} H^*\varphi$ is the far field pattern of the single layer potential, thus $\frac{1}{4\pi} H^*\varphi = GS\varphi$. Solving for G and substituting into $F = -GH$ yields the assertion.

In light of the two previous theorems one is now faced with the problem of expressing the range of H^* by using only properties of the operator F. This is a purely functional analytic problem. We were able to prove the following result:

Theorem 7. *Let $X \subset U \subset X^*$ and $Y \subset V \subset Y^*$ be Gelfand triples*[2] *with dense imbeddings, $F : Y^* \to Y$, $H : Y^* \to X$, $T : X \to X^*$ linear and bounded operators with $F = H^* T H$ and, furthermore,*

(i) H is one-to-one and compact and has dense range.
(ii) $\mathrm{Re}\, T := (T + T^*)/2 = C + K$ with coercive operator $C : X \to X^*$ and compact operator $K : X \to X^*$. With "coercive" we mean that $C^* = C$ and there exists $c > 0$ such that $\langle Cx, x\rangle \geq c\|x\|^2$ for all $x \in X$, where $\langle \cdot, \cdot \rangle$ denotes the dual form in $\langle X^*, X\rangle$.
(iii) $\mathrm{Im}\, T := (T - T^*)/(2i)$ is non-negative on X, i.e. $\mathrm{Im}\,\langle Tx, x\rangle \geq 0$ for all $x \in X$.
(iv) $\mathrm{Re}\, T$ is one-to-one or $\mathrm{Im}\, T$ is positive on X (i.e. $\mathrm{Im}\,\langle Tx, x\rangle > 0$ for all $x \in X$, $x \neq 0$).

Then $F_\# := |\mathrm{Re}\, F| + \mathrm{Im}\, F$ is positive and

$$\mathcal{R}(H^*) = \mathcal{R}(F_\#^{1/2}).$$

The operators $|\mathrm{Re}\, F|$ and $F_\#^{1/2}$ of the self-adjoint operators $\mathrm{Re}\, F$ and $F_\#$, respectively, are, as usual, defined by the spectral decomposition.

For a proof we refer to [54].

Combining the previous three theorems we arrive at the main result of this section.

Theorem 8. *Assume that k^2 is not an eigenvalue of $-\Delta$ with respect to the Dirichlet boundary condition and, in addition, with respect to the boundary condition of the scattering problem (A-a) – (A-c). Then we have:*

$$z \in \Omega \iff \phi_z \in \mathcal{R}(F_\#^{1/2}) \tag{21}$$

where $F_\# := |\mathrm{Re}\, F| + \mathrm{Im}\, F$.

Remarks:

In special cases where the scattering matrix is unitary, e.g. in the cases of Dirichlet- or Neumann boundary conditions, $F_\#$ can be replaced by $(F^*F)^{1/2}$, see [51].

[2] i.e. X and Y are reflexive Banach spaces, U and V Hilbert spaces with their duals being identified with the spaces themselves.

The condition $\phi_z \in \mathcal{R}(F_{\#}^{1/2})$ can be reformulated using an eigensystem $\{\mu_j, \psi_j : j \in \mathbb{N}\}$ of the self-adjoint and positive operator $F_{\#}$. Indeed, by simple arguments from spectral theory, we can reformulate (21) as

$$z \in \Omega \iff \phi_z \in \mathcal{R}(F_{\#}^{1/2}) \iff \sum_{j=1}^{\infty} \frac{|(\phi_z, \psi_j)|^2}{\mu_j} < \infty . \tag{22}$$

3 Numerical Methods for the Inverse Scattering Problem

We recall that we want to solve the inverse scattering problem to determine Ω from the knowledge of the far field pattern $u^\infty(\hat{x}; \hat{\theta})$ for all $\hat{x}, \hat{\theta} \in S^{d-1}$ (or for all $\hat{x} \in S^{d-1}$ and only some $\hat{\theta}$). In Figs. 5-6, the contour lines of the real and the imaginary parts of the far field pattern corresponding to two different obstacles in $\mathbb{R}^2$ are shown. The pair shown in Fig. 5 belong to a disc while the pair shown in Fig. 6 belong to a more complicated region. The inverse scattering problem is to determine the shape of Ω from these pictures.

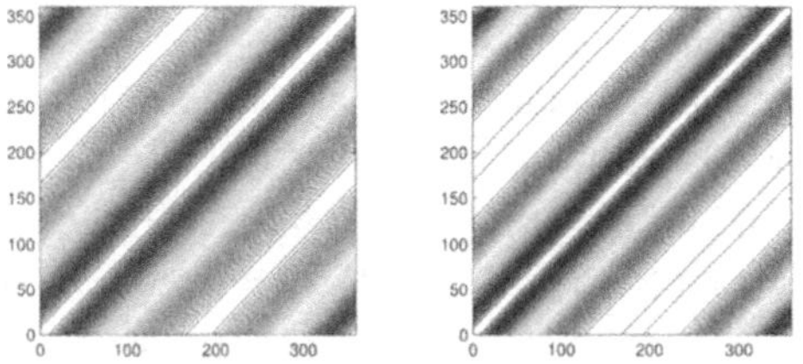

Fig. 5. Real and imaginary parts of far field for a disc

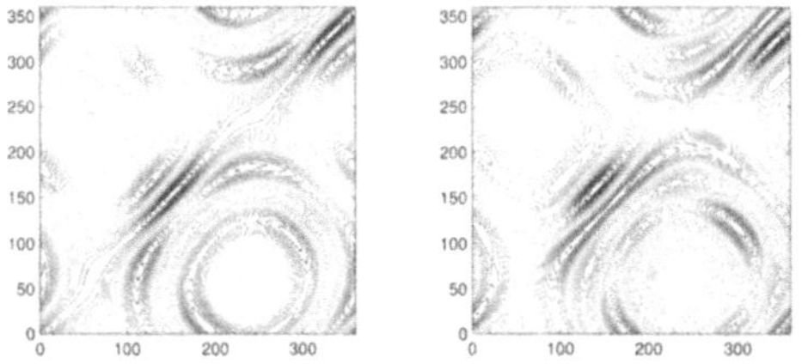

Fig. 6. Real and imaginary parts of far field for more complicated object

In this section we again restrict ourselves to the case of an impenetrable scatterer. We divide the methods into two classes. Motivated by the factorisation method, in the first class we collect all methods which avoid solving a sequence of direct scattering problems. The second class contains iterative methods such as Newton type methods.

3.1 Direct Methods

Factorisation Methods:

Before we formulate the case of finitely many data points we recall the functional space setting: With the far field operator, given by

$$(F\psi)(\hat{x}) = \int_{S^{d-1}} u^\infty(\hat{x};\hat{\theta})\,\psi(\hat{\theta})\,ds(\hat{\theta})\,, \quad \hat{x} \in S^{d-1}\,,$$

we define the self-adjoint operator $F_\# := |\mathrm{Re}\,F| + \mathrm{Im}\,F$ (or $F_\# = (F^*F)^{1/2}$), and, for z from some region R which is known to contain the unknown obstacle Ω, the function

$$\phi_z(\hat{x}) = e^{-ikz\cdot\hat{x}}\,, \quad \hat{x} \in S^{d-1}\,.$$

Then we know from (22) that

$$z \in \Omega \iff W(z) := \left[\sum_{j=1}^{\infty} \frac{|\langle \phi_z, \psi_j\rangle|^2}{\mu_j}\right]^{-1} > 0\,, \tag{23}$$

where $\{\mu_j, \psi_j : j \in \mathbb{N}\}$ is an eigensystem of $F_\#$. Therefore, the sign of $W(z)$ is just the characteristic function of Ω.

In practice, only a finite number of measurements is available. For our two-dimensional numerical experiments we assume that $u^\infty(\hat{x}_i;\hat{\theta}_j)$ for $i,j = 1,\dots,N$ is given where $\hat{\theta}_j = \big(\cos(j\cdot 2\pi/N), \sin(j\cdot 2\pi/N)\big) \in S^1$, $j = 1,\dots,N$, and $\hat{x}_i = \big(\cos(i\cdot 2\pi/N), \sin(i\cdot 2\pi/N)\big) \in S^1$, $i = 1,\dots,N$. We define the matrix $\mathbf{F} = \big(u^\infty(\hat{x}_i;\hat{\theta}_j)\big)_{i,j=1,\dots,N} \in \mathbb{C}^{N\times N}$ and compute $\mathbf{F}_\# := |\mathrm{Re}\,\mathbf{F}| + \mathrm{Im}\,\mathbf{F}$ (or $\mathbf{F}_\# = (\mathbf{F}^*\mathbf{F})^{1/2}$, respectively) and an eigensystem $\{\mu_j, \boldsymbol{\psi}_j : j = 1,\dots,N\}$ of $\mathbf{F}_\#$. Then, with

$$\boldsymbol{\varphi}_z = \left(e^{-ik\,z\cdot\hat{x}_1}, \dots, e^{-ik\,z\cdot\hat{x}_N}\right)^\top \in \mathbb{C}^N,$$

we plot the function

$$W_N(z) = \left[\sum_{j=1}^{N} \frac{|\boldsymbol{\varphi}_z^*\boldsymbol{\psi}_j|^2}{\mu_j}\right]^{-1}, \quad z \in R\,.$$

Actually, we chose an initial grid $\mathcal{G}$ in R, plotted $W_N(z)$ for every grid point z and refined the grid iteratively in the neighbourhood of all points where $W_N(z)$ was strictly positive. The plots shown in Fig. 7 are generated in this way.

We observe that this method is able to reconstruct obstacles with more than one component–without knowing *a priori* that there are multiple objects.

Closely related to the factorisation methods are the *Linear Sampling Methods* (see [11–16, 22–24]) which work with $F_\# = F$ and use a regularisation of the equation $Fg = \phi_z$ of the first kind. The theory, however, is not as complete and satisfying as for the factorisation method in the author's opinion.

Other sampling methods have also been proposed. We mention the *point source method* [70] by Potthast or *the probe method* [43] by Ikehata.

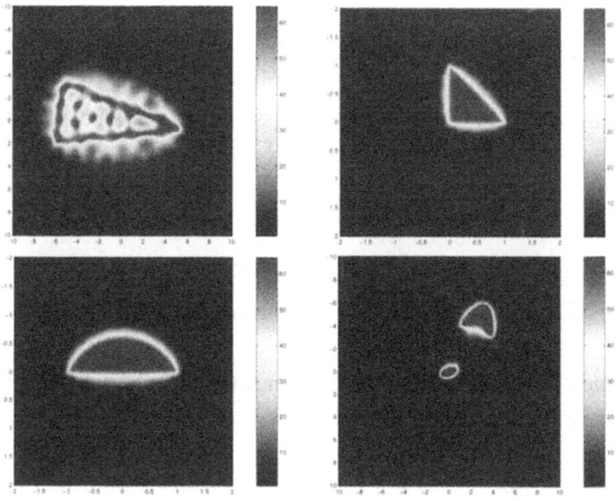

Fig. 7. Reconstructions obtained using factorisation method

Analytic Continuation Methods:

These methods usually consist of two steps. In the first step one continues the scattered field from ∞ to the domain. This step is highly ill-posed but linear. In the second step one determines $\partial\Omega$ as the set where the boundary conditions are satisfied. The advantage of these methods is that they require data for only one (or a few) incident direction. In this section we restrict ourselves to Dirichlet boundary conditions to keep the exposition readable.

Again, we formulate this approach in the function space setting, i.e. before discretisation. We describe one of these methods, proposed in [55] (see also [18]).

We assume that we are given $f(\hat{x})$ for all $\hat{x} \in S^{d-1}$ where we think of f as being an approximation of $u^\infty(\cdot;\hat{\theta})$ for some fixed $\hat{\theta} \in S^{d-1}$.

Step 1: We choose a domain $D \subset \mathbb{R}^d$ with $\overline{D} \subset \Omega$ (a priori information) and "solve" the integral equation of the first kind

$$\gamma_d \int_{\partial D} \varphi(y)\, e^{-ik\,\hat{x}\cdot y}\, ds(y) \;=\; f(\hat{x})\,, \quad \hat{x} \in S^{d-1}\,. \tag{24}$$

Note that the integral operator is just the adjoint H^* of the Herglotz operator of the previous section. If k^2 is not a Dirichlet eigenvalue of $-\Delta$ in D then H^* is one-to-one and compact and has dense range. Therefore, the solution of (24) requires regularisation.

Note: If φ is a solution of this equation, i.e. $\gamma_d H^*\varphi \;=\; f$ then, by Rellich's Lemma and analytic continuation,

$$\tilde{S}\varphi(x) \;:=\; \int_{\partial D} \varphi(y)\,\Phi(x,y)\,ds(y) \;=\; u^s(x)\,, \quad x \notin \Omega\,.$$

In this sense $\tilde{S}\varphi$ is an approximation of u^s.

Step 2: Once an approximation $\tilde{S}\varphi$ of u^s has been found, one determines the contour Γ (in $\mathbb{R}^2$) or the surface Γ (in $\mathbb{R}^3$) with $\tilde{S}\varphi \approx -u^{inc}$ on Γ.

In practice, one has to choose a class of parametrisation for the admissible contours Γ. In this paper we assume that every Γ has the form

$$x = \rho(\hat{x})\,\hat{x}\,, \quad \hat{x} \in S^{d-1} \quad \text{with } \rho \in U\,, \tag{25}$$

where $U \subset \{\rho \in C^{1,\tau}(S^{d-1}) : \rho > 0 \text{ on } S^{d-1}\}$ is compact with respect to the topology of the Hölder space $C^{1,\tau}(S^{d-1})$. We write Γ_ρ for the contour parametrised by $\rho \in U$.

We combine steps 1 and 2 and define for some fixed coupling parameter $\beta > 0$ and a regularising parameter $\alpha > 0$ the functional

$$J(\varphi, \rho, \alpha) := \left\|\gamma_d H^* \varphi - f\right\|^2_{L^2(S^{d-1})} + \alpha\|\varphi\|^2_{L^2(\partial D)} + \beta\|u^{inc} + \tilde{S}\varphi\|^2_{L^2(\Gamma_\rho)}$$

for $\varphi \in L^2(\partial D)$ and $\rho \in U$. Then the following results can be proven (see [18] for details).

Theorem 9. *For every $\alpha > 0$ there exists $\hat{\rho} \in U$ and $\hat{\varphi} \in L^2(\partial D)$ which minimises $J(\cdot, \cdot, \alpha)$ over $L^2(\partial D) \times U$.*

Theorem 10. *Let the exact contour $\partial\Omega$ have the form $\partial\Omega = \Gamma_{\hat{\rho}}$ for some $\hat{\rho} \in U$. Then*

$$J^\alpha := \min\{J(\varphi, \rho, \alpha) : \varphi \in L^2(\partial D)\,,\ \rho \in U\} \longrightarrow 0\,,\ \alpha \to 0\,.$$

Theorem 11. *Let the assumption of Theorem 10 hold and u be the exact solution corresponding to $\partial\Omega$. Let $\alpha_j \to 0$, $j \to \infty$, and (φ_j, ρ_j) be optimal for α_j. Then there exist accumulation points ρ of (ρ_j), and u vanishes on every such accumulation point Γ_ρ.*

Finally, we state a result on finite dimensional approximations: Let $X_{n-1} \subset X_n \subset L^2(\partial D)$ be a nested sequence of finite dimensional subspaces such that $\bigcup_n X_n$ is dense in $L^2(\partial D)$. Furthermore, let $U_{n-1} \subset U_n \subset U$ be a nested sequence of compact sets such that $\bigcup_n U_n$ is dense in U. Then we have:

Theorem 12. *Let $\alpha > 0$ be fixed. Then*

$$J^\alpha_n := \min\{J(\varphi, \rho, \alpha) : \varphi \in X_n\,,\ \rho \in U_n\} \longrightarrow J^\alpha\,, \quad n \to \infty\,.$$

If $(\varphi_n, \rho_n) \in X_n \times U_n$ is optimal for J^α_n, then there exists a subsequence $\rho_n \to \rho$, and, for some φ, (φ, ρ) is optimal for J^α.

This method can easily be modified to treat the limited aperture problem. Also, measurements of the far field for more than one angle of incidence can be incorporated. Instead of adding the corresponding discrepancies in the far field, Zinn in [77] suggested to use suitable linear combinations of incident plane waves. Numerical results are reported in, e.g. [55, 77].

Finally, we want to mention that we may replace the analytic continuation part 1 by any other convenient method. For example, Angell, Kleinman and Roach [4] used an expansion with respect to radiating spherical wave functions, Angell, Jiang and Kleinman [1] point sources. For numerical examples we refer to [2, 3, 47, 48].

A method which also falls into this class has been suggested by Colton and Monk in a series of papers (see [19–21]). The idea is to construct a superposition of the incident fields such that the scattered field is approximately that of a point source. Actually, this method was the starting point of the Linear Sampling Methods. We do not describe these methods here but refer to the original literature (see also [18]). A comparison between these two methods has been made in [57].

3.2 Iterative Methods–Abstract Theory

We can formulate the inverse scattering problem as the problem of solving a nonlinear equation of the type

$$K(\rho) \;=\; f\,, \tag{26}$$

where again f is an approximation of the exact far field pattern $u^\infty(\cdot;\hat\theta)$ and K is the nonlinear operator which maps the parametrisation ρ into the far field pattern of the solution of the scattering problem (A-a) – (A-c) outside of Γ_ρ. There exists an extended literature on the numerical treatment of this kind of nonlinear and ill-posed equations. In this part we recall some approaches. For a comprehensive overview of the state of the art we refer to [28, 74].

To state the abstract framework, let X and Y be Hilbert spaces, $\overline{y} \in Y$, and $K : X \supset \mathcal{D}(K) \to Y$ a (possibly nonlinear) operator. We consider the task to construct a suitable approximation of the exact solution $\overline{x}$ of $K(\overline{x}) = \overline{y}$. Usually, only an approximation $y^\delta \in Y$ of $\overline{y}$ is known with measurement error $\|y^\delta - \overline{y}\| \le \delta$. Therefore, instead of $K(x) = \overline{y}$ one can only construct (approximate) solutions of $K(x) = y^\delta$.

The following classes of methods have been studied so far:

- Nonlinear Tikhonov-regularisation methods,
- methods of Landweber-type (i.e. steepest descent methods),
- methods of Gauss–Newton–type,
- second order methods.

We will briefly recall a Landweber method and a Gauss–Newton method.

Landweber Method:

First, note that $\nabla_x \|K(x) - y^\delta\|^2 = 2\,K'(x)^*\big(K(x) - y^\delta\big)$, where $K'(x) : X \to Y$ denotes the Fréchet derivative of K at x. Therefore, the steepest descent method starts with some $x_0^\delta := \hat{x}$ and constructs the sequence $\big(x_n^\delta\big)$ by the recursion formula

$$x_{n+1}^\delta \;:=\; x_n^\delta \;-\; \omega_n\, K'\big(x_n^\delta\big)^*\big(K(x_n^\delta) - y^\delta\big)\,,\; n = 0, 1, \ldots \tag{27}$$

where $\omega_n \le 1/\sup_{x\in\mathcal{D}(K)} \|K'(x)\|^2$ are scaling factors or step-size controls.

One of the central results for constant $\omega_n = \omega$, scaled to $\omega = 1$, i.e. $\sup_{x \in \mathcal{D}(K)} \|K'(x)$ 1, needs the following assumptions (see [36]):

$(A1_L)$ $\mathcal{D}(K)$ is open and convex,
$(A2_L)$ K is continuously Fréchet-differentiable on $\mathcal{D}(K)$, and
$(A3_L)$ there exists $\rho > 0$ and $\eta < 1/2$ and $\overline{x} \in B_\rho(\hat{x})$ with $K(\overline{x}) = \overline{y}$ and

$$\|K(x) - K(z) - K'(z)(x - z)\| \leq \eta \, \|K(x) - K(z)\| \quad \text{for all } x, z \in B_{2\rho}(\hat{x}) .$$

The central tool of the convergence analysis is contained in the following lemma.

Lemma 3. *Let* $(A1_L) - (V3_L)$ *hold. If* $\tau > (2 + 2\eta)/(1 - 2\eta)$ *and*

$$\|K(x_n^\delta) - y^\delta\| \geq \tau\delta \quad \textit{for } n = 0, 1, \ldots, n_* - 1 ,$$

then

$$n_*(\tau\delta)^2 \leq \sum_{n=0}^{n_*-1} \|K(x_n^\delta) - y^\delta\|^2 \leq \frac{\tau\|\overline{x} - \hat{x}\|^2}{(1 - 2\eta)\tau - 2(1 + \eta)} .$$

We can draw the following conclusions:

If $\delta = 0$, i.e. the data contain no error, then we can let n_* tend to infinity and arrive at

$$\sum_{n=0}^{\infty} \|K(x_n^\delta) - y^\delta\|^2 < \infty .$$

If $\delta > 0$ then the following *stopping rule* is well defined:

For $\tau > (2 + 2\eta)/(1 - 2\eta)$ there exists a smallest n_* with

$$\|K(x_n^\delta) - y^\delta\| \geq \tau\delta , \; n = 0, \ldots, n_* - 1 , \quad \text{and} \quad \|K(x_{n_*}^\delta) - y^\delta\| < \tau\delta . \tag{28}$$

Theorem 13. *Let* $(A1_L) - (V3_L)$ *hold and* $\tau > (2 + 2\eta)/(1 - 2\eta)$.

(a) If $\delta = 0$ *then the sequence* (x_n) *constructed by (27) converges to the solution* $\overline{x}$ *of* $K(x) = \overline{y}$ *as* $n \to \infty$.
(b) If $\delta > 0$ *then, with the stopping rule* $n_* = n_*(\delta)$ *of (28) we have* $x_{n_*(\delta)}^\delta \to \overline{x}$, $\delta \to 0$, *where again* $K(\overline{x}) = \overline{y}$.

This theorem proves only convergence as $\delta \to 0$. *Rates of convergence* require additional assumptions. The common assumption is the following, known under the name of "source condition":

$(A4_L)$ There exists $\nu \in (0, 1/2]$ and $w \in X$ with

$$\overline{x} - \hat{x} = \left(K'(\overline{x})^* K'(\overline{x})\right)^\nu w .$$

Further technical assumptions are needed. We refer to Hanke, Neubauer and Scherzer [36] or Deuflhard, Engl and Scherzer [27]. Under these assumptions the following can be shown:

Theorem 14. *Let* $(A1_L)-(A4_L)$ *hold, let* $\tau > (2+2\eta)/(1-2\eta)$ *and let* $n_* = n_*(\delta)$ *be determined by the stopping rule (28), i.e.*

$$\|K(x_n^\delta) - y^\delta\| \ge \tau\delta\,,\; n = 0,\dots,n_*-1\,, \quad \textit{and} \quad \|K(x_{n_*}^\delta) - y^\delta\| < \tau\delta\,.$$

Then, if $\|w\|$ *is sufficiently small, there exists* $c > 0$ *with*

$$\left\|x_{n_*(\delta)}^\delta - \overline{x}\right\| \le c\,\delta^{2\nu/(2\nu+1)}\,.$$

Gauss–Newton–Type Methods:

The classical Newton iteration for solutions of $K(x) = y$ is given by

$$x_{n+1} = x_n - K'(x_n)^{-1}\big(K(x_n) - y\big)\,, \quad n = 0,1,\dots,$$

provided $K'(x)$ is boundedly invertible. For ill-posed problems $K'(x)$ is compact. Therefore, the Newton equation $K'(x_n)z = K(x_n) - y$ is a linear equation of the first kind and has to be regularised. The classical *Levenberg-Marquardt method* uses Tikhonov regularisation, i.e. solves the equation $\big[\alpha_n I + (K'(x_n)^* K'(x_n)\big]z = K'(x_n)^*\big(K(x_n) - y\big)$ instead. This leads to the sequence $x_0^\delta := \hat{x}$ and

$$x_{n+1}^\delta := x_n^\delta - \big[\alpha_n I + K'(x_n^\delta)^* K'(x_n^\delta)\big]^{-1} K'(x_n^\delta)^*\big(K(x_n^\delta) - y^\delta\big)$$

for $n = 0,1,\dots$ Here α_n are appropriately chosen regularisation parameters. There is a different interpretation of this formula: x_{n+1}^δ minimises the functional

$$x \mapsto \|K(x_n^\delta) - y^\delta + K'(x_n^\delta)(x - x_n^\delta)\|^2 + \alpha\|x - x_n^\delta\|^2$$

with respect to $x \in X$. This method has been investigated by Hanke in [33].

Instead of the Tikhonov-regularisation in the Newton equation, a conjugate gradient method is also possible, see [34]. However, only convergence results are known but not rates of convergence.

A slightly different method is the *iteratively regularised Levenberg-Marquardt method* proposed by Bakushinskii in [5]: With $x_0^\delta := \hat{x}$ one defines

$$x_{n+1}^\delta = x_n^\delta - \big[\alpha_n I + K'(x_n^\delta)^* K'(x_n^\delta)\big]^{-1}\big[K'(x_n^\delta)^*\big(K(x_n^\delta) - y^\delta\big) + \alpha_n(x_n^\delta - \hat{x})\big]\,,$$

$n = 0,1,\dots$. Now x_{n+1}^δ minimises the functional

$$x \mapsto \|K(x_n^\delta) - y^\delta + K'(x_n^\delta)(x - x_n^\delta)\|^2 + \alpha\|x - \hat{x}\|^2$$

over $x \in X$. The following *assumptions* are similar to those for the Landweber method.

$(A1_N)$ There exists $\rho > 0$ and $\overline{x} \in B_\rho(\hat{x})$ with $K(\overline{x}) = \overline{y}$, and K' is Lipschitz continuous in $B_\rho(\hat{x})$.

$(A2_N)$ There exists $\nu \geq 1/2$ and $w \in X$ with

$$\overline{x} - \hat{x} = \left(K'(\overline{x})^* K'(\overline{x})\right)^{\nu} w \,.$$

$(A3_N)$ There exists $r > 1$ with

$$\alpha_n > 0\,, \quad 1 \leq \frac{\alpha_n}{\alpha_{n+1}} \leq r\,, \quad \lim_{n\to\infty} \alpha_n = 0\,.$$

For $\delta = 0$ the following convergence rate is known: There exists $c > 0$ with

$$\|x_n - \overline{x}\| \leq c\,\alpha_n^{\nu}\,.$$

For $\delta > 0$ and the discrepancy principle (28) as stopping rule one can show existence of $c > 0$ such that

$$\|x^{\delta}_{n_*(\delta)} - \overline{x}\| \leq c\,\delta^{2\nu/(2\nu+1)} \quad \text{if } 0 \leq \nu \leq \frac{1}{2}\,.$$

Extensions of his result have been given by Hohage in [41, 42] and Deuflhard, Engl, and Scherzer in [27].

3.3 Application to Inverse Scattering Problems

In this part we come back to the inverse scattering problem: Given a wave number $k > 0$ and incident field $u^{inc}(x) = \exp\left(ik\,\hat{\theta} \cdot x\right)$, $x \in \mathbb{R}^d$, and a far field pattern $f(\hat{x})$ for all $\hat{x} \in S^{d-1}$, determine the obstacle Ω.

The direct scattering problem defines the operator $K : \rho \mapsto u^{\infty}$ which maps the parametrisation ρ of Γ_ρ onto the far field pattern of the solution u^s of the scattering problem (A-a) – (A-c). The inverse scattering problem is to solve $K(\rho) = f$ for ρ.

To apply the previously described methods one has to compute $K(\rho)$ as well as the Fréchet derivative $K'(\rho)$. As mentioned already, the computation of $K(\rho)$ means essentially solving a direct scattering problem. $K'(\rho)$ is related to the *domain derivative*. Let $\Omega \subset \mathbb{R}^d$ be fixed with boundary $\partial\Omega = \Gamma_\rho$ and total field $u = u^{inc} + u^s$.

For sufficiently smooth vector fields $h : \partial\Omega \to \mathbb{R}^d$ define "parallel" surfaces

$$\partial\Omega_\varepsilon := \left\{x + \varepsilon\, h(x) : x \in \partial\Omega\right\}, \quad \varepsilon \in [0, \varepsilon_0]\,.$$

For sufficiently small $\varepsilon > 0$ the surfaces $\partial\Omega_\varepsilon$ are boundaries of domain Ω_ε, and one defines the domain derivative $K'(\Omega; h)$ by

$$K'(\Omega; h) := \left.\frac{d}{d\varepsilon} K(\partial\Omega_\varepsilon)\right|_{\varepsilon=0}\,.$$

For the particular K of the scattering problem the domain derivative $K'(\Omega; h)$ is computable (see [26, 38, 49, 67–69]) and, for the Dirichlet boundary conditions, it has the following representation: It is $K'(\Omega; h) = v^{\infty}$ were v^{∞} is the far field pattern of the solution v of the exterior boundary value problem

$$\Delta v + k^2 v = 0 \text{ in } \mathbb{R}^2 \setminus \overline{\Omega}, \quad v = -(h \cdot \nu)\frac{\partial u}{\partial \nu} \text{ on } \partial\Omega,$$

and v satisfies Sommerfeld's radiation condition (4). From this a representation of the Fréchet derivative $K'(\rho)$ can easily be derived. We note that the computation of $K'(\rho)\eta$ involves the solution of a boundary value problem for the same differential equation and on the same region as for the evaluation of $K(\rho)$ itself – only for different boundary values. This makes the computation of $K'(\rho)\eta$ fast. Also, we note that the adjoint $K'(\rho)^*$ has simple representations. We refer to, e.g. [35, 49] for details.

The iterative methods from the previous section have been successfully applied to inverse scattering problems. However, it is not clear if the theoretical assumptions $(A3_L)$, $(A4_L)$, or $(A2_N)$ are valid for these applications. Some progress in this direction has been made by Hohage [41] and Potthast [71].

Hettlich and Rundell proposed a second order method in [39]. We are grateful to them for providing the plots shown in Figs. 8-9 taken from [40]. Figure 8 shows the best, an average and the worst result obtained using Newton's method and the second order method. The number of iterations in each case is indicated. The maximal number of iterations was limited by 20.

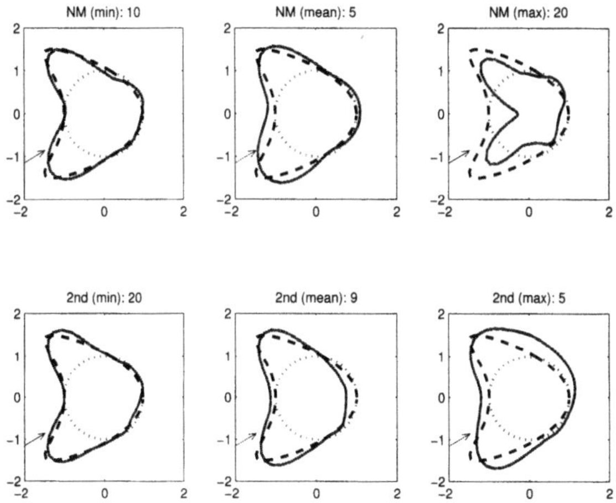

Fig. 8. Comparison of Newton method and second order method of [39]. The true curve is given by the dashed line ("kite"). The dotted circle denotes the initial curve. The angle of incidence is also indicated by an arrow

Figure 9 shows the initial iterate and the best result for Newton's method and for the second degree method, respectively, for a smaller initial circle. We observe the improvement of the second degree method over Newton's method after the first iterate. The figure also shows the errors in the data and the scattering obstacles, respectively, as a function of the iteration number. For Newton's method we clearly observe that one has to take the stopping rule into account since otherwise the results become worse again.

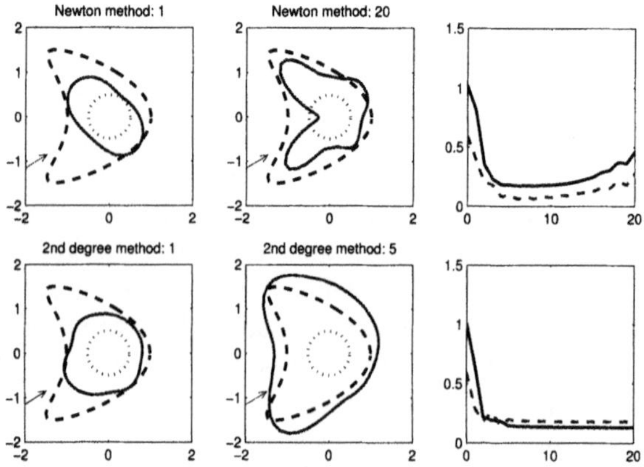

Fig. 9. Initial iterate and best result obtained using Newton's method and second order method for smaller initial circle. The true curve is given by the dashed line ("kite"). The dotted circle denotes the initial curve. The angle of incidence is indicated by an arrow. The final plots show the errors in the data and the scattering obstacles as a function of iteration number

4 Conclusions

In this article we tried to give a short overview on the classical theory of direct and inverse scattering theory for time-harmonic waves as well as on some areas of current research. We concentrated on the presentation of two approaches for solving the inverse scattering problem which seem to be the most successful ideas at present. As an example of the first approach we described the factorisation method from the class of "direct" approaches for solving the inverse problem. The second class of methods contain iterative methods of Newton type. By using characterisations of the domain derivative the cost of these method can be reduced such that also three-dimensional problems can be solved. There is still a gap between the abstract convergence analysis of these methods in Hilbert spaces and the applications to inverse scattering problems.

References

1. T.S. Angell, X. Jiang and R.E. Kleinman: On a numerical method for inverse acoustic scattering. Inverse Problems 13 (1997), 531–545.
2. T.S. Angell, R.E. Kleinman, B. Kok and G.F. Roach: A constructive method for identification of an impenetrable scatterer. Wave Motion 11 (1989), 185–200.
3. T.S. Angell, R.E. Kleinman, B. Kok and G.F. Roach: Target reconstruction from scattered far field data. Ann. des Télécommunications 44 (1989), 456–463.
4. T.S. Angell, R.E. Kleinman and G.F. Roach: An inverse transmission problem for the Helmholtz equation. Inverse Problems 3 (1987), 149–180.
5. A.B. Bakushinskii: The problem of the convergence of the iteratively regularized Gauss-Newton method. Comput. Meth. Math. Phys. 32 (1992), 1353–1359.

6. H. Brakhage and P. Werner: Über das Dirichletsche Aussenraumproblem für die Helmholtzsche Schwingungsgleichung. Arch. Math. 16 (1965), 325–329.
7. M. Brühl: Gebieterkennung in der elektrischen Impedanztomographie. Dissertation thesis, University of Karlsruhe, 1999.
8. M. Brühl: Explicit characterization of inclusions in electrical impedance tomography. SIAM J. Math. Anal. 32 (2001), 1327–1341.
9. M. Brühl and M. Hanke: Numerical implementation of two noniterative methods for locating inclusions by impedance tomography. Inverse Problems 16 (2000), 1029–1042.
10. K. Bryan and L.F. Caudill: An inverse problem in thermal imaging. SIAM J. Appl. Math. 56 (1996), 715–735.
11. F. Cakoni and D. Colton: On the mathematical basis of the Linear Sampling Method. Preprint 2002.
12. F. Cakoni, D. Colton and H. Haddar: The linear sampling method for anisotropic media. J. Comp. Appl. Math. 146 (2002), 285–299.
13. F. Cakoni, D. Colton and P. Monk: The direct and inverse scattering problems for partially coated obstacles. Inverse Problems 17 (2001), 1997–2015.
14. F. Cakoni and H. Haddar: The linear sampling method for anisotropic media: Part II. Preprint, 2002.
15. D. Colton, K. Giebermann and P. Monk: A regularized sampling method for solving three-dimensional inverse scattering problems. SIAM J. Sci. Comput. 21 (2000), 2316–2330.
16. D. Colton and A. Kirsch: A simple method for solving inverse scattering problems in the resonance region. Inverse Problems 12 (1996), 383–393.
17. D. Colton and R. Kress: Integral equations in scattering theory. Wiley-Interscience Publications 1983.
18. D. Colton and R. Kress: Inverse acoustic and electromagnetic scattering problems. 2nd edition, Springer-Verlag, New-York, 1998.
19. D. Colton and P. Monk: A novel method for solving the inverse scattering problem for time-harmonic acoustic waves in the resonance region. SIAM J. Appl. Math. 45 (1985), 1039–1053.
20. D. Colton and P. Monk: A novel method for solving the inverse scattering problem for time-harmonic acoustic waves in the resonance region II. SIAM J. Appl. Math. 46 (1986), 506–523.
21. D. Colton and P. Monk: The numerical solution of the three dimensional inverse scattering problem for time-harmonic acoustic waves. SIAM J. Sci. Stat. Comp. 8 (1987), 278–291.
22. D. Colton and P. Monk: A linear sampling method for the detection of leukemia using microwaves. SIAM J. Appl. Math. 58 (1998), 926–941.
23. D. Colton and P. Monk: A linear sampling method for the detection of leukemia using microwaves II. SIAM J. Appl. Math. 60 (2000), 241–255.
24. D. Colton, M. Piana and R. Potthast: A simple method using Morozov's discrepancy principle for solving inverse scattering problems. Inverse Problems 13 (1997), 1477–1493.
25. G. Dassios and R. Kleinman: Low frequency scattering. Oxford Science Publications, 2000.
26. M.C. Delfour and J.-P. Zolésio: Shapes and geometries. Analysis, differential calculus, and optimization. SIAM, Philadelphia, 2001.
27. P. Deuflhard, H. Engl and O. Scherzer: A convergence analysis of iterative methods for the solution of nonlinear ill-posed problems under affinely invariant conditions. Inverse Problems 14 (1998), 1081–1106.
28. H. Engl, M. Hanke and A. Neubauer: Regularization of inverse problems. Kluwer Academic Publishers, Dordrecht, 1996.

29. T. Gerlach and R. Kress: Uniqueness in inverse obstacle scattering with conductive boundary condition. Inverse Problems 12 (1996), 619–625.
30. D. Gilbarg and N.S. Trudinger: Elliptic partial differential equations. 2nd edition, Springer-Verlag, New-York, 1983.
31. N. Grinberg: On the Inverse obstacle scattering problem with robin or mixed boundary condition: Application of the modified Kirsch factorization method. University of Karlsruhe, Department of Mathematics, Preprint 02/4.
32. N. Grinberg and A. Kirsch: The linear sampling method in inverse obstacle scattering for impedance boundary conditions. Journal of Inverse and Ill-Posed Problems, to appear.
33. M. Hanke: A regularized Levenberg-Marquardt scheme with applications to inverse groundwater filtration problems. Inverse Problems 13 (1997), 79–95.
34. M. Hanke: Regularizing properties of a truncated Newton-cg algorithm for nonlinear inverse problems. Numer. Funct. Anal. Opt. 18 (1998), 971–993.
35. M. Hanke, F. Hettlich and O. Scherzer: The Landweber iteration for an inverse scattering problem. In: Proc. of the 1995 design engineering technical conferencees, Vol 3 Part C, K.-W. Wang et al. (eds.), The American Society of Mechanical Engineers, New York, 1995, 909–915.
36. M. Hanke, A. Neubauer and O. Scherzer: A convergence analysis of the Landweber iteration for nonlinear ill-posed problems. Numer. Math. 72 (1995), 21–37.
37. F. Hettlich: On the uniqueness of the inverse conductive scattering problem for the Helmholtz equation. Inverse Problems 10 (1994), 129–144.
38. F. Hettlich: Fréchet derivatives in inverse obstacle scattering. Inverse Problems 11 (1995), 371–382.
39. F. Hettlich and W. Rundell: A second degree method for nonlinear inverse problems. SIAM J. Numer. Anal. 37 (2000), 587–620.
40. F. Hettlich: Personal communications.
41. T. Hohage: Logarithmic convergence rates of the iteratively regularized Gauss-Newton method for an inverse potential and an inverse scattering problem. Inverse Problems 13 (1997), 1279–1299.
42. T. Hohage: Regularization of exponentially ill-posed problems. Numer. Funct. Anal. Optim. 21 (2000), 439–464.
43. M. Ikehata: Reconstruction of an obstacle from the scattering amplitude at fixed frequency. Inverse Problems 14 (1998), 949–954.
44. V. Isakov: On the uniqueness in the inverse transmission scattering problem. Comm. Part. Diff. Equa. 15 (1990), 1565–1587.
45. V. Isakov: Inverse problems for partial differential equations. Springer Verlag, New-York, 1998.
46. D.S. Jones: Integral equations for the exterior acoustic problem. Q. J. Mech. Appl. Math. 27 (1974), 129–142.
47. D.S. Jones and X.Q. Mao: The inverse problem in hard acoustic scattering. Inverse Problems 5 (1989), 731–748.
48. D.S. Jones and X.Q. Mao: A method for solving the inverse problem in soft acoustic scattering. IMA J. Appl. Math. 44 (1990), 127–143.
49. A. Kirsch: The domain derivative and two applications in inverse scattering theory. Inverse Problems 9 (1993), 81–96.
50. A. Kirsch: Introduction to the mathematical theory of inverse problems. Springer-Verlag, New-York, 1996.
51. A. Kirsch: Characterization of the shape of a scattering obstacle using the spectral data of the far field operator. Inverse Problems 14 (1998), 1489–1512.

52. A. Kirsch: Factorization of the far field operator for the inhomogeneous medium case and an application in inverse scattering theory. Inverse Problems 15 (1999), 413–429.
53. A. Kirsch: New characterizations of solutions in inverse scattering theory. Applicable Analysis 76 (2000), 319–350.
54. A. Kirsch: The MUSIC-algorithm and the factorization method in inverse scattering theory for inhomogeneous media. Inverse Problems 18 (2002), 1025–1040.
55. A. Kirsch and R. Kress: A numerical method for an inverse scattering problem. In *Inverse Problems* (H. Engl and C. Groetsch, eds.) Academic Press, Orlando (1987), 279–290.
56. A. Kirsch and R. Kress: Uniqueness in inverse obstacle scattering. Inverse Problems 9 (1993), 285–299.
57. A. Kirsch, R. Kress, P. Monk and A. Zinn: Two methods for solving the inverse acoustic scattering problem. Inverse Problems 4 (1988), 749–770.
58. A. Kirsch and L. Päivärinta: On recovering obstacles inside inhomogeneities. Math. Meth. Appl. Sci. 21 (1998), 619–651.
59. A. Kirsch and S. Ritter: A linear sampling method for inverse scattering from an open arc. Inverse Problems 16 (2000), 89–105.
60. R.E. Kleinman and G.F. Roach: On modified Green's functions in exterior problems for the Helmholtz equation. Proc. Royal Soc. London A383 (1982), 313–332.
61. P.D. Lax and R.S. Phillips: Scattering theory. Academic Press, New York, 1967.
62. R. Leis: Zur Dirichletschen Randwertaufgabe des Aussenraums der Schwingungsgleichung. Math. Z. 90 (1965), 205–211.
63. R. Leis: Initial boundary value problems in mathematical physics, Teubner, 1986.
64. C. Miranda: Partial differential equations of elliptic type. Springer Verlag, 1970.
65. J.-C. Nédélec: Acoustic and electromagnetic equations. Springer Verlag, 2001.
66. O.I. Panich: On the question of the solvability of the exterior boundary-value problems for the wave equation and Maxwell's equations. Usp. Mat. Nauk 20A (1965), 221–226 (in Russian).
67. O. Pironneau: Optimal shape design for elliptic systems. Springer Verlag, Berlin, Heidelberg, New York, 1984.
68. R. Potthast: Fréchet differentiability of boundary integral operators. Inverse Problems 10 (1994), 431–447.
69. R. Potthast: Domain derivatives in electromagnetic scattering. Math. Meth. Appl. Sci. 19 (1996), 1157–1175.
70. R. Potthast: A point-source method for inverse acoustic and electromagnetic obstacle scattering problems. IMA J. Appl. Math. 61 (1998), 119–140.
71. R. Potthast: On the convergence of Newton's method in inverse scattering. Inverse Problems 17 (2001), 1419–1434.
72. F. Rellich: Über das asymptotische Verhalten der Lösungen von $\Delta u + \lambda u = 0$ in unendlichen Gebieten. Jber. Deutsch. Math. Verein. 53 (1943), 57–65.
73. B.P. Rynne and B.D. Sleeman: The interior transmission problem and inverse scattering from inhomogeneous media. SIAM J. Math. Anal. 22 (1991), 1755–1762.
74. A.N. Tikhonov, A.V. Goncharrsky, V.V. Stepanov and A.G. Yagola: Numerical methods for the solution of ill-posed problems. Kluwer Academic Publishers. Dordrecht, 1990.
75. F. Ursell: On the exterior problems of acoustics. Math. Proc. Cambridge Philos. Soc. 74 (1973), 117–125.
76. F. Ursell: On the exterior problems of acoustics II. Math. Proc. Cambridge Philos. Soc. 84 (1978), 545–548.
77. A. Zinn: On an optimisation method for the full- and limited-aperture problem in inverse acoustic scattering for a sound-soft obstacle. Inverse Problems 5 (1989), 239–253.

Herglotz Wave Functions in Inverse Electromagnetic Scattering Theory*

David Colton and Peter Monk

Department of Mathematical Sciences, University of Delaware, Newark, DE 19716, USA
colton@math.udel.edu, monk@math.udel.edu

1 Introduction

Ever since the invention of radar during the Second World War, scientists and engineers have strived not only to detect but also to identify unknown objects through the use of electromagnetic waves. Indeed, as pointed out in [19], "Target identification is the great unsolved problem. We detect almost everything; we identify nothing". A significant step forward in the resolution of this problem occurred in the 1960's with the invention of synthetic aperture radar (SAR) and since that time numerous striking successes have been recorded in imaging by electromagnetic waves using SAR [1], [7]. However, as the demands of radar imaging have increased, the limitations of SAR have become increasingly apparent. These limitations arise from the fact that SAR is based on the "weak scattering" approximation and ignores polarisation effects. Indeed, such incorrect model assumptions have caused some scientists to ask "how (and if) the complications associated with radar based automatic target recognition can be surmounted" ([1], p. 5).

Until recently, the alternative to SAR in electromagnetic imaging was the use of nonlinear optimisation techniques [15], [16], [23]. Although this approach has achieved success in some areas, in general it is far too computationally expensive for many practical applications. However, in the past few years an alternative to SAR and nonlinear optimisation techniques has been introduced. This alternative method is called the *linear sampling method* and, like SAR, is a linear method but, unlike SAR, makes no restrictive model assumptions on the scattering process. The main purpose of this paper is to describe this new method in electromagnetic inverse scattering theory and to provide some numerical examples of its practicality. Although overcoming some of the limitations of both SAR and nonlinear optimisation techniques, we hasten to mention that the linear sampling method is no panacea. In particular, its implementation requires a large amount of multi-static data (i.e. for each of many directions of the incident wave, the scattered field is measured at many dif-

* This research was supported in part by a grant from the Air Force Office of Scientific Research.

ferent observation points) and only the support, but not the material properties, of the scatterer is determined. Nevertheless, the linear sampling method provides an alternative to existing methods in electromagnetic imaging and one which we feel holds considerable promise for surmounting some of the current problems in target recognition.

As with any other reconstruction algorithm, the linear sampling method cannot be viewed in isolation but rather as part of an overall view of the inverse scattering problem for electromagnetic waves. In particular, issues of uniqueness and continuous dependence are inseparable from the issue of reconstruction. Furthermore, such issues for the inverse problem cannot be resolved without a firm mathematical basis for the direct or forward scattering problem. To clarify these observations, we note that in many (if not most!) situations neither the shape nor the material properties of the object being imaged are known and hence neither the shape nor boundary conditions are known. A typical example of this occurs when the target has been (partially) coated by an unknown material in order to avoid detection. Hence in general the desired uniqueness result for the inverse problem should not depend on knowing the boundary condition a priori. Such results exist and are based on the assumption that the direct scattering problem, whatever it may be, depends continuously on the boundary data. In a different direction, since the inverse scattering problem is ill-posed, in order to restore stability some type of a priori information is needed and an estimate on the noise level is in general more realistic than the knowledge of, for example, an a priori bound on the curvature of the scattering object.

Keeping the above ideas in mind, the plan of our paper is as follows. We begin by considering the direct scattering problem for two representative situations, the first being when the scatterer is a partially coated perfect conductor (including the case of possibly no coating at all!) and the second the case when the scatterer is a penetrable, isotropic, inhomogeneous medium. In each case we establish existence, uniqueness and continuous dependence on the data with respect to an appropriate norm. We then consider the corresponding inverse problems and first establish uniqueness. These uniqueness results are then followed by a presentation of the linear sampling method which is based on a factorisation of the far field operator F in the form $F = \mathcal{B}\gamma\mathcal{H}$ where $\mathcal{B}$ maps the incident field onto the electric far field pattern, γ is a trace operator and $\mathcal{H}$ maps tangential vector fields on the unit sphere onto the class of vector Herglotz wave functions [11]. The validity of the linear sampling method is then established by showing that $\mathcal{B}$ and γH are injective with dense range.

Surprisingly, one of the theorems proved during our analysis of the linear sampling method also is useful in the analysis of a numerical scheme for solving the direct scattering problem. In Sect. 5 we outline the Ultra Weak Variational Formulation of Maxwell's equations due to Cessenat and Després [5] and show how to prove convergence in the case where the mesh is fixed and the order of approximation is increased element by element.

We conclude our paper with the presentation of some numerical examples. In these examples we stabilise the inverse scattering problem by assuming that an a priori estimate of the noise level is available and then making use of Tikhonov regularisation and the Morozov discrepancy principle.

In what follows the word "proof" in fact means "outline of proof", i.e. our aim is simply to give the reader an idea of how the proof proceeds together with appropriate references where full details can be found.

2 Electromagnetic Scattering Problems

In this section we will discuss the two scattering problems that we will study in this paper and show that in both cases the forward problem is well posed. In addition we will establish the well-posedness of two related interior problems which turn out to be basic in our discussion of the corresponding inverse problems.

We begin by describing the direct scattering problem for a partially coated perfect conductor. Let $D \subset \mathbb{R}^3$ be a bounded region with boundary Γ such that $D_e := \mathbb{R}^3 \setminus \overline{D}$ is connected. Each connected component of D is assumed to be a Lipschitz curvilinear polyhedron with smooth faces. We assume that the boundary $\Gamma = \Gamma_D \cup \Pi \cup \Gamma_I$ is split into two disjoint parts Γ_D and Γ_I (it may be that Γ_I is the empty set) having Π as their possible common boundary in Γ and assume that Γ_D and Γ_I can be written as the union of a finite number of open smooth faces having unit outward normal ν. Let k denote the positive wavenumber of the radiation. The first direct scattering problem we are interested in is to find an electric field E and magnetic field H such that

$$\nabla \times E - ikH = 0 \tag{1a}$$

$$\nabla \times H + ikE = 0 \tag{1b}$$

in D_e satisfying the following boundary conditions

$$\nu \times E = 0 \quad \text{on } \Gamma_D \tag{2a}$$

$$\nu \times \nabla \times E - i\lambda(\nu \times E) \times \nu = 0 \quad \text{on } \Gamma_I \tag{2b}$$

where $\lambda > 0$ is the surface impedance which is assumed to be a (possibly different) constant on each connected subset of Γ_I. The total fields E and H are given in terms of the incident fields and the scattered fields E^s, H^s by

$$E = E^i + E^s \tag{3a}$$

$$H = H^i + H^s. \tag{3b}$$

The scattered fields E^s, H^s satisfy the Silver–Müller radiation condition

$$\lim_{r \to \infty} (H^s \times x - rE^s) = 0 \tag{4}$$

uniformly in $\hat{x} = x/|x|$ where $r = |x|$ and the incident field E^i, H^i is given by

$$E^i(x) := \frac{i}{k} \nabla \times \nabla \times p e^{ikx \cdot d} = ik(d \times p) \times d e^{ikx \cdot d} \tag{5a}$$

$$H^i(x) := \nabla \times p e^{ikx \cdot d} = ikd \times p e^{ikx \cdot d}. \tag{5b}$$

Here d is a unit vector giving the direction of propagation and p is the (constant) polarization vector.

We now need to be more precise concerning the regularity assumptions satisfied by E and H. To this end, let $L_t^2(\Gamma_I)$ denote the space of square integrable tangential vector fields defined on $\Gamma_I, H^{-1/2}(\Gamma_D)$ the Sobolev space of negative half-integer order defined on Γ_D and

$$H(\text{curl}, D) := \left\{u \in (L^2(D))^3 : \nabla \times u \in (L^2(D))^3\right\}.$$

We then define

$$X(D, \Gamma_I) := \left\{u \in H(\text{curl}, D) : \nu \times u\big|_{\Gamma_I} \in L_t^2(\Gamma_I)\right\}$$

equipped with the norm

$$\|u\|^2_{X(D,\Gamma_I)} := \|u\|^2_{H(\text{curl},D)} + \|\nu \times u\|^2_{L^2(\Gamma_I)}.$$

with $H_{loc}(\text{curl}, D_e)$ and $X_{loc}(D_e, \Gamma_I)$ defined in the obvious way. We furthermore introduce the trace space on Γ_D by

$$Y(\Gamma_D) := \Big\{f \in (H^{-1/2}(\Gamma_D))^3 : \exists u \in H_0(\text{curl}, B_R),$$
$$\nu \times u\big|_{\Gamma_I} \in L_t^2(\Gamma_I) \text{ and } f = \nu \times u\big|_{\Gamma_D}\Big\}$$

where $B_R := \{x : |x| < R\} \supset \overline{D}$ and $H_0(\text{curl}, B_R)$ is the space of functions $u \in H(\text{curl}, B_R)$ such that $\nu \times u\big|_{\partial B_R} = 0$. It is easy to show that $Y(\Gamma_D)$ is a Banach space with respect to the norm

$$\|f\|^2_{Y(\Gamma_D)} := \inf \left\{\|u\|^2_{H(\text{curl},B_R)} + \|\nu \times u\|^2_{L^2(\Gamma_I)}\right\}$$

where the infimum is taken over all functions $u \in H_0(\text{curl}, B_R)$ such that $\nu \times u\big|_{\Gamma_I} \in L_t^2(\Gamma_I)$ and $f = \nu \times u|_{\Gamma_D}$.

We can now formulate the following *exterior mixed boundary value problem* for Maxwell's equations which includes (1)-(5) as a special case: given $f \in Y(\Gamma_D)$ and $h \in L_t^2(\Gamma_I)$ find $E \in X_{loc}(D_e, \Gamma_I)$ and $H = (ik)^{-1}\nabla \times E$ such that

$$\nabla \times \nabla \times E - k^2 E = 0 \quad \text{in } D_e \tag{6a}$$
$$\nu \times E = f \quad \text{on } \Gamma_D \tag{6b}$$
$$\nu \times \nabla \times E - i\,\lambda(\nu \times E) \times \nu = h \quad \text{on } \Gamma_I \tag{6c}$$
$$\lim_{r\to\infty} (H \times x - rE) = 0. \tag{6d}$$

We will also need to consider the corresponding *interior mixed boundary value problem* where we are given $f \in Y(\Gamma_D)$ and $h \in L_t^2(\Gamma_I)$ and wish to find $E \in X(D, \Gamma_I)$ such that

$$\nabla \times \nabla \times E - k^2 E = 0 \quad \text{in } D \tag{7a}$$

$$\nu \times E = f \quad \text{on } \Gamma_D \tag{7b}$$

$$\nu \times \nabla \times E - i\lambda(\nu \times E) \times \nu = h \text{ on } \Gamma_I. \tag{7c}$$

We first prove the existence of a unique solution to the interior mixed boundary value problem.

Theorem 1. *Assume that either Γ_I is not empty or that k is not a Maxwell eigenvalue of D. Then the interior mixed boundary value problem (7) has a unique solution which satisfies*

$$\|E\|_{X(D,\Gamma_I)} \leq C(\|f\|_{Y(\Gamma_D)} + \|h\|_{L^2(\Gamma_I)})$$

for some positive constant C.

Proof. If $\Gamma_I \neq \emptyset$ uniqueness follows by using Green's first vector theorem and the unique continuation principle (for details see [4]). If $\Gamma_I = \emptyset$ uniqueness follows by the definition of a Maxwell eigenvalue [11].

To prove existence we consider the variational formulation of (7): find $E \in X(D,\Gamma_I)$ satisfying $\nu \times E = f$ on Γ_D such that

$$\int_D (\nabla \times E \cdot \nabla \times \overline{\phi} - k^2 E \cdot \overline{\phi})\, dV + i\lambda \int_{\Gamma_I} E_T \cdot \overline{\phi}_T\, dS = -\int_{\Gamma_I} h \cdot \overline{\phi}_T dS \tag{8}$$

for every test function $\phi \in \tilde{X}$ where

$$\tilde{X} := \left\{ u \in H(\text{curl}, D) : \nu \times u\big|_{\Gamma_D} = 0 \text{ and } \nu \times u\big|_{\Gamma_I} \in L^2_t(\Gamma_I) \right\}$$

and $E_T := (\nu \times E) \times \nu$. From the definition of the space $Y(\Gamma_D)$ there exists a function $U \in X(D,\Gamma_I)$ such that $\nu \times U\big|_{\Gamma_D} = f$. By subtracting from both sides of (8) the expression

$$\int_D (\nabla \times U \cdot \nabla \times \overline{\phi} - k^2 U \cdot \overline{\phi})\, dV + i\lambda \int_{\Gamma_I} U_T \cdot \overline{\phi}_T\, dS$$

we are now led to the problem of finding $w \in \tilde{X}$ such that for every $\phi \in \tilde{X}$ we have

$$a(w,\phi) = \langle h, \phi \rangle - a(U,\phi) \tag{9}$$

where the sesquilinear form $a : \tilde{X} \times \tilde{X} \to \mathbb{C}$ is defined by

$$a(u,\psi) = (\nabla \times u, \nabla \times \psi) - k^2(u,\psi) + i\lambda\, \langle u_T, \psi_T \rangle$$

with $(\cdot,\cdot)$ denoting the $L^2(D)$ scalar product and $\langle \cdot,\cdot \rangle$ the $L^2(\Gamma_I)$ scalar product. The sesquilinear form $a(u,\psi)$ is systematically studied in Chap. 5 of [24] for the case when Γ_D and Γ_I are closed manifolds and the extension to the case when Γ_D and Γ_I are open subsets of the boundary Γ is carried out in [4]. In particular, it is shown in [4] that (9) leads to the problem of inverting a Fredholm operator of index zero and the theorem follows.

The exterior mixed boundary value problem can be treated in a similar manner but in the domain $D_e \cap B_R$. The boundary condition on the artificial surface ∂B_R is imposed by using the capacity operator (see, e.g. [25]). By Rellich's lemma uniqueness holds even if $\Gamma_I = \emptyset$.

Theorem 2. *The exterior mixed boundary value problem (6) has a unique solution which satisfies*

$$\|E\|_{X(D_e \cap B_R, \Gamma_I)} \leq C(\|f\|_{Y(\Gamma_D)} + \|h\|_{L^2(\Gamma_I)})$$

for some positive constant C (depending on R).

The second class of scattering problems we shall be concerned with in this paper is connected with the scattering of time harmonic electromagnetic plane waves by an isotropic inhomogeneous medium having bounded support D where D has a connected complement and the boundary Γ of D is in class C^2. The direct scattering problem in this case can be formulated (under appropriate simplifying assumptions) as the exterior problem of finding an electric field E and a magnetic field H such that $E, H \in C^1(\mathbb{R}^3)$ and

$$\nabla \times E - ikH = 0 \tag{10a}$$
$$\nabla \times H + ikn(x)E = 0 \tag{10b}$$

in $\mathbb{R}^3$ where n is a complex valued function such that $n \in C^{1,\alpha}(\mathbb{R}^3)$ for $0 < \alpha < 1$, Re $n > 0$, Im $n \geq 0$ and $n(x) = 1$ for $x \in D_e$. In addition

$$E = E^i + E^s \tag{11a}$$
$$H = H^i + H^s \tag{11b}$$

where E^i, H^i are given by (5) and E^s, H^s satisfy the Silver–Müller radiation condition

$$\lim_{r \to \infty} (H^s \times x - rE^s) = 0 \tag{12}$$

uniformly in $\hat{x} = x/|x|$. It is not difficult to show that E and $H = (ik)^{-1}$ curl E is a solution of (10)–(12) if and only if E satisfies the integral equation (See Sect. 9.2 of [11])

$$\begin{aligned} E(x) = E^i(x) - k^2 \int_{\mathbb{R}^3} \Phi(x,y) m(y) E(y)\, dV(y) \\ + \text{grad} \int_{\mathbb{R}^3} \frac{1}{n(y)} \text{grad}\, n(y) \cdot E(y) \Phi(x,y)\, dV(y) \end{aligned} \tag{13}$$

for $y \in \mathbb{R}^3$ where $m := 1 - n$ and

$$\Phi(x,y) := \frac{1}{4\pi} \frac{e^{ik|-y|}}{|x-y|}, \; x \neq y. \tag{14}$$

The uniqueness of a solution to (10)–(12) follows from an application of Gauss' divergence theorem and the unique continuation principle (Theorem 9.4 of [11]). The following theorem now follows by applying the Riesz–Fredholm theory to the integral equation (13) (Theorem 9.5 of [11]).

Theorem 3. *Let B_R be a ball containing D. Then there exists a unique solution E, H of (10)–(12) and this solution depends continuously on E^i, H^i with respect to the maximum norm over $\overline{B}_R$.*

As in the case of the previously considered problem of scattering by an obstacle, we will also need for future purposes to consider an interior problem corresponding to the exterior problem (10)–(12). In particular, for $z \in D$ consider the electric dipole with polarisation q defined by

$$E_e(x, z, q) := \frac{i}{k} \nabla_x \times \nabla_x \times q\Phi(x, z) \tag{15a}$$

$$H_e(x, z, q) := \nabla_x \times q\Phi(x, z) \tag{15b}$$

The *interior transmission problem* is to find fields $E, E_0, H, H_0 \in C^1(D) \cap C(\overline{D})$ such that

$$\left.\begin{aligned} \nabla \times E - ikH = 0 \\ \nabla \times H + ikn(x)E = 0 \end{aligned}\right\} \quad \text{in} \quad D \tag{16a}$$

$$\left.\begin{aligned} \nabla \times E_0 - ikH_0 = 0 \\ \nabla \times H_0 + ikE_0 = 0 \end{aligned}\right\} \quad \text{in} \quad D \tag{16b}$$

$$\left.\begin{aligned} (E - E_0) \times \nu = E_e \times \nu \\ (H - H_0) \times \nu = H_e \times \nu \end{aligned}\right\} \quad \text{on} \quad \Gamma \tag{16c}$$

where ν is again the unit outward normal to Γ. We shall proceed to showing the existence of a weak solution to the interior transmission problem by following the ideas of [12] (see [17]).

We begin by defining the volume potential $T_m E$ for $E \in (L^2(D))^3$ by the integral operator on the right-hand side of (13) and note that the interior transmission problem (16) can be written in integral form as

$$E - E_0 = T_m E \quad \text{in } D \tag{17a}$$

$$T_m E = E_e \quad \text{in } D_e. \tag{17b}$$

We now define the Hilbert space

$$L^2_m(D) := \left\{ f : D \to \mathbb{C}^3 \text{ measurable and } \int_D |f|^2 |m| \, dV < \infty \right\}$$

with scalar product

$$(f, g)_{|m|} := \int_D f \cdot \overline{g} \, |m| \, dV$$

where we assume that $|m| > 0$ in D. We also introduce the subspace $\mathbb{H} \subset L^2_m(D)$ defined by

$$\mathbb{H} := \text{span}\,\{M_n^m,\ \nabla \times M_n^m : n = 1, 2, \cdots,\ m = -n, \cdots, n\}$$

where $M_n^m = \nabla \times (x j_n(k|x|) Y_n^m(\hat{x}))$, Y_n^m is a spherical harmonic and j_n is a spherical Bessel function. We denote the closure of $\mathbb{H}$ in $L^2_m(D)$ by $\overline{\mathbb{H}}$.

Definition: *A pair* (E_1, E_0) *is said to be a weak solution to the interior transmission problem if* $(E_1, E_0) \in L^2_m(D) \times \overline{\mathbb{H}}$ *and satisfies (17).*

Finally, in order to prove the existence of a unique weak solution to the interior transmission problem we need to define the sesquilinear form $(\cdot,\cdot)_m$ on $L^2_m(D) \times L^2_m(D)$ by

$$(f, g)_m := \int_D f \cdot \bar{g} m \, dV$$

and denote by $\overline{\mathbb{H}}^{\perp m}$ the orthogonal complement of $\overline{\mathbb{H}}$ in $L^2_m(D)$ with respect to $(\cdot,\cdot)_m$. If Im $n > 0$ in D it can be shown that $L^2_m(D) = \overline{\mathbb{H}}^{\perp m} \oplus \overline{\mathbb{H}}$ with respect to $(\cdot,\cdot)_m$ and that the projection operator $P : L^2_m(D) \to \overline{\mathbb{H}}^{\perp m}$ defined by this decomposition is bounded [17].

Theorem 4. *Assume that* $n \in C^{1,\alpha}(\mathbb{R}^3)$ *for some* $\alpha, 0 < \alpha < 1$, $n(x) = 1$ *in* $\mathbb{R}^3 \setminus \overline{D}$, *Im* $n > 0$ *in* D *and there exists a positive constant* M *such that* $|\nabla n| \leq M\sqrt{|n - 1|}$. *Then there exists a unique weak solution to the interior transmission problem.*

Remark. The condition Im $n > 0$ is needed for uniqueness and the estimate on ∇n is needed in order to ensure the compactness of the operator $T_m : L^2_m(D) \to L^2_m(D)$.

Proof. Uniqueness is proved by using a limiting argument in connection with Green's first vector theorem [17]. To prove existence [17] we first assume without loss of generality that $q = (0, 0, 1)$ and $z = 0 \in D$ in (15) and note that condition (17b) is equivalent to

$$\nabla \times T_m E = \nabla \times E_e \quad \text{in } D_e. \tag{18}$$

Since

$$\nabla \times T_m E(x) = -k^2 \nabla \times \int_D \Phi(x, y) E(y) m(y) \, dV(y)$$

we see from the addition formula for Φ (Theorem 6.27 of [11]) that in order to ensure that (18) is valid we need to construct a solution E of (17a) such that $E \in \mathbb{H}_0^{\perp m}$ where

$$\begin{aligned} \mathbb{H}_0 := \text{span}\,[&\{M_n^m : n = 1, 2, \cdots,\ m = -n, \cdots, n\} \\ &\cup \{\nabla \times M_n^m : n = 1, 2, \cdots,\ m = -n, \cdots, n, (m, n) \neq (0, 1)\}] \end{aligned}$$

and such that $(E, \nabla \times M_1^0)_m = \gamma$ for γ an appropriately chosen constant. To this end we note that since $\mathbb{H}_0^{\perp m} \cap \overline{\mathbb{H}} \neq \emptyset$ there exists $\psi \in \mathbb{H}_0^{\perp m} \cap \overline{\mathbb{H}}$ such that

$(\psi, \nabla \times M_1^0)_m \neq 0$. By the uniqueness of weak solutions to the interior transmission problem, the boundedness of the projection operator P defined above and the compactness of T_m we can conclude that for any constant c there exists a unique solution to the operator equation

$$(I - PT_m)E = c\psi. \tag{19}$$

Now consider $E \in L_m^2(D)$ satisfying (19) and set $E_0 := (I - P)T_mE - c\psi$. By construction $E_0 \in \overline{\mathbb{H}}$ and E_0, E satisfy (17). Furthermore, a short calculation shows that $(E, \phi)_m = 0$ for $\phi \in \mathbb{H}_0$ and $(E, \nabla \times M_1^0)_m = c(\psi, \nabla \times M_1^0)_m$. Hence c can be chosen such that $(E, \nabla \times M_1^0)_m = \gamma$ and the proof is complete (we note that the analysis in [17] is based on using a magnetic dipole instead of the electric dipole (15)).

3 Inverse Problems

There are many inverse problems in scattering theory. However, the inverse problems we are concerned with in this paper are to 1) determine the support D of a partially coated obstacle from far field data and 2) to determine the support D of an isotropic inhomogeneous medium from far field data. We will show in this section that in each case D is uniquely determined from the far field data. We will also present in this section a heuristic discussion of the linear sampling method for reconstructing the support in each case, delaying a mathematical justification of this approach to the next section of our paper.

We began with establishing uniqueness. From either(1)–(5) or (10)–(12) it can easily be shown [11] that in each case the scattered electric field E^s has the asymptotic behaviour

$$E^s(x) = \frac{e^{ikr}}{r}\left(E_\infty(\hat{x}, d, p) + O\left(\frac{1}{r}\right)\right) \tag{20}$$

as $r = |x| \to \infty$ where E_∞ is known as the *electric far field pattern* and is an infinitely differentiable tangential vector field on the unit sphere Ω. The inverse scattering problem we are interested in is to determine D from a knowledge of $E_\infty(\hat{x}, d, p)$ for $\hat{x}, d \in \Omega$ and $p \in \mathbb{R}^3$. We note that E is a linear function of p and, in the case of (1)–(5), no a priori knowledge is assumed of Γ_I, Γ_D or λ.

As usual for inverse problems, the first question to ask is concerned with uniqueness and, in the case of (1)–(5) we base our proof on the ideas of Kirsch and Kress [21] (see Theorem 7.1 in [11]).

Theorem 5. *Let E_∞ be the electric far field pattern corresponding to the exterior mixed boundary value problem (1)–(5). Then D is uniquely determined by $E_\infty(\hat{x}, d, p)$ for $\hat{x}, d \in \Omega, p \in \mathbb{R}^3$.*

Remark. Since $E_\infty(\hat{x}, d, p)$ is an analytic function of $\hat{x}$ and d on Ω, it suffices to know E_∞ for $\hat{x}$ and d on an open subset of Ω. Furthermore, since $E_\infty(\hat{x}, d, p)$ is linear in p, it suffices to know E_∞ for three linearly independent vectors p_1, p_2, p_3.

Proof. Assume that there are two different domains D_1 and D_2 giving rise to the same far field pattern E_∞. By Rellich's lemma, and the fact that an entire solution to Maxwell's equations satisfying the Silver–Müller radiation condition must be identically zero, we can restrict our attention to the case when $\overline{D_1} \cap \overline{D_2} \neq \emptyset$. Finally, by using the mixed reciprocity relation of Potthast [28], we can assume that the scattered fields $E^s_{1,e}(x,z,p)$ and $E^s_{2,e}(x,z,p)$ corresponding to the scattering of electric dipoles by D_1 and D_2 coincide for all x, z in the unbounded component G of $\overline{D_1} \cup \overline{D_2}$ and all polarisations p.

Since $D_1 \neq D_2$, without loss of generality, there exists $x^* \in \partial G$ such that $x^* \in \Gamma_1$ and $x^* \notin \Gamma_2$ where Γ_j is the boundary of D_j, $j = 1, 2$. Of course, we can assume without loss of generality that x^* is not on the boundary Π_1 between Γ_{1D} and Γ_{1I}. In particular, we have that $z_n := x^* + n^{-1}\nu(x^*) \in G$ for n sufficiently large. Then, using Theorem 2, we have that $E^s_{2,e}(x^*, z_n, p)$ remains bounded in $X(D_{2,e} \cap B_R, \Gamma_{2,I})$ as $n \to \infty$ but on the other hand $E^s_{1,e}(x^*, z_n, p)$ cannot remain bounded in $X(D_{1,e} \cap B_R, \Gamma_{1,I})$ as $n \to \infty$ due to the dipole at $x = x^*$. But this contradicts $E^s_{1,e} = E^s_{2,e}$ in G and therefore $D_1 = D_2$.

We now turn our attention to uniqueness for the inverse problem associated with (10)–(12). The proof of the following theorem is quite technical and hence we only give the key steps in the proof. The result is due to Colton and Päivärinta [13] with a subsequent simplification being given by Hähner [18]. Generalisations to the case of variable permeability have been considered by Ola, Päivarinta and Somersalo [26], Ola and Somersalo [27] and Sun and Uhlmann [29].

Theorem 6. *Let E_∞ be the electric far field pattern corresponding to the exterior transmission problem (10)–(12). Then the index of refraction $n(x)$ is uniquely determined by $E_\infty(\hat{x}, d, p)$ for $\hat{x}, d \in \Omega, p \in \mathbb{R}^3$.*

Remark. The remark after the statement of Theorem 5 also holds in this case.

Proof. The key steps in the proof are as follows. Full details can be found in [13].

1. It is shown that the set of all solutions to the exterior transmission problem for $d \in \Omega$, $p \in \mathbb{R}^3$, is complete in the closure in $L^2(B)$ of all solutions to

$$\nabla \times E - ikH = 0 \tag{21a}$$

$$\nabla \times H + ikn(x)E = 0 \tag{21b}$$

   in B where B is a ball containing the support of $m = 1 - n$.
2. If there exist two refractive indices n_1 and n_2 having the same electric far field pattern, then it is shown, using Step 1, that

$$\int_{\mathbb{R}^3} E_1(x) \cdot (n_1(x) - n_2(x)) E_2(x)\, dV(x) = 0$$

   where E_j, H_j is any solution of (21) with $n = n_j, j = 1, 2$.

3. A solution E, H of (21) is constructed such that E has the form

$$E(x) = e^{i\zeta\cdot x}\left[\eta + R_\zeta(x)\right] \tag{22}$$

where $\zeta, \eta \in \mathbb{C}^3$, $\eta \cdot \zeta = 0$ and $\zeta \cdot \zeta = k^2$. A problem here is that R_ζ does *not* tend to zero as $|\zeta| \to \infty$.
4. Choose E_j to be of the form (22) where $\zeta = \zeta_j$ with $\zeta_1 + \zeta_2 = \xi \in \mathbb{R}^3$. By choosing $\eta_j = \eta(\zeta_j)$ appropriately and substituting E_j into (21) we have, letting $|\zeta_j| \to \infty$, that

$$\int_{\mathbb{R}^3} e^{i\xi\cdot x}(n_1(x) - n_2(x))\, dV(x) = 0.$$

Hence, by the Fourier integral theorem, $n_1(x) = n_2(x)$ for all $x \in \mathbb{R}^3$.

As mentioned above, Theorem 6 is *not* in general true for anisotropic media, i.e. in the case when n is a matrix. However in this case it can be shown that the support D of $I - n$ is uniquely determined [3]. The proof is based on the ideas of the proof of Theorem 5 together with an analysis of a particular interior transmission problem.

Having established uniqueness, the next step in studying the inverse scattering problem is to derive a reconstruction algorithm for determining D. In particular, we want an algorithm that does not require the a priori knowledge of knowing that the electric far field pattern is associated with (1)–(5) or (10)–(12). A method for doing this is the *linear sampling method* which was first introduced for acoustic waves by Colton and Kirsch [10] and Colton, Piana and Potthast [14] and for electromagnetic waves by Potthast [28], Kress [22] and Colton, Haddar and Monk [9]. Here we present a heuristic introduction to the method, delaying its mathematical justification and numerical implementation to the next section of this paper. We begin by defining the *far field operator* $F : L^2_t(\Omega) \to L^2_t(\Omega)$ by

$$(Fg)(\hat{x}) := \int_{\Omega} E_\infty(\hat{x}, d, g(d))\, dS(d). \tag{23}$$

where E_∞ is the electric far field pattern of either (1)–(5) or (10)–(12). We further define an *electromagnetic Herglotz pair with kernel* g by

$$E_g(x) := \int_{\Omega} e^{ikx\cdot d} g(d)\, dS(d) \tag{24a}$$

$$H_g(x) := \frac{1}{ik}\nabla \times E_g(x). \tag{24b}$$

We note that by superposition we have that Fg is the electric far field pattern for either (1)–(5) or (10)–(12) corresponding to the electromagnetic Herglotz pair with kernel ikg as incident field. Finally, we define the *far field equation* by

$$Fg(\hat{x}) = E_{e,\infty}(\hat{x}, z, q) \tag{25}$$

where

$$E_{e,\infty}(\hat{x}, z, q) = \frac{ik}{4\pi}(\hat{x} \times q) \times \hat{x}\, e^{ik\hat{x}\cdot z} \tag{26}$$

is the electric far field pattern of the electric dipole (15).

We are now in a position to introduce the basic idea of the linear sampling method. Suppose that for each $z \in D$ there exists $g(\cdot, z) \in L_t^2(\Omega)$ such that the far field equation (25) is satisfied. Then by Rellich's lemma

$$\int_\Omega E^s(x, d, g(d))\, dS(d) = E_e(x, z, q) \tag{27}$$

for $x \in D_e$ and in particular for $x \in \Gamma$. As $z \to x \in \Gamma$ we have that $E_e(x, z, q) \to \infty$ and hence from (27) and the regularity of E^s we must have that

$$\lim_{\substack{z \to x \in \Gamma \\ z \in D}} \|g(\cdot, z)\|_{L_t^2(\Omega)} = \infty. \tag{28}$$

Under the above assumptions it is also possible to conclude that the electric field of the electromagnetic Herglotz pair with kernel g becomes infinite as $z \to \Gamma$. Hence Γ is characterised by points where the solution of the far field equation become unbounded as a function of z.

The above argument is purely heuristic since in general there is no solution $g \in L_t^2(\Omega)$ of the far field equation! Indeed it can be shown that a solution to the far field equation exists if and only if, in the case of (1)–(5), the solution of the interior mixed boundary value problem (7) for f, g the appropriate traces of the electric dipole is the electric field of an electromagnetic Herglotz pair and, in the case of (10)–(12), the solution E_0, H_0 of the interior transmission problem (16a)–(16c) is an electromagnetic Herglotz pair [11]. In general this is not true in either case and hence a solution to the far field equation does not exist. A second problem is that even the above heuristic argument breaks down if $z \in D_e$. In particular, it is not at all clear how, or even if, the linear sampling method can be given a mathematical justification. Of course, such a justification can indeed be given (otherwise we would not be writing this paper!) and details of how this can be done will be given in the next section.

4 The Linear Sampling Method

In this section we will give a mathematical justification of the linear sampling method that was described heuristically at the end of the last section. As mentioned in the Introduction, the basic idea is to factor the far field operator F in the form $F = \mathcal{B}\gamma\mathcal{H}$ where $\mathcal{B}$ maps the incident field onto the electric far field pattern, γ is a trace operator and $\mathcal{H}$ maps tangential fields on the unit sphere Ω onto the class of vector Herglotz wave functions. The desired result then follows by showing that $\mathcal{B}$ and γH are injective with dense range and $E_{e,\infty}(\cdot, z, q)$ is in the range of $\mathcal{B}$ if and only if $z \in D$. We will concentrate our attention on the scattering problem (1)–(5) and then briefly describe how a similar approach is valid for the scattering problem (10)–(12).

Recalling the space $Y(\Gamma_D)$ defined in Sect. 2, we define the operator $\mathcal{H} : L^2_t(\Omega) \to C^\infty(\mathbb{R}^3)$ by $\mathcal{H}g = E_g$ where E_g is the electric field of an electromagnetic Herglotz pair with kernel g. The trace operator $\gamma : C^\infty(\mathbb{R}^3) \to Y(\Gamma_D) \times L^2_t(\Gamma_I)$ is defined by

$$\gamma E := \begin{cases} \nu \times E & \text{on } \Gamma_D \\ \nu \times \nabla \times E - i\lambda\nu \times (E \times \nu) & \text{on } \Gamma_I \, . \end{cases}$$

Noting that $E_g(x) = 0$ for $x \in D$ if and only if $g = 0$ we see from Theorem 1 that $\gamma\mathcal{H}$ is injective provided $\Gamma_I \neq \emptyset$.

Theorem 7. *Assume that $\Gamma_I \neq \emptyset$. Then the range of $\gamma\mathcal{H}$ is dense in $Y(\Gamma_D)\times L^2_t(\Gamma_I)$.*

Proof. We need to show that the dual operator $(\gamma\mathcal{H})^T : Y(\Gamma_D)' \times L^2_t(\Gamma_I) \to L^2_t(\Omega)$ is injective. A simple computation shows that if $(a_1, a_2) \in Y(\Gamma_D)' \times L^2_t(\Gamma_I)$ then

$$(\gamma\mathcal{H})^T[a_1, a_2] = d \times \left\{ \int_{\Gamma_D} e^{ikx\cdot d}(a_1 \times \nu)\, dV(x) \right.$$

$$\left. + d \times \int_{\Gamma_I} e^{ikx\cdot d}(a_2 \times \nu)\, dS(x) - i\lambda \int_{\Gamma_I} e^{-ikx\cdot d}\, [\nu \times (a_2 \times \nu)]\; dS(x) \right\} \times d$$

and that this is the far field pattern of

$$P(z) = \nabla \times \nabla \times \int_{\Gamma_D} \Phi(x,z)(a_1, \times\nu) dS(x) + k^2 \nabla \times \int_{\Gamma_I} \Phi(x,z)(a_2 \times \nu) dS(x)$$
$$-i\lambda\, \nabla \times \nabla \times \int_{\Gamma_I} \Phi(x,z)\, [\nu \times (a_2 \times \nu)]\, dS(x). \tag{29}$$

Now assume that $(\gamma\mathcal{H})^T[a_1, a_2] = 0$. It can then be shown using Rellich's lemma and the discontinuity properties of surface potentials with densities in a Sobolev space that

$$\begin{aligned} \nu \times P^- \big|_{\Gamma_D} &= 0 \\ \left[\nu \times \nabla \times P^- - i\lambda\nu \times (P^- \times \nu)\right] \big|_{\Gamma_I} &= 0 \end{aligned} \tag{30}$$

where the minus sign indicates the limits from inside D and the boundary data (30) is understood in the L^2 limit sense. Since $\nabla \times \nabla \times P - k^2 P = 0$ in D, one can now conclude that $P = 0$ in D and by the discontinuity properties of surface potentials we now see from (29) that $a_1 = a_2 = 0$. Hence $(\gamma\mathcal{H})^T$ is injective and this concludes the proof. For details we refer the reader to [4].

Corollary 1. *Assume that $\Gamma_I \neq \emptyset$. Then the electric field of the solution to the interior mixed boundary value problem (7) can be approximated by the electric field of an electromagnetic Herglotz pair with respect to the $X(D, \Gamma_I)$ norm.*

Proof. The result is a consequence of Theorem 7 and the a priori estimate of Theorem 1.

We now define the operator $\mathcal{B} : Y(\Gamma_D) \times L^2_t(\Gamma_I) \to L^2_t(\Omega)$ which maps the boundary data (f, h) onto the electric far field pattern $E_\infty \in L^2_t(\Omega)$ of the exterior mixed boundary value problem (6). It is easy to see that $\mathcal{B}$ is an injective, compact linear operator.

Theorem 8. *The range of $\mathcal{B}$ is dense in $L^2_t(\Omega)$.*

Proof. As in the proof of Theorem 7, we need to show that the dual operator $\mathcal{B}^T : L^2_t(\Omega) \to Y(\Gamma_D)' \times L^2_t(\Gamma_I)$ is injective. By a straightforward but lengthy calculation [4] it is possible to show that

$$4\pi\mathcal{B}^T g = \begin{cases} \nu \times (\nabla \times E_g - \nabla \times \tilde{E}) \times \nu & \text{on } \Gamma_D \\ \nu \times (E_g - \tilde{E}) \times \nu & \text{on } \Gamma_I \end{cases} \tag{31}$$

where E_g is again the electric field of an electromagnetic Herglotz pair with kernel g and $\tilde{E} \in X_{loc}(D_e, \Gamma_I)$ is the solution of the exterior mixed boundary value problem (9) with $f = \nu \times E_g$ and $h = \nu \times \nabla \times E_g - i\lambda(\nu \times E_g) \times \nu$. Now suppose $\mathcal{B}^T g = 0$. Then (31) and the boundary condition satisfied by $\tilde{E}$ imply that

$$\left.\begin{aligned} \nu \times \tilde{E} &= \nu \times E_g \\ \nu \times \nabla \times \tilde{E} &= \nu \times \nabla \times E_g \end{aligned}\right\} \quad \text{on } \Gamma$$

and hence $\tilde{E}$ and $\tilde{H} = (ik)^{-1}\nabla \times \tilde{E}$ can be extended to a solution of Maxwell's equation in all of $\mathbb{R}^3$. But, since $\tilde{E}, \tilde{H}$ satisfy the Silver–Müller radiation condition, $\tilde{E} = 0$ which implies $E_g = 0$ in D and hence $g = 0$. Thus $\mathcal{B}^T$ is injective and the theorem is proved.

We are now in a position to give a mathematical justification of the linear sampling method for the case of the scattering problem (1)–(5). We first observe that the far field equation (25) can be written in the form

$$\mathcal{B}\gamma\mathcal{H}g = -\frac{1}{ik}E_{e,\infty}(\cdot, z, q). \tag{32}$$

Suppose $z \in D$. In this case it is easy to see that $E_{e,\infty}(\cdot, z, q)$ is in the range of $\mathcal{B}$. Let $E \in X(D, \Gamma_I)$ be the solution of the interior mixed boundary value problem (7) for $f = (ik)^{-1}\nu \times E_e\big|_{\Gamma_D}$ and $h = (ik)^{-1}\left(\nu \times \nabla \times E_e - i\lambda(\nu \times E_e) \times \nu\right)\big|_{\Gamma_I}$. Then if $\Gamma_I \neq \emptyset$ we have from Theorem 7 that for every $\epsilon > 0$ there is a $g_\epsilon = g_\epsilon(\cdot, z, q) \in L^2_t(\Omega)$ such that $\mathcal{H}g_\epsilon$ satisfies

$$\|\gamma(E - \mathcal{H}g_\epsilon)\|_{Y(\Gamma_D)\times L^2_t(\Gamma_I)} < \epsilon \tag{33}$$

and hence

$$\|\mathcal{B}\gamma\mathcal{H}g_\epsilon + \frac{1}{ik}E_{e,\infty}(\cdot, z, q)\|_{L^2_t(\Omega)} < C\epsilon$$

for some positive constant C. But from (33) we have that

$$\lim_{z\to\Gamma} \|\gamma\mathcal{H}g_\epsilon(\cdot,z)\|_{Y(\Gamma_D)\times L^2_t(\Gamma_I)} = \infty$$

and hence

$$\lim_{z\to\Gamma} \|(\mathcal{H}g_\epsilon)(\cdot,z)\|_{X(D,\Gamma_I)} = \infty$$

and

$$\lim_{z\to\Gamma} \|g_\epsilon(\cdot,z)\|_{L^2_t(\Omega)} = \infty.$$

Now assume that $z \in D_e$. In this case $E_{e,\infty}(\cdot,z,q)$ is not in the range of $\mathcal{B}$. However, using Theorem 8 and Tikhonov regularisation we can find functions $(f_z^\alpha, h_z^\alpha) \in Y(\Gamma_D) \times L^2(\Gamma_I)$ corresponding to a parameter $\alpha = \alpha(\delta)$ chosen by a regular regularisation strategy (see, e.g. [11]) such that

$$\|\mathcal{B}(f_z^\alpha, h_z^\alpha) + \frac{1}{ik}E_{e,\infty}(\cdot,z,q)\|_{L^2_t(\Omega)} < \delta \tag{34}$$

for an arbitrarily small $\delta > 0$ and

$$\lim_{\alpha\to 0}(\|f_z^\alpha\|_{Y(\Gamma_D)} + \|h_z^\alpha\|_{L^2(\Gamma_I)}) = \infty. \tag{35}$$

Note that $\alpha \to 0$ as $\delta \to 0$. Again assuming that $\Gamma_I \neq \emptyset$, for every $\epsilon > 0$ we can use Theorem 7 to find $g_{\alpha,\epsilon}(\cdot,z,q) \in L^2_t(\Omega)$ such that

$$\|\mathcal{B}\gamma\mathcal{H}g_{\alpha,\epsilon} - \mathcal{B}(f_z^\alpha, h_z^\alpha)\|_{L^2_t(\Omega)} < \epsilon \tag{36}$$

and hence from (34) and (36) we have that

$$\|\mathcal{B}\gamma\mathcal{H}g_{\alpha,\epsilon} + \frac{1}{ik}E_{e,\infty}(\cdot,z,q)\|_{L^2_t(\Omega)} < \epsilon + \delta.$$

Furthermore, (38) implies that

$$\lim_{\alpha\to 0} \|(\mathcal{H}g_{\alpha,\epsilon})(\cdot,z)\|_{X(D,\Gamma_I)} = \infty$$

and

$$\lim_{\alpha\to 0} \|g_{\alpha,\epsilon}(\cdot z)\|_{L^2_t(\Omega)} = \infty.$$

We summarise these results in the following theorem which gives a mathematical justification of the linear sampling method for the case of the scattering problem (1)–(5) [4].

Theorem 9. *Assume that $\Gamma_I \neq \emptyset$. Then if F is the far field operator corresponding to the scattering problem (1)–(5) we have the following results.*

1. *If $z \in D$ then for every $\epsilon > 0$ there exists a solution $g_\epsilon(\cdot,z) = g_\epsilon(\cdot,z,q) \in L^2_t(\Omega)$ satisfying the inequality*

$$\|Fg_\epsilon(\cdot,z) - E_{e,\infty}(\cdot,z,q)\|_{L^2_t(\Omega)} < \epsilon.$$

Moreover, this solution satisfies

$$\lim_{z\to\Gamma} \|\mathcal{H}g_\epsilon(\cdot,z)\|_{X(D,\Gamma_I)} = \infty \text{ and } \lim_{z\to\Gamma} \|g_\epsilon(\cdot,z)\|_{L^2_t(\Omega)} = \infty.$$

2. *If* $z \in D_e$ *then for every* $\epsilon > 0$ *and* $\delta > 0$ *there exists a solution* $g_{\delta,\epsilon}(\cdot, z) = g_{\delta,\epsilon}(\cdot, z, q) \in L^2_t(\Omega)$ *of the inequality*

$$\|Fg_{\delta,\epsilon}(\cdot, z) - E_{e,\infty}(\cdot, z, q)\|_{L^2_t(\Omega)} < \epsilon + \delta$$

such that

$$\lim_{\delta \to 0} \|\mathcal{H}g_{\delta,\epsilon}(\cdot, z)\|_{X(D,\Gamma_I)} = \infty \text{ and } \lim_{\delta \to 0} \|g_{\delta,\epsilon}(\cdot, z)\|_{L^2_{t(\Omega)}} = \infty.$$

We remark that if $\Gamma_I = \emptyset$, i.e. D is a perfect conductor, then Theorem 9 holds provided k is not a Maxwell eigenvalue for D. In this case it is in fact possible to eliminate the restriction that k is not a Maxwell eigenvalue by replacing the far field operator F by the *combined far field operator* $F_c : L^2_t(\Omega) \to L^2_t(\Omega)$ defined by

$$(F_c g)(\hat{x}) := \gamma \int_\Omega E_\infty(\hat{x}, d, g(d))\, dS(d) + \mu \int_\Omega H_\infty(\hat{x}, d, g(d)) \times d)\, dS(d)$$

where $H_\infty := \hat{x} \times E_\infty$ is the magnetic far field pattern and $\gamma > 0,\ \mu < 0$ are real numbers [2].

It is also possible to consider limited aperture far field data, in which case in the above theorem $L^2_t(\Omega)$ is replaced by $L^2_t(\Omega_0)$ where $\Omega_0 \subset \Omega$ [2].

A theorem analogous to Theorem 9 can also be established for the scattering problem (10)–(12). The idea of the proof is similar to Theorem 9. The operator $\mathcal{B}$ is replace by $\mathcal{F}T_m(I - T_m)^{-1} : \overline{\mathbb{H}} \to L^2_t(\Omega)$ where $\mathcal{F}E$ denotes the electric far field pattern associated with a radiating solution E to Maxwell's equations in $\mathbb{R}^3 \setminus \overline{D}$ and the trace operator γ is now the identity operator on $L^2_m(D)$. The density result corresponding to Theorem 7 is trivial, since elements of $\mathbb{H}$ are in fact Herglotz wave functions, and the density result corresponding to Theorem 8 follows from the existence of a weak solution to the interior transmission problem (Theorem 4). For details we refer the reader to [17].

5 Numerical Approximation of the Forward Problem

In the numerical results we shall show in the next section, the data for the inverse problem is the far field pattern E_∞. This is provided by computing an approximation to E, H satisfying (1)–(5) or (10)–(12). In this section we shall outline how this is done and prove a new convergence result for the method via Theorem 7. Since the method, termed the Ultra Weak Variational Formulation or UWVF by its inventors Cessenat and Després [5], [6], is quite novel we shall describe it in some detail.

The first step in applying the UWVF to our problem is to introduce an artificial boundary surface Σ (assumed to be the boundary of a Lipschitz polyhedron and connected) containing D in its interior. We apply a boundary condition on Σ motivated by the Silver–Müller radiation condition (4). To approximate (1)–(5) we define the computational domain R to be the domain inside Σ and outside D (to approximate

(10)–(12) the domain R is just the interior of Σ). Then we approximate the solution (E, H) of (1)–(5) by $(\tilde{E}, \tilde{H})$ that satisfy

$$\left.\begin{aligned} \nabla \times \tilde{E} - ik\tilde{H} = 0 \\ \nabla \times \tilde{H} + ik\tilde{E} = 0 \end{aligned}\right\} \text{ in } R, \tag{37a}$$

together with the boundary conditions

$$\nu \times \tilde{E} = 0 \text{ on } \Gamma_D, \tag{37b}$$

$$\nu \times \nabla \times \tilde{E} - i\lambda(\nu \times \tilde{E}) \times \boldsymbol{\nu} = 0 \text{ on } \Gamma_I, \tag{37c}$$

$$\tilde{H} \times \nu - (\nu \times \tilde{E}) \times \nu = -(H^i \times \nu - (\nu \times E^i) \times \nu) \text{ on } \Sigma, \tag{37d}$$

where ν is the unit outward normal to D on Γ_I and Γ_D and the outward normal to R on Σ. This is just an interior mixed problem and hence existence and uniqueness of $(\tilde{E}, \tilde{H})$ is guaranteed by Theorem 1 since $\Sigma \neq \emptyset$. A similar problem is also obtained from (10)–(12). Both these problems can be written in a form suitable for the UWVF as the problem of finding $(\tilde{E}, \tilde{H})$ such that

$$\left.\begin{aligned} \nabla \times \tilde{E} - ik\tilde{H} \qquad &= 0 \\ \nabla \times \tilde{H} + ikn(x)\tilde{E} &= 0 \end{aligned}\right\} \text{ in } R, \tag{38a}$$

$$\tilde{H} \times \nu - \sigma(\nu \times \tilde{E}) \times \nu = Q(\tilde{H} \times \nu + \sigma(\nu \times \tilde{E}) \times \nu) + g \text{ on } \partial R \tag{38b}$$

where ν is the unit outward normal to R and Q and σ are piecewise constant functions on ∂R. The function g is a given tangential vector field. In addition $\sigma > 0$. For example to obtain (37a)–(37d) we choose $n = 1$, $Q = 0$, $\sigma = \lambda/k$ and $g = 0$ on Γ_I, together with $Q = -1$, $\sigma = 1$ and $g = 0$ on Γ_D. Finally we also need $Q = 0$, $\sigma = 1$, and $g = -(H^i \times \nu - E^i_T)$ on Σ.

The UWVF applied to (38) is, according to [5], constructed as follows: First R is subdivided to disjoint subdomains $K_1, \ldots, K_M$ each having connected complement. In our implementation of the method $\{K_j\}_{j=1}^M$ is a regular tetrahedral finite element mesh of Ω. Hence we must assume that R is a Lipschitz polyhedron (not curvilinear). For technical reasons we also need to assume that the index of refraction n is piecewise constant on this partition. For simplicity we shall describe the UWVF with only two subdomains K_1 and K_2. It is then relatively straightforward to extend the method to $M > 2$ domains. We denote by $\Sigma_{1,2}$ the common surface $\overline{K}_1 \cap \overline{K}_2$ with normal ν_1 pointing out of K_1 and by $\Sigma_{2,1}$ the same surface viewed as a portion of ∂K_2 with normal ν_2 pointing out of K_2. We denote by $(\tilde{E}_j, \tilde{H}_j)$ the restriction of $(\tilde{E}, \tilde{H})$ to K_j, $j = 1, 2$. Then $(\tilde{E}_j, \tilde{H}_j)$, $j = 1, 2$, satisfies

$$\left.\begin{aligned} \nabla \times \tilde{E}_j - ik\tilde{H}_j = 0 \\ \nabla \times \tilde{H}_j + ikn(x)\tilde{E}_j = 0 \end{aligned}\right\} \text{ in } K_j, \tag{39a}$$

$$\begin{aligned} &\tilde{H}_j \times \nu_j - \sigma(\nu_j \times \tilde{E}_j) \times \nu_j = \\ &Q\left(\tilde{H}_j \times \nu_j + \sigma(\nu \times \tilde{E}_j) \times \nu\right) + g \text{ on } \partial K_j \cap \partial R \end{aligned} \tag{39b}$$

and the interface conditions

$$\tilde{H}_1 \times \nu_1 - \sigma_{12}(\nu_1 \times \tilde{E}_1) \times \nu_1 = -\tilde{H}_2 \times \nu_2 - \sigma_{12}(\nu_2 \times \tilde{E}_2) \times \nu_2, \tag{40a}$$

$$\tilde{H}_1 \times \tilde{\nu}_1 + \sigma_{12}(\nu_1 \times \tilde{E}_1) \times \nu_1 = -\tilde{H}_2 \times \nu_2 + \sigma_{12}(\nu_2 \times \tilde{E}_2) \times \nu_2, \tag{40b}$$

on Σ_{12} where σ_{12} is a positive parameter chosen to be

$$\sigma_{12} = \sqrt{|n_1|\,|n_2|} \text{ on } \Sigma_{1,2}.$$

Obviously (40a) and (40b) guarantee the continuity of $\nu \times E$ and $\nu \times H$ across $\Sigma_{1,2}$. Hence (39a)–(40b) is equivalent to (38). More subdomains simply imply more interfaces and corresponding interface conditions.

The UWVF is based on the following "Isometry Lemma" from [5] for which we need to define the space

$$X_j = \{(u, v) \in H(\text{curl}; K_j) \times H(\text{curl}; K_j) \mid \\ v \times \nu_j + \sigma_j(\nu_j \times u) \times \nu_j \in L_t^2(\partial K_j)\}$$

where $\sigma_j = \sigma$ on $\partial\Omega_j \setminus \Sigma_{12}$ and $\sigma_j = \sigma_{12}$ on Σ_{12}.

Lemma 1. *Let $(\tilde{E}_j, \tilde{H}_j) \in X_j$ satisfy (39a) for $j = 1, 2$, and assume in addition that (40a)–(40b) are satisfied in $L_t^2(\Sigma_{12})$. Suppose furthermore that $(\phi_j, \psi_j) \in X_j$ satisfy the adjoint problem*

$$\left.\begin{array}{ll} \nabla \times \phi_j - ik\psi_j & = 0 \\ \nabla \times \psi_j + ik\overline{n}(x)\phi_j & = 0 \end{array}\right\} \text{ in } K_j,\ j = 1, 2,$$

with sufficient regularity that

$$\psi_j \times \nu_j - \sigma_j(\nu_j \times \phi_j) \times \nu_j \in L_t^2(\partial K_j).$$

Then

$$\int\limits_{\partial K_j} \frac{1}{\sigma_j}\left(\tilde{H}_j \times \nu_j + \sigma_j\left(\nu_j \times \tilde{E}_j\right) \times \nu_j\right) \cdot \left(\overline{\psi_j \times \nu_j + \sigma_j\left(\nu_j \times \phi_j\right) \times \nu_j}\right) dS$$
$$= \int\limits_{\partial K_j} \frac{1}{\sigma_j}\left(\tilde{H}_j \times \nu_j - \sigma_j\left(\nu_j \times \tilde{E}_j\right) \times \nu_j\right) \cdot \left(\overline{\psi_j \times \nu_j - \sigma_j\left(\nu_j \times \phi_j\right) \times \nu_j}\right) dS. \tag{41}$$

Proof. This result follows from integration by parts. By expanding both sides of the inequality we see that the left-hand and right-hand sides of (41) differ by

$$2\int_{\partial K_j} \tilde{H}_j \times \nu_j \cdot (\nu_j \times \overline{\phi}_j) \times \nu_j - (\overline{\psi}_j \times \nu_j) \cdot \left(\nu_j \times \tilde{E}_j\right) \times \nu_j \, dS$$
$$= -2\int_{K_j} \nabla \times \tilde{H}_j \cdot \overline{\phi}_j - \tilde{H}_j \cdot \nabla \times \overline{\phi}_j + \nabla \times \overline{\phi}_j \cdot \tilde{E}_j - \overline{\phi}_j \dot{\nabla} \times \tilde{E}_j \, dV$$

and using the equations for $(\tilde{E}_j, \tilde{H}_j)$ and (ϕ_j, ψ_j) proves that this vanishes.

Using the Isometry Lemma the UWVF is easy to formulate. Let

$$\mathcal{X}_j = \tilde{H}_j \times \nu_j + \sigma\left(\nu_j \times \tilde{E}_j\right) \times \nu_j \qquad \text{on } \partial K_j, \tag{42}$$

and

$$\mathcal{Y}_j = \psi_j \times \nu_j + \sigma(\nu_j \times \phi_j) \times \nu_j \qquad \text{on } \partial K_j. \tag{43}$$

Then using the transmission conditions (40a) and (40b) and the boundary condition (39b) in (41) we obtain the problem of finding $(\mathcal{X}_1, \mathcal{X}_2) \in L^2_t(\partial K_1) \times L^2_t(\partial K_2)$ such that

$$\begin{aligned}\int\limits_{\partial K_1} \frac{1}{\sigma_1}\mathcal{X}_1 \cdot \overline{\mathcal{Y}}_1 \, dS = &- \int_{\Sigma_{12}} \frac{1}{\sigma_{12}}\mathcal{X}_2 \cdot \overline{F_1(\mathcal{Y}_1)}\, dS \\ &+ \int_{(\partial K_1)\setminus\Sigma_{12}} \left(\frac{1}{\sigma_1} Q\mathcal{X}_1 + g\right) \cdot \overline{F_1(\mathcal{Y}_1)}\, dS\end{aligned} \tag{44a}$$

$$\begin{aligned}\int\limits_{\partial K_2} \frac{1}{\sigma_2}\mathcal{X}_2 \cdot \overline{\mathcal{Y}}_2 \, dS = &- \int_{\Sigma_{21}} \frac{1}{\sigma_{12}}\mathcal{X}_1 \cdot \overline{F_2(\mathcal{Y}_2)}\, dS \\ &+ \int_{(\partial K_2)\setminus\sigma_{12}} \frac{1}{\sigma_2} (Q\mathcal{X}_2 + g) \cdot \overline{F_2(\mathcal{Y}_2)}\, dS\end{aligned} \tag{44b}$$

for all $(\mathcal{Y}_1, \mathcal{Y}_2) \in L^2_t(\partial K_1) \times L^2_t(\partial K_2)$ where $F_j : L^2_t(\partial K_j) \to L^2_t(\partial K_j)$, $j = 1, 2$, is the operator such that if $\mathcal{Y}_j = x_j \times \nu_j + \sigma_j(\nu_j \times \phi_j) \times \nu_j$ on ∂K_j then $F_j(\mathcal{Y}_j) = x_j \times \nu_j - \sigma_j(\nu_j \times \phi_j) \times \nu_j$.

Cessenat [5] proves that (44a)–(44b) has a unique solution, and that $\mathcal{X}_1$ and $\mathcal{X}_2$ indeed is the impedance trace of $(\tilde{E}, \tilde{H})$, in the sense of (42).

To discretise (44a)–(44b) we need to choose subspaces of $L^2_t(\partial K_j)$ such that the map F_j is easy to compute. In this context the plane wave basis is particularly attractive. Following Cessenat, we start by choosing p_j directions $d_{l,j}$, $1 \le l \le p_j$, on the unit sphere together with p_j polarisations $P_{l,j}$ such that $P_{l,j} \neq 0$ and $d_{l,j} \cdot P_{l,j} = 0$ for all l. In order to obtain a more sparse discrete matrix we then define, for $1 \le l \le p_j$, $j = 1, 2$,

$$\begin{aligned} F_{l,j} &= (P_{l,j} + iP_{l,j} \times d_{l,j}), \\ G_{l,j} &= (P_{l,j} - iP_{l,j} \times d_{l,j}). \end{aligned}$$

Then we can define two families of solutions of Maxwell's equations (39a) with n replaced by $\overline{n}$ as follows:

$$\begin{aligned} \phi^F_{l,j} &= F_{l,j} \exp\left(ik\sqrt{\overline{n}_j} d_{l,j} \cdot x\right), & \psi^F_{lj} &= \frac{1}{ik}\nabla \times \phi_{l,j}, \\ \phi^G_{l,j} &= G_{l,j} \exp\left(ik\sqrt{\overline{n}_j} d_{l,j} \cdot x\right), & \psi^G_{lj} &= \frac{1}{ik}\nabla \times \phi_{l,j}. \end{aligned}$$

Using these functions we can define the discrete spaces $V_j \subset L^2_t(\partial K_j)$ by

$$V_j = \text{span}\left\{\phi^F_{l,j} \times \nu_j + \sigma_j(\nu_j \times \phi^F_{l,j}) \times \nu_j, \text{ and } \phi^G_{l,j} \times \nu_j + \sigma_j(\nu_j \times \phi^G_{l,j}) \times \nu_j, \; 1 \le l \le p_j\right\}.$$

The discrete problem is to find $\mathcal{X}_1^p \in V_1$ and $\mathcal{X}_2^p \in V_2$ such that (44a) and (44b) are satisfied for all test functions $\mathcal{Y}_1^p \in V_1$ and $\mathcal{Y}_2^p \in V_2$. Obviously the resulting linear system consists of $2p_1 + 2p_2$ equations in $2p_1 + 2p_2$ unknowns and it is possible to show that this system has a unique solution. The convergence analysis of $(\mathcal{X}_1^p, \mathcal{X}_2^p)$ to $(\mathcal{X}_1, \mathcal{X}_2)$ is not as well developed as for the finite element method. Cessenat [5] shows that if Q is constant and $|Q| \le \delta < 1$ on ∂R then

$$\begin{aligned}&\sqrt{\left\|\frac{1}{\sigma_1^{1/2}}(\mathcal{X}_1 - \mathcal{X}_1^p)\right\|^2_{L^2(\partial R\cap\partial K_1)} + \left\|\frac{1}{\sigma_2^{1/2}}(\mathcal{X}_2 - \mathcal{X}_2^p)\right\|^2_{L^2(\partial R\cap\partial K_2)}} \\ &\le \frac{2}{\sqrt{1-\delta^2}}\sqrt{\left\|\frac{1}{\sigma_1^{1/2}}(\mathcal{X}_1 - \phi_1^p)\right\|^2_{L^2_t(\partial K_1)} + \left\|\frac{1}{\sigma_2^{1/2}}(\mathcal{X}_1 - \phi_2^p)\right\|^2_{L^2_t(\partial K_2)}}\end{aligned} \tag{45}$$

for any $\phi_j^p \in V_j, j = 1, 2$. A similar estimate holds for multiple elements $K_j, j = 1, \cdots, K_M$, and via interpolation error estimates for plane waves, it is possible to prove that there is a choice of directions such that the UWVF converges as the mesh size tends to zero [5].

Another way of obtaining convergence is to fix the mesh (in our example, fix K_1 and K_2) and increase the number of directions $p_j, j = 1, 2$, in order to obtain a more accurate solution. We shall now prove the following convergence result.

Theorem 10. *Suppose Q is constant, $|Q| \le \delta < 1$ and $p_1 = p_2 = p$. Given $\epsilon > 0$ there exists a number of directions P_ϵ such that for all $p > P_\epsilon$ there is a choice of directions $d_1, \cdots, d_p$ on each element such that*

$$\sqrt{\|\mathcal{X}_1 - \mathcal{X}_1^p\|^2_{L^2_t(\partial R\cap\partial K_1)} + \|\mathcal{X}_2 - \mathcal{X}_2^p\|^2_{L^2_t(\partial R\cap\partial K_2)}} < \epsilon.$$

Remark. The proof shows that the directions $d_1, \cdots, d_p$ for this theorem should be chosen to correspond to a sufficiently accurate quadrature scheme on the unit sphere. For quadrature on the sphere see for example [30].

Proof. We need only show that there is a function in $V_j, j = 1, 2$, with the desired accuracy and use (45). Consider K_1. From Theorem 7 there is a Herglotz kernel $g \in L^2_t(\Omega)$ such that

$$\left\|\frac{1}{\sigma_1^{1/2}}\left(\mathcal{X}_1 - \frac{1}{ik}\nabla \times \mathcal{H}g \times \nu - i\sigma_1(\mathcal{H}g)_T\right)\right\|_{L^2_t(\partial K_1)} \le \frac{\sqrt{1-\delta^2}}{12}\epsilon$$

where $\mathcal{H}$ is the Herglotz operator defined before Theorem 7 and $(\mathcal{H}g)_T = (\nu_j \times \mathcal{H}g) \times \nu_j$. Because $(C^\infty(\Omega))^3 \cap L^2_t(\Omega)$ is dense in $L^2_t(\Omega)$ there is then a smooth kernel $\tilde{g} \in (C^\infty(\Omega))^3 \cap L^2_t(\Omega)$ such that

$$\left\| \frac{1}{\sigma_1} \left(\frac{1}{ik} \nabla \times \mathcal{H}(g - \tilde{g}) \times \nu + i\sigma_1 \mathcal{H}(g - \tilde{g})_T \right) \right\|_{L_t^2(\partial K_1)} \leq \frac{\sqrt{1-\delta^2}}{12} \epsilon.$$

But by definition

$$\mathcal{H}\tilde{g} = \int_{\Omega} \tilde{g}(d) \exp(ikx \cdot d)\, dS(d).$$

Since the integrand is smooth there is a quadrature scheme which when applied to the previous integral gives

$$\mathcal{H}^Q \tilde{g} = \sum_{j=1}^{p_1} \tilde{g}(dj) w_j \exp\left(ikx \cdot d_j\right)$$

where d_j, $j = 1, \ldots, p_1$, are quadrature points on Ω and w_j, $j = 1, \cdots, p_1$, are quadrature weights such that

$$\left\| \frac{1}{\sigma_1} \left(\frac{1}{ik} \nabla \times (\mathcal{H} - \mathcal{H}^Q)\tilde{g} \times \nu + i\sigma_1 (\mathcal{H} - \mathcal{H}^Q \tilde{g}_T) \right) \right\|_{L_t^2(\partial K_1)} \leq \sqrt{\frac{1-\delta^2}{12}} \epsilon.$$

Using these estimates and the triangle inequality shows that

$$\left\| \frac{1}{\sigma_1} \left(\mathcal{X}_1 - \frac{1}{ik} \nabla \times \mathcal{H}^Q \tilde{g} \times \nu - i\sigma_1 (\mathcal{H}^Q \tilde{g})_T \right) \right\|_{L_t^2(\partial K_1)} \leq \frac{\sqrt{1-\delta^2}}{4} \epsilon.$$

Repeating the estimate on K_2 and taking $p = p_1 + p_2$ shows that the desired estimate holds.

It is unfortunate that no estimate of the order of convergence can be obtained by this method. If the Herglotz kernel g could be characterised more precisely it might be possible to prove such an estimate.

Now that we have described the UWVF, its potential advantages compared to the finite element method should be clear. Because it is a discontinuous method, the basis can be changed from element to element with ease. This allows the number of basis functions to be chosen depending on the size of the element (see [20]). A high order scheme can be obtained by simply increasing the number of directions of the plane waves per element. Perhaps more important, the discrete problem can be solved using the bi-conjugate gradient scheme.

However the UWVF also has some disadvantages. First of all it is quite complicated! Secondly the absorbing boundary condition is low order and hence a large computational domain is needed to approximate a scattering problem. This problem is currently being attacked by coupling the UWVF to an integral equation formulation of the exterior problem. A third problem is that the method can suffer from ill-conditioning. This can be controlled by a careful choice of the number of directions per element [20]. In addition the scheme only provides an approximation to the field on the faces of the tetrahedra in the mesh. Although the theory does not guarantee convergence away from ∂R, our experience is that the solution throughout R

is approximated by the solution obtained by the UWVF scheme. However, from the point of view of computing far field data, it is sufficient to know $\mathcal{X}$ on the boundary of R and use the usual integral representation for E_∞ to obtain an approximate far field pattern.

All the results we report in the next section are obtained using far field data computed from the near field solutions obtained by the UWVF.

6 Numerical Examples for the Inverse Problem

The discretisation of (25) is relatively straightforward provided attention is paid to the fact that it is a severely ill-posed linear integral equation. We start by applying numerical quadrature to the integral in (25). In particular we use a finite element approximation to Ω to compute the necessary quadrature weights. It is shown in [9] that if $(\hat{x}, e_1(\hat{x}), e_2(\hat{x}))$ is an orthonormal basis for $\mathbb{R}^3$ then the resulting discrete equations for an approximate kernel g_a are

$$\sum_{j=1}^{N} \omega_j E_\infty(-d_j, -\hat{x}, e_l(\hat{x})) \cdot g_a(d_j, z, q) = e_l(\hat{x}) \cdot E_{e,\infty}(\hat{x}, z, q), \tag{46}$$

for $l = 1, 2$ where $d_1, d_2, \cdots, d_N$ are N quadrature points on Ω (and also the directions of the N incident fields used in the inverse problem). The quadrature weights are $\omega_1, \omega_2, \cdots, \omega_N$. The equations given by (46) are required to hold for $\hat{x}$ in a suitable discrete set giving the position of measurements on Ω. In this study we use $\hat{x} = d_n$, $1 \le n \le N$, so that the same vectors are used for observation and incident directions. We thus have $2N$ equations for the $2N$ tangential components of g_a. More precisely we write

$$g_a(d_j, z, q) = g_1(d_j, z, q)e_1(d_j) + g_2(d_j, z, q)e_2(d_j)$$

and define

$$A_{l,m}(d_j, d_n) = \omega_j E_\infty(-d_j, -d_n, e_l(d_n)) \cdot e_m(d_j)$$

for $1 \le l, m \le 2$, and $1 \le j, n \le N$, to obtain the following discrete equations for $\{g_n(d_j, z, q)\}_{n=1,2}^{j=1,\ldots,N}$

$$\sum_{j=1}^{N} \sum_{m=1}^{2} A_{l,m}(d_j, d_n) g_m(d_j, z, q) = e_l(d_n) \cdot E_{e,\infty}(d_n, z, q) \tag{47}$$

for $1 \le l \le 2$, and $1 \le n \le N$. This can be written in matrix form as the $2N \times 2N$ linear system

$$A_\infty \vec{g}(z, q) = \vec{b}(z, q). \tag{48}$$

Since the original linear equation (25) is ill-posed, this system is ill-conditioned as N increases. We thus need to apply techniques suitable for ill-conditioned matrix equations to solve the problem. In particular, we stabilise the problem using Tikhonov regularisation. If $\gamma > 0$ denotes the Tikhonov regularisation parameter, the regularised solution of (48), denoted by $\overrightarrow{g_\alpha}(z,q)$, satisfies

$$\gamma \overrightarrow{g_\alpha} + A_\infty^* A_\infty \overrightarrow{g_\alpha} = A_\infty^* \overrightarrow{b} \tag{49}$$

where A_∞^* is the conjugate transpose of the matrix A_∞. The parameter γ is chosen using Morozov's discrepancy principle via the singular value decomposition of A_∞ and an estimate δ of the error in the data for the inverse problem [8].

In the upcoming examples the data for the inverse problem is computed using the UWVF technique outlined in the previous section. The output from the UWVF code is an estimate for A_∞. In order to avoid inverse crimes we also perturb this matrix by noise. If A_∞ denotes the output from the UWVF forward code, we actually solve (49) using the perturbed matrix A_∞^ϵ given by

$$(A_\infty^\epsilon)_{l,m} = (A_\infty)_{l,m}\,(1 + \epsilon R_{1,l,m} + i\epsilon R_{2,l,m})$$

for $1 \le l, m \le 2N$, where R_1 and R_2 are $2N \times 2N$ matrices of numbers from a random number generator producing random numbers uniformly distributed in $(-1,1)$. The Morozov discrepancy principle requires an estimate of the error δ in A_∞^ϵ. We neglect the error due to the UWVF and simply take $\delta = \|A_\infty - A_\infty^\epsilon\|$ (in the spectral norm). In the examples we shall show we take $\epsilon = 0.01$ which results in approximately 0.75% relative error in the spectral norm.

We choose $N = 42$ directions, and the wavenumber k is always such that the obstacles are roughly the same order of size as the wavelength of the incident field.

The linear sampling method proceeds by computing $\overrightarrow{g_\alpha}(z,q)$ for given z and $q = e_1, e_2, e_3$, the three unit vectors. The sampling point z is usually chosen to lie on a regular grid in the region under inspection. For this paper we use a uniform grid of $51 \times 51 \times 51$ points. We plot either isosurfaces of the function $G(z)$ defined by

$$G(z) = \frac{1}{3}\left(\|\overrightarrow{g_\alpha}(z,e_1)\|^{-1} + \|\overrightarrow{g_\alpha}(z,e_2)\|^{-1} + \|\overrightarrow{g_\alpha}(z,e_3)\|^{-1}\right)$$

or contour maps of $G(z)$ on planes cutting the sampling region. The isosurface values plotted are usually approximately 0.1–0.3 of the maximum value of G. In [9] we showed that using all polarisations q is essential for obtaining an accurate reconstruction.

We shall present two preliminary examples of the use of the linear sampling method for mixed problems. Our first example D is the unit sphere and $k = 4$. The sampling region for z is the cube $[-2,2]^3$. For comparison we have computed results for a perfectly conducting sphere (see Fig. 1 for the surface mesh on D used by the UWVF scheme) and for a partially coated sphere. The coated region is the irregular surface shown in the right hand panel of Fig. 1. In Fig. 2 we show the reconstruction of the unit sphere using $k = 4$ and perfectly conducting boundary conditions. This should be compared to the results in Fig. 3 where the reconstruction of the coated obstacle is shown. The isosurface reconstructions (plotting $G(z) = 0.2 \max_z G(z)$) are almost indistinguishable. But the contour maps of $G(z)$ do show that the mixed boundary data has influenced the computation of $\overrightarrow{g}_\infty$.

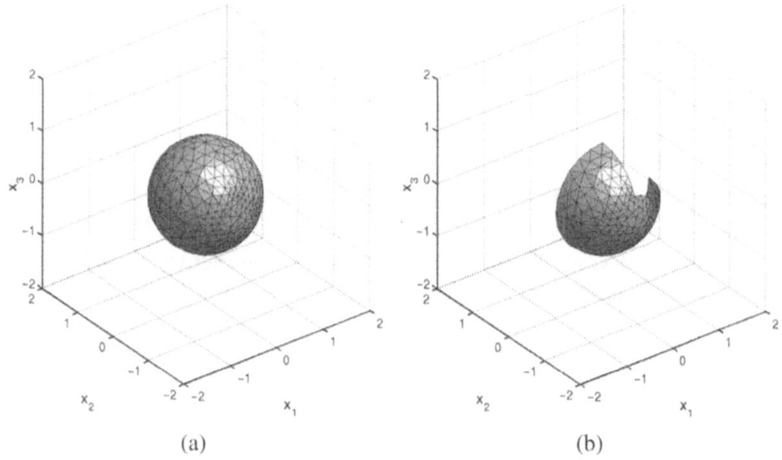

Fig. 1. The left hand figure shows the surface mesh used by the UWVF scheme to compute the far field data. For the coated sphere example, the segment of the boundary shown in the right hand figure supports an impedance boundary condition with $\lambda = k$ (the remainder of the sphere is perfectly conducting)

In our second example the domain D is two balls each having a radius 0.25 centred at ± 0.5. The wavenumber $k = 4$ so that the wavelength of the incident field is 1.57. Thus the spheres are smaller than 1/3 of a wavelength in diameter and separated by 2/3 of a wavelength centre to centre. On each sphere the hemisphere $x_2 > 0$ supports an impedance boundary condition with $\lambda = k$. The forward problem was computed using a mesh of 65776 tetrahedra refined around the boundary interface between impedance and perfect conducting data with six directions per tetrahedron resulting in a total of 789312 unknowns.

The reconstruction is shown in Fig. 4 where the domain for z is $[-1, 1]^3$. The effect of the coating is barely visible in the bottom left contour plot, but not visible in the reconstruction in the bottom right panel. In terms of computer time we note that the forward problem took about 4 days on a 400MHz Silicon Graphics Origin 2000 whereas the inverse problem took 4 minutes on a 300MHz Silicon Graphics Octane.

7 Conclusion

We have considered the inverse electromagnetic scattering problem of determining the shape of a partially coated obstacle or an isotropic inhomogeneous medium in $\mathbb{R}^3$ from multi-static far field data at fixed frequency in the resonance region, i.e. no assumptions are made on the wavenumber being large or small. Uniqueness and existence for the direct problem are established as well as uniqueness for the inverse problem. We prove the mathematical validity of the linear sampling method for solv-

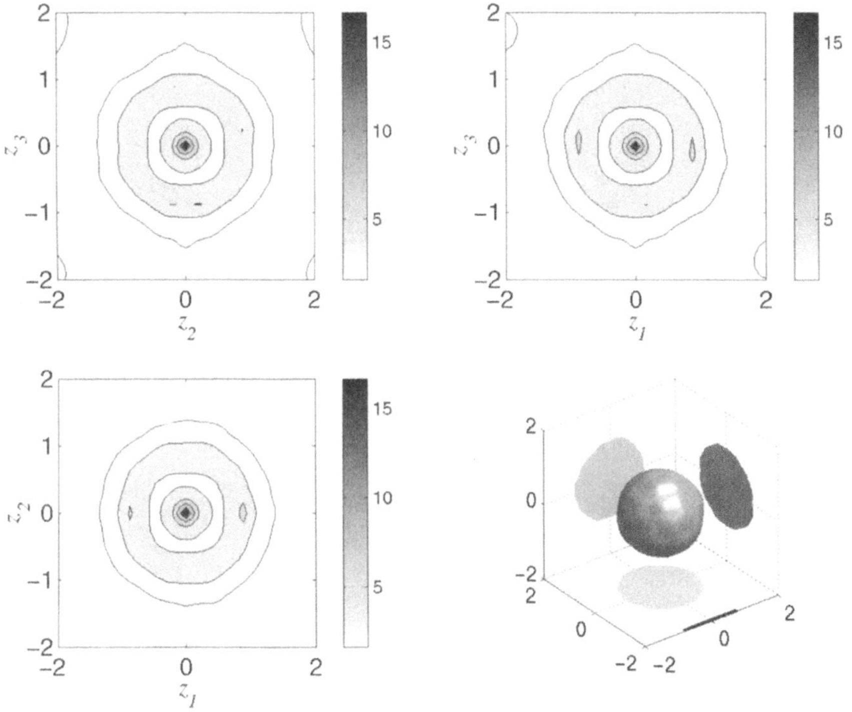

Fig. 2. Here we show the results of reconstructing the perfectly conducting unit sphere using 42 directions and $k = 4$. The top two panels and the bottom left panel show contour plots of $G(z)$ in the z_1, z_2 and z_3 planes. A good reconstruction is shown in the bottom right hand panel using the isosurface $G(z) = 0.2 \max_z G(z)$. This result should be compared to Fig. 3. The dark bar in the lower right panel represents the wavelength of the field

ing the inverse problem. Synthetic far field data is generated by using the ultra weak variational formulation of Maxwell's equations due to Cessenat and Després and numerical examples are given for the inverse problem in different cases.

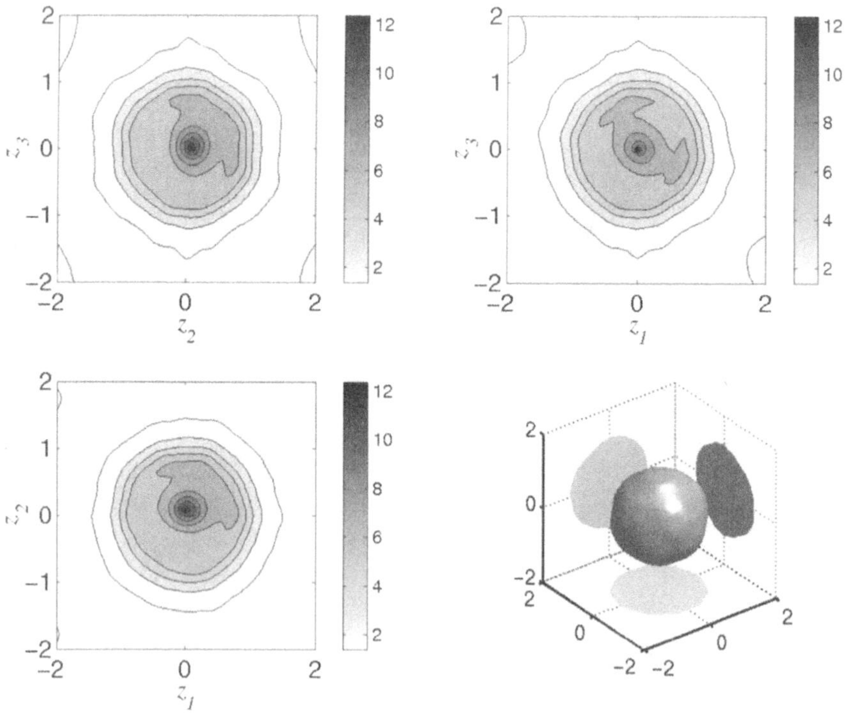

Fig. 3. Results of reconstructing the unit sphere with mixed boundary data and the same parameters as for the perfectly conducting sphere in Fig. 1. The portion of the sphere shown in the right-hand panel of Fig. 1 supports an impedance boundary condition with $\lambda = k$. As in Fig. 2 (to which this figure should be compared), the top two panels and the bottom left panel show contour plots of $G(z)$ in the z_1, z_2 and z_3 planes. These plots clearly show the effect of the mixed boundary condition. However the reconstruction in the bottom right hand panel using the isosurface $G(z) = 0.2 \max_z G(z)$ is very similar to that shown in Fig. 2

References

1. B. Borden, *Radar Imaging of Airborne Targets*, IOP Publishing, Bristol, 1999.
2. F. Cakoni and D. Colton, Combined far field operators in electromagnetic inverse scattering theory, *Math. Methods Applied Science*, to appear.
3. F. Cakoni and D. Colton, A uniqueness theorem for an inverse electromagnetic scattering problem in inhomogeneous anisotropic media, *Proc. Edin. Math. Soc.*, to appear.
4. F. Cakoni, D. Colton and P. Monk, The electromagnetic inverse scattering problem for partially coated Lipschitz domains, to appear.
5. O. Cessenat, *Application d'une nouvelle formulation variationnelle aux équations d'ondes harmoniques. Problèmes de Helmholtz 2D et de Maxwell 3D*, PhD thesis, Université Paris IX Dauphine, 1996.

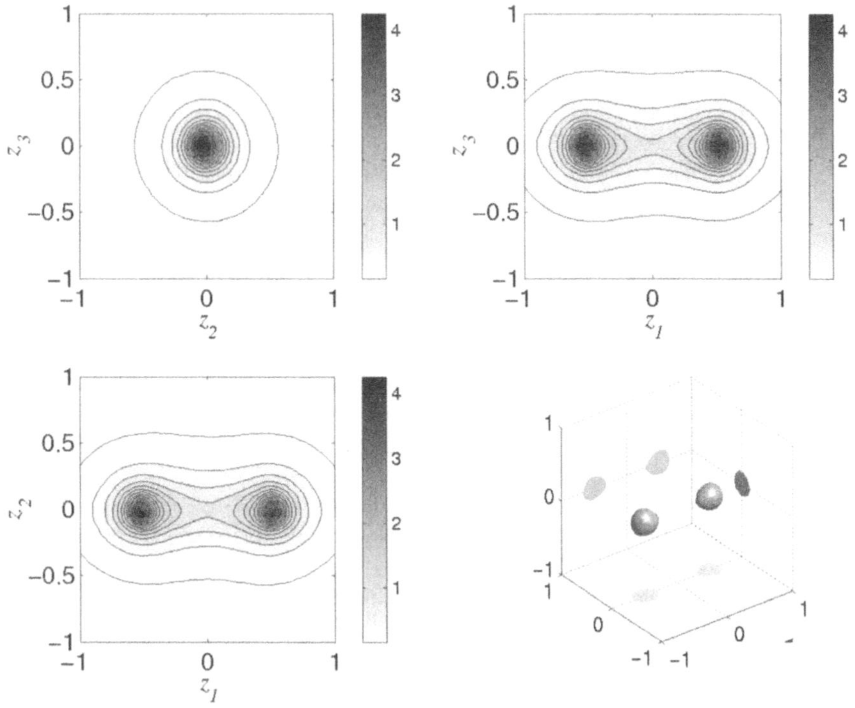

Fig. 4. Here we show the results of reconstructing the two balls with mixed boundary data (the hemisphere $x_2 > 0$ of each ball is coated with an impedance boundary condition having $\lambda = k$). The top left panel shows a contour map for $z_1 = -0.5$. The remaining contour plots are at $z_2 = 0$ and $z_3 = 0$. The bottom left contour plot ($z_3 = 0$) is slightly asymmetric and shows the influence of the mixed data. However the reconstruction shown in the bottom right hand panel using the isosurface $G(z) = 0.3 \max_z G(z)$ does not show an obvious effect of the mixed boundary data

6. O. Cessenat and B. Després, Application of the ultra-weak variational formulation of elliptic PDEs to the 2-dimensional Helmholtz problem, *SIAM J. Numer. Anal.*, **35** (1998), 255–299.
7. M. Cheney, A mathematical tutorial on synthetic aperture radar, *SIAM Review* **43** (2001), 301-312.
8. D. Colton, J. Coyle, and P. Monk, Recent developments in inverse acoustic scattering theory, *SIAM Review*, **42** (2000), 369-414.
9. D. Colton, H. Haddar and P. Monk, The linear sampling method for solving the electromagnetic inverse scattering problem, *SIAM J. Sci. Comput.*, **12** (2002), 719-731.
10. D. Colton and A. Kirsch, A simple method for solving inverse scattering problems in the resonance region, *Inverse Problems*, **12** (1996), 383-393.
11. D. Colton and R. Kress, *Inverse Acoustic and Electromagnetic Scattering Theory*, Springer-Verlag, New York, Second Edition, 1998.

12. D. Colton and L. Päivärinta, Far field patterns and the inverse scattering problem for electromagnetic waves in an inhomogeneous medium, *Math. Proc. Comb. Phil. Soc.* **103** (1988), 561-575.
13. D. Colton and L. Päivärinta, The uniqueness of a solution to an inverse scattering problem for electromagnetic waves, *Arch. Rational Mech. Anal.* **119** (1992), 59-70.
14. D. Colton, M. Piana and R. Potthast, A simple method using Morozov's discrepancy principle for solving inverse scattering problems, *Inverse Problems*, **13** (1997), 1477-1493.
15. O. Dorn, H. Bertete-Aguirre, J.G. Berryman and G. C. Papanicolaou, A nonlinear inversion method for 3D-electromagnetic imaging using adjoint fields, *Inverse Problems* **15** (1999), 1523-1558.
16. M. Haas, W. Rieger, W. Rucker and G. Lehner, Inverse $3D$ acoustic and electromagnetic obstacle scattering by iterative adaption, in *Inverse Problems of Wave Propagation and Diffraction*, G. Chavent and P. Sabatier, eds., Springer-Verlag, Heidelberg, 1997, 204-215.
17. H. Haddar and P. Monk, The linear sampling method for solving the electromagnetic inverse medium problem, *Inverse Problems* **18** (2002), 891-906.
18. P. Hähner, Electromagnetic wave scattering: theory, in *Scattering*, R. Pike and P. Sabatier, eds., Academic Press, London, 2002, 211-229.
19. A. E. Hooper and H. N. Hambric, Unexploded ordinance (UXO)): The problem, in *Detection and Identification of Visually Obscured Targets*, C. E. Baum, ed., Taylor and Francis, Philadelphia, 1999 1-8.
20. T. Huttunen, P Monk and J.P. Kaipio, Computational Aspects of the Ultra Weak Variational Formulation, *J. Comput. Phys,* **182** (2002), 27-46.
21. A. Kirsch and R. Kress, Uniqueness in inverse obstacle scattering, *Inverse Problems*, **9** (1993), 285-299.
22. R. Kress, Electromagnetic wave scattering: theory, in *Scattering*, R. Pike and P. Sabatier, eds., Academic Press, London, 2002, 191-210.
23. P. Maponi, M. Recchioni and F. Zirilli, The use of optimization in the reconstruction of obstacles from acoustic or electromagnetic scattering data, in *Large Scale Optimization with Applications, Part I: Optimization in Inverse Problems and Design*, L. Biegler, et. al. eds., IMA Volumes in Mathematics and Its Applications, Vol. 92, Springer, New York, 1997, 81-100.
24. P. Monk, *Finite Element Methods for Maxwell's Equations*, Oxford University Press, Oxford, 2002.
25. J. C. Nédélec, *Acoustic and Electromagnetic Equations*, Springer-Verlag, New York, 2001.
26. P. Ola, L. Päivärinta, and E. Somersalo, An inverse boundary value problem in electrodynamics, *Duke Math. Jour.* **70** (1993) 617-653.
27. P. Ola and E. Somersalo, Electromagnetic inverse problems and generalized Sommerfeld potentials, *SIAM J. Appl. Math.* **56** (1996), 1129-1145.
28. R. Potthast, *Point Sources and Multipoles in Inverse Scattering Theory*, Research Notes in Mathematics, Vol. 427, Chapman and Hall/CRC, Boca Raton, Florida, 2001.
29. Z. Sun and G. Uhlmann, An inverse boundary value problem for Maxwell's equations, *Arch. Rat. Mech. Anal.* **119** (1992), 71-93.
30. R. Womersley and I. Sloan, How good can polynomial interpolation on the sphere be?, *Advances in Computational Mathematics*, 14 (2001), 195–226.

Appendix: Colour Figures

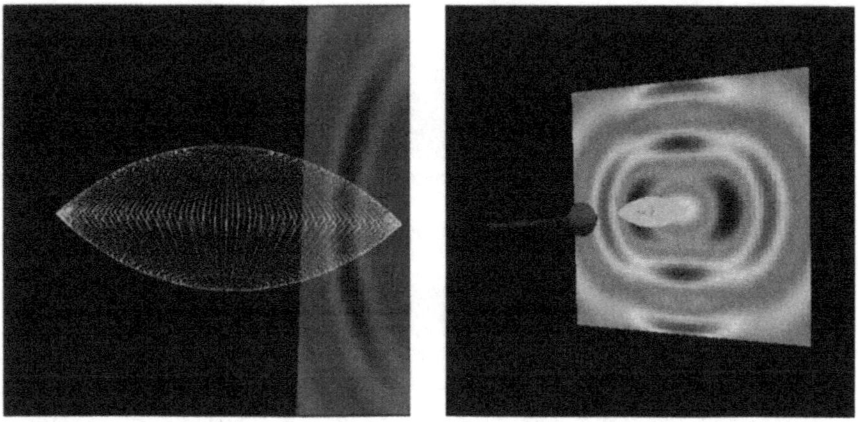

Fig. 1. Left: Flying saucer. **Right:** Electromagnetic scattering from a two-wavelength diameter flying saucer, with maximum far field errors of $3.0 \cdot 10^{-5}$

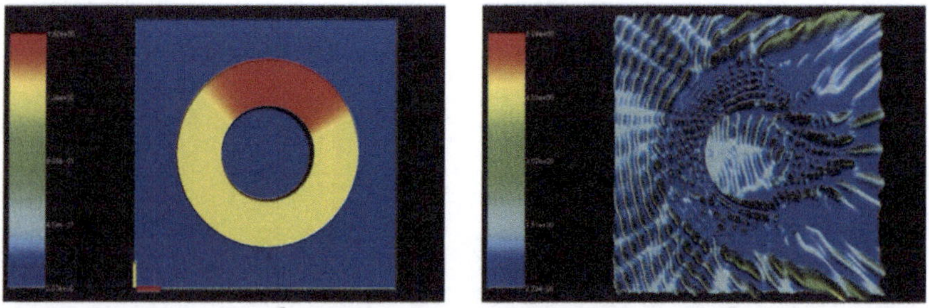

(a) Scatterer ($q = -m = n^2 - 1$) (b) Near Field Intensity ($|u|^2$)

Fig. 2. Two-dimensional scatterer. Diameter $= 10\lambda$. Computations using N unknowns

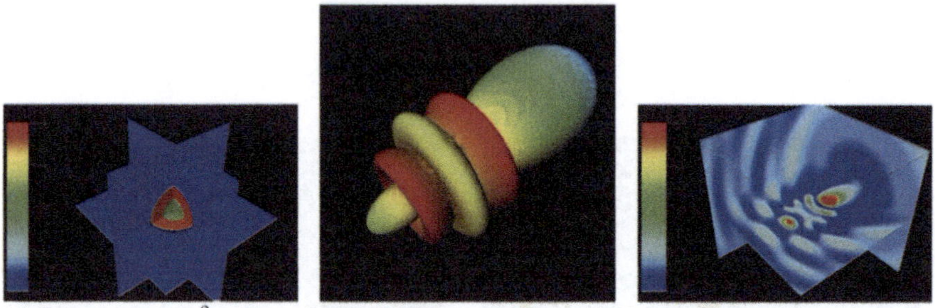

(a) Scatterer ($q = n^2 - 1$) (b) Far Field Intensity (c) Near Field Intensity

Fig. 3. Layered Sphere of radius a– $ka = 4$

(a) Scatterer ($q = n^2 - 1$) (b) Far Field Intensity (c) Near Field Intensity

Fig. 4. Array of Smooth Scatterers (Potentials) – $6\lambda \times 6\lambda \times 6\lambda$

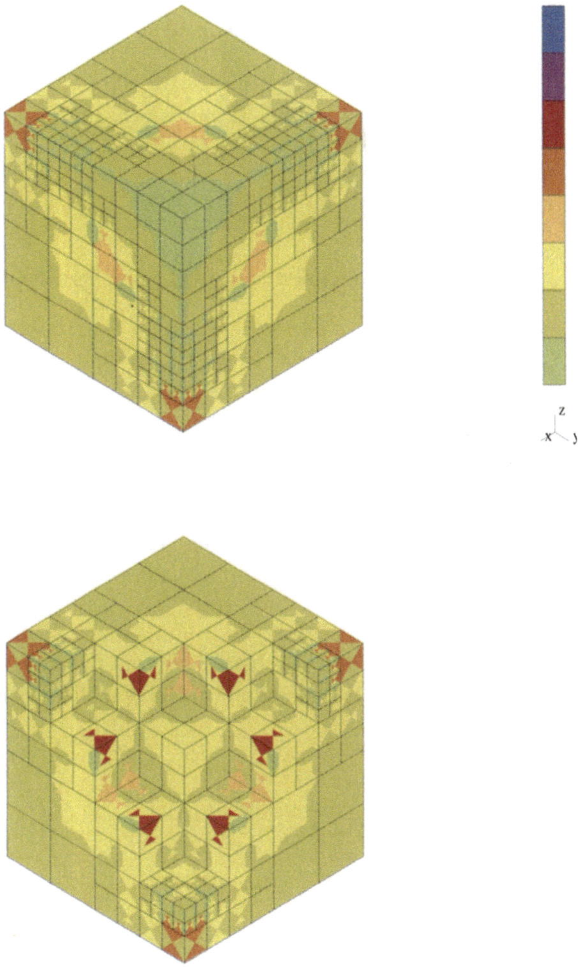

Fig. 5. An automatically generated 3D *hp* mesh for an elliptic problem. View of the mesh from outside and inside (notice the symmetry)

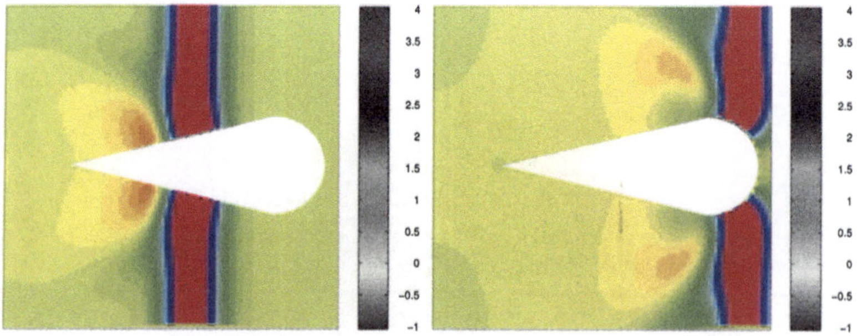

Fig. 6. The solution at two successive instants (see p. 238)

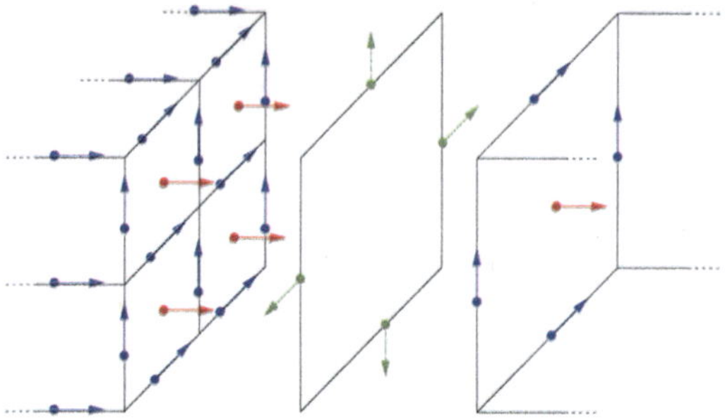

Fig. 7. The degrees of freedom matching coarse and fine grids for the Yee scheme *with* a Lagrange multiplier (see p. 253): the electric field in blue, the magnetic field in red, the electric current in green

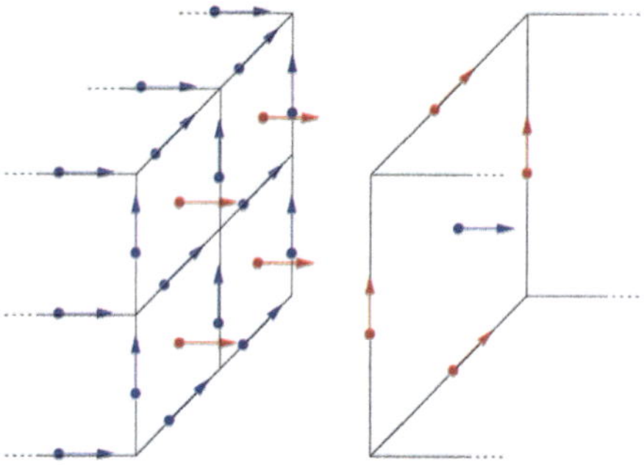

Fig. 8. As previous figure but *without* a Lagrange multiplier (see p. 256): the electric field in blue, the magnetic field in red

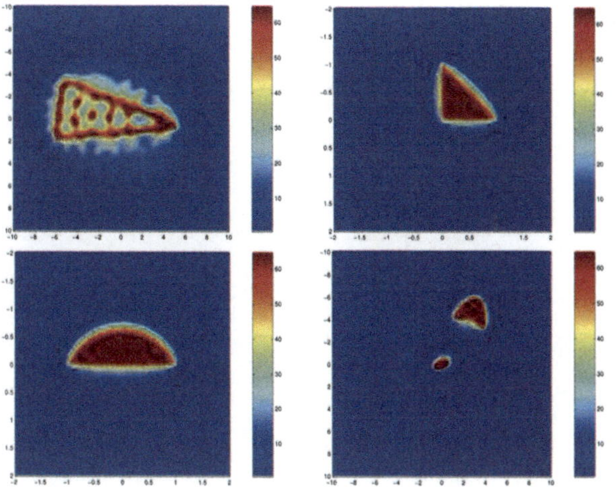

Fig. 9. Reconstructions obtained using factorisation method

Editorial Policy

§1. Volumes in the following three categories will be published in LNCSE:

i) Research monographs
ii) Lecture and seminar notes
iii) Conference proceedings

Those considering a book which might be suitable for the series are strongly advised to contact the publisher or the series editors at an early stage.

§2. Categories i) and ii). These categories will be emphasized by Lecture Notes in Computational Science and Engineering. **Submissions by interdisciplinary teams of authors are encouraged.** The goal is to report new developments – quickly, informally, and in a way that will make them accessible to non-specialists. In the evaluation of submissions timeliness of the work is an important criterion. Texts should be well-rounded, well-written and reasonably self-contained. In most cases the work will contain results of others as well as those of the author(s). In each case the author(s) should provide sufficient motivation, examples, and applications. In this respect, Ph.D. theses will usually be deemed unsuitable for the Lecture Notes series. Proposals for volumes in these categories should be submitted either to one of the series editors or to Springer-Verlag, Heidelberg, and will be refereed. A provisional judgment on the acceptability of a project can be based on partial information about the work: a detailed outline describing the contents of each chapter, the estimated length, a bibliography, and one or two sample chapters – or a first draft. A final decision whether to accept will rest on an evaluation of the completed work which should include

- at least 100 pages of text;
- a table of contents;
- an informative introduction perhaps with some historical remarks which should be accessible to readers unfamiliar with the topic treated;
- a subject index.

§3. Category iii). Conference proceedings will be considered for publication provided that they are both of exceptional interest and devoted to a single topic. One (or more) expert participants will act as the scientific editor(s) of the volume. They select the papers which are suitable for inclusion and have them individually refereed as for a journal. Papers not closely related to the central topic are to be excluded. Organizers should contact Lecture Notes in Computational Science and Engineering at the planning stage.

In exceptional cases some other multi-author-volumes may be considered in this category.

§4. Format. Only works in English are considered. They should be submitted in camera-ready form according to Springer-Verlag's specifications. Electronic material can be included if appropriate. Please contact the publisher. Technical instructions and/or TEX macros are available via http://www.springer.de/math/authors/help-momu.html. The macros can also be sent on request.

General Remarks

Lecture Notes are printed by photo-offset from the master-copy delivered in camera-ready form by the authors. For this purpose Springer-Verlag provides technical instructions for the preparation of manuscripts. See also *Editorial Policy*.

Careful preparation of manuscripts will help keep production time short and ensure a satisfactory appearance of the finished book.

The following terms and conditions hold:

Categories i), ii), and iii):
Authors receive 50 free copies of their book. No royalty is paid. Commitment to publish is made by letter of intent rather than by signing a formal contract. Springer-Verlag secures the copyright for each volume.

For conference proceedings, editors receive a total of 50 free copies of their volume for distribution to the contributing authors.

All categories:
Authors are entitled to purchase further copies of their book and other Springer mathematics books for their personal use, at a discount of 33,3 % directly from Springer-Verlag.

Addresses:

Timothy J. Barth
NASA Ames Research Center
NAS Division
Moffett Field, CA 94035, USA
e-mail: barth@nas.nasa.gov

Michael Griebel
Institut für Angewandte Mathematik
der Universität Bonn
Wegelerstr. 6
D-53115 Bonn, Germany
e-mail: griebel@iam.uni-bonn.de

David E. Keyes
Computer Science Department
Old Dominion University
Norfolk, VA 23529-0162, USA
e-mail: keyes@cs.odu.edu

Risto M. Nieminen
Laboratory of Physics
Helsinki University of Technology
02150 Espoo, Finland
e-mail: rni@fyslab.hut.fi

Dirk Roose
Department of Computer Science
Katholieke Universiteit Leuven
Celestijnenlaan 200A
3001 Leuven-Heverlee, Belgium
e-mail: dirk.roose@cs.kuleuven.ac.be

Tamar Schlick
Department of Chemistry
Courant Institute of Mathematical Sciences
New York University
and Howard Hughes Medical Institute
251 Mercer Street
New York, NY 10012, USA
e-mail: schlick@nyu.edu

Springer-Verlag, Mathematics Editorial IV
Tiergartenstrasse 17
D-69121 Heidelberg, Germany
Tel.: *49 (6221) 487-8185
e-mail: peters@springer.de
http://www.springer.de/math/peters.html

Lecture Notes in Computational Science and Engineering

Vol. 1 D. Funaro, *Spectral Elements for Transport-Dominated Equations.* 1997. X, 211 pp. Softcover. ISBN 3-540-62649-2

Vol. 2 H. P. Langtangen, *Computational Partial Differential Equations.* Numerical Methods and Diffpack Programming. 1999. XXIII, 682 pp. Hardcover. ISBN 3-540-65274-4

Vol. 3 W. Hackbusch, G. Wittum (eds.), *Multigrid Methods V.* Proceedings of the Fifth European Multigrid Conference held in Stuttgart, Germany, October 1-4, 1996. 1998. VIII, 334 pp. Softcover. ISBN 3-540-63133-X

Vol. 4 P. Deuflhard, J. Hermans, B. Leimkuhler, A. E. Mark, S. Reich, R. D. Skeel (eds.), *Computational Molecular Dynamics: Challenges, Methods, Ideas.* Proceedings of the 2nd International Symposium on Algorithms for Macromolecular Modelling, Berlin, May 21-24, 1997. 1998. XI, 489 pp. Softcover. ISBN 3-540-63242-5

Vol. 5 D. Kröner, M. Ohlberger, C. Rohde (eds.), *An Introduction to Recent Developments in Theory and Numerics for Conservation Laws.* Proceedings of the International School on Theory and Numerics for Conservation Laws, Freiburg / Littenweiler, October 20-24, 1997. 1998. VII, 285 pp. Softcover. ISBN 3-540-65081-4

Vol. 6 S. Turek, *Efficient Solvers for Incompressible Flow Problems.* An Algorithmic and Computational Approach. 1999. XVII, 352 pp, with CD-ROM. Hardcover. ISBN 3-540-65433-X

Vol. 7 R. von Schwerin, ***M**ulti **B**ody **S**ystem **SIM**ulation.* Numerical Methods, Algorithms, and Software. 1999. XX, 338 pp. Softcover. ISBN 3-540-65662-6

Vol. 8 H.-J. Bungartz, F. Durst, C. Zenger (eds.), *High Performance Scientific and Engineering Computing.* Proceedings of the International FORTWIHR Conference on HPSEC, Munich, March 16-18, 1998. 1999. X, 471 pp. Softcover. 3-540-65730-4

Vol. 9 T. J. Barth, H. Deconinck (eds.), *High-Order Methods for Computational Physics.* 1999. VII, 582 pp. Hardcover. 3-540-65893-9

Vol. 10 H. P. Langtangen, A. M. Bruaset, E. Quak (eds.), *Advances in Software Tools for Scientific Computing.* 2000. X, 357 pp. Softcover. 3-540-66557-9

Vol. 11 B. Cockburn, G. E. Karniadakis, C.-W. Shu (eds.), *Discontinuous Galerkin Methods.* Theory, Computation and Applications. 2000. XI, 470 pp. Hardcover. 3-540-66787-3

Vol. 12 U. van Rienen, *Numerical Methods in Computational Electrodynamics.* Linear Systems in Practical Applications. 2000. XIII, 375 pp. Softcover. 3-540-67629-5

Vol. 13 B. Engquist, L. Johnsson, M. Hammill, F. Short (eds.), *Simulation and Visualization on the Grid.* Parallelldatorcentrum Seventh Annual Conference, Stockholm, December 1999, Proceedings. 2000. XIII, 301 pp. Softcover. 3-540-67264-8

Vol. 14 E. Dick, K. Riemslagh, J. Vierendeels (eds.), *Multigrid Methods VI.* Proceedings of the Sixth European Multigrid Conference Held in Gent, Belgium, September 27-30, 1999. 2000. IX, 293 pp. Softcover. 3-540-67157-9

Vol. 15 A. Frommer, T. Lippert, B. Medeke, K. Schilling (eds.), *Numerical Challenges in Lattice Quantum Chromodynamics.* Joint Interdisciplinary Workshop of John von Neumann Institute for Computing, Jülich and Institute of Applied Computer Science, Wuppertal University, August 1999. 2000. VIII, 184 pp. Softcover. 3-540-67732-1

Vol. 16 J. Lang, *Adaptive Multilevel Solution of Nonlinear Parabolic PDE Systems.* Theory, Algorithm, and Applications. 2001. XII, 157 pp. Softcover. 3-540-67900-6

Vol. 17 B. I. Wohlmuth, *Discretization Methods and Iterative Solvers Based on Domain Decomposition.* 2001. X, 197 pp. Softcover. 3-540-41083-X

Vol. 18 U. van Rienen, M. Günther, D. Hecht (eds.), *Scientific Computing in Electrical Engineering.* Proceedings of the 3rd International Workshop, August 20-23, 2000, Warnemünde, Germany. 2001. XII, 428 pp. Softcover. 3-540-42173-4

Vol. 19 I. Babuška, P. G. Ciarlet, T. Miyoshi (eds.), *Mathematical Modeling and Numerical Simulation in Continuum Mechanics.* Proceedings of the International Symposium on Mathematical Modeling and Numerical Simulation in Continuum Mechanics, September 29 - October 3, 2000, Yamaguchi, Japan. 2002. VIII, 301 pp. Softcover. 3-540-42399-0

Vol. 20 T. J. Barth, T. Chan, R. Haimes (eds.), *Multiscale and Multiresolution Methods.* Theory and Applications. 2002. X, 389 pp. Softcover. 3-540-42420-2

Vol. 21 M. Breuer, F. Durst, C. Zenger (eds.), *High Performance Scientific and Engineering Computing.* Proceedings of the 3rd International FORTWIHR Conference on HPSEC, Erlangen, March 12-14, 2001. 2002. XIII, 408 pp. Softcover. 3-540-42946-8

Vol. 22 K. Urban, *Wavelets in Numerical Simulation.* Problem Adapted Construction and Applications. 2002. XV, 181 pp. Softcover. 3-540-43055-5

Vol. 23 L. F. Pavarino, A. Toselli (eds.), *Recent Developments in Domain Decomposition Methods.* 2002. XII, 243 pp. Softcover. 3-540-43413-5

Vol. 24 T. Schlick, H. H. Gan (eds.), *Computational Methods for Macromolecules: Challenges and Applications.* Proceedings of the 3rd International Workshop on Algorithms for Macromolecular Modeling, New York, October 12-14, 2000. 2002. IX, 504 pp. Softcover. 3-540-43756-8

Vol. 25 T. J. Barth, H. Deconinck (eds.), *Error Estimation and Adaptive Discretization Methods in Computational Fluid Dynamics.* 2003. VII, 344 pp. Hardcover. 3-540-43758-4

Vol. 26 M. Griebel, M. A. Schweitzer (eds.), *Meshfree Methods for Partial Differential Equations.* 2003. IX, 466 pp. Softcover. 3-540-43891-2

Vol. 27 S. Müller, *Adaptive Multiscale Schemes for Conservation Laws.* 2003. XIV, 181 pp. Softcover. 3-540-44325-8

Vol. 28 C. Carstensen, S. Funken, W. Hackbusch, R. H. W. Hoppe, P. Monk (eds.), *Computational Electromagnetics.* Proceedings of the GAMM Workshop on "Computational Electromagnetics", Kiel, Germany, January 26-28, 2001. 2003. X, 209 pp. Softcover. 3-540-44392-4

Vol. 29 M. A. Schweitzer, *A Parallel Multilevel Partition of Unity Method for Elliptic Partial Differential Equations.* 2003. V, 194 pp. Softcover. 3-540-00351-7

Vol. 30 T. Biegler, O. Ghattas, M. Heinkenschloss, B. van Bloemen Waanders (eds.), *Large-Scale PDE-Constrained Optimization.* 2003. approx. 344 pp. Softcover. 3-540-05045-0

Vol. 31 M. Ainsworth, P. Davies, D. Duncan, P. Martin, B. Rynne (eds.) *Topics in Computational Wave Propagation.* Direct and Inverse Problems. 2003. VIII, 399 pp. Softcover. 3-540-00744-X

Vol. 32 H. Emmerich, B. Nestler, M. Schreckenberg (eds.) *Interface and Transport Dynamics.* Computational Modelling. 2003. XV, 432 pp. Hardcover. 3-540-40367-1

Vol. 33 H. P. Langtangen, A. Tveito (eds.) *Advanced Topics in Computational Partial Differential Equations.* Numerical Methods and Diffpack Programming. 2003. XIX, 659 pp. Softcover. 3-540-01438-1

Texts in Computational Science and Engineering

Vol. 1 H. P. Langtangen, *Computational Partial Differential Equations.* Numerical Methods and Diffpack Programming. 2nd Edition 2003. XXVI, 855 pp. Hardcover. ISBN 3-540-43416-X

Vol. 2 A. Quarteroni, F. Saleri, *Scientific Computing with MATLAB.* 2003. IX, 257 pp. Hardcover. ISBN 3-540-44363-0

For further information on these books please have a look at our mathematics catalogue at the following URL: `http://www.springer.de/math/index.html`

Printed by Printforce, the Netherlands